Statistical Modeling With R

Statistical Modeling With R

A dual frequentist and Bayesian approach for life scientists

PABLO INCHAUSTI

Centro Universitario Regional del Este, Universidad de la República, Uruguay

Great Clarendon Street, Oxford, OX2 6DP,
United Kingdom

Oxford University Press is a department of the University of Oxford.
It furthers the University's objective of excellence in research, scholarship,
and education by publishing worldwide. Oxford is a registered trade mark of
Oxford University Press in the UK and in certain other countries

Impression: 1

Published in the United States of America by Oxford University Press
198 Madison Avenue, New York, NY 10016, United States of America

British Library Cataloguing in Publication Data
Data available

Library of Congress Control Number: 2022937827

ISBN 978–0–19–285901–3 (hbk)

ISBN 978–0–19–285902–0 (pbk)

DOI: 10.1093/oso/9780192859013.001.0001

Printed and bound by
CPI Group (UK) Ltd, Croydon, CR0 4YY

Cover image: John Lund/Getty Images.

For Joana,
because two is so much more than one . . .

Preface

A preface is the closest an author may have to the "letter of marque" used by European governments in the seventeenth century to authorize piracy with the sovereign's tacit consent. An author can write just about anything in the preface with the resigned permission of the editor.

Having finished the book, I hereby give myself permission to write in the first person singular. I have spent many thousands of hours endlessly writing and rewriting this book over the last 24 months. It has been a challenge, a pleasure, a thrill, and a burden at the same time.

At this point, I have the following unsettling mixture of feelings:

* Relief and joy: Finally it is over.
* Pride and accomplishment: I have done it! I have done it! I have done it!
* Uncertainty: Was it worth the effort? Will it be well/badly received?
* Insecurity: Did I double-check everything? Does it have any embarrassing errors?

It is now time for the book to sink or swim on its own at the hands of its readers.

A long time ago, my undergraduate adviser sent me to fetch some rainfall data from the Venezuelan Ministry of the Environment. I reluctantly went, got lost inside the ugly building, failed to gather the data, but stumbled by chance into a nearly empty and shabby Ministry bookstore. There I found a gem: the original 1975 Spanish edition of the book *Areografía: estrategias geográficas de las especies* by Eduardo Rapoport. He was a clever and original Argentinian ecologist who anticipated what later became known as macroecology. The Pergamon Press translation (*Areography: The Geographic Strategies of Species*) missed the preface of the Spanish original, easily the best preface of a scientific book I have ever read. I read Rapoport's preface at the bookstore, and right away I bought the book using all the money I had, including the return bus fare. I walked the 4.5 km home, reading the book between traffic lights. Recognizing the absence of rules for what the preface of a scientific book should contain, Rapoport aimed to bring a humanized depiction of his CV, so that "science books would leave home and be taken to the dentist waiting room," and you knew what to say if you ever meet the author. Following the master Eduardo Rapoport, here is my attempt.

Things that I love

* The dazzling imaginations of Gabriel García Márquez, Julio Cortázar, and Alejo Carpentier, the dignity of Primo Levi, the wisdom of Umberto Eco, the humanity of Italo Calvino, *The Magus* by John Fowles.
* The music and shining smile of Louis Armstrong, the saxophone of John Coltrane (his mission: "a masterpiece by midnight"), and Art Tatum playing the piano; the Beatles, Eric Clapton, Sting, and Mark Knopfler; the blues of Taj Mahal, Keb'Mo, and Buddy

Guy; the stirring voices of Annie Lennox, Ute Lemper, and Madeleine Peyroux; the lyrics of Leonard Cohen and Bob Dylan.
* Two scientific heroes: John (JBS) Haldane and Richard Feynman.
* All the Monty Python films and the original BBC series.
* All Picasso, except the pink period; Gustav Klimt and Vincent van Gogh.
* Let's share a full Uruguayan barbecue, including mollejas (thymus), kidneys, and sweet blood sausages. It would be glorious to enjoy some French goat cheese, a fresh green salad with endives and cherry tomatoes, red wine of course, and a mango or passionfruit mousse to bring my soul closer to earthly paradise.
* And, above all, let's talk, exchanging stories, books, and anecdotes. The folders of my memory hold countless megabytes of historical, literary, and scientific information, some of which may even interest or entertain you for a while. And I can swiftly change my opinion on any issue under the sun as many times as you can manage to convince me with good arguments, sensible evidence, and a modicum of straight reasoning.

Things I hate

* Social injustice in any shape or form.
* The stupidity of the military and all its cheerleaders.
* Social, racial, and sexual discrimination under any disguise or shade.
* All totalitarian ideologies and forms of thought.
* The stiffness, conservatism, intolerance, and backwardness of the traditional Catholic Church and of many recently created protestant churches.
* Pineapple on pizza: a horrendous mix that spoils two great things.

My story

My own biography is pretty ordinary. I will recall just a few events that might perhaps inspire others to believe in themselves. I was born in Uruguay, a small country that lies sandwiched between Argentina and Brazil at the bottom left of the world map. After starting public primary school there, I followed my father to Venezuela. I started university wishing to become an electrical engineer, but finally managed to graduate in biology at the tardy age of 26.

I desperately wanted to study more and become a scientist. With my then partner, we managed to get admitted to the State University of New York at Stony Brook (now Stony Brook University) by sheer luck. In September 1992, we gathered all our money and belongings, got some family loans, and traveled to New York. We landed there perfectly unaware of everything, including how to get to the university from the airport. We had expected to pay the first year of university fees, betting that our good background would allow us to obtain good grades that might lead to some financial support. But it turned out that we did not have to pay any university fees at all!

And even better, one day before the start of my first semester, another graduate student chose to take care of her ill grandmother and declined her teaching assistant position. I was offered it, and of course took it: $750 a month minus taxes amounted to touching heaven. The next day I went to teach some very basic biology to 25 puzzled American students. During my fourth day in an English-speaking country, I barely understood 40% of what the students said. But I had an inspired idea that saved me. I shamelessly told them of a hearing disability that required them to speak slowly and very loudly for me to understand them. And they did it to such an extent that my strange disability miraculously disappeared after a few weeks. We quickly bought a TV to help train my wooden ears. At first, the only program that I could understand was the (British) Prime Minister's Questions that was broadcast on CSPAN very late at night just for the (dis)pleasure of insomniacs. This very theatrical, ceremonial, and mostly pointless weekly exercise of British politics was my door to understanding spoken English, and the starting of an anglophilia that only Brexit recently and definitely managed to cure.

At Stony Brook, I met Lev Ginzburg by chance while eating sandwiches at the Department of Ecology and Evolution. This very intelligent and witty Russian mathematician became my PhD supervisor. At first, it was nearly impossible to understand what this man was talking so quickly about. I used to share a 6 m^2 office in front of his. Лев often called me into his office using mundane excuses to spend many hours talking and teaching me on a one-to-one basis as if I was a medieval apprentice. These interactions over the years shaped me into a scientist, and affected my brain more than anything since gastrulation. The wide range of topics of these conversations included mathematics, ecology, classical physics, dynamical systems, risk analysis, philosophy of science, the latest books we were reading, and who knows what else ... I still vividly recall two entire Friday afternoons that Лев devoted to teaching me the puzzling basics of quantum mechanics (including the Schrödinger wave equation) using a small green blackboard and white chalk. It was an indescribable pleasure to have received such a gift of human knowledge from you, мой дорогой друг и наставник.

I graduated in 1998, and my Italian passport (life lesson for the young: you can never have too many passports; acquire as many as possible since some may open unexpected doors) got me an EU fellowship for a postdoc with John Lawton at Imperial College, UK. I later moved to France where I lived and worked for nine years. Other moves following a non-traditional and hardly straight path took me back to Uruguay, where I live now.

I will not bother the reader with further details of my academic past. There is hardly any merit involved in it. Like you, I have 23 pairs of chromosomes in every cell, blood the same color as yours, and a genome that differs from yours and from Mandela's, Einstein's, Himmler's, and Stalin's by about six million DNA bases (~0.06%, an irrelevant difference since only about 2% of our DNA is translated into proteins). Therefore, rest assured that there is nothing special, unique, or even good about me. You can easily do better than me if you wish.

Just trust me on this one. Most people who succeed in life are those that seriously apply their heart and mind and energy long enough to pursue their dreams with stubborn determination. I am convinced that life (or the universe, or the gods) rewards persistence and single-mindedness over apparent leaps of inspired genius. However, for that you first need to hold dreams and ambitions for yourself. Nobody can teach you to dream and aspire to a higher future than your present. Dreaming turns out to be a spontaneous and personal affair. I gleaned the next quote (out of context, and oddly enough due to Lenin) from a Julio Cortázar book that summarizes well what I wish to convey:

> The rift between dreams and reality causes no harm if only the person dreaming believes seriously in his dream, if he attentively observes life, compares his observations with his castles in the air, and if, generally speaking, he works conscientiously for the achievement of his fantasies. If there is some connection between dreams and life then all is well.

I have been helped beyond the call of duty by the staff of Oxford University Press. Ian Sherman, senior editor of Life Sciences, incredibly remembered me after a 19-year hiatus and, even more surprisingly, believed in and liked the idea of this book. He variously guided, prompted, kept quiet, and encouraged me, and I cannot thank him enough for all this and more. I must also thank Katie Lakina for putting the production of this book back on track, Karen Moore for her diligent and dedicated work during the transformation of many files into a finished book, and Richard Hutchinson for his attentive and careful copyediting that greatly improved the quality of the text that you are reading.

The free and open software R and the many packages used in this book stem from the fantastic and creative work of many generous scientists and programmers around the world. Their incredible work has created the collective property of statistical knowledge that made this book possible. While I lack the means to thank you all, let me at least raise a glass to toast you with endless gratitude. If there is any informatics god, its blessings should also extend to the creators and maintainers of Linux Ubuntu and LibreOffice.

Sebastián Aguiar, Marc Kéry, Enrique Lessa, Daniel Naya, and Matías Schrauf kindly read, commented on, and corrected different chapters of this book. Their input and feedback prompted changes that led to improvements and hopefully fewer embarrassing mistakes. The stubborn errors, plain inconsistencies, and straight omissions that might remain are, of course, mine only. Melina Aranda, Javier García, Daniel Naya, Alicia Ponce, and Agustín Sáez kindly provided data from their published papers that are used as either case studies or problems at the end of some chapters. I thank Alexandra Elbakyan for allowing me to access an enormous amount of essential information that I could not otherwise have ever dreamed to read and use in this book: Өте көп.

I refuse to indulge in the tacky final sentences that end the prefaces of many scientific books: "Last but not least, I want to thank . . . for their patience and . . . for the many hours I spent . . ." Oh no, please not that again! But I will say this: over the last 12 years, I have been blessed beyond deserving by the earthly gods to share my life with Joana Gagliardi. She is my magnificent partner, my passionate lover, my close friend, and a truly great and beautiful woman with a shiny soul enveloped by a large smile and almond-shaped eyes. I have also had the privilege to share these years with Fiamma (24) and Iahel (20), Joana's bright daughter and son, whom I have seen grow into two beautiful adults who are the better angels of my soul.

This is enough now. You did not buy the book to read this babble. You want some stats, and that is what you will find starting on the next page. Should you have any comments, complaints, remarks, or suggestions, or have spotted any small or large errors, I want to hear from you, so please write to pablo.inchausti.f@gmail.com.

With warm regards,
Pablo
34°54′39″ S, 54°52′21″ W

Contents

PART II Applying the Generalized Linear Model to Varied Data Types

PART I

The Conceptual Basis for Fitting Statistical Models

CHAPTER 1

General Introduction

1.1 The purpose of statistics

The first article of the first issue of *Annual Review of Statistics* was entitled "What is statistics?" (Fienberg 2014). It started by listing eight different and only partly overlapping definitions. It is hard to imagine that chemists or physicists would provide as much variety when defining their own trades. The American Statistical Association offers a very inclusive definition: "Statistics is the science of learning from data, and of measuring, controlling and communicating uncertainty" (https://www.amstat.org/asa-newsroom). While not every statistician would agree with this, it serves to highlight that statistics is a kind of meta-discipline aiming to extract real-world insights from data gathered within other realms of knowledge (Wild et al. 2011). Statistics is a meta-discipline because, in dealing with the fuzziness, imprecision, and vagaries of real-world data, it pushes its practitioners to formulate "theoretical scaffolds" that can be used on other areas of knowledge.

Obtaining insights from statistics involves specifying hypotheses, gathering data relevant to a problem, modeling data with quantitative methods, and interpreting quantitative findings within the specific context of the scientific hypotheses that motivated the research. These activities do not, and cannot, take place as an intellectual abstraction aiming to solve problems within the clearly defined boundaries of applied mathematics where statistics is sometimes placed. Mathematicians often need to (over-)simplify the context of the initial problem to better define a narrower, more interesting, and hopefully solvable research question. In contrast, in statistics the context is the key to interpreting the findings of computer printouts of tables and graphs and to transforming data into insights in terms of the research problem and hypotheses that motivated the gathering of evidence. The practice of statistics is (or rather should be) something far more subtle and interesting than a quasi-mechanical quest to contrast and reject hypotheses whenever $p < 0.05$, as you might have learned in undergraduate courses.

"Statisticians are engaged in an exhausting but exhilarating struggle with the biggest challenge that philosophy makes to science: how do we translate information into knowledge?" (Senn 2003 p. 3). Taken at face value, how can this last statement fail to excite you? Statisticians deal with the excruciating messiness of real-world data. By that we mean the uncertainty in the measurements of variables, the pervasive variability of the world, and the often foggy relations between the variables that we aim to uncover in order to claim empirical support for a scientific hypothesis. Statistics has to tackle the chance and contingency that lie entangled within real-world data, and whose influence can be as pervasive as that of the signal related to the main patterns that we wish to reliably retrieve. The statistical holy grail is to uncover an approximate statistical model that could have plausibly generated (and hence fits acceptably well) the available evidence. But this is not

Statistical Modeling With R. Pablo Inchausti, Oxford University Press. © Pablo Inchausti (2023).
DOI: 10.1093/oso/9780192859013.003.0001

all. The magnitudes of the estimated parameters of such a well-fitting model should allow the evaluation of a statistical hypothesis and have a tangible, real-world interpretation in the research context that prompted the design of the experiment, the gathering of data, and its analysis.

1.2 Statistics in a schizophrenic state?

Over the last century, statistics has fully developed two theoretical frameworks (frequentist and Bayesian, to be explained in Chapters 2 and 3) that have contended to become "the right and appropriate" way of analyzing data. You will not find practitioners in other scientific disciplines spilling so many barrels of ink fighting each other without ever achieving complete victory. These two frameworks largely stem from two different views of probability that have coexisted since the seventeenth century, and their proponents and defenders have engaged in acrimonious and protracted disputes during most of the twentieth century. The currently dominant frequentist framework is an incoherent blend that arose from the protracted clash between R. Fisher on one side and J. Neyman and E. Pearson on the other. It is likely that Fisher and Neyman/Pearson only agreed on their strong dislike and distrust of the use of prior information (again, to be explained in Chapter 2) as a subjective and arbitrary component of the Bayesian framework that they wanted uprooted from statistics. Aiming for objectivity and conclusions that are independent of whoever analyzes the data, most of the practice of statistics championed under the frequentist framework has turned into a quasi-mechanized procedure aiming to reject statistical hypotheses.

It is currently fair to say that a clear majority of scientists have been educated in courses based on (and hence only use) frequentist methods. However, being in (a rapidly growing) minority does not suggest, or even less proves, that the champions of the Bayesian framework are "wrong" by any stretch of the imagination. The struggle for primacy between proponents of these two statistical frameworks has been largely inconclusive thus far. At present, scientists have a more ecumenical or pragmatic view of using what seems appropriate, and what they know best, to solve the problem at hand. Scientists needing to employ the other framework almost need to relearn from scratch. This book explains, discusses, and applies both the frequentist and Bayesian statistical frameworks to analyze the different types of data that are commonly gathered by research scientists and students.

The book in your hands aims to present material in an informal, approachable, and progressive manner suitable for research scientists and graduate students with a modicum of previous training. The book covers all the material in a theoretically rigorous manner, focusing on the practical applications of all methods to actual research data. It aims to provide just enough theoretical background for you to understand the basic underpinnings of the statistical models explained here. Every important formula will be "translated" into words to provide a clear, non-intimidating description to readers with only a basic background in mathematics and inferential statistics. In contrast to books laden with more theory, this is a "how-to" book. It emphasizes teaching by learning to compute using R, and to thoroughly interpret the results from the viewpoint and needs of research scientists and students.

1.3 How is this book organized?

It is unthinkable to carry out statistical analysis of meaningful amounts data of even moderate complexity without a computer. This book will make extensive use of the R programming environment (http://www.r-project.org/). This is an open-source (one can access and edit the code of all the R functions and save a revised version in one's computer), interpreted (it does not require compilation to be executed) programming language environment for statistical computing and graphics. R runs on Linux, Windows, and macOS, among others, and is the brainchild of its creators Ross Ihaka and Robert Gentleman. It is now supported by the R Foundation for Statistical Computing (Thieme 2018). R has experienced phenomenal growth since August 1993 to become one of the most popular and fastest growing programs for statistical analysis and graphics worldwide. Being a programming language, R can be easily extended by writing functions and extensions. There is a growing and very active R community creating packages (more than 17,500 packages in April 2021) and providing answers in terms of code and explanations in many active and fast-reacting mailing lists. R code is mostly written in the R language itself, although advanced users can link it to other computer languages such as C, C++, FORTRAN, Java, and Python using specific commands to assist in the execution of computer-intensive tasks.

Most statistics books using R aim for standalone use by providing brief (and by necessity incomplete) introductory chapters about the installation and basic use of R, including the basic commands to generate graphics. This introductory material about R can take up several chapters, often 10 to 20 percent of the overall length of many statistical textbooks. There are many books and companion websites that cover both the basic steps for using R and producing graphs: see Beckerman et al. (2017), Lander (2017), Petchey et al. (2021), and Teetor (2017) for the basics of R; Horton and Kleinman (2011) and Kabacoff (2011) for simple graphics, and Abedin and Mittal (2015), Chang (2012), and Teutonico (2015) for `ggplot2` graphics. We felt it unwise to provide the same material in print yet again. The companion website (www.oup.com/companion/InchaustiSMWR) contains detailed information about the installation of R in Windows, macOS, and Linux, along with the basic syntax for using and manipulating R objects. The website also provides detailed explanations for making basic plots in R using the package `ggplot2` (Wickham 2016), which is rapidly becoming the dominant approach to producing graphics in R. From here on, all R code in the book will be shown `in this font and highlighted in gray`. While the code necessary for each statistical analysis will be thoroughly explained in each chapter, the code used to make all the figures can be found on the companion website to avoid distracting you from understanding the main ideas. You will also find all the data sets and scripts (i.e., text files with commands) for each chapter in the companion website.

R has a rather minimalist interface in which the user types commands and obtains statistical and graphical results. RStudio (https://rstudio.org) has become a very popular graphical interface that manages the interaction between the user and R with great flexibility. The installation and basic use of this free graphical interface is also explained on the companion website. Nonetheless, all statistical and graphical analyses described in this book are independent of whether one uses a graphical interface such as RStudio.

This book is organized in three parts. Part I will provide the fundamental definitions of probability that underlie the frequentist and Bayesian frameworks, and develops the notion of parameter estimation as the main goal of statistical inference (Chapter 2). Chapter 3 then covers the basic underpinnings of the frequentist and Bayesian methods of parameter estimation (i.e., maximum likelihood, and the Markov chain and Hamiltonian Monte Carlo algorithms) that will be used in the data analyses of all the chapters of Parts II and III.

Part II represents the bulk of this book. It covers the analysis of the main types of data gathered in social and natural sciences from both frequentist and Bayesian perspectives. Each data set will be analyzed with both frameworks. Readers may choose to focus on separate, largely self-contained chapters depending on the type of response variable. However, the single effects of numerical and categorical explanatory variables (Chapters 4 to 6) should be examined as basic foundational aspects. Chapter 7 covers the theoretical basis of model selection (and a few other things), again for both frequentist and Bayesian frameworks. Chapter 8 reviews the conceptual basis of the generalized linear models that allow viewing most of the analyses explained in separate chapters of Part II as special cases. The assessment of statistical significance of parameter estimates, the calculation of confidence intervals, and the assessment of model goodness of fit are also covered. The rest of Part II covers, in separate chapters, the analysis of different types of data commonly encountered in scientific research involving binary, count, proportions, and other real-valued outcome variables. The quality of fit of all the statistical models to the data will be assessed with residual analysis and related methods, all of which will be explained in detail.

Part III builds on the understanding gained in Part II to incorporate random or population-level effects (Chapter 13). This enables the incorporation of structure in the data imposed by experimental and survey designs (Chapter 14). It is at this point that the book reaches its highest level of complexity, generality, and usefulness. As in all chapters of Part II, the emphasis is placed on formulating the starting statistical model, fitting the model using either the frequentist or Bayesian framework, interpreting and understanding the model outputs, assessing the goodness of fit to the data, and translating into words and figures the statistical findings for interpretation.

The book was structured and written assuming an imaginary reader interested in acquiring a broad and comprehensive understanding of univariate statistical analysis after a basic undergraduate course as taught in most engineering and science faculties around the world. These single-semester courses provide a basic understanding of descriptive statistics (mean, variance, quartiles), the basic notions of probability theory, a working knowledge of some probability distributions (e.g., normal, binomial), how to calculate the confidence intervals of at least the population mean, the basis (i.e., types of statistical errors, the notion of statistical significance) for testing statistical hypotheses about the differences between two means, and hopefully simple linear regression. The book starts slowly to progressively build a basic understanding of the main concepts and ideas that will be used in subsequent chapters.

1.4 How to use this book

In 1963 the Argentinian writer Julio Cortázar published the remarkable book *Hopscotch* (or *Rayuela* for those who can read it in the Spanish original). This novel has 155 mostly short chapters, 99 of which were considered "expendable" by its author. Even more, Julio Cortázar proposed several alternative ways in which his book could be read as if the chapters were pieces of many different possible puzzles to be assembled at will by its readers. Following Cortazar's lead, here are a few suggested paths for using this book:

- If you lack a reasonable knowledge of R and how to make graphics, you should definitely start by reading the introductory material about R and R graphics on the companion website.
- Should you not be interested in the historical roots and the conceptual basis of the frequentist and Bayesian frameworks over which statisticians have spilled so much ink, you may skip Chapters 2 and 3. However, please have a look at the final table

of Chapter 3 highlighting the main differences between the Bayesian and frequentist approaches that are worth knowing even if just for basic statistical literacy.

- If you are just interested in a specific data analysis (say, logistic regression, factorial analysis of variance, count regression), Table 2.1 points to the chapters you need depending on the probability distribution appropriate for modeling each type of response variable. Beware that you may need to have a look at parts of Chapter 8 to understand certain key features of the generalized linear models such as the link function. The main aspects of incorporating numerical and/or categorical explanatory variables in models are covered in Chapters 4 to 6, and they are valid for all models covered in this book.
- If you wish to learn either frequentist or Bayesian statistics, you may only read selected parts of specific chapters and simply dismiss the other half. But again, at this point in the twenty-first century it is becoming essential for scientists to possess at least a broad understanding of the theoretical/conceptual basis of both frequentist and Bayesian frameworks as discussed in Chapter 3. You will need the basics just to avoid getting lost and being fooled while reading papers.
- Readers only interested in Bayesian statistics may find it frustrating there there is no single chapter devoted to priors, the perennially debated feature of this framework. Starting in Chapter 4, the setting of priors is progressively built up in complexity in different chapters. There is a summary of the many non-exclusive steps or approaches to defining priors in the different chapters on page 323.
- Should you be interested in model selection in either the frequentist or Bayesian framework, you need to read parts of Chapter 7 to acquire at least a flavor of how it is done in either framework. Please read this chapter before doing any model selection with your specific data type, as unwritten and oral traditions have plagued too much of statistical model selection carried out by life scientists. Although the book has limited emphasis on model selection issues, there are specific examples in Chapters 11 and 12.
- Readers with data stemming from specific experimental designs should first read the chapter dealing with the type of data in Part II, then have at least a quick read on the theoretical basis of the mixed models (Chapter 13), and then carry out the data analysis perhaps inspired by one of the several examples given in the chapters of Part III.
- Finally, for readers wishing to acquire a broad and reasonably exhaustive overview of univariate statistics, the author suggests starting with Chapters 4 to 6, jumping to Chapter 8 to cover the basic theory of generalized linear models, and then going straight to the chapter(s) dealing with the types of data according to Table 2.1.

Whichever of the suggested (or other) paths you take through this book, it is very likely that you will have to flip back and forth to improve or check your understanding of a concept, an idea, or the interpretation of model results, or simply the code for an analysis or a figure. In this regard, while each chapter is self-contained, the book is heavily cross-referenced to allow you to find your way back and forth between chapters as needed.

References

Abedin, J. and Mittal, H. (2015). *R Graphs Cookbook*, 2nd edn. Packt Publishing, Birmingham.

Beckerman, A., Childs, D., and Petchey, O. (2017). *Getting Started with R: An Introduction for Biologists*. Oxford University Press, Oxford.

Chang, W. (2012). *R Graphics Cookbook*, 2nd edn. CRC Press / Chapman and Hall, New York.

Fienberg, S. (2014). What is statistics? *Annual Review of Statistics and Applications*, 1, 1–19.

Horton, N. and Kleinman, K. (2011). *Using R for Data Management Statistical Analysis and Graphics*. CRC Press / Chapman and Hall, New York.
Kabacoff, R. (2011). *R in Action*. Manning Publications, New York.
Lander J. (2017). *R for Everyone: Advanced Analytics and Graphics*, 2nd edn. Addison-Wesley, New York.
Petchey, O. Beckerman, A., Childs, D., et al. (2021). *Insights from Data with R: An Introduction for the Life and Environmental Sciences*. Oxford University Press, Oxford.
Teetor, P. (2017). *R Cookbook*. O'Reilly Publishing, New York.
Teutonico, D. (2015). *ggplot2 Essentials*. Packt Publishing, Birmingham.
Senn, S. (2003). *Dicing with Death: Chance, Risk and Healing*. Cambridge University Press, Cambridge.
Thieme, N. (2018). The R generation. *Significance*, 15, 14–20.
Wickham, H. (2016) *ggplot2: Elegant Graphics for Data Analysis*. Springer, New York.
Wild, C., Pfannkuch, M., and Horton, N. (2011). Towards more accessible conceptions of statistical inference. *Journal of the Royal Statistical Society A*, 174, 247–295.

CHAPTER 2

Statistical Modeling

A short historical background

2.1 What is a statistical model?

Using data to test statistical hypotheses, to fit empirical relations, or to explore suggestive patterns requires formulating statistical models. All statistical tests of hypotheses and statistical estimators of parameters are derived from statistical models. In very general terms, a statistical model can be defined as a mathematical equation(s) having at least one variable exhibiting stochastic (i.e., probabilistic) variation to represent the inherent uncertainty of observing its potential values.

The statistical models considered in this book contain a single response variable Y reflecting the effect of, or the variation associated with, the explanatory variables X. The latter can be any number of numerical variables, categorical variables denoting groups, or combinations thereof (i.e., interactions between explanatory variables). In all the models considered in this book, the response variable is a random variable with an associated probability distribution whose parameters embody both the effect of the explanatory variables and the variability of its potential values. Statistical models are thus equations that can be seen as data-generating mechanisms. They contain explicit assumptions that may reproduce the data for some combination of their parameters and values of the explanatory variables.

You might recall from previous introductory courses the existence of probability mass functions (PMFs) and probability density functions (PDFs) that are associated with discrete and continuous random variables, respectively. PMFs and PDFs are collectively also termed "probability distributions," and sometimes both are also subsumed under the term PDF. The names of some probability distributions that may spring to mind are binomial, Poisson, normal, and perhaps others. Which probability distribution could or should be used for each statistical model essentially depends on the main attributes of its response variable. Rather than showing a bestiary of the probability distributions that will be considered in this book along with their equations and their different shapes according to particular parameter values, we simply list them in relation to the type of data to which they apply (i.e., the domain of the response variable) in Table 2.1, and defer further details to the respective chapters where the analysis of each data type is explained. In addition, you can find such bestiaries of probability distributions in almost any statistics book on the shelf of the library of your institute, as well as on the internet.

Yet, why must the response variable Y of all statistical models be a random variable? There are several lines of argumentation for this (Blitzstein and Hwang 2014). One line of reasoning is that the randomness of the outcome variables results from the epistemic uncertainty (a fancy way of saying limited knowledge), and the measurement errors

Statistical Modeling With R. Pablo Inchausti, Oxford University Press. © Pablo Inchausti (2023).
DOI: 10.1093/oso/9780192859013.003.0002

Table 2.1 List of probability functions considered in this book.

Nature of the response variable	Probability distribution	Chapter
Dichotomous (or binary)	binomial	9
Unbounded counts: positive integers	Poisson, negative binomial, zero-inflated, or zero-augmented	10, 11
Proportions: real values in [0,1]	beta	12
Real values, strictly positive	gamma	12
Real values, positive and/or negative	normal	4–6

described before prevent us from precisely predicting them before actually measuring or estimating them during the data collection. Whenever we repeatedly perform simple experiments and measure or estimate the values of an outcome variable that characterize its outcome, we inevitably observe that variability is a pervasive feature. Every time you drive the same car 5 km at a constant speed it takes a different amount of time to reach its destination. After giving the same amount of water to identical genetic clones from the same original plant you will observe that plant height will vary among the pots. Variability, be it due to the uncertainty of the variables affecting an outcome or to the vagaries of measurement, is as important a part of reality as are the main trends observed in data. There would be no need for statistics in the absence of variability since, barring measurement error, a given set of inputs would then always render the same set of observable outputs.

We are interested in statistics because we need to understand how to explore, analyze, and interpret the data at hand in the context of our current research, or because we wish to understand some of the main principles involved in designing experiments and surveys to gather the data. True enough, the statistics involved in the exploration, analysis, and interpretation of data and in designing experiments and surveys requires a decent minimal background in probability theory. This much you already knew, which is why you acquired such basic knowledge before picking up this book. This book need not pretend to be a self-contained encyclopedia by repeating the introductory material on probability whose retelling has become an enduring ritual of statistics textbooks. Should you wish to refresh these fundamental concepts and ideas, consider consulting Wasserman (2004), Westfall and Henning (2014), and Blitzstein and Hwang (2014) among many, many others.

More interesting and useful to the goals of this book would be to recall the main interpretations of probability. This is because these interpretations of probability underlie and gave origin to the frequentist and Bayesian frameworks of statistical inference that are the subject matter of this book.

2.2 What is this thing called probability?

Probability is a principled way of quantifying uncertainty by assigning plausibility or credibility to a set of mutually exclusive possibilities or results of an experiment or observation. The concept of probability has a long, interesting, and convoluted history (see Tabak 2004, Stigler 1986, and Weisberg 2014). The origin of modern probability stems from Antoine Gombaud's (he was also known as Chevalier de Méré) question to Blaise Pascal (1623–1652) in a Paris salon regarding the fair division of stakes of an interrupted card game accounting for the previous and potential gains of each player. Gombaud's questions led to a brief correspondence exchange between Pascal and Pierre de Fermat (1607–1655). Pascal and Fermat were mostly concerned with evaluating gambles and equity, not with evaluating either evidence (i.e., data) or truth in arguments. Their correspondence would probably have vanished from public view were it not for Christian Huygens' 1657 book *De Ratiociniis in Ludo Aleae* (*On Reasoning in Games of Dice*). While still exclusively focusing on analyzing games of chance, Huygens' book remained the reference for probability for the next 50 or so years.

Jacob Bernoulli's posthumously published book *The Art of Conjecturing* (1713) marked a turning point in the history of probability for several reasons. First, Bernoulli (1654–1705) showed how to calculate probabilities as a frequency obtained by the ratio of the number of favorable events to the total number of events. By so doing, Bernoulli defined forever probability as an index of uncertainty that is bounded by 0 and 1. He also related the calculation of (some aspects of) probability to data, and made the crucial link between probability and the long-term frequency of an event, later called the law of large numbers. Second, Bernoulli applied probability to model uncertainty in areas other than gambling, such as human mortality and criminal justice, and by so doing he created what came to be known as "subjective probability." By making the crucial intellectual leap of viewing human existence as an existential lottery akin to a game of chance, Bernoulli was able to calculate mortality odds in order to price the lifetime yearly payment given by the state to the lenders of money to European states at the time. Bernoulli was probably the first to practically apply probability theory outside of games of chance.

Following the chronological line, Thomas Bayes' paper published posthumously in 1763 became the next key contribution to what in the twentieth century became known as statistical theory. Here's the historical context. The Scottish philosopher David Hume had argued in 1748 that "causes and effects are discoverable not by reason, but by experience." Hume stated that we can never be certain about the cause of a given effect as either or both of them may be due to an as yet unknown ultimate cause of both. In Hume's view, inductive inference embraced the uncertainty of inferring causes from effects by referring to probable rather than to definite causes. Being a Presbyterian minister and trained in mathematics, Bayes (1702–1761) wanted to counter Hume's view by finding a mathematical way based on probability theory to reliably infer the cause from an observed effect. His solution came to be known as Bayes' theorem or rule. It allows us to compute so-called "inverse probabilities," i.e., the chances of inferring a cause from its observed effects.

Bernoulli's monumental contribution was continued by Pierre-Simon Laplace (1749–1827) with two major works published in 1774 and 1814. In them, Laplace not only further developed the two views of probability contained in Bernoulli's book, but also independently reached the same result as Bayes, using a clearer and more thorough analysis. It is Laplace's results that form the current basis of what is called Bayes' rule in modern statistics (Chapter 3).

At this point, it is useful to reconsider the two main interpretations of probability considered by Bernoulli and Laplace in more detail to help synthesize the major ideas. You should also be aware that there are other classifications and accounts of the historical developments and interpretations of probability (e.g., Tabak 2004, Howie 2004, Zabell 2005, Stigler 1986, Weisberg 2014), but for the sake of brevity it suffices to consider here the two main, broad interpretations of probability.

The first interpretation of probability is sometimes called *aleatory probability* (Spiegelhalter 2019). It describes either a chance or experimental setup involving the process of obtaining uncertain observations, or the intrinsic uncertainty in nature. Let's also recall that the unpredictable chance events at subatomic levels can only be characterized using probability. Aleatory probability describes the propensity of the occurrence of events referring to an objective reality that is independent of an observer's knowledge, and of the amount of information they possess to describe it. This interpretation of probability includes the frequentist interpretation present in the books of Bernoulli and Laplace. In it, the probability of an event is the long-run proportion of times that it occurs within a set of infinitely many identical potential repetitions of an experiment or observation. Calculating that proportion thus requires defining a reference set of hypothetical, replicated experiments whose cumulative results would reflect the true tendency or propensity to observe any of all the possible outcomes. Aleatory probability requires an actual or hypothesized chance mechanism capable of generating a set of uncertain results whose frequencies we could count in a large or infinite set of identical trials or observations. The objective or frequentist interpretation of probability was later formalized in John Venn's (1866) *The Logic of Chance*. In Venn's view, probability is objective, literal, and not a conceptual or personal belief. As a consequence, Venn dogmatically dismissed the use of probability to refer to single events or anything unrelated to frequency (Howie 2004). The frequentist view of probability gained traction in the natural and social sciences in the late nineteenth century as scientists inspired by Francis Galton and Adolphe Quetelet were amassing ever larger amounts of empirical data (see Clayton 2021). Fisher (1925, p. 25) credited Venn with "developing the concept of probability as an objective fact, verifiable by observations of frequency," which came to be the dominant view for most scientists.

The other main interpretation of probability can be called *epistemic probability* (Spiegelhalter 2019). Here, probability refers to a measure of a personal degree of belief in some proposition. Therefore, epistemic probability is a mental construct that does not directly apply to actual events, but to our imperfect knowledge of reality that may be progressively modified by information. By "knowledge of reality" we refer to any observable event such as flipping a coin, rolling a die, or observing the results of experiments. First considered by both Bernoulli and Laplace as a universal model of rationality, this view of probability was later viewed by Augustus De Morgan and George Boole as a logical relationship between evidence and belief that measures our ignorance about the true state of affairs in the world (Howie 2004). This view of probability as a measure of a degree of belief was later championed by John Maynard Keynes (1921), Frank Ramsey (1931), and Bruno de Finetti (1933). Jeffreys (1939) wrote: "The essence of the present theory is that no probability, direct, prior, or posterior, is simply a frequency." In contrast to aleatory probability, epistemic probability quantifies the amount of information we possess regarding the occurrence of an event, or the degree of truth of a statement, or the degree of (un)certainty about an event. Additional information would generally decrease our ignorance and hence reduce our epistemic uncertainty.

Consider the perennial example of tossing a fair coin. The adjective "fair" to describe a coin could come from the long-term sequence of tosses yielding similar numbers of heads and tails. But it could just as well stem from considering the physical constitution of the coin that allows an equal chance of obtaining a heads or a tails. Or from the absence of additional knowledge that warrants you believing otherwise. Or from our a priori personal considerations that it is an even or fair bet. The probability of heads will be the same regardless of one's interpretation of probability. Following William Feller (1967, p. 3), "we shall no more attempt to explain the 'true meaning' of probability than the modern physicist dwells on the 'real meaning' of mass and energy or the geometer discusses the nature of a point."

2.3 Linking probability with statistics

Our short historical account of probability ended sometime in 1920s. This is when Ronald A. Fisher formulated a novel framework for a frequency-based general theory of parametric statistical inference. Fisher's framework included, among other things, maximum likelihood, tests of significance, randomization methods, sampling theory, analysis of variance, and experimental design. In 1922 Fisher published one of the most influential papers in the history of statistics that fundamentally changed its theory and methods forever. In this paper he single-handedly coined fundamental concepts such as "parameter," "statistic," "variance," "sufficiency," "consistency," "information," "estimation," "maximum likelihood estimate," "efficiency," and "optimality." He was also the first to use Greek letters for unknown parameters and Latin letters for their estimates. Much like in classical physics, the founding father of modern statistics was also an ill-tempered genius from Cambridge. In the words of Hald (1999, p. 1): "There are three revolutions in parametric statistical inference due to Laplace (1774), Gauss and Laplace in 1809–1812, and Fisher (1922)." The first revolution formally introduced the method of inverse probability, the second developed linear statistical methods based on the normal distribution, and the third introduced maximum likelihood as the workhorse method for statistical inference (Hald 1999). Indeed, it might be said that statistics as the child of probability theory was born with Bayes' posthumous 1763 paper, and was brought to maturity by Laplace who used inverse probability, via the now-standard Bayes' theorem (Pawitan 2001). The second revolution involved the development of a theory of errors by Gauss (1809). It was inspired by the need to adjust and summarize observational data from astronomy or surveying. Gauss proposed use of the normal distribution and the principle of least squares as a general method of estimation (Fig. 2.1).

Century	**XVII**	**XVIII**	**XIX**	**XX**		
Theory of probability	1654 B. Pascal & P. Fermat exchange 1657 C. Huygens' book	1713 J. Bernoulli's book	1866 J. Venn book	Books by 1921 J. Keynes 1925 F. Ramsey 1933 B. de Finetti 1933 A. Kolmogorov		
Frequentist statistics				1922, 1925, 1935 R. Fisher books and papers	1932, 1934 J. Neyman & E. Pearson papers	
Bayesian statistics		1763 T. Bayes paper	1812 P. Laplace book	1939 H. Jeffries book	1953 Metropolis et at paper	1990s Rediscovery of MCMC

Fig. 2.1 Dates and theoretical landmarks.

Prior to the third, Fisherian revolution, statistics mostly consisted of a collection of semi-independent, discipline-specific methods (Efron 1998) developed to analyze data in biology, agronomy, psychology, astronomy, etc. These methods included the least-squares method of estimation, linear regression and correlation, chi-square tables, and the t-test (see Stigler 1986, 1999 and Hald 1999 for historical overviews). These methods were applied to the large data sets that were amassed throughout the nineteenth century in the social and natural sciences by compulsive data-gatherers such as Adolphe Quetelet and Francis Galton. Throughout that century and until the 1930s, Bayes' theorem was widely taught and used for statistical inference using Laplace's methods by key statisticians such as K. Pearson and F. Edgeworth (Howie 2004, Zabell 2005). When developing the method of least squares, Gauss explicitly invoked Laplace's Bayesian perspective of using a uniform prior distribution for the unknown parameters (Stigler 2007). Fisher himself had "learned [Bayes' theorem] at school as an integral part of the subject, and for some years saw no reason to question its validity" (Fisher, 1936, p. 248).

Fisher's (1922) paper has been called "arguably the most influential article on [theoretical statistics] in the twentieth century" (Stigler 2005). In this paper, Fisher developed the method of maximum likelihood (to be explained in detail in Chapter 3), and also defined and evaluated desirable properties of the estimates of the parameters of statistical models. This provided a whole new conceptual framework for the science of statistics. Fisher (1922) explicitly defined probability as a frequency ratio in a "hypothetical infinite population" from which the actual sample is a circumstantial random realization now in the hands of the researcher (Chapter 3). And he also gave a new, technical meaning to the word "likelihood" to express the confidence that parameters have the specified values in a statistical population (another term that he coined in this paper) given the sample. Although there were earlier versions of likelihood (Howie 2004), Fisher (1922) was the first to clearly present it as a non-Bayesian approach; he also evaluated the properties and calculated the precision of parameter estimators. To summarize, Fisher proposed a new, universal method for parameter estimation applicable even to samples of moderate size and without the restrictive assumptions of other existing methods of parameter estimation (such as Pearson's chi square and method of moments), obtained the precision of parameter estimates (i.e., standard errors), and developed the modern method of statistical hypothesis testing. And all that in a single paper!

In 1925, Fisher published his most important book, *Statistical Methods for Research Workers*. This book contains a collection of non-mathematical prescriptions for the statistical analysis of biological and agronomic data using methods developed by Fisher himself (Hald 1999). It was later followed by *The Design of Experiments* (Fisher 1935), which also had a profound influence on the planning and analysis of comparative experiments in agriculture, biology, industry, psychology, and other areas. Fisher's monumental contribution thus combined the proper conduct and design of experiments with the proper analytical methods of data based on maximum likelihood into a coherent research program. A strict frequentist interpretation of probability underlay Fisher's theory and methods, whose strong influence dealt an almost killer blow to inverse probability (Howie 2004). Nonetheless, Fisher's views on probability evolved throughout his life and became somewhat more tolerant to the epistemic view of probability (Aldrich 2005).

Fisher (1925) actually created a new paradigm of statistical hypothesis testing. Fisher's book aimed "to put into the hands of research workers, and especially of biologists, the means of applying statistical tests accurately to numerical data" (Fisher 1925, p. 16). Specifically, Fisher's procedure involved a single statistical hypothesis that needed not be a null hypothesis (i.e., zero difference, or no relation). In his words, "every experiment may be said to exist only to give the facts a chance of disproving the null hypothesis" (Fisher 1935, p. 16). Provisionally assuming this single hypothesis to be true, Fisher's procedure then used the data to calculate a *p*-value. This is the probability of observing similar or more extreme data in potential random samples from the same hypothetically infinite statistical population. In a sense, the *p*-value signals an outlier indicating that "either an exceptionally rare chance has occurred, or the theory [hypothesis] of random distribution is not true" (Fisher 1956, p. 42). Fisher stated: "It is usual and convenient for experimenters to take 5 per cent as a standard level of significance, in the sense they are prepared to ignore all results which fail to reach this standard" Fisher (1935, p. 15). He also warned that "no scientific worker has a fixed level of significance at which from year to year, and in all circumstances, he rejects hypotheses; he rather gives his mind to each particular case in the light of his evidence and his ideas" (Fisher 1956, p. 42).

Why does the significance level of 5% now seem to be cast in stone? At that time, calculating the exact *p*-value involved long and tedious numerical integration of the PDF. The best tables available in 1925 were those previously published by Karl Pearson. These tables constituted an important source of income for Pearson's biometry laboratory. Unsurprisingly, after their acrimonioous public dispute Fisher could not obtain a waiver of Pearson's copyright to reproduce these tables in full in his 1925 book (Stigler 2006). He then resorted to the practical solution of calculating and printing just the 5% and 1% quantiles for several probability distributions in his book, and recommended comparing the value of the estimated test statistic with his 5% and 1% quantiles. Oddly enough, this commercial dispute might be the likely start of the 5% criterion of statistical significance that in the eyes of many became a sacred number. The current prevalence of the 5% is a testament to the extraordinary influence of Fisher's book.

In the 1930s, Jerzy Neyman and Egon Pearson aimed to build a theory of statistical decision-making by improving on Fisher's approach. Both theories were strongly anchored on the likelihood principle, and on a frequentist interpretation of probability (Lehmann 2011). While Fisher viewed the testing of a statistical hypothesis as an aspect of inductive inference, Neyman and Pearson considered theirs mostly as on an inductive behavioral basis. Neyman & Pearson (1928, 1933) set up two mutually exclusive statistical hypotheses, defined two types of decision error (type I and type II), found the existence of a trade-off between them, and discussed that the choice of which error to minimize would depend on the empirical consequences of the decision. Their approach also used Fisher's significance test, whose *p*-values were interpreted as the long-term error rates in a sequence of repeated random samplings from a finite statistical population.

A long, acrimonious public debate ensued in the 1930s between Fisher on one side and Neyman and Pearson on the other about the framing and interpretation of the significance tests. Public statements such as "a threat to the intellectual freedom of the West" (Fisher to Neyman) and "worse than useless" (Neyman to Pearson) did not help its settlement, which only occurred on Fisher's death. Textbooks following the Second World War contained a hybrid (not to be confused with a synthesis) of Fisher's, Neyman's, and Pearson's views that has been, and still is, taught in courses around the world. This hybrid includes Fisher's significance testing and the 5% significance level, and Neyman and Pearson's two statistical hypotheses and their associated decision errors, confidence intervals, and the use of statistical power to aid the planning of experiments (Chapter 14).

2.4 The early Bayesian demise during the 1930s

Fisher developed a forceful view rejecting "inverse probability" (as Bayesian methods were known at the time) despite it having been the main methodology for statistical inference for nearly 150 years. Howie (2004, p. 55) traced Fisher's rejection of inverse probability to his over-reaction to Soper et al.'s (1917) comments on Fisher's (1915) exact distribution of the correlation coefficient, considered a misapplication of inverse probability by Fisher. Being a lifelong devotee of frequentist probability, Fisher was incensed by the description of his method as "Bayesian" (in today's terms). Fisher contended that it was Soper et al. (1917) who did not understand probability, because they mistakenly applied the concept to both the results of an experiment and to statistical hypotheses, which are distinct things, each requiring a separate measure of uncertainty (Howie 2004). Pearson's refusal to publish Fisher's response in *Biometrika* started a feud that that only ended on Pearson's death. Nonetheless, Fisher even had to "plead guilty in my original statement of the Method of Maximum Likelihood [Fisher 1912] to having based my argument upon the principle of inverse probability" (Fisher 1922, p. 326). Fisher did not mince words when he called inverse probability "fundamentally false and devoid of foundation" (Fisher 1930, p. 528), "that the theory of inverse probability is founded upon an error, and must be wholly rejected" (Fisher 1925, p. 10), and "a mistake, perhaps the only mistake to which the mathematical world has so deeply committed itself" (Fisher 1922, p. 325).

Fisher's other main objection was that the flatness of prior probability densities was not a "property of noninformativity" or absence of knowledge about the values of model parameters (Schweder and Hjort 2016). As we discuss in Chapter 3, assuming uniform prior distributions (known as Laplace's "principle of insufficient reason") of model parameters when using Bayesian methods was often the only means to obtain analytical solutions in the pre-computer days. Fisher forcefully argued (which turned out to be true, at least in some cases) that leaving open the choice of prior distributions introduced an arbitrary and subjective component in statistical inference. Although these would seem to be unnecessary technicalities at this point, the modest goal here is to give you at least the gist of the polemics leading to the early demise of Bayesian methods at the time.

While Fisher and others were developing the frequentist framework, Harold Jeffreys' (1939) book *Theory of Probability* contained the first modern exposition of Bayesian methods (Zabell 1989). This book remained for many years the only systematic explanation of how to apply Bayesian methods to scientific problems. Unlike Fisher's strict frequentist interpretation, Jeffreys considered probability as an appropriate measure for all types of uncertainty. Jeffreys believed that Fisher's (1922) maximum likelihood method was basically Bayesian (Zabell 1989), and his own theory basically fused Fisher's likelihood principle with Bayes' theorem and subjective probability as the mechanisms for achieving inferential coherence (Lindley 1965). In reviewing Jeffreys' (1939) book, Fisher pointed out that Jeffreys made "a logical mistake on the first page (i.e., Bayes rule) which invalidates all the 395 formulae in his book" (Aldrich 2005).

The sheer personality and strong influence of Fisher in the development of theoretical statistics and experimental methods based on a frequentist probability in the 1920s and 1930s led to the early demise of inference based on the Laplacian inverse probability method. In the early 1930s, Fisher and Jeffreys publicly debated the meaning of probability and methods for statistical inference. Although their debate essentially ended inconclusively, in practice Jeffreys and the Bayesians lost (McGrayne 2010). Fisher's views were seconded by the vast majority of statisticians in the following 30 years. In his review of Jeffreys' (1939) book, Wilks (1941, p. 194) was prophetic in thinking it "doubtful that there will be many scholars thoroughly familiar with the system of statistical thought initiated by R. A. Fisher and extended by Neyman, Pearson, Abraham Wald and others who will abandon this system in favor of the one proposed by Jeffreys in which inverse probability plays the central role."

The statistical theory and methods pioneered by Fisher, Neyman, Pearson, and Wald have been labeled "frequentist" and "classical" in the statistical literature, and those based on the Laplacian inverse probability as "Bayesian." It was Nagel (1936) who coined the term "frequentist" to refer to theory and methods that were also called "classical" by

Neyman (1937). At that time, the label "classical" attached to frequentist statistics was an oxymoron since the theory and methods based on inverse probability (what we would now call Bayesian) were at least 200 years older than those created just 15 years earlier by Fisher and others. Ironically, it appears that Fisher himself was the first to use the label "Bayesian" (Stigler 2006).

Until its revival in the early 1990s, what was to become the Bayesian framework barely survived in small, scattered research groups and university departments, mostly in the US and the UK (Stigler 2008). Nevertheless, the true renaissance of the Bayesian framework leading to its current popularity had to wait until the rediscovery of Monte Carlo Markov chain methods (a class of algorithms to numerically estimate parameters of statistical models; see Chapter 3) in the late 1980s that made possible computations that could only be dreamed of a few decades earlier. The story of the recent Bayesian renaissance, along with a fuller explanation of frequentist and Bayesian frameworks, comes next in Chapter 3.

References

Aldrich, J. (2005). Fisher and regression. *Statistical Science*, 20, 401–417.

Blitzstein, J. and Hwang, J. (2014). *Introduction to Probability*. CRC Press / Chapman and Hall, New York.

Clayton, A. (2021). *Bernoulli's Fallacy: Statistical Illogic and the Crisis of Modern Science*. Columbia University Press, New York.

de Finetti, B. (1933). *Theory of Probability: A Critical Introductory Treatment*. John Wiley and Sons, New York.

Efron, B. (1998). R. A. Fisher in the 21st century. *Statistical Science*, 13, 95–122.

Feller, W. (1967). *An Introduction to Probability Theory and Its Applications*, vol. 1. John Wiley and Sons, New York.

Fisher, R. (1912). On an absolute criterion for fitting frequency curves. *Messenger of Mathematics*, 41, 155.

Fisher, R. (1915). Frequency distribution of the values of the correlation coefficient in samples from an indefinitely large population. *Biometrika*, 10, 507–521.

Fisher, R. (1922). On the mathematical foundations of theoretical statistics. *Proceedings of the Royal Society A*, 222, 309–368.

Fisher, R. (1925). *Statistical Methods for Research Workers*. Oliver and Boyd, Edinburgh.

Fisher, R. (1930). Inverse probability. *Mathematical Proceedings of the Cambridge Philosophical Society*, 26, 528–535.

Fisher, R. (1935). *The Design of Experiments*. Oliver and Boyd, Edinburgh.

Fisher, R. (1936). Uncertain inference. *Proceedings of the American Academy of Arts and Science*, 71, 245–258.

Fisher, R. (1956). *Statistical Methods for Research Workers*, 3rd. edn. Hafner Press, New York.

Gauss, C. (1809). *Theory of the Motion of Heavenly Bodies Moving about the Sun in Conic Sections* (English translation by C. H. Davis), reprinted 1963, Dover, New York.

Hald, A. (1999). On the history of maximum likelihood in relation to inverse probability and least squares. *Statistical Science*, 14, 214–222.

Howie, D. (2004). *Interpreting Probability Controversies and Developments in the Early 20th Century*. Cambridge University Press, Cambridge.

Jeffreys, H. (1939). *Theory of Probability*. Oxford University Press, Oxford.

Keynes, J. M. (1921). *A Treatise on Probability*. Macmillan and Co., London.

Laplace, P. (1774). Mémoire sur la probabilité des causes par les événements. *Mémoires de l'Academie Royale des Sciences Presentés par Divers Savan*, 6, 621–656.

Laplace, P. (1814). *Essai philosophique sur les probabilités*. Translated by Truscott, F. and Emory, F. Cosimo Classics, New York.

Lehmann, E. (2011). *Fisher, Neyman, and the Creation of Classical Statistics*. Springer, New York.
Lindley, D. (1965) *Introduction to Probability and Statistics from a Bayesian Viewpoint*. Cambridge University Press, Cambridge.
McGrayne, S. (2010). *The Theory That Would Not Die*. Yale University Press, New Haven.
Nagel, E. (1936). The meaning of probability. *Journal of the American Statistical Association*, 31, 10–31.
Neyman, J. (1937). Outline of a theory of statistical estimation based on the classical theory of probability. *Philosophical Transactions of the Royal Society A*, 236, 333–380.
Neyman, J. and Pearson, E. (1928). On the use and interpretation of certain test criteria. *Biometrika*, 20, 175–240.
Neyman, J. and Pearson, E. (1933). On the problem of the most efficient tests of statistical hypotheses. *Philosophical Transactions of the Royal Society A*, 231, 289–337.
Pawitan, J. (2001). *In All Likelihood: Statistical Modeling and Inference Using Likelihood*. Oxford University Press, Oxford.
Ramsey, F. (1931). Truth and probability In F. Brainwaite (ed.), *The Foundations of Mathematics and other Logical Essays*, pp. 156–198. Harcourt, Brace and Co., London.
Schweder, T. and Hjort, N. (2016). *Confidence, Likelihood, Probability: Statistical Inference with Confidence Distributions*. Cambridge University Press, Cambridge.
Soper, H., Young, A., Cave, B. et al. (1917). On the distribution of the correlation coefficient in small samples. Appendix II to the papers of "Student" and R. A. Fisher. A cooperative study. *Biometrika*, 11, 328–413.
Spiegelhalter, D. (2019). *The Art of Statistics: Learning From Data*. Penguin Publishers, London.
Stigler, S. (1986). *The History of Statistics: The Measurement of Uncertainty Before 1900*. Harvard University Press, Harvard.
Stigler, S. (1999). *Statistics on the Table*. Harvard University Press, Harvard.
Stigler, S. (2005). Fisher in 1921. *Statistical Science*, 20, 32–49.
Stigler, S. (2006). Fisher and the 5% level. *Significance*, 21, 12.
Stigler, S. (2006). When did Bayesian inference became "Bayesian?" *Bayesian Analysis*, 1, 1–40.
Stigler, S. (2007). The epic story of maximum likelihood. *Statistical Science*, 4, 598–620.
Tabak, J. (2004). *Probability And Statistics: The Science Of Uncertainty*. Checkmark Books, New York.
Venn, J. (1866). *The Logic of Chance, An Essay on the Foundations and Province of the Theory of Probability, with Especial Reference to Its Application to Moral and Social Science*. Macmillan and Co., London.
Wasserman, L. (2004). *All of Statistics: A Concise Course in Statistical Inference*. Springer, New York.
Weisberg, H. (2014). *Willful Ignorance: The Mismeasure of Uncertainty*. John Wiley and Sons, New York.
Westfall, P. and Henning, K. (2014). *Understanding Advanced Statistical Methods*. CRC Press / Chapman and Hall, New York.
Wilks, S. (1941). Theory of Probability by Harold Jeffreys. *Biometrika*, 32, 192–194.
Zabell, S. (1989). R. A. Fisher on the history of inverse probability. *Statistical Science*, 4, 247–263.
Zabell, S. (2005) *Symmetry and its Discontents: Essays on the History of Inductive Probability*. Cambridge University Press, Cambridge.

CHAPTER 3

Estimating Parameters

The main purpose of statistical inference

R packages needed in this chapter:
`packages.needed<-c("ggplot2","reshape2","gridExtra","mvtnorm")`
`lapply(packages.needed,FUN=require,character.only=T)` # loads these packages in the working session.

3.1 Introduction

Statistical inference aims to estimate the values of the (unknown) parameters in a statistical model. Every statistical model is a theoretical construct framed as a mathematical equation that could generate the data in hand for some combination of its parameters and values of the explanatory variables. Estimating the values of the model parameters allows the use of a statistical model as a data-generating mechanism whose predictions can be compared with available evidence.

The goal in statistics is rarely to reach conclusions or make statements about the actual data in hand. Rather, the latter are considered a small subset obtained from a much larger collection of potential values of the response and explanatory variables that could have potentially been recorded for the same hypothetical data-generating process. The statistical population is defined as the full set of values of the response variable that could potentially be included in a sample. The information embodied by a sample (from the point of view of statistical inference, the parameter estimates) is nothing more than the transitory means to the larger end of reaching reliable conclusions about the statistical population. In other words, statistical estimation is typically about the process thought to have generated your actual data or, similarly, about "all" data sets it could potentially have produced.

Formally speaking, the statistical population imagined to define a statistical model can be considered as having a hypothetically infinite size. The study population (set of all patients, plots of land, etc.) can be comprised of a potentially large number of elements, of which only a small number are sampled. In the rare case of a complete census, we can obviate the use of statistics since the sampled data will allow us to know all that can possibly be known about the values of the parameters (e.g., means, variances) of the statistical model with infinite precision or zero estimation error. In all other cases, we need statistical induction, i.e., reaching general conclusions about the statistical population from the limited information contained in the data. This is unavoidably uncertain and only approximately valid under specific circumstances. One of these circumstances involves random sampling in which every element of a statistical population has the same chance

Statistical Modeling With R. Pablo Inchausti, Oxford University Press. © Pablo Inchausti (2023).
DOI: 10.1093/oso/9780192859013.003.0003

of being included in our sample. Random sampling is then a generic circumstance permitting the use of the parameter estimates obtained from the sample to infer the values of parameters of the study population.

3.2 Least squares: A theory of errors and the normal distribution

The first approach to parameter estimation can actually be found in Bayes' (1763) paper. It was independently and substantially developed by Laplace (1814) into what we now call Bayesian methods. Bayes and Laplace both used inverse probability via the now standard Bayes' theorem to estimate the parameter(s) of a statistical model from data. The Bayes–Laplace method was the only available method of statistical inference until Gauss developed the method of least squares in 1809 in connection with the normal distribution. He thus generated a theory of errors when fitting curves and surfaces to measurements in geodesy and astronomy. In 1814 Laplace proved the first central limit theorem, whereby the additive composition of independent and identically distributed random variables has a normal distribution. He thus justified the use of the method of least squares for large sample sizes. The nineteenth century witnessed the increasing gathering of large data sets in social and natural sciences that led to the development of analytical methods such as linear regression and correlation. The least squares method played no small part in the analyses of those large data sets of quantitative variables that could be assumed to follow a normal distribution (Chapter 2).

Least squares is a parameter estimation method in linear models where the response variables follow a normal distribution with constant variance (this is known as homoscedasticity; for more on this see Chapters 4 to 6). However, many response variables are not, and cannot be transformed to be, normally distributed. Therefore, the least squares method can only be used in a relatively narrow set of circumstances.

3.3 Maximum likelihood

3.3.1 *The basic concepts*

This is a general method for the estimation of the parameters in statistical models that was single-handedly created by R. A. Fisher between 1912 and 1922 (Chapter 2). Maximum likelihood can estimate the values of model parameters for any probability distribution adopted to describe a response variable. Maximum likelihood became an essential, general-purpose tool for inference in frequentist statistics, as it allowed the evaluation of statistical significance, calculation of confidence intervals, model assessment and comparison, and prediction (Millar 2011, Pawitan 2001). Importantly, maximum likelihood estimates for a normal response variable coincide with the corresponding least-squares estimates.

Likelihood, a very old term dating to at least the fourteenth century, is loosely synonymous with probability, but it acquired a new meaning in statistics after Fisher's work. Fisher (1912) initially derived the ideas of maximum likelihood from the "principle of inverse probability" (Chapter 2). Bayes' theorem framed in the context of parameter estimation states:

$$\Pr(\text{parameter} \mid \text{data}) \times \Pr(\text{data}) = \Pr(\text{data} \mid \text{parameter}) \times \Pr(\text{parameter}). \tag{3.1}$$

In words, this theorem is a statement based on the definition of conditional probability that allows inverting the first term on each side of the equality sign. It relates the probability of the parameter having a certain value given the data, Pr(parameter | data), with the probability of observing the data given a certain value of the parameter, Pr(data | parameter). We will delve further into Bayes' theorem when covering the Bayesian method of parameter estimation.

Before delving into maximum likelihood, let's recall the use of a probability distribution in undergraduate courses with the example of coin tossing. We would write that the response variable Y = "# of heads in $k = 10$ tosses" follows a binomial distribution with parameters (k, π), where π is the probability of observing a head in each toss. To do so, we assumed that there were two mutually exclusive results, and that successive tosses are mutually independent. Now, assuming a certain value of the parameter π, we could easily calculate the probability of observing any possible result (i.e., $Y = 0, 1, \ldots, 10$ heads) in $k = 10$ tosses. These calculations assumed the binomial distribution as adequate, and that we knew the value of π from some external knowledge.

In statistics, however, we always have the inverse problem: given some data (e.g., the number of heads or tails observed in $k = 10$ tosses), and assuming a statistical model (here, Binomial(k, π)), we wish to estimate the "best" value to assume for the parameter π. The goal of maximum likelihood is to estimate the values of the parameters of a statistical model that are deemed most likely to have generated the observed data (Fig. 3.1).

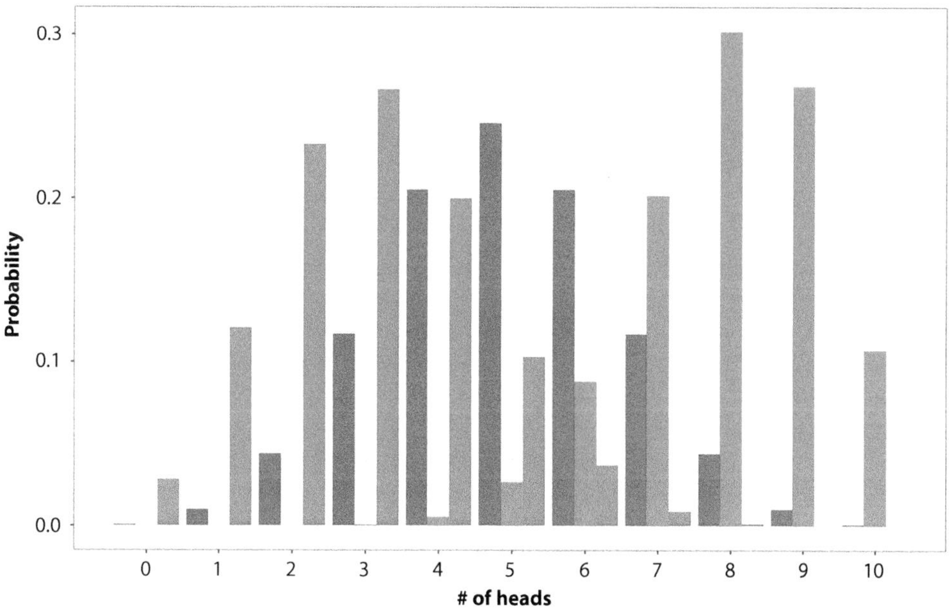

Fig. 3.1 Probabilities of observing 0–10 heads in $k = 10$ coin tosses for $\pi = 0.3$ (blue), $\pi = 0.5$ (red), and $\pi = 0.8$ (green).

Fisher (1922, p. 310) defined "the likelihood that any parameter (or set of parameters) should have any assigned value(s) is proportional to the probability that if this were so, the totality of the observations should be that observed." That is, Likelihood(data | parameters) $\propto$ Pr(parameters | data), i.e., the left-hand side of Eq. (3.1), with the constant of proportionality being arbitrary and dependent on the statistical model being fitted.

Although Fisher viewed both probability and likelihood as measures of uncertainty, for him they were on a very different footing. He stated that we can speak of the probability of observing any value of the response variable, given a statistical model (e.g., binomial) and its parameter estimate (say $\pi = 0.5$), but we speak of the likelihood of a statistical hypothesis (i.e., a statement about parameter values) given the observed data. The Fisherian likelihood method would then allow the estimation of the parameter π given the data (e.g., 6 heads in 10 coin tosses) and the statistical model (binomial).

Importantly, the likelihood function is not a probability function. Hence, unlike probability, the likelihood does not obey the various axioms of probability such as adding up to one. Instead, the likelihood assesses the relative plausibility of different values of a parameter given the observed data. The Fisherian Likelihood(data | parameters) views the observed data set as one of the infinitely many possible random realizations of the probabilistic data-generation mechanism embodied by the statistical model. Intriguingly, Fisher (1922) also defined likelihood as a "measure of our rational degree of belief in a conclusion." In the context of the above example, it is the belief that the parameter π is equal to (say) 0.38 given the data. The latter definition is akin to the epistemic view of probability (Chapter 2). However, Fisher always consistently aligned with the frequentist framework of probability (Chapter 2), and was vehemently opposed to the use of probability as denoting uncertainty about specific statements or hypothesis (such as $\pi = 0.38$). In Fisher's view, such statements can be either true or false, but do not admit a frequentist underpinning of probability. In keeping with the frequentist view of probability, Fisher consistently reserved the term likelihood for evaluating and comparing statistical models and hypotheses given the data.

3.3.2 *Obtaining maximum likelihood estimates*

Maximum likelihood estimates of model parameters are obtained by maximizing the likelihood function with respect to its parameters. This will render the parameter values that are most likely to have produced the observed data, given the PMF or PDF used to model the response variable. Thus, the likelihood function neither comes out of the blue, nor does its definition require divine inspiration. Rather, it is defined by the probability density or mass function of the response variable in the statistical model.

So, how can we obtain the values of the parameters that maximize that function? In some cases, maximum likelihood estimates of model parameters can be obtained by analytically calculating the derivative of the likelihood function with respect to its parameters, and then setting it to zero to determine its maximum. When we do this, we end up with simple formulae that allow us to obtain the maximum likelihood parameter estimate directly from the data. In most other cases, maximizing the likelihood function is achieved using numerical methods, many of which have been implemented in R. We will illustrate both approaches.

Let's start with the analytic estimation of parameters for the coin tossing example by maximum likelihood. We define the response variable Y = # heads and assume we obtained 6 heads in 10 independent tosses. We further assume that the binomial distribution is a reasonable statistical model such that we can assume $Y \sim \text{Binomial}(k, \pi)$. Now, our goal is to estimate that value of π that is most likely to have generated the observed data (i.e., 6 heads in 10 tosses). We know that the Binomial(k, π) PMF is

$$\Pr(Y = y \mid \pi) = \binom{k}{y} \pi^y (1 - \pi)^{k-y}, \tag{3.2}$$

where the first term counts the number of ways in which we can obtain y heads in k tosses. By substituting $y = 6$ and $k = 10$, Eq. (3.2) would yield the probability of obtaining the observed data for a given value of the parameter π. But we don't know the value of π; we will estimate that value of π that is most likely to have give risen to the entire data set of 6 heads in 10 tosses. We already know that by Fisher's definition the likelihood $L(\pi \mid \text{data})$ is proportional to $\Pr(Y \mid \pi)$, and hence

$$L\left(Y = y \mid \pi\right) \approx \binom{k}{y} \pi^{y}(1-\pi)^{k-y}. \tag{3.3}$$

Taking logarithms in Eq. (3.3) will transform a product of three terms into a sum of three terms that is easier to manipulate. Also, because logarithm is a monotonic function, the value of π that maximizes L will also maximize the log-likelihood log L. Therefore,

$$\log L\left(Y = y \mid \pi\right) \approx \log\binom{k}{y} + \log\left(\pi^{y}\right) + \log\left((1-\pi)^{k-y}\right), \tag{3.4}$$

And, applying the properties of logarithms,

$$\log L\left(Y = y \mid \pi\right) \approx \log\binom{k}{y} + y_i \log\left(\pi\right) + (k-y)\log\left((1-\pi)\right). \tag{3.5}$$

To maximize log L with respect to π, we take its derivative and set it to zero:

$$\begin{aligned} \frac{d\log L\left(Y = y \mid \pi\right)}{d\pi} &\approx \frac{d\left(\log\binom{k}{y} + y\log\left(\pi\right) + (k-y)\log\left((1-\pi)\right)\right)}{d\pi} = 0 \\ &\approx 0 + \frac{y}{\pi} - \frac{(k-y)}{1-\pi} = 0. \end{aligned} \tag{3.6}$$

Note that the first term in Eq. (3.6) does not contain the parameter π, and hence it is a constant whose derivative with respect to π is equal to zero. Solving the second line of Eq. (3.6) gives $\pi = \frac{y}{k}$, which in words says that the value of π that is most likely to have produced the observed data (6 heads in 10 tosses) is precisely 6/10. Of course, you knew this result in advance. We only show it here to illustrate obtaining the maximum likelihood estimate using an analytical (i.e., "pen and paper") procedure.

Turning now to the numerical estimation of π, let's start by recalling that Eq. (3.4) will give the log-likelihood of obtaining a given number of heads in $k = 10$ tosses for a value of π. The log-likelihood in Eq. (3.4) is a convenient and equivalent way of implementing Eq. (3.3) that is numerically more accurate for very small numbers. We could now implement either Eq. (3.3) or (3.4) in R as a function, but R has the binomial density function as a shortcut, so we could calculate the probability of obtaining 6 heads in 10 tosses if π were equal to 0.5 as follows:

```
> dbinom(x=6, size=10, prob=0.5)
 [1] 0.205078
```

We could implement Eq. (3.4) to calculate the log-likelihood of 6 heads in 10 tosses as:

```
> log(dbinom(x=6, size=10, prob=0.5))
 [1]-1.58
```

or as:

```
> dbinom(x=6, size=10, prob=0.5, log=TRUE)
 [1]-1.58
```

Of course, we claim at this point that we were ignoring the true value of the MLE of π; otherwise we would not be doing any of this!

What we need to do now is to evaluate the log-likelihood of the data (6 heads in 10 tosses) for a plausible range of values of π to identify the value that is most likely to have generated the observed data. We first generate a vector of values for the parameter to be evaluated:

```
DF4 = data.frame(prob = seq(from = 0.001, to = 0.999, by = 0.001))
```

Then, we can use the binomial PMF in R with its log argument set to `TRUE` to calculate the log-likelihood for different values of π given by the variable `prob`:

```
Lbinom=dbinom(prob, x, size, log=TRUE)
```

We can use this function to yield the log-likelihood of the data for a given value of its argument `prob`:

```
> LBinom(0.3, x=6, size=10)
[1] -3.3
> LBinom(0.8, x=6, size=10)
[1] -2.43
```

The candidate value of $\pi = 0.3$ is less likely than the $\pi = 0.8$ for `x` = 6 heads in `size` = 10 tosses. We can calculate the log-likelihoods of many different possible values of π, and store the results inside a newly created variable `Lik` in `DF4` by:

```
DF4$Lik=LBinom(prob=DF4$prob, x=6, size=10)
```

Let's examine the first six (of the 999) rows of `DF4`:

```
> head(DF4)
   prob   Lik
1 0.001 -36.1
2 0.002 -31.9
3 0.003 -29.5
4 0.004 -27.8
5 0.005 -26.5
6 0.006 -25.4
```

We can then plot these log-likelihood values to visually (and crudely, or by "brute force") find the value of π that is most likely to have produced the observed data and which we are going to call the MLE.

We could locate the value of π associated with the maximum of the log-likelihood function (Fig. 3.2) by using the R function `which.max` that gives the row number of the maximum value of a vector:

```
> which.max(DF4$Lik)
[1] 600
```

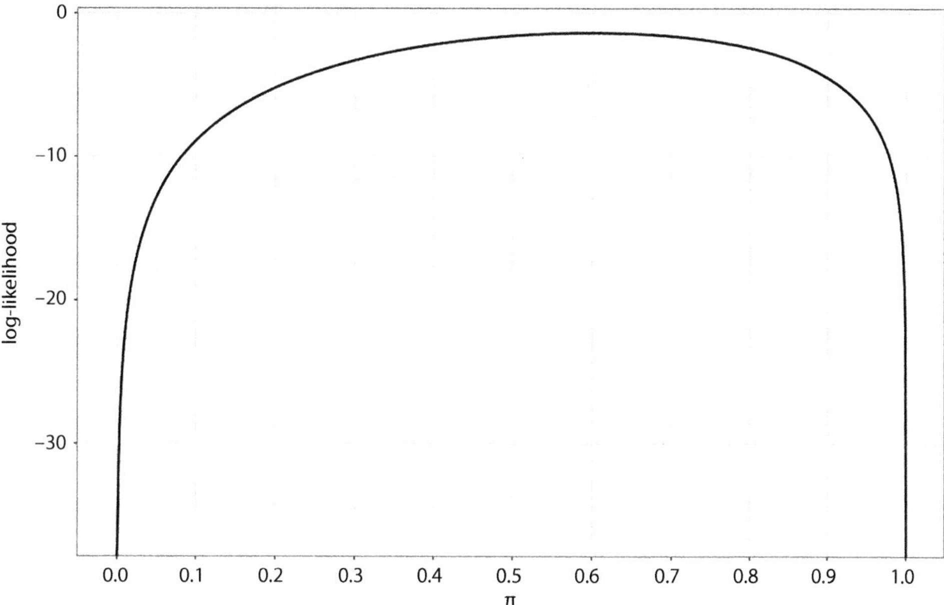

Fig. 3.2 The log-likelihood of observing 6 heads in 10 tosses for a set of plausible values of π. The maximum log-likelihood estimate of π appears to be around 0.6.

Then, recalling the syntax for handling data frames:

```
> DF4[which.max(DF4$Lik),]
    prob        Lik
600  0.6 -1.383009
```

Thus, $\pi = 0.6$ is the maximum log-likelihood (−1.38309) value among the 999 values of π evaluated and stored in `DF4`. Because we could have considered more (or fewer) values of π to obtain more (or less) accuracy in its estimation, we must do better than this to attain a more reliable result. One option is the function `mle2` of the package `bbmle`, as follows:

```
prob.estimated=mle2(minuslogl=LBinom,data=list(x=6, size=10),
                start=list(prob=0.2))
```

This calls the command `mle2` to use our log-likelihood function (`LBinom`), the data (6 heads in 10 tosses), and provides a starting value for the optimization routine (such as the Newton–Raphson method) employed to numerically calculate the maximum of `LBinom` for the data set, and stores the result in `prob.estimated`. The result is:

```
> summary(prob.estimated)
Maximum likelihood estimation
Call:
mle2(minuslogl = LBinom, start = list(prob=0.2),data = list(x = 6, size = 10))
Coefficients:
     Estimate Std. Error z value     Pr(z)
prob  0.60000    0.15492   3.873 0.0001075
-2 log L: 2.766018
```

We obtained the maximum likelihood estimate and its standard error (the rest of the output is of no interest for our purposes here). We can obtain the 95% confidence interval of π by:

```
> confint(prob.estimated)
Profiling...
 2.5% 97.5%
0.299 0.854
```

The 95% CI was obtained by finding the lower and upper limits of the interval of values of π that encompass 95% of the (inverted) likelihood profile (Fig. 3.3).

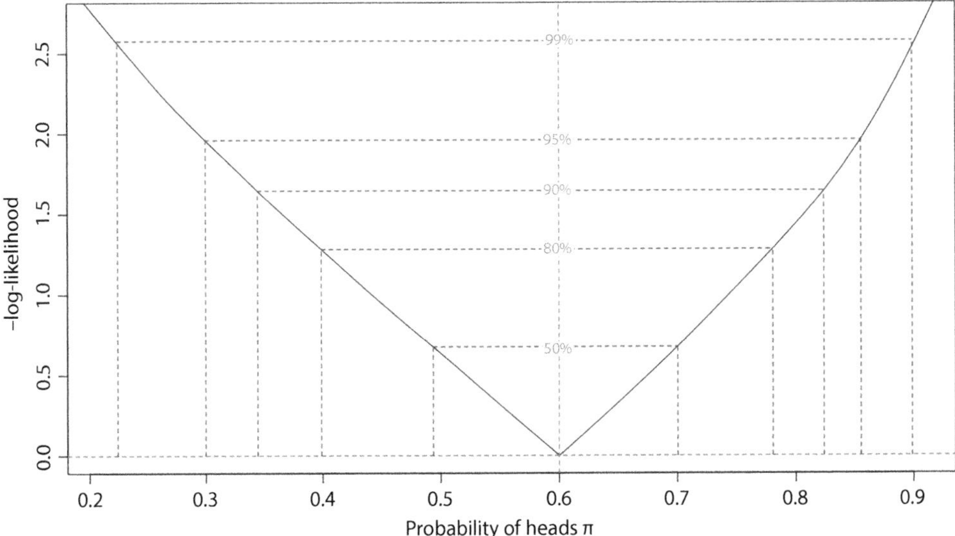

Fig. 3.3 (Inverted) log-likelihood profile showing the maximum likelihood estimate (bottom of the curve) and confidence intervals at different levels for the parameter π. Note that we show the negative log-likelihood profile here, since function optimisers such as `optim` and `nlm` in R always find the minimum and not the maximum of a function; hence, we work with the negative log-likelihood in practice.

Should a statistical model have two or more parameters (for instance, $Y \sim \text{Normal}(\mu_Y, \sigma_Y^2)$), the method of maximum likelihood allows us to simultaneously estimate the combination of all model parameters that is most likely to have generated the observed data. What the clever optimization algorithms in R do is to search for the combination of values in the parameter space that jointly maximizes the likelihood for the observed data. When statistical models have many parameters, it becomes more likely that there might be several local maxima. As we know, there are multiple peaks in a mountain region, one of which is the highest. Identification of the maximum likelihood parameter estimates becomes a more complicated numerical problem then. What happens is that by starting the numerical search at different sets of initial parameter values, the optimization algorithm may end up reaching different local maxima rather than the single, global maximum. There are several solutions to this well-known challenge for maximum likelihood estimation. They involve changing the optimization algorithm and trying out different initial parameter values to evaluate the consistency of the resulting numerical solutions.

The correctness of the parameter estimates obtained by maximum likelihood is contingent on having postulated the correct, or at least an adequate, statistical model. We will discuss methods and approaches to assess the adequacy of fit of statistical models to data in each of the subsequent chapters. In the remainder of the book we will not carry out any direct numerical estimation of model parameters by maximum likelihood. However, we emphasize that it is useful to have a feeling for how this estimation is carried out under the hood, by R, for models fitted under the frequentist perspective.

3.3.3 *Using maximum likelihood estimates in statistical inference*

Fisher (1922) conceived the notion of likelihood as a method for objective reasoning with data. The use of likelihood in statistical inference either to assess the degree of support that data provides to a single hypothesis or to discern between hypotheses is based on the axiom of likelihood (Etz 2018, Reid 2010), which has two components. First is the likelihood principle, which posits that the likelihood function, Eq. (3.3), captures all the information in the data about a certain model parameter that is relevant for the evaluation of statistical evidence for a hypothesis (Edwards 1992). Second is the likelihood axiom, which states that "within the framework of a statistical model, a particular set of data supports one statistical hypothesis better than another if the likelihood of the first hypothesis [given] the data exceeds the likelihood of the second hypothesis" (Edwards 1992, p. 30).

Unlike probability, the likelihood function (or any precise value of it such as its maximum, see Fig. 3.3) has no meaning or interest per se, unless it is compared with another value of likelihood calculated for the same data. Because `LBinom(0.3, x = 6, size = 10) = -3.3` is smaller than `LBinom(0.8, x = 6, size = 10) = -2.43`, we can confidently say that $\pi = 0.8$ is more likely to have produced the data of 6 heads in 10 coin tosses than $\pi = 0.3$. This can be more clearly appreciated when comparing the ratio of actual likelihoods (and not of their log-likelihoods) of the two hypotheses:

```
> exp(LBinom(prob=0.8, x=6, size=10))/exp(LBinom(prob=0.3, x=6, size=10))
[1] 2.396295
```

Recall that the likelihood of a parameter given the data, Likelihood(data | π), is proportional to the probability of the parameter given the data, Pr(π | data), so the above ratio of the two likelihoods (Likelihood(data | $\pi = 0.8$) / Likelihood(data | $\pi = 0.3$)) corresponds to the ratio Pr($\pi = 0.8$ | data) / Pr($\pi = 0.3$ | data). Hence, the data lend 2.396 times more support to the hypothesis $\pi = 0.8$ than to the other hypothesis $\pi = 0.3$. Although the likelihood ratio does not say which (if any) of these two hypotheses is "true," it does say that the available evidence renders the first hypothesis more likely than the other.

Maximum likelihood became the method of choice in statistics to estimate parameters of statistical models after Neyman and Pearson's (1933) lemma proved that likelihood ratio tests are the uniformly most powerful (i.e., they have the highest probability of correctly finding a true effect among all tests with the same level of significance) hypothesis tests about a parameter. This lemma provides the crucial link between having a maximum likelihood parameter estimate and testing a statistical hypothesis in the frequentist framework.

The process of statistical inference in the frequentist framework is based on comparing the maximum likelihood estimates from the actual data with a long-run, hypothetical distribution of parameter estimates known as the sampling distribution that would be obtained if we were to extract repeated samplings from the same population having fixed but unknown parameter values. The bridge between probability models and statistical inference is then provided by a sampling distribution of maximum likelihood parameter estimates (Wakenfield 2013). With larger sample sizes and hence more knowledge about the statistical population, the standard error of maximum likelihood parameter estimates decreases and the confidence intervals become narrower. Therefore, the parameter estimate gets ever closer to the fixed, true values of the parameters.

Fisher (1922) proved that maximum likelihood estimates have many optimal and desirable properties, such as being:

- sufficient (provide complete information about the parameter being estimated);
- consistent (asymptotically, i.e., for ever larger sample sizes they converge to the true parameter value);
- efficient (asymptotically have the lowest possible variance (= uncertainty) among comparable estimators);
- invariant to reparametrization (the same result is obtained with different parametrizations).

And, as if that were not enough for a single paper, Fisher (1922) also showed that (the inverse of the negative of the square root of) the second derivative of the likelihood function with respect to a parameter (which, in our coin example, is obtained by differentiating Eq. (3.3) again) is actually the standard error (SE) of the model parameter π. In practice, the derivatives of the likelihood function to obtain the parameter estimates and their standard errors are calculated numerically in R via the so-called Hessian matrix (you can do `?optim`). To sum up, the method of maximum likelihood not only allows estimation of the value of the parameter(s) that most likely generated the data given the statistical model, but these point estimates also have several optimality properties and there is an associated method to obtain their standard errors.

Once we have maximum likelihood estimates and their standard errors, there are two main inferential procedures within the framework of frequentist statistics. These are the likelihood ratio test and the Wald test. The *likelihood ratio test* permits comparing the degree of empirical support that two statistical models receive from data (Pawitan 2001, Millar 2011). It is based on the large-sample-size results of Wilks (1938), who showed that twice the logarithmic ratio of the likelihoods of two hypotheses—in the above example, $2\log\left(\frac{\text{Likelihood}(\pi=0.8|\text{data})}{\text{Likelihood}(\pi=0.3|\text{data})}\right)$—follows a chi-squared ($\chi^2$) distribution with one degree of freedom. It should be noted that Wilks' test requires calculating the maximum likelihood of the two models being compared, which may not be trivial for very complex statistical models, and it relies on a quadratic approximation of the likelihood functions around their maximum that is reasonable for large sample sizes (Fig. 3.4; Pawitan 2001).

The Wilks test can compare two statistical models (i.e., two statistical hypotheses) fitted to the same data but differing in only one parameter. However, this comparison requires that the two models are nested such that the simpler model is a special case of the more complex one (Fig. 3.5).

The *Wald test* allows the assessment of whether a given maximum likelihood parameter estimate differs significantly from zero (or from any other hypothesized value) by

$$\frac{\theta_{\text{full}} - \theta_{\text{restr}}}{\text{SE}\left(\theta_{\text{full}}\right)} \sim \text{Normal}\left(0, 1\right), \tag{3.7}$$

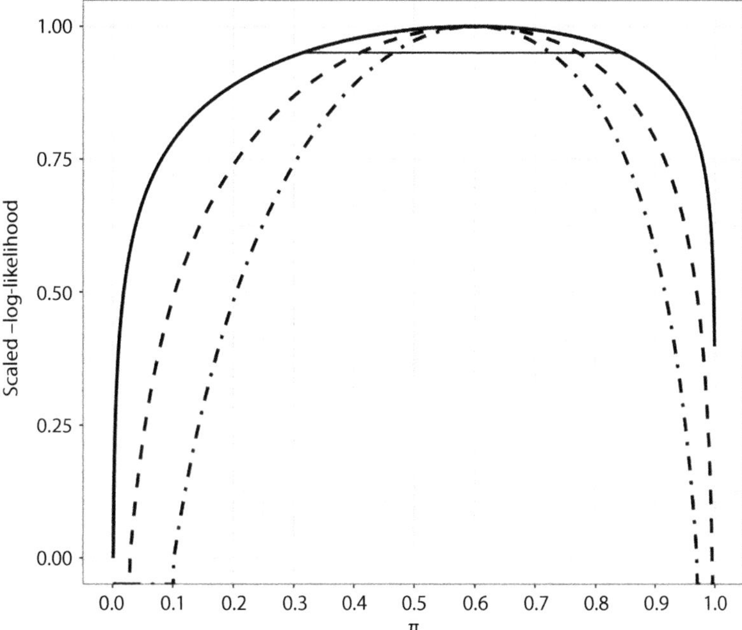

Fig. 3.4 Log-likelihood functions of observing 6 heads in 10 tosses (continuous), 300 heads in 500 tosses (dashed), and 600 heads in 100 tosses (dot-dashed) for a set of plausible values of π. In all cases, the maximum log-likelihood estimate of π appears to be around 0.6. Note that the log-likelihood function becomes closer to a parabola (recall the quadratic approximation involved in Wilks' test) as the sample size increases. The log-likelihood functions were scaled to show them in the same graph. Note also that the 95% confidence interval (thin horizontal line) becomes narrower (i.e., more precise) as the sample size increases.

where θ_{full} is the estimate under the more complex model, and θ_{restr} is either a desired value for testing the parameter θ (such as zero), or the parameter estimate for the restricted model that is a special case of the full model. $\text{SE}(\theta_{\text{full}})$ is the standard error associated with the maximum likelihood estimate. The Wald test also assumes a quadratic approximation to the log-likelihood function (Pawitan 2001). Compared to the likelihood-ratio test, the Wald test only requires the estimation of the full model. This lowers the computational burden compared to Wilks' test (Etz 2018). Although seldom used in this way, the Wald test can be used to test in one operation whether two models differ in several parameters (Millar 2011).

We make extensive use of Wilks' test as one of the procedures involved in model selection, and of the Wald test (and also of a more robust alternative, the parametric bootstrap) to assess the statistical significance of model parameters across this book. We discuss the parametric bootstrap in Part III.

3.4 Bayesian parameter estimation: The basics

The starting point to explain Bayesian parameter estimation has to be Bayes' theorem or rule: $\Pr(A \mid B) \times \Pr(B) = \Pr(B \mid A) \times \Pr(A)$. We begin with a clear-cut example to keep things simple. Let's assume that the author of this book suspects that he might suffer from prostatic cancer, and his doctor recommends carrying out the prostate-specific antigen

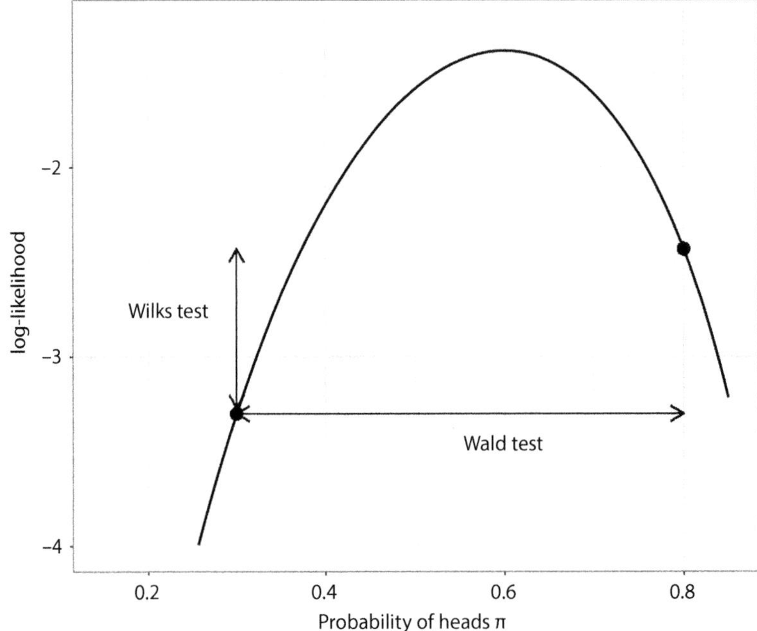

Fig. 3.5 Log-likelihood functions of observing 6 heads in 10 tosses. The dots correspond to the log-likelihoods for $\pi = 0.3$ and $\pi = 0.8$. The Wald test corresponds to the horizontal separation between $\pi = 0.3$ and $\pi = 0.8$ in SE units, and the log-likelihood ratio test to the vertical distance in the log-likelihoods for $\pi = 0.3$ and $\pi = 0.8$.

(PSA) test to quantify the levels of a protein in the blood. To apply Bayes' rule, let A be the true but unknown state of the patient (cancer, no cancer), and B the positive result of the PSA test, such that

$$\Pr(\text{cancer} \mid +) \times \Pr(+) = \Pr(+ \mid \text{cancer}) \times \Pr(\text{cancer}).$$

The + result of the test is a fixed result: it is the data that we have in hand. What the patient wants to know is the probability of actually having cancer given that the test was positive:

$$\Pr(\text{cancer} \mid +) = \frac{\Pr(+ \mid \text{cancer}) \times \Pr(\text{cancer})}{\Pr(+)}. \tag{3.8}$$

Before making any harsh decisions, it is crucial to recall that all clinical tests can yield correct assessments and also make mistakes by yielding false positives (test + but patient fine, i.e., a misdiagnosis) and false negatives (test – but patient ill, potentially leading to death). All clinical tests have what is called their sensitivity (i.e., the probability of correctly detecting a suffering patient) and their specificity (i.e., the probability of correctly detecting that a patient is not ill). Specificity and sensitivity are properties of the test or clinical diagnosis method, and thus they are unaffected by, and independent of, the true but unknown state (ill, not ill) of the patient. Of course, we would ideally like to have a test whose specificity and sensitivity are very close to 1 so that the probabilities of mistakes—(1 – specificity) or Pr(false +), and (1 – sensitivity) or Pr(false –)—become close to 0. In the case of the PSA test, the specificity and sensitivity are 0.33 and 0.14 (Vickers et al. 2010). This means that 67% of the positive and 86% of the negative results of the PSA test are erroneous diagnoses.

The denominator Pr(+) in Eq. (3.8) is called the "total probability of the data," which in this case covers all the ways in which one can obtain a positive PSA test. This can happen either because of a correct (true +) diagnosis of a sufferer, or an incorrect (false +) diagnosis of a non-sufferer:

$$\Pr(+) = \Pr(+ \mid \text{cancer}) \times \Pr(\text{cancer}) + \Pr(+ \mid \text{no cancer}) \times \Pr(\text{no cancer})\,. \tag{3.9}$$

Substituting in Eq. (3.8), we obtain

$$\Pr(\text{cancer} \mid +) = \frac{\Pr(+ \mid \text{cancer}) \times \Pr(\text{cancer})}{\Pr(+ \mid \text{cancer}) \times \Pr(\text{cancer}) + \Pr(+ \mid \text{no cancer}) \times \Pr(\text{no cancer})}. \tag{3.10}$$

To apply Eq. (3.10), we need to know Pr(cancer), or the annual incidence rate of prostate cancer. What do we know of this disease? We know that it only affects men, and mostly when they are 40 years or older. The current US population is about 320 million people, with an approximately even (50:50) sex ratio, and the proportion of people older than 40 years is about 40%. Therefore, we expect about 320 × 0.5 × 0.4 = 64 million men older than 40 years in the US. In addition, there were 174,650 diagnosed cases of prostate cancer in the US in 2018.[1] Therefore, the annual incidence rate, Pr(cancer), of prostate cancer in the US is about: $\frac{175{,}000}{64 \text{ million}} \approx \frac{1}{365}$. Substituting the incidence rate (1/365) of prostate cancer in the USA and the specificity (0.33) of the PSA test into Eq. (3.10) yields

$$\Pr(\text{cancer} \mid +) = \frac{0.33 \times \frac{1}{365}}{0.33 \times \frac{1}{365} + (1 - 0.33) \times \left(1 - \frac{1}{365}\right)} = 0.00135 \approx \frac{1}{740}.$$

That is, the chances of being ill (the potential cause of the + PSA test) for the average 40-year-old man currently living in the US are a bit less than one in a thousand. The point here is not to linger on this hypothetical case. Rather, we wanted to to illustrate the use of Bayes' theorem to calculate the "inverse probability," namely the probability of the cause (cancer) of an observed effect or result (the + result in the PSA test). Here, the "direct probability" is the very uninteresting probability of having a + PSA test result given that the patient has cancer (which, of course, we don't know, otherwise the patient wouldn't have undergone the PSA test).

Although it was not explicitly identified, the previous example had a binary response variable (Y = {cancer, no cancer}), and to estimate the probability π of the single (k = 1) patient being ill we had to resort to Bayes' theorem. We will now employ the same approach in a slightly more complex example involving the counting of plants in k = 50

[1] https://www.cancer.net/cancer-types/prostate-cancer/statistics

square sample units of 1 m^2. Some of these plots will contain 0, 1, 2, etc. plants. Here are the data in tabular form showing the number of sampling units with 0, 1, ..., 7, 8 plants:

```
> table(DF5$plants)
 0  1  2  3  4  5  6  7  8
 3  5 11 15  5  7  1  1  2
```

Given that the response variable Y represents counts (Chapter 10), we may as a first approximation assume $Y \sim$ Poisson (μ_Y) as a plausible statistical model: $\Pr(Y = y) = \frac{e^{-\mu_Y}(\mu_Y^y)}{y!}$. We of course ignore the value of parameter μ_Y and wish to estimate it using Bayesian methods. Now, in the initial statement of conditional probabilities, A is the unknown value of the parameter μ_Y and B is the observed data, and thus Eq. (3.8) becomes:

$$\Pr(\mu_Y \mid \text{data}) = \frac{\Pr(\text{data} \mid \mu_Y) \times \Pr(\mu_Y)}{\Pr(\text{data})}. \tag{3.11}$$

Let's start by identifying the terms in Eq. (3.11):

- $\Pr(\mu_Y \mid \text{data})$ is the posterior distribution. It reflects our uncertainty about the parameter μ_Y *after* seeing the data, i.e., *including* the information about it contained in the data set. Uncovering the entire posterior distribution is the ultimate goal of Bayesian statistical inference. This distribution contains all the knowable information about the parameter μ_Y of the Poisson statistical model for plant counts.
- $\Pr(\mu_Y)$ is the prior distribution of μ_Y. It reflects our knowledge about this parameter *before* seeing the data, i.e., *not including* the information they carry about the parameter. Whether and how to express any existing prior knowledge in the statistical analysis has been the key point of dispute between frequentist and Bayesian statistics. We will obviously discuss this important issue at length later.
- $\Pr(\text{data} \mid \mu_Y)$ is the likelihood, Eq. (3.3). It tells us how plausible it is to have observed the actual count data, given a potential value of the parameter μ_Y.
- Pr(data) is the probability of observing the actual count data for all possible values of the parameter μ_Y under our model.

Calculating the denominator is very hard because it requires considering all possible values of μ_Y. Let's first understand why this is so.

We know in advance that the parameter μ_Y, the mean number of plants in the Poisson statistical model, must be a positive, real number. To simplify things, let's for a moment pretend that μ_Y could only be either 1.32 or 1.33. This is false, of course, but creating an artificially simple case helps us to gain an intuitive understanding of complex situations or equations. If μ_Y could only be {1.32, 1.33}, substituting these two values in the denominator of Eq. (3.11) would lead to

$$\Pr(\mu_Y \mid \text{data}) = \frac{\Pr(\text{data} \mid \mu_Y) \times \Pr(\mu_Y)}{\Pr(\text{data} \mid \mu_Y = 1.32) \times \Pr(\mu_Y = 1.32) + \Pr(\text{data} \mid \mu_Y = 1.33) \times \Pr(\mu_Y = 1.33)}. \tag{3.12}$$

Let's first notice that the denominator in Eq. (3.12) corresponds to a weighted sum of the likelihoods, $\Pr(\text{data} \mid \mu_Y)$, of observing the data given either of the two possible values of μ_Y.

As weights we have the prior probability that μ_Y assumes either of these two values. The correct calculation of the weighted sum of likelihoods in Eq. (3.11) must in fact consider that μ_Y could assume any positive positive, real number, not just these hypothetical two values. This correct calculation must thus involve the sum of an infinite set of plausible values of μ_Y, which amounts to integration,

$$\Pr(\mu_Y \mid \text{data}) = \frac{\Pr(\text{data} \mid \mu_Y) \times \Pr(\mu_Y)}{\int \Pr(\text{data} \mid \mu_Y) \times \Pr(\mu_Y)\, d\mu_Y}, \tag{3.13}$$

where the $\int$ sign indicates that we are summing weighted likelihoods, $\Pr(\text{data} \mid \mu_Y) \times \Pr(\mu_Y)$, and $d\mu_Y$ indicates the variable (μ_Y) whose values are being summed. The denominator in Eq. (3.13) is the total likelihood of the data, in complete analogy to Eq. (3.12).

Let's now imagine another statistical model with three parameters (θ_1, θ_2, θ_3) instead of just one as in our Poisson model. The posterior distribution will now be a three-dimensional (one axis per parameter) function that will convey the joint plausibility for each triplet (θ_1, θ_2, θ_3) of values of the model parameters given the data. Equation (3.13) would become

$$\Pr(\theta_1, \theta_2, \theta_3 \mid \text{data}) = \frac{\Pr(\text{data} \mid \theta_1, \theta_2, \theta_3) \times \Pr(\theta_1)\Pr(\theta_2)\Pr(\theta_3)}{\int\int\int \Pr(\text{data} \mid \theta_1, \theta_2, \theta_3)\Pr(\theta_1)\Pr(\theta_2)\Pr(\theta_2)\, d\theta_1 d\theta_2 d\theta_3}.$$

The denominator in this threatening-looking expression means that we would need to calculate the total likelihood of the data considering all triplets of possible values of the parameters θ_1, θ_2, and θ_3, weighted by their prior probabilities $\Pr(\theta_i)$. Now we have a triple integral (one per model parameter) involving three infinite sums over the parameter space, or sets of plausible values of the three model parameters. A brute-force implementation of such a triple integral in a computer would have to consider 1,000 plausible real values for each parameter θ_i (which would make $1{,}000^3 = 10^9$ triplets or combinations to evaluate), and then perform a rough calculation of the average likelihood for each. We can then appreciate that the magnitude of the task of computing the denominator of Eq. (3.13) quickly increases with the number of model parameters (this is known as the "curse of dimensionality"). Virtually every model that will be considered and fitted in this book will have more than three parameters. This would make the computation of the denominator involving multiple integrals a perennially thorny and time-consuming task. As Lambert (2018) puts it, "the devil (of Bayesian models) lies in the denominator."

To appreciate why we can avoid the devilish denominator of Eq. (3.13), let's return for a moment to the artificial case where the true value of μ_Y in the Poisson model could (magically) only be either 1.32 or 1.33. Because μ_Y may only take one of two particular values here, we could evaluate Eq. (3.12) separately for each value,

$$\begin{aligned}&\Pr(\mu_Y = 1.32 \mid \text{data})\\ &= \frac{\Pr(\text{data} \mid \mu_Y = 1.32) \times \Pr(\mu_Y = 1.32)}{\Pr(\text{data} \mid \mu_Y = 1.32) \times \Pr(\mu_Y = 1.32) + \Pr(\text{data} \mid \mu_Y = 1.33) \times \Pr(\mu_Y = 1.33)}\end{aligned}$$

(and likewise for $\mu_Y = 1.33$). We can now hopefully see that the denominator in the above expression is just a simple scaling that divides each value, $\Pr(\mu_Y = 1.32 \mid \text{data})$, by the sum of both possible values of μ_Y, i.e., $\Pr(\mu_Y = 1.32 \mid \text{data}) + \Pr(\mu_Y = 1.33 \mid \text{data})$. Any scaling of a function by dividing with a constant will only change the height, but not the shape, of the function. This means that the shape of Eq. (3.11) is preserved when we get rid of the "devilish" hard-to-compute denominator by writing

$$\text{Posterior}\,(\mu_Y \mid \text{data}) \propto \text{Likelihood}\,(\text{data} \mid \mu_Y) \times \text{Prior}\,(\mu_Y)\,. \tag{3.14}$$

Equation (3.14) is clearly by far *THE* most important equation for Bayesian inference, so let's put it in words: *The posterior distribution of a parameter,* $\Pr(\mu_Y \mid \text{data})$*, is proportional to the product of the likelihood,* $\Pr(\text{data} \mid \mu_Y)$*, and the prior distribution of the parameter,* $\Pr(\mu_Y)$*.* The posterior distribution thus calculated will not be a "true probability distribution" in the sense of integrating to 1. However, the posterior distribution that we will obtain with Eq. (3.14) will have the exact same shape and will be proportional to the true and correct posterior distribution that would be obtained if we had computed the "evil" integral in the denominator of Eq. (3.13). Because of this proportionality, all descriptive statistics (mean, mode, ranges, standard deviation, etc.) from the posterior distribution in Eq. (3.14) will be identical to those that could be obtained from Eq. (3.13). Virtually all statistical inference in the Bayesian framework is based on analyzing the posterior distribution thus obtained.

The likelihood $\Pr(\text{data} \mid \mu_Y)$ in Eq. (3.14) denotes the values of the parameter that are most likely to have produced the observed data. In the Bayesian framework, the likelihood is the means by which the information contained in the data enters into the process of parameter estimation. The information contained in the data modifies the previous knowledge about the parameter embodied in the prior distribution $\Pr(\mu_Y)$ by giving "more weight" to those parameter values receiving more empirical support from the evidence. If we have very few data (a relatively wide likelihood; cf. Fig. 3.5) or if the data contain very little new information, the posterior, $\Pr(\mu_Y \mid \text{data})$, and the prior, $\Pr(\mu_Y)$, distributions would resemble one another. We could interpret the latter result as implying that either we learned almost nothing new from the data, or that the available evidence confirmed what we already knew about the parameter μ_Y. If, based on previous evidence or any source of inside knowledge, we had suspected the values of μ_Y ought to be in a narrow range (for instance, [1.46, 1.84]) with a specific mean (for example, 1.62), it would take a lot of data with substantially new evidence to shift such a tight prior distribution.

In a qualitative sense, it is possible to visualize two caricatured extremes involved in Eq. (3.14). At one extreme, very large data sets might overpower a vague or diffuse prior distribution, so that the posterior distribution would resemble the likelihood. At the other, overly precise prior distributions for model parameters might overwhelm the information harbored by the data, and hence the prior and posterior distributions would resemble each other. Needless to say, the overwhelming majority of cases will lie somewhere between these black-and-white extremes.

The use of prior knowledge in the estimation of parameters of a statistical model has been, is, and seemingly will always be, a contentious issue. The main founders of the frequentist statistical framework (Fisher, Neyman, Pearson, and Wald) repeatedly voiced their strong rejection of the use of Laplace's uniform prior distributions (or "flat priors" in the Bayesian jargon) that was the standard and only approach up to the 1930s (see Chapter 2). In fact, Bayesian estimation using flat, uniform priors will obtain identical results to a maximum likelihood estimation (e.g., Gelman et al. 2014, Kéry and Schaub 2012, Kruschke 2015). But beware: seemingly flat, uninformative priors depicting that "everything is possible" can be quite informative in other contexts (e.g., Hobbs and Hooten 2015; see Chapter 9). While Bayes, Laplace, and others had to use uniform priors to obtain analytic posterior distributions in the few cases when this was possible, there have been conceptual advances in defining less arbitrary prior distributions. Jeffreys (1946) addressed an important critique of Fisher's by developing a method to propose priors that are invariant to reparametrization. Raifa and Schlaifer (1961) proposed conjugate priors whereby the posterior distribution would have the same algebraic form as the prior, but with different parameter values reflecting the influence of the data. Conjugate priors are tremendously important in analytically obtaining the denominator of Eq. (3.13) as part of the full posterior distribution (Hobbs and Hooten 2015). While conjugate priors constituted a great leap forward in Bayesian modeling, they are insufficient to analytically obtain posterior distributions of complex statistical models. Their compulsive use ended up constraining the flexibility needed to define priors in statistical models with many parameters (Lambert 2018).

The frequentist methods based on maximum likelihood developed mostly by Fisher, Neyman, and Wald were deemed "objective"; by contrast, Bayesian methods were negatively labeled as "subjective." The adjective "objective" was justified on the two grounds. First, on the strict view of probability as describing the aleatory uncertainty of events in nature on the basis of their relative frequency of occurrence (Chapter 2). Second, on using parameter estimation methods independent of the views of the data analyst and, above all, devoid of the arbitrariness involved in defining prior distributions (Chapter 2). All statistical methods use prior knowledge to choose relevant explanatory variables, to design experiments and surveys, and to interpret results (Gelman 2012). But the founders of the frequentist framework abhored the encoding of any prior knowledge in the prior distributions of model parameters. In fairness, it is often unclear how much and which prior knowledge is used to accomplish these tasks in frequentist data analysis. Regardless of what we may think of Bayesian data analysis, at least the prior knowledge is always explicitly communicated and is open to public scrutiny.

During the first half of the twentieth century, prior distributions were often chosen for analytical convenience to obtain the posterior distribution by the applied mathematicians and statisticians in whose hands Bayesian statistics remained. Empirical justifications and understanding regarding the choice of the priors became an issue of largely secondary importance. Bayesian methods endured the frequentist crush for most of the twentieth century, and until recently were viewed as a sort of acquired taste for certain hard problems tackled by eccentric, die-hard specialists.

Bayesian statistical modeling was released from the "tyrannic need of analytical solutions" requiring certain specific priors in the 1980s by the increasing use of numerical methods to obtain the posterior distribution for just about any chosen priors. We show in detail the main numerical method, Markov chain Monte Carlo (MCMC), and a recent improvement (Hamiltonian Monte Carlo) later in this chapter. The MCMC algorithm and its variants are by far the main factor responsible for the growing use of Bayesian methods in applied statistics and in the social and natural sciences (e.g., Gelman et al. 2014, Hobbs and Hooten 2015, Kruschke 2015, Lambert 2018).

At this point, it is important to appreciate that enlarging the scope away from the analytical convenience of the uniform priors and of Jeffreys' (1939) conjugate priors opened a whole new world of possibilities for portraying prior knowledge about model parameters. On the one hand, many other probability distributions can now be used for the priors, provided that these choices are justified and accepted by other practitioners. On the other hand, the same probability distribution can be a more or less informative prior depending on how we set it up. The prior distributions of parameters $\beta \sim \text{Normal}(0, 1)$ or $\beta \sim \text{Normal}(0, 10)$ would give 95% ranges of plausible parameter values in [−1.96, 1.96] and [−19.60, 19.60]. The latter probability distribution is flatter and more diffuse than the former. Instead of just putting different values (1 and 10) in the variance of the prior distribution for β, we could have defined $\beta \sim \text{Normal}(0, \sigma_\beta^2)$, where σ_β^2 is now called a hyperparameter whose value is to be estimated as part of the analysis. The latter approach is the basis of Bayesian hierarchical models (see Part III). Importantly, the degree of informativeness of a prior is not an absolute feature since it is contingent on the model complexity and the data at hand (Gelman et al. 2008).

The Bayesian literature and textbooks contain an array of many fuzzily defined adjectives to label the degree of informativeness of priors, including "vague," "strong," "uninformative," "weak," "diffuse," "informative," "low-information," "weakly informative," and "very/strongly informative." It would probably be hopeless to attempt any precise equivalence of these fuzzily defined categories. Rather than continuing to drown in cloud of ink, we will discuss in detail the choice of prior distributions for the parameters of the main classes of models in the respective chapters. In addition, we will assess in each case the sensitivity of the results (i.e., the posterior distributions of model parameters) to the seemingly irreducible arbitrariness of defining priors. The drawback of this approach is that you may not find a single chapter containing "all you wanted to know about priors and never dared to ask." We summarize different non-exclusive approaches that can be used to gauge the effects of priors on page 323.

So where do we stand now? At this point, the goal is for you to have understood that:

- The two main "ingredients" involved in Bayesian parameter estimation (the likelihood function and the prior distribution) somehow combine to produce "the result" (the posterior distribution) that contains all attainable knowledge about the model parameter(s).
- Setting up priors remains a contentious issue for many (though not for others, who think that the importance of the issue is greatly overblown) that is discussed later in more detail.
- Equation (3.14) is the fundamental equation involved in Bayesian parameter estimation.

So, let's finally see how to numerically solve Eq. (3.14) to obtain Bayesian parameter estimates from data under some statistical model.

3.5 Bayesian methods: Markov chain Monte Carlo to the rescue

Bayesian statistics underwent a revolution in the 1990s (Gelfand and Smith 1990) when a clever computer algorithm developed in the 1950s and used in a few specialized areas (e.g., image processing) started to be used to solve Eq. (3.14) by simulation. The algorithm in question, Markov chain Monte Carlo, was invented by Metropolis et al. (1953) at Los Alamos National Laboratory, New Mexico within the secrecy of the Cold War, and was published in the *Journal of Chemical Physics*. The problem that Metropolis et al. (1953) were tackling was how to simulate a liquid in equilibrium with its gas phase (Geyer 2011). The exact details need not concern us here. What Metropolis et al. (1953) realized and proved was that they did not need to simulate the exact dynamics since a simulated Markov chain having the same equilibrium distribution would allow them to extract the same conclusions. That is, a clever numerical simulation allowed them to obtain an approximate but correct solution to a problem that had previously been insoluble by analytical means.

Let's dissect what lies behind the name MCMC and then understand how it works. MCMC contains two features: Markov chain and Monte Carlo. A Markov chain is sequence of random elements whose values $\{X_1, X_2, \ldots, X_k\}$ are obtained using some probabilistic method whose conditional distribution of X_{k+1} given $X_1, \ldots, X_k$ only depends on the previous value X_k. This is just the definition of the first-order Markov chain that is used in all MCMC algorithms. For a fuller and more general description of Markov chains, see Norris (1997). This Markov chain is defined by a probabilistic rule that determines how to obtain the value X_{k+1} given the previous value X_k. In turn, Monte Carlo refers to a broad class of algorithms that use random sampling to obtain numerical results to problems that cannot, or that are too hard to, be solved by exact methods. The main uses of Monte Carlo methods are in optimization, sampling, and estimation (Kroese et al. 2014). Monte Carlo methods are usually traced to Stanislaw Ulam and John von Neumann's work at Los Alamos in 1946. Ulam had the original idea while trying to find a practical way of solving an intractable combinatorial computation (calculating the probability of winning at the solitaire card game) during his convalescence from tuberculosis. The approach was first adopted by von Neumann for implementation with direct applications to neutron diffusion, and later given a code name by Nicholas Metropolis in reference to Ulam's uncle's fondness for gambling (Metropolis 1987).

MCMC is a numerical adaptation of a random walk through plausible values in the parameter space by obtaining random samples from the posterior distribution whose analytical expression, shape, and main features are unknown in advance. The temporal sequence of values of each population parameter obtained by MCMC constitutes a chain. It would seem mysterious and magical to generate samples from a probability distribution whose algebraic expression is unknown to us (and, actually, we do not even need to know it!), but wait, hang in there, and the mysterious fog will hopefully start to clear shortly. We will start with the simplest and oldest MCMC algorithm, known as the Metropolis algorithm, and then explain a more general version (Metropolis–Hastings) and a special case (the Gibbs algorithm) that is particularly useful in models with many parameters (i.e., in the vast majority of cases).

To discuss in detail the Metropolis MCMC algorithm, let's recall the data of plant counts that we encountered a few pages ago (Fig. 3.6). We hypothesized that Y, the number of plants per sampling unit, may have a Poisson distribution whose parameter μ_Y we need to estimate from data. We also know that the parameter μ_Y will be a positive real number. This means that its plausible values will be in the interval $(0, +\infty)$.

We might consider two extreme approaches to obtain the posterior distribution of μ_Y using Eq. (3.14), neither of which will give us its posterior distribution (Lambert 2018). One extreme option would be letting the computer do a random walk to blindly choose plausible values for μ_Y in this interval, and then evaluating their likelihood to generate the observed count data. This random drunkard search of parameter values would be very wasteful and inefficient given the infinite number of values of μ_Y to be considered. In addition, it would offer no guarantees of uncovering the posterior distribution of μ_Y in a finite number of evaluations. The "curse of dimensionality" again warns us that this inefficiency will quickly become far worse for statistical models with more parameters. The other extreme option would be to focus only on a few values depending on their likelihood of producing the observed count data. This would relatively quickly reach the maximum likelihood estimate of μ_Y but would not uncover its posterior distribution. While the likelihood function for simple statistical models is often unimodal, this is far from being common in more complex models. Hence, every optimization method must

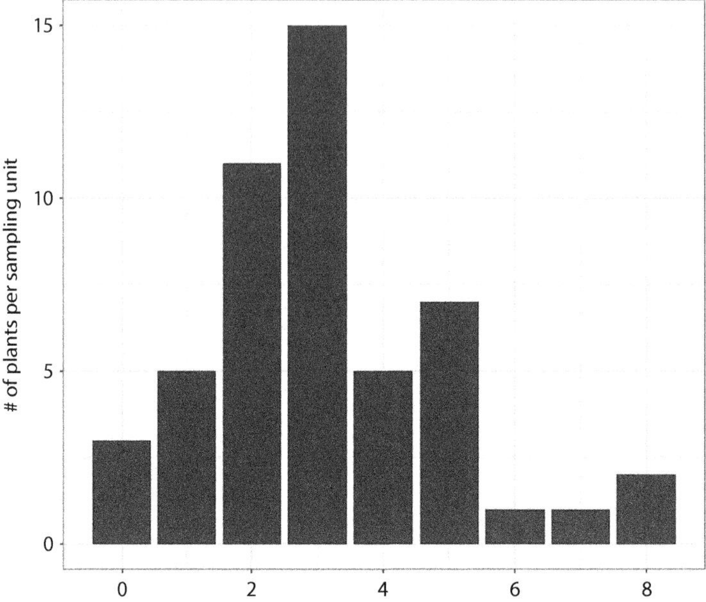

Fig. 3.6 Simulated data of plant counts per sampling unit.

beware the dangers of local maxima. Intuitively, our method to solve Eq. (3.14) should ideally combine some random wandering in the parameter space (here, the set of plausible values of μ_Y) with a guided basis for selecting the most suitable values of μ_Y on the basis of the likelihood of having produced the observed counts. And this is what MCMC does.

These are the six steps involved in the basic Metropolis MCMC algorithm:

1. At iteration 1, choose an initial (set of) model parameter value(s) A.
2. Propose a new (set of) model parameter value(s) B.
3. Evaluate the posterior probabilities of both A and B, Pr(A) and Pr(B).
4. Decision rule:
   a. If Post(B) > Post(A), always jump $A \rightarrow B$; add B to the chain of estimates.
   b. If Post(B) $\leq$ Post(A), randomly jump $A \rightarrow B$ in proportion of how much bigger Pr(B) is than Pr(A). If jumping, add B to the chain of estimates.
5. Either keep or, when needed, refresh the current value of the chain.
6. Repeat steps 2 to 5 many (never less than a few thousand) times.

Steps 3 and 4 deserve and require further elaboration and clarification. Let's start with the decision rule in step 4, which can also be stated as

$$R = \min\left(1, \frac{\text{Likelihood}\left(\text{data} \mid \mu_Y^B\right)\text{Prior}\left(\mu_Y^B\right)}{\text{Likelihood}\left(\text{data} \mid \mu_Y^A\right)\text{Prior}\left(\mu_Y^A\right)}\right), \tag{3.15}$$

with A and B being the initial and the new proposed value of μ_Y, and Likelihood(data | μ_Y) being the product of the likelihood of observing the count data given a proposed value of μ_Y, and Pr(μ_Y), the prior probability of μ_Y. The latter product is proportional to the posterior probability (μ_Y| data) density for each plausible value of μ_Y, cf. Eq. (3.14). Hence, the criterion for accepting a new proposal value μ_Y (and if so, adding a new parameter estimate to the chain) reflects a local rule that compares how much better the new proposed value of the parameter Posterior(μ_Y^B | data) is compared to the current value in the chain, Posterior(μ_Y^A | data). If Posterior(μ_Y^B | data) > Posterior(μ_Y^A | data), the new parameter estimate μ_Y^B will always be added to the chain; otherwise, the new value μ_Y^B may be probabilistically accepted depending on how much greater its posterior probability would be compared with that of the current value μ_Y^A. In practical terms, the decision is taken by comparing the value of R in Eq. (3.15) with a value randomly extracted (i.e., a variate) from the uniform distribution in the range [0, 1]. The total probability of data—i.e., the "devilish" denominator in Eq. (3.13)—takes no part in the acceptance of parameter values since its value would cancel in the decision rule stated in Eq. (3.14).

Before deciding on whether to accept a new proposal value μ_Y^B using the acceptance rule, we obviously need a rule to generate the new proposal from its current value μ_Y^A. This transition rule is known as the proposal distribution in the Bayesian jargon (e.g., Hobbs and Hooten 2015, and many others). The original Metropolis MCMC algorithm uses a symmetric rule whereby $\Pr(\mu_Y^A \rightarrow \mu_Y^B) = \Pr(\mu_Y^B \rightarrow \mu_Y^A)$ (see below for the Metropolis–Hastings algorithm). A simple way to implement a symmetric proposal distribution is to define a normal distribution centered at μ_Y^A with variance σ^2 defined by the user. Therefore, there will be a 95% chance that the next plausible parameter value will be within 1.96 × sqrt(variance) of the current value. As we might guess, as this variance becomes smaller, the next proposed value μ_Y^B is likely to be more similar to the current one. If this variance is too small, the algorithm will explore a very small set of similar parameter values and there will be a slow convergence to a solution. In contrast, if the variance is too big, there may be large jumps in parameter values resulting in a very coarse exploration of the parameter space with a high rejection rate. In many implementations of the MCMC algorithm, this variance is predefined so as to generate a rejection rate of about 25% of the proposed parameter values. Experience suggests that this acceptance rate leads to efficient exploration of the parameter space for many statistical models (e.g., Gelman et al. 1996).

The combined effect of the proposal distribution and the mixed deterministic–stochastic acceptance rule makes the MCMC algorithm somewhat sub-optimal in the sense of occasionally accepting parameter values that do not necessarily have a high support from the data. This feature turns out to be essential to uncover the full posterior distribution and avoid getting stuck in those parts of the parameter space with a high likelihood of being able to generate the data. The actual "magic" of the MCMC algorithm is that by sampling candidate parameter values and evaluating their suitability using the rejection rule in Eq. (3.15), the resulting chain will contain parameter values in proportion to the posterior distribution probability that will be uncovered in the process.

Does the MCMC algorithm work? Algorithms are specific, numerical, and approximate solutions to much harder problems. We thus need assurances that not only do they reach a solution, but also that it is a correct solution to the problem posed. The Markov chain used in MCMC requires three conditions to converge to a solution: it must be irreducible or ergodic, not transient, and aperiodic (Gelman et al. 2014). In practical terms, these conditions mean that the entire set of values in the parameter space may eventually be visited by the random walk of the chain and their values evaluated by the MCMC rejection rule. These conditions are met by most Markov chains, including those typically used in all MCMC algorithms. The Markov chain of MCMC converges to a solution in a finite number of steps, and its sequence of parameter values forms a stationary distribution of parameter estimates. By "stationary distribution" we mean that the mean, variance, shape, and other features of the posterior distribution barely change if we obtain additional parameter estimates by computational brute force. If so, the computer does not need to keep running. So MCMC leads to a stable solution (the stationary distribution of parameter estimates), but is this solution the posterior distribution that we are seeking? It is not too hard to show (see Gelman et al. 2014) that the stationary distribution reached by MCMC *is* actually the posterior distribution that we are seeking. So yes, MCMC attains a solution, and this solution is what we are seeking: the posterior distribution of the parameters.

Let's finally show a step-by-step example of the workings of the MCMC algorithm with the plant count data. A few pages ago we stated that the response variable Y: # plants per sampling unit may have a Poisson distribution whose parameter μ_Y (a positive real number) needs to be estimated. Equation (3.14) states that Posterior $\propto$ Likelihood × Prior or, better, log(Posterior) $\propto$ log(Likelihood) + log(Prior), and here the likelihood will be the Poisson distribution, $\Pr(Y = y) = \frac{e^{-\mu_Y}\left(\mu_Y^y\right)}{y!}$. We now need to specify the prior distribution for μ_Y. In order to avoid a further digression, we are going to use, without much explanation, the lognormal distribution to model our knowledge (or lack thereof) of the parameter μ_Y before looking at the data. For the time being (see Chapter 4), it suffices to know that the range of the lognormal distribution is $(0, +\infty)$ and it has a variety of shapes depending on its parameters, the mean and variance (Fig. 3.7).

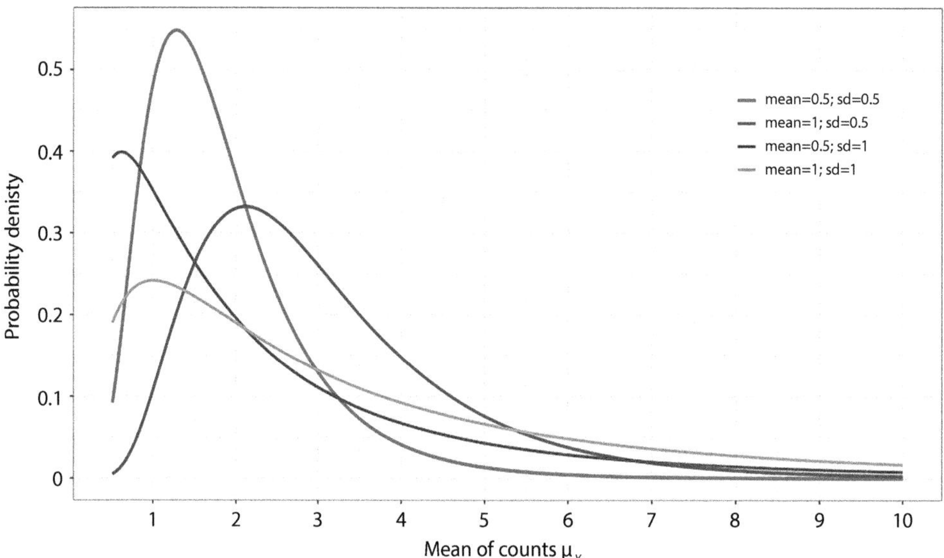

Fig. 3.7 Using the lognormal distribution as the prior distribution for the parameter μ_Y, showing four different combinations of the mean and standard deviation of the lognormal distribution.

We will (arbitrarily, at this point) use Lognormal(mean = 0.5, sd = 0.5) as the prior distribution of the parameter μ_Y. We are now in a position to start filling the chain by applying the MCMC algorithm by hand at first. To simplify matters in this first approximation to the MCMC algorithm we will not implement the proposal distribution (see below) that governs the jumps between consecutive possible parameter values in a chain. Instead, we now take five candidate values of the Poisson parameter μ_Y in the vector `cand` of the data frame `MCMC` as if they were magically given:

```
MCMC=data.frame(cand=c(1.32,0.96,2.34,1.18,1.55))
```

Table 3.1 The first five iterations of our MCMC algorithm that aims to estimate the mean plant density μ_Y of the response variable plant count that is modeled using the Poisson distribution. We show five candidate values of μ_Y, their log-prior density, log-likelihood, and log-posterior density (the logarithmic transformation of Eq. (3.14)), and the R criterion, Eq. (3.15), for deciding whether to accept each candidate parameter value and incorporate it into an MCMC chain.

Iteration	Candidate value of μ_Y	Log-prior	Log-likelihood	Log-posterior	R	Accept?
1	1.32	−0.602	−143	−144	1.000	—
2	0.96	−0.770	−175	−175	1.000	Accept
3	2.34	−1.321	−105	−106	0.604	Accept
4	1.18	−0.615	−153	−154	1.000	Accept
5	1.55	−0.672	−129	−130	0.844	Reject

A few explanations are in order to understand the calculations of Table 3.1. First, we do the calculations using log-probability and log-likelihood to avoid rounding and numerical problems involved in dealing with very small numbers. If we do exp[log-likelihood + log(Prior)] we would of course obtain the same value of the posterior probability that we would have obtained by doing Posterior = Likelihood × Prior. Second, we need to create a function to calculate the log-likelihood of obtaining the entire plant count data for each candidate value of the parameter μ_Y. That will be assessed by MCMC. This function will be very similar to the function `Lbinom` that we used in Section 3.4:

```
Lpoisson=function(lambda,data=DF5$plants){sum(log(dpois(data,lambda)))}
```

This function has two arguments: the data (all the plant counts stored in `DF5$plants`), and the parameter `lambda` (the mean) of the Poisson distribution used in the model likelihood, and it will return the overall likelihood that each value of the mean of the Poisson distribution could have generated the entire data set.

To obtain the log-likelihood for the first candidate value of the parameter μ_Y:

```
> LPoisson(lambda=MCMC$cand[1],data=DF5$plants)
[1] -143
```

To calculate the log-likelihoods for all candidate values of μ_Y we could either enter these values one by one (by changing the index inside the square brackets), or use the R function `sapply` which will apply the function `LPoisson` to every element of a vector much faster, and yield the output as a vector that is stored in an object we will call `MCMC`:

```
> MCMC$Lik=sapply(MCMC$cand, LPoisson)
[1] -143 -175 -105 -153 -129
```

Now for the logs of the prior probabilities, the command `dlnorm(MCMC$cand, meanlog = 0.5, sdlog = 0.5, log = T)` will calculate the log of the prior probability for each candidate value of the Poisson parameter:

```
> MCMC$Prior=dlnorm(MCMC$cand,meanlog=0.5,sdlog=0.5, log=T)
[1] -0.602 -0.770 -1.321 -0.615 -0.672
```

We can now add the log-likelihood and the log of the prior to obtain the log of the posterior and store the results with `MCMC$Posterior = MCMC$Lik + MCMC$Prior`, giving

```
> MCMC$Posterior
[1] -144 -175 -106 -154 -130
```

Having obtained the log of the posterior for the five candidate values of μ_Y, we can now decide on whether to accept them and write them as estimates on our short chain using the criterion of Eq. (3.15). Although for five values one could do the calculations one by one using the above formulae, using just a tiny bit of R programming will allow us to generate some code that will be used later with a more realistic example:

```
1.for (i in 2:nrow(MCMC)){
2.MCMC$R[i]=min(1,exp(MCMC$Posterior[i])/exp(MCMC$Posterior[i-1]))
   MCMC$Accept[i]=ifelse(MCMC$R[i]>runif(1,min=0, max=1),"Accept","Reject") }
```

What the `for` loop does is (starting with the second candidate value of the Poisson parameter and continuing up to the last value held in the last row of `MCMC`; line 1) calculate the R criterion from Eq. (3.15) on line 2 using exp(Posterior) to revert our calculation of log(Posterior), and reach the decision of keeping or not each candidate parameter value by comparing its R value with a random deviate from a uniform distribution between zero and one, that is, a proportion. In R, every `for` loop starts and ends with curly brackets.

```
> MCMC
  cand Posterior  Prior  Lik      R   Accept?
1 1.32      -144 -0.602 -143     NA    <NA>
2 0.96      -175 -0.770 -175 1.000   Accept
3 2.34      -106 -1.321 -105 0.604   Accept
4 1.18      -154 -0.615 -153 1.000   Accept
5 1.55      -130 -0.672 -129 0.844   Accept
```

Here, it is critical to note a very important point: if we run the code again with the same count data and the same candidate values for the Poisson parameter, we might just by sheer chance alter the decision concerning the third and fifth candidate values whose values of the criterion R are smaller than 1. This is precisely the somewhat sub-optimal nature of the MCMC algorithm that allows uncovering the entire posterior distribution rather than focusing only on that (or those) value(s) of the Poisson parameter receiving the highest support from the data.

Obviously, our first approach to generate a short Markov chain with just five estimates of μ_Y is not enough to obtain the posterior distribution of the Poisson parameter. We need to modify the above code to make a more realistic MCMC algorithm by including the proposal distribution that will dictate the jumps between consecutive plausible parameter values in the chain. The modified algorithm will evaluate 5,000 candidate values of the Poisson parameter:

```
1. MCMC1=
    data.frame(matrix(nrow=5000,ncol=1));names(MCMC1)="mean"
2. MCMC1$mean[1]=3.5
3. for(i in 2:5000){
4.    current=MCMC1$mean[i-1]
5.    prop=rnorm(n=1,mean=current, sd=0.5)
6.    post.prop=sapply(prop,LPoisson)+
      dlnorm(prop,meanlog=0.5,sdlog=0.5, log=T)
7.    post.curr=sapply(current, LPoisson)+
      dlnorm(current,meanlog=0.5,sdlog=0.5, log=T)
8.    R=min(1,exp(post.prop)/exp(post.curr))
9.    MCMC1$mean[i]=ifelse(R >runif(n=1,min=0,max=1),prop,
      current)
10.  }
```

Line 1 creates an empty data frame `MCMC1` to hold the estimates of μ_Y and names the single column (with a default name of `V1`). Line 2 simply gives a starting value to the variable `mean`. The `for` loop starts on line 3 and ends with the curly bracket on line 10. Then we give values to the current value of `mean` in the chain (line 4) and to the proposed candidate value `prop` (line 5). The proposal distribution is normal, with mean the current value of the chain. Its standard deviation (set somewhat arbitrarily to 0.5) indicates how far from the mean the most likely values of `prop` will lie on average. "Arbitrarily" here just means that we tried several values and settled on 0.5 to avoid generating negative candidate values of μ_Y that will generate an error when calculating the log-likelihood and the log of the prior; we will proceed differently when dealing with count models in Chapter 11, but let's keep it simple and roughly adequate for now. So, now we need to obtain log(Posterior) for the `current` and `prop` parameter values (lines 6 and 7) that are compared by calculating the value of `R` (line 8) as before. The decision of whether to accept the `prop` value or to keep the `current` value of the Poisson parameter is made on line 9. The result will be a data frame `MCMC1` with 5,000 estimates of the Poisson parameter μ_Y stored in the variable `mean`. This is probably the simplest hand-made MCMC algorithm that we can write. These 10 lines of code may hopefully suffice to understand the gist of the idea behind the MCMC algorithm.

Here's the resulting Posterior distribution in the data frame `MCMC1`:

```
> head(MCMC1)
      mean
1 3.500000
2 3.466757
3 3.047261
4 3.047261
5 2.973406
6 2.973406
```

We could characterize the distribution of 5,000 estimates with summary statistics:

```
> summary(MCMC1$mean)
  Min. 1st Qu.  Median    Mean 3rd Qu.    Max.
 2.267   2.906   3.063   3.068   3.219   4.113
```

We can calculate an interval containing 95% of its values by:

```
> quantile(MCMC1$mean, probs=c(0.025,0.975))
    2.5%    97.5%
2.619477 3.559602
```

This interval is called the 95% credible interval in Bayesian statistics. It tells us that the probability (since it is calculated from our posterior distribution that is proportional to the true posterior probability distribution) that μ_Y lies between 2.619 and 3.559 is 0.95. The credible interval of a parameter is analogous, but *not* identical, to a confidence interval.

Many of us have struggled in introductory statistics courses with the twisted interpretation of the confidence interval of a parameter. Let's recall what it is. If we were to extract many (say, 1,000) random samples from the same statistical population, and for each sample we calculated the statistics needed to compute its 95% confidence interval (we would now have 1,000 confidence intervals, one per random sample), then 95% of these intervals would contain the true but unknown value of the parameter. All the parameter values bracketed inside the confidence interval are, of course, equally plausible. Given that we sampled only once and hence have only one confidence interval, we cannot know whether the true parameter value lies in our interval. But if we were to repeat the random sampling and generate a hypothetical sampling distribution of its estimate, we could say how often we expect the true value of the parameter to be bracketed by the CI: 95% of the time, or in 95% of those hypothetical intervals that will never be calculated. But just about every student learning statistics will incorrectly interpret the CI by stating that there is a 95% chance that the interval contains the true parameter value. Wrong: that would be the interpretation of the Bayesian credible interval, not of the frequentist confidence interval. It is perhaps best to interpret the confidence interval by saying that it is a range of plausible parameter values that is compatible or consistent with a statistical model or a statistical hypothesis.

We can describe the posterior distribution just obtained using a density plot (Fig. 3.8).

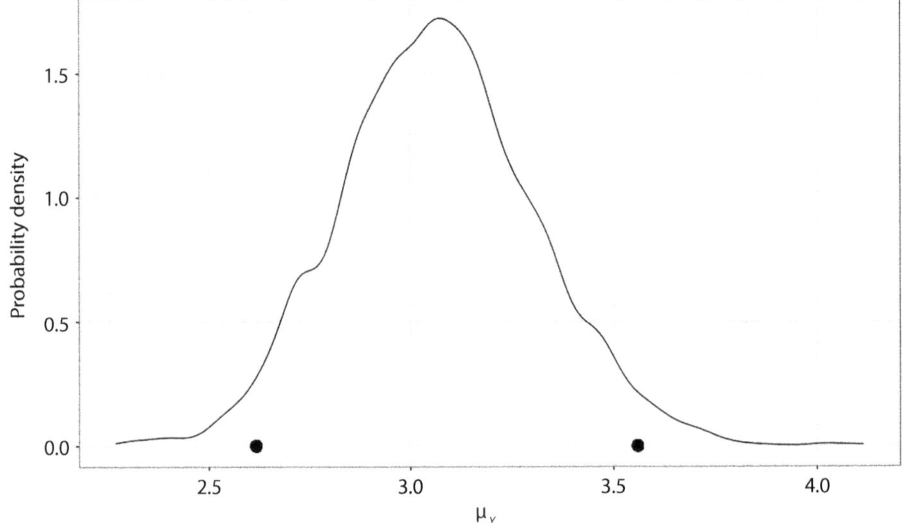

Fig. 3.8 Density plot of the posterior distribution of the Poisson parameter μ_Y. The wiggly lines show that this is not a theoretical but a numerically calculated distribution. The dots correspond to the lower and upper limits of the 95% credibility interval shown in the text.

The basic ideas just explained about using the MCMC algorithm for a Poisson model with a single parameter are extended to more complex and realistic statistical models throughout this book. We will not need to write the R code to carry out parameter estimation with MCMC. Rather, we will use the optimized algorithms of a standalone program (Stan) that is connected with R by several dedicated packages. We will discuss the workings of these packages in detail, starting in Chapter 4. We do not need to master the basics of microelectronics to use a cell phone. However, in the case of statistics, it really helps to have a basic understanding of the algorithms and numerical methods used to estimate parameters for two main reasons. First, to avoid fooling oneself first (and later others) by obtaining unreliable statistical results that lead to incorrect inferences. Second, to be able to tweak and troubleshoot the many options of these algorithms and numerical methods on an informed basis when aiming to fit complex models with a limited amount of data.

3.6 Quality control for the algorithms of Bayesian methods

So, we have crafted and run our little MCMC algorithm, and we then quickly went on to characterize the posterior distribution of the Poisson parameter. We were too fast and too reckless, but this may be understandable after having swallowed many abstract ideas related to MCMC. There are two main aspects of the quality control that should be done before we look at the parameter estimates: assessing the convergence of the MCMC algorithm and the degree of independence of the parameter estimates obtained.

Because MCMC is a numerical algorithm, we first need to verify that it has converged to a solution, and that this solution is a stationary distribution as we have been promised by the basic theory of Markov chains discussed above. We can picture a Markov chain as a blind explorer walking with a short memory (it only remembers the previous value in the chain) through the parameter space and evaluating candidate parameter values to assess their worth, Eq. (3.14). Even more, we should use more than one (and ideally no less than three) chains to assess if they all eventually converge to the same stationary distribution of parameter values. Each Markov chain explorer will laboriously and independently wander through the parameter space. What we would ideally like is that all Markov chains examine large ranges of the parameter space without getting stuck in specific parts of it. This is achieved by visual examination of the so-called trace plots showing the sequence of parameter estimates of each chain (we will do so starting in Chapter 4).

Gelman and Rubin (1992) proposed the numerical metric R to characterize the degree of convergence of multiple chains for each parameter. The R metric compares the average variance of sampled parameter values within each chain (W) with the variance of sampled parameter values calculated between chains (B). This is similar to the logic of the one-way analysis of variance that compares two estimators of variance (Chapter 5) to decide whether there are differences in means between groups (here it would be differences between chains). The formula for R is $\sqrt{\frac{W+\frac{1}{length}(B-W)}{W}}$ (see the details in Gelman et al. 2014) and this metric is routinely calculated by nearly all Bayesian software, including the R packages that we use later. In practical terms, often $R > 1.1$ is taken as a rule of thumb to indicate inadequate convergence of the chains. The convergence of the chains can and must also be examined by visually comparing the degree of overlap of the posterior distributions of each chain for all model parameters. We start doing this in Chapter 4 onwards.

It turns out that a certain number of the parameter values of the initial iterations of each chain will be affected by their initial values, and thus may not be representative of the posterior distribution. This undesirable effect is easily avoidable by discarding the first 1,000 values of each chain (called the burn-in period in the Bayesian jargon). After the burn-in period, the proposal distribution used in the MCMC algorithm always generates the next plausible candidate parameter value from the previously accepted value in the Markov chain. So what? If we have a sequence of values of some variable X generated by $X_{t+1} = X_t$ + random shift (this is the simplest random walk, and it is also known as the autoregressive model of order one, or AR(1) in the time series literature) analogous to our proposal distribution, it should be intuitive that consecutive parameter values in the chain will necessarily be correlated. Hence, the 5,000 values in our chain for the Poisson parameter were not serially independent estimates. Fortunately, we can use the time series literature to obtain the effective number of independent parameter estimates in a chain by $\frac{length}{1+2\sum\rho_i}$, where *length* is the nominal length of the Markov chain, and ρ_i is the autocorrelation coefficient at lag i that measures the Pearson correlation coefficient calculated for pairs of parameter estimates separated by i steps. What this formula says is that the stronger the serial autocorrelation (i.e., the degree of similarity) between values in the chain, the smaller the number of effectively independent draws.

We need to take into account the non-independence of parameter estimates when calculating the standard error that is used to describe the precision of the mean of the parameter estimates. The effective number of parameter estimates will be routinely estimated for each parameter by any software doing Bayesian analysis, using this simple expression (although Stan uses an expression based on the Fourier transform; Lambert 2018). A common solution to having autocorrelated parameter estimates is to thin the chain by only storing one of every k accepted values (this is known as the thinning rate in the Bayesian jargon), meaning that the other $k - 1$ parameter estimates drawn will be discarded and never seen by the user. Therefore, the computer will not only discard the burn-in part, but also, when $k = 20$, 19 out of 20 (95%) of accepted posterior draws. The posterior distribution assessment available to the user will contain (say) 5,000 draws, not counting the burn-in and thinning rate applied. These incredibly wasteful procedures that discard so much information are feasible only because computer power is cheap and getting cheaper all the time. We comment on the convergence, autocorrelation, and the effective sample size of chains just discussed for every Bayesian model fitted in this book.

3.7 More general MCMC variations: Metropolis–Hastings and Gibbs algorithms

The Metropolis MCMC algorithm explained in the previous section employs a symmetric proposal distribution, whereby the probability that the chain jumps between two parameter values A and B is identical in either direction: $\Pr(A \to B) = \Pr(B \to A)$. This symmetric proposal distribution does not work well whenever there are boundaries of feasible values of model parameters. Examples include a variance that must be strictly positive and a correlation coefficient that must lie in [–1, 1]. Hastings (1970) proposed a change involving a potentially asymmetric proposal distribution that includes and improves on the original Metropolis algorithm as a special case. In keeping with the historical precedence, the main basic MCMC algorithm is now known as the Metropolis–Hastings algorithm. The main difference between the original and the more general version is in the acceptance criterion, Eq. (3.15), which now includes the asymmetric proposal distribution and becomes

$$R = \min\left(1, \frac{\text{Likelihood}\left(\text{data} \mid Par^B\right)\text{Prior}\left(Par^B\right)\Pr\left(Par^A \to Par^B\right)}{\text{Likelihood}\left(\text{data} \mid Par^A\right)\text{Prior}\left(Par^A\right)\Pr\left(Par^B \to Par^A\right)}\right). \tag{3.16}$$

The only difference from Eq. (3.15) is the appearance of a third term (the probability of the chain jumping between the two parameter values A and B) in the numerator and denominator. Otherwise, the algorithms are identical. This seemingly slight but important change proposed by Hastings (1970) is critical to dealing with the bounded parameter spaces that frequently appear in many statistical models.

Another variation of the original Metropolis MCMC algorithm is called a Gibbs algorithm or Gibbs sampler that can be shown to be a special case of the Metropolis MCMC algorithm (Gelman et al. 2014). The Gibbs algorithm was proposed by Geman and Geman (1984) and it is especially useful for complex models with many parameters. It has given rise to several popular platforms for carrying out Bayesian estimation such as BUGS (an acronym standing for "Bayesian analysis using Gibbs sampling"; Gilks et al. 1994) and JAGS ("Just another Gibbs sampler"; Plummer 2003). In particular, BUGS can be credited as the main program responsible for allowing non-statisticians to carry out MCMC estimation at the start of the Bayesian revival during the early 1990s. There have been several variations and improvements of the BUGS software over time (Lunn et al. 2012). It is fair to say that the more recent and faster algorithms and platforms for Bayesian analysis (among them, JAGS) available today have superseded the original WinBUGS and then OpenBUGS software.

To describe the Gibbs algorithm, let's first define a simple statistical model to describe a real-valued response variable Y with a normal distribution, $Y \sim \text{Normal}(\mu_Y, \sigma_Y^2)$, whose parameters mean μ_Y and variance σ_Y^2 need to be estimated. Equation (3.14) now becomes

$$\text{Posterior}\left(\mu_Y, \sigma_Y^2 \mid \text{data}\right) \propto \text{Likelihood}\left(\text{data} \mid \mu_Y, \sigma_Y^2\right) \times \text{Prior}\left(\mu_Y, \sigma_Y^2\right). \qquad (3.17\text{a})$$

The posterior distribution will now be a two-dimensional distribution, with a probability density for each pair of values of the parameters μ_Y and σ_Y^2. The idea behind Gibbs sampling is to factorize Eq. (3.17a) using the fact that assuming the independence of these parameters gives a simpler equation:

$$\text{Posterior}\left(\mu_Y, \sigma_Y^2 \mid \text{data}\right) \propto \text{Likelihood}\left(\text{data} \mid \mu_Y, \sigma_Y^2\right) \times \text{Prior}\left(\mu_Y\right)\text{Prior}\left(\sigma_Y^2\right). \qquad (3.17\text{b})$$

In words, Eq. (3.17b) means that we do not need to generate pairs of values of μ_Y and σ_Y^2 from a joint prior distribution (a much harder problem) but that we can do it from two independent prior distributions (a much simpler problem).

The steps of the Gibbs algorithm for our current problem are:

1. Give initial values to the parameters μ_Y and σ_Y^2, and at each iteration:
2. (optional) Randomly choose which parameter we update first (say, σ_Y^2).
3. Sample a new value for μ_Y (keeping the value of σ_Y^2 constant) from its conditional posterior distribution using the most up-to-date values of μ_Y and σ_Y^2.
4. Repeat step 3 for the other parameter(s) ($\sigma_Y^2$for this iteration).
5. Repeat steps 2–4 many times until the convergence of the MCMC chains.

The cornerstone of the Gibbs algorithm involves factoring the complex Eq. (3.17a) into a collection of univariate prior distributions, Eq. (3.17b), that can be sampled one at a time using updated information. Randomizing the order of the parameters to be updated at each iteration is, however, optional (Fig. 3.9).

The "factorization trick" in Eq. (3.17b) seems to work reasonably well provided that the parameters are not too strongly correlated, which is often (but not always) the case for many statistical models (Lambert 2018, Hooten and Hefley 2019). The goal here has been just to provide intuitive, informal explanations of the main ideas and the rationale behind the basic Metropolis, Metropolis–Hastings, and Gibbs algorithms that remain the workhorse for the Bayesian analysis of many statistical models. There are more advanced and complex versions of these algorithms that are better suited for more complex (i.e.,

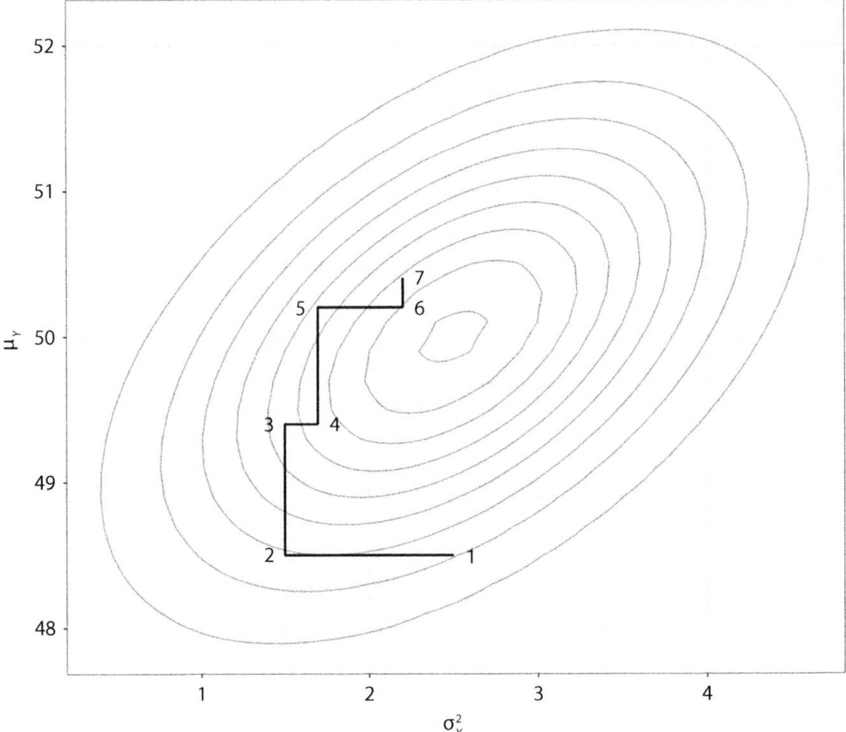

Fig. 3.9 Representation of the first seven steps of the Gibbs algorithm to estimate the joint posterior distribution of the mean μ_Y and variance σ^2_Y whose projection of the (unknown) posterior distribution is shown as a gray contour map. At each step of the algorithm, only one randomly chosen parameter is updated using the Metropolis–Hastings acceptance rule and incorporated into the bivariate Markov chain, hence the straight lines connecting the successive steps of the algorithm.

high-dimensional) models with highly correlated parameters such as the next MCMC algorithm.

3.8 Recent advances in Bayesian methods: Hamiltonian Monte Carlo

Hybrid or Hamiltonian Monte Carlo (HMC) is the third and last MCMC algorithm that we cover in this book. HMC is a recent addition to the arsenal of MCMC algorithms (see Neal 2011 for a historical overview). It uses a different way of generating candidate parameter values that is more efficient and versatile than the Metropolis–Hastings and Gibbs algorithms (Monnahan et al. 2017, Hooten and Hefley 2019). "Efficient" here means that for the same data set, HMC converges faster to the posterior distribution than the Metropolis–Hastings and Gibbs algorithms, and produces samples that typically have less autocorrelation and hence contain more information about a posterior (Monnahan et al. 2017). And by "versatile" we mean that HMC can tackle and numerically solve complex problems with many correlated parameters that either cannot be solved, or would take too long to solve, by the Metropolis–Hastings and Gibbs algorithms (Hooten and Hefley 2019).

These desirable features of HMC are due to the way in which it generates proposals using the paradigm of Hamiltonian dynamics from physics (Carpenter et al. 2016). We do not need to grasp the details of Hamiltonian dynamics to understand HMC. Let's start by recalling that the parameter proposals of the Metropolis–Hastings algorithm were obtained by performing a random walk through the parameter space. As a consequence, the proposed parameter values were often likely to have a posterior probability similar to the current values. The Gibbs algorithm essentially did the same but just updating one parameter at a time, rather than all the parameters at once as in Metropolis–Hastings. Inevitably, these algorithms may make a slow and inefficient exploration of high-dimensional parameter spaces, particularly when the model parameters are correlated (Hooten and Hefley 2019).

What HMC does is to make longer jumps from the current to the new proposed values, thereby exploring faster (with fewer rejections) and providing a more thorough examination of the parameter space than MCMC. These longer jumps are based on the evaluation of both the actual value and the gradient (or steepness) of the posterior probability of parameter values after a short-term trajectory or move over the parameter space. While Metropolis–Hastings makes a single jump from current to proposal parameter values, HMC makes a wiggly trajectory (that depends on both the posterior probability and its gradient and some random kicks) over the parameter space from current proposed values (Hooten and Hefley 2019).

The essence of the HMC algorithm (based on Lambert 2018) is:

1. Select initial parameter values (let's call them A) and calculate their posterior probability Post(A).
2. Calculate the gradient of the posterior probability for each parameter from a probability distribution—let's call them Grad(A). The gradient is a vector of partial derivatives (read rates of change) of the probability distribution for each parameter.
3. Use some algorithm to generate the short-term wiggly trajectory over the parameter space (leapfrog is often used, but here it really does not matter which is used).
4. At the end of this wiggly trajectory at the parameter set B, calculate the posterior probability, Post(B), gradient of the posterior probability for each parameter, Grad(B).
5. Calculate the criterion R:

$$R = \min\left(1, \frac{\text{Likelihood}\left(\text{data} \mid Par^{B}\right)\text{Prior}\left(Par^{B}\right)\Pr\left(Par^{A} \to Par^{B}\right)\text{Grad}\,(B)}{\text{Likelihood}\,(\text{data} \mid Par^{A})\,\text{Prior}\,(Par^{A})\Pr\,(Par^{B} \to Par^{A})\,\text{Grad}\,(A)}\right). \tag{3.18}$$

6. Decision rule:
   a. If Post(B) > Post(A), always jump $A \to B$; add B to the chain of estimates.
   b. If Post(B) $\leq$ Post(A), randomly jump $A \to B$ by comparing R with a uniform deviate in the range [0, 1] (just as we did before!).
7. Either keep or, when needed, refresh the current value of the chain.
8. Repeat steps 1 to 7 many times until convergence.

We may hopefully appreciate that there is a great similarity between Metropolis–Hastings MCMC and the HMC algorithms. But, there are some important differences between them. The main one is that the proposal distribution in HMC involves a short deterministic trajectory (and not a random walk of a single step) based on the gradient of the posterior distribution (Hooten and Hefley 2019). As a consequence, HMC may undertake big jumps and provide a more efficient exploration of the values across the parameter space.

Implementation of the HMC algorithm is more complex than for Metropolis–Hastings, and HMC requires the fine-tuning of other key details such as the calculation of fast gradients and the generation of the wiggly trajectory (by numerical integration of a differential equation, if we care to know it).

Some words of relief are needed and overdue here. After reading this far, it would seem that we would need to be a combination of a card-carrying statistician and a high-level computer programmer to carry out any Bayesian analysis. Rest assured: you do not! Some people feel that "you must code your MCMC yourself to make sure that you know what the program is doing." We do not need to reinvent the wheel by spending many hours to produce sub-optimal computer code for each problem we need to solve. What we will do is to use R to write the statistical models to analyze each type of data, then employ optimized black-box MCMC or HMC software to do the Bayesian calculations for us, and later retrieve the results into R to analyze and validate the statistical results.

We have already mentioned BUGS and JAGS as the main programs to implement the Metropolis–Hastings and Gibbs algorithms. Carpenter et al. (2016) developed Stan, a probabilistic model language that implements HMC and produces an output that can be analyzed and post-processed by R packages. Stan has very clearly written documentation posted on the web (https://mc-stan.org/). It also has a very active and responsive community of users that is of great help in troubleshooting most problems that arise during the climbing of the learning curve involved in fitting complex statistical models. This is not to say that Stan is either the ultimate or universal MCMC algorithm capable of optimally solving every estimation problem. Besides JAGS, we can also mention NIMBLE (De Valpine et al. 2017) and INLA (Rue et al. 2017) as two flexible and efficient platforms for Bayesian parameter estimation capable of dealing with all the statistical models covered in this book. We explain how to install Stan for Windows, Mac, and Linux on the companion web site. We will interact with Stan through R-Studio using specific R packages in the coming chapters. No doubt in a few years there will be better theoretical approaches and yet more efficient and more general MCMC algorithms that will supersede what we have explained here.

3.9 Bayesian hypothesis tests

In the Bayesian framework, the null hypothesis statistical test is accomplished by the calculation of the Bayes factor. Given two statistical hypotheses about a parameter θ being equal or not to a certain value, H_0: $\theta = 0$ and H_1: $\theta \neq 0$, we could use Eq. (3.14) to write

$$\frac{\text{Posterior}\,(\theta\,\text{for H}_1 \mid \text{data})}{\text{Posterior}\,(\theta\,\text{for H}_0 \mid \text{data})} \approx \frac{\text{Likelihood}\,(\text{data} \mid \theta\,\text{for H}_1)}{\text{Likelihood}\,(\text{data} \mid \theta\,\text{for H}_0)} \times \frac{\text{Prior}\,(\theta\,\text{for H}_1)}{\text{Prior}\,(\theta\,\text{for H}_0)}, \tag{3.19}$$

which reflects how the ratio of the prior beliefs of H_1 relative to H_0 are modified by the data (the ratio of their likelihoods) to yield the ratio of posterior beliefs in these hypotheses. Intuitively, if the data provided equal support to each hypothesis, we would not change our prior beliefs on these hypotheses. We could then rewrite Eq. (3.19) as

$$\frac{\text{Posterior}\,(\theta\,\text{for H}_1 \mid \text{data})}{\text{Posterior}\,(\theta\,\text{for H}_0 \mid \text{data})} \approx \text{Bayes factor} \times \frac{\text{Prior}\,(\theta\,\text{for H}_1)}{\text{Prior}\,(\theta\,\text{for H}_0)}. \tag{3.20}$$

Table 3.2 Two sets of criteria for the use of Bayes factors smaller than unity to decide the degree of empirical support for the hypothesis H_0 in a Bayesian hypothesis test.

Bayes factor	Strength of evidence against H_0	
	Jeffreys (1939)	Goodman (1999)
1 to 1/3	Bare mention	—
1/3 to 1/10	Substantial	Weak to moderate
1/10 to 1/30	Strong	Moderate to strong
1/30 to 1/100	Very strong	Strong
1/100 to 1/300	Decisive	Very strong

The Bayes factor (Jeffreys 1939) quantifies the change from prior to posterior odds brought about by the data (Kass and Raftery 1995, Morey et al. 2016, Ly et al. 2020). If the Bayes factor were equal to (say) 7, H_1 would be seven times more likely than H_0 given the data and their priors (Table 3.2). Just like the original derivation of the p-values, the Bayes factor can be used as a continuous measure of the relative support that two hypotheses receive from the data (Goodman 1999, Held and Ott 2018). The Bayes factor has been proposed as a suitable criterion for Bayesian hypothesis tests. And of course, people have come up with totally arbitrary (but perhaps useful) rules of thumb for how its values can be translated into discrete conclusions (Table 3.2).

Three comments here. First, calculating the Bayes factor is not as straightforward as it may seem at first sight. When the two hypotheses are nested (i.e., when one is a special case of the other), the Bayes factor can be calculated using the Savage–Dickey density ratio (Wagenmakers et al. 2010). This approximation is, however, highly dependent on the priors used (Kruschke and Liddell 2017, Etz et al. 2018, Heck 2019). Hence, employing the Bayes factor in hypothesis tests is bound to be subject to the seemingly eternal bickering about the choice of priors in the Bayesian framework. Second, there might seem to be a way out of the dependence of Bayes factors on prior distributions. Sellke et al. (2001), Bayarri et al. (2016), and Held and Ott (2019) have derived relationships between minimum Bayes factors and p-values (which, being obtained within the frequentist framework, do not depend on prior distributions) in two-sided hypothesis tests. At this point, the generality of these relationships seems unclear for most common statistical models covered in this book. Third, and most important, the discretization of the Bayes factor, a continuous measure of relative evidence for two hypotheses, entails exactly the same problems encountered by the analogous procedure in the frequentist framework. There is no point in having "universal thresholds" to make yes/no decisions that can be mechanically applied by researchers in all problems and contexts. We should rather strive to estimate the magnitudes of the effects and relationships between variables, quantifying their uncertainty using statistical models that fit the data reasonably well, and make decisions based on the contingent factual evidence (Armheim et al. 2019).

3.10 Summary of the main differences between maximum likelihood and Bayesian methods

In closing, it is useful to summarize and highlight the main conceptual differences between the frequentist (based on maximum likelihood) and Bayesian frameworks that will be employed in this book. While some ideas were already mentioned in the chapter, they bear repeating to emphasize them. The two key features that differentiate these frameworks are their fundamental interpretation of probability, and the explicit use of prior knowledge in the process of estimation. All the differences featured in Table 3.3 stem from these two fundamental aspects.

After taking probability strictly and only as a relative frequency of events obtained with respect to a hypothetical statistical population, a frequentist inevitably views data as just a chance event out of many other samples that could have been drawn from the same population. The sampling distribution of parameter estimates is a central aspect of frequentist inference that reflects what values of the parameter estimates could have been obtained ... although actually they didn't in our current sample. *P*-values and confidence intervals are all derived from the sampling distributions of potential parameter estimates. The frequentist framework seeks to find those parameter estimates (and their standard errors) that maximize the chances of having produced the data we have collected.

Bayesian statisticians often criticize sampling distributions, *p*-values, and confidence intervals by referring to them as statements or figments about potential data that have neither occurred nor will ever occur. Inference in the frequentist framework focuses on the probability of obtaining the collected data given a model or hypothesis defined by the parameter values, Pr(data | hypothesis). In contrast, inference in the Bayesian framework is based on the posterior distribution, Pr(hypothesis | data). The strong dislike of Fisher, Neyman, and Pearson of prior distributions prompted them to develop methods devoid of the perceived arbitrariness of the prior distributions that are an essential ingredient in any Bayesian analysis. In a literal sense, the world of knowledge is created anew every time a frequentist analyzes their data because no previously acquired knowledge explicitly enters into the analysis. Employing uniform prior distributions for every model parameter to signify our complete ignorance about their values in a Bayesian analysis is exactly equivalent to a frequentist analysis based on maximum likelihood. The von Mises–Bernstein theorem states that the posterior distribution converges to a multivariate normal distribution centered at the maximum likelihood estimates for large enough sample sizes, and under some general regularity conditions.

Many Bayesian statisticians in turn view data analysis as a progressive learning process involving the successive refinement of knowledge embodied in parameter values by the incorporation of new evidence in subsequent applications of Bayes' rule. New data allows updating our previous knowledge that will now become embodied in the prior distribution of the latest analysis. The use of prior distributions is fundamental in Bayesian statistics. And it has been, and may probably remain, its most contentious and divisive aspect. Though progress has been made and new theory and algorithms have clarified many standing issues, how to define prior distributions remains the Bayesians' very weak and soft spot for many. Some frequentist statisticians (e.g., Dennis 1996) view the use of prior distributions as an unacceptable Faustian bargain to attain statistical knowledge. There is indeed a terminological maze of confusion when defining the degree of informativeness of prior distributions that can be disquieting. Bayesian inference may be hard for many users due to its reliance on clever algorithms. It may also (to some at least) demand a stronger background in statistics and computer programming than does frequentist statistics. As we (should) know, there is no free lunch, ever.

As with every human endeavor, the frequentist and Bayesian frameworks would seem to make some seemingly outrageous assumptions (i.e., the sampling distribution) or take

Table 3.3 Main differences between the frequentist and Bayesian frameworks of statistical inference.

	Frequentist	Bayesian
The goal of inference is to obtain:	The maximum likelihood point estimates (and their standard errors) of model parameters.	The joint posterior probability distribution of model parameters using Bayesian algorithms (MCMC).
View of probability	A relative frequency of occurrence based on the actual or potential repetition of the same sampling event.	A degree of plausibility of occurrence of an event. It may also include a frequentist interpretation.
View of available data	A single realization of a random process out of many possible data that could be obtained from the same population.	Taken as fixed and given once data were obtained. New data can be used to progressively refine our knowledge of parameter values in separate analyses.
Statistical parameters are:	Fixed constants whose precise values are both unknown and unknowable but can be estimated with a given precision that essentially depends on the sample size.	Fixed constants whose values remain unknown. The uncertainty about their values is expressed using probability. Because the latter is used as a measure of knowledge about a parameter, they are treated equivalently to random variables.
Role of the likelihood function	As the central and only feature of statistical inference. The likelihood reflects the plausibility of obtaining the data given the values of parameters.	As the means by which the information contained in the data modifies our previous knowledge of model parameters according to Bayes' rule.
Inference refers to:	Comparing the actual parameter estimates with a sampling distribution that could be potentially obtained by repeated sampling from the same population.	Strictly based on the available data. Consists of characterizing and comparing the posterior distributions of model parameters.
Can we state that a statistical hypothesis (e.g., $\beta = 0$) is "true"?	No. We can only state the chances of potentially obtaining the current, or more extreme, data if a statistical hypothesis were true.	Yes. We can calculate the probability that a statistical hypothesis were true, given the available data and our previous knowledge of the parameters. This is a direct consequence of the Bayesian use of probability as a measure of uncertainty about parameters.
Measures of precision of parameters	Confidence intervals, interpreted as the proportion of times that a parameter would lie in range if one were to obtain many random samples and calculate the intervals from the same statistical population.	Credible intervals, interpreted as the probability that a parameter actually lies in range.
Use of previous knowledge	Included only at the discretion of the analyst. Previous knowledge does not affect current inference.	An essential part of the analysis. Depicted as prior distributions for each model parameter.

wild leaps of faith and imagination (i.e., the prior distributions) to be able to crack the difficult problem of statistical inference. Most assumptions are never exactly true and strictly valid in every conceivable case. The issue at stake is not truth and validity but usefulness: what can be achieved by making these assumptions that could not have been possible otherwise? The disputes between adherents to these statistical frameworks were often acronymous for most of the twentieth century until the 1990s. At the time of writing, while there is no conceivable theoretical synthesis between the frequentist and Bayesian frameworks, there is an agreed sense of peaceful and ecumenical coexistence. For scientists like us who feel no deep attachment to schools of thought and philosophical traditions, they are the conceptual basis of essential tools of data analysis. As such, they need to be well understood, used, and interpreted as an essential part of the hard and uncertain problem of scientific inference from data.

References

Amrheim, V., Greenland, G., and McShane, B. (2019). Retire statistical significance. *Nature*, 567, 305–308.

Bayarri, M., Benjamin, D., Berger, J. et al. (2016). Rejection odds and rejection ratios: A proposal for statistical practice in testing hypotheses. *Journal of Mathematical Psychology*, 72, 90–103.

Bayes, T. (1763). An essay towards solving a problem in the doctrine of chances. *Philosophical Transactions of the Royal Society*, 53, 370–418.

Carpenter, R., Gelman, A., Hoffman, M. et al. (2016). Stan, a probabilistic programming language. *Journal of Statistical Software*, 76, 1–32.

De Valpine, P., Turek, D., Paciorek, C. et al. (2017). Programming with models: Writing statistical algorithms for general model structures with NIMBLE. *Journal of Computational and Graphical Statistics*, 26, 403–413.

Dennis, B. (1996). Should ecologists become Bayesians? *Ecological Applications*, 6, 1093–1105.

Edwards, A. (1992). *Likelihood*. The John Hopkins University Press, Baltimore.

Etz, M. (2018). Introduction to the concept of likelihood and its applications. *Advances in Methods and Practices in Psychological Science*, 1, 60–69.

Etz, M., Gronau, Q., Dablander, F. et al. (2018). How to become a Bayesian in eight easy steps: An annotated reading list. *Psychonomic Bulletin Review*, 25, 219–234.

Fisher, R. (1912). On an absolute criterion for fitting frequency curves. *Messenger of Mathematics*, 45, 155–160.

Fisher, R. (1922). On the mathematical foundations of theoretical statistics. *Proceedings of the Royal Society A*, 222, 309–368.

Gelfand, A. and Smith, A. (1990). Sampling-based approaches to calculating marginal densities. *Journal of American Statistical Association*, 85, 398–410.

Gelman, A. (2012). Ethics and the statistical use of prior information. *Chance*, 25, 52–55.

Gelman, A., Hwang, J., and Vehtari, A. (2014). Understanding predictive information criteria for Bayesian models. *Statistical Computing*, 24, 997–1016.

Gelman, A., Jakulin, A., Pittau, M. et al. (2008) A weakly informative default prior distribution for logistic and other regression models. *Annals of Statistics*, 2, 1360–1383.

Gelman A., Meng X., and Stern H. (1996). Posterior predictive assessment of model fitness via realized discrepancies. *Statistica Sinica*, 7, 733–807.

Gelman, A. and Rubin, D. (1992). Inference from iterative simulation from multiple sequences. *Statistical Science*, 7, 457–511.

Geman, S. and Geman, D. (1984). Stochastic relaxation, Gibbs distributions, and the Bayesian restoration of images. *IEEE Transactions on Pattern Analysis and Machine Intelligence*, 6, 721–741.

Geyer, C. (2011). Introduction to Markov chain Monte Carlo. In S. Brooks et al. (eds), *Handbook of Markov Chain Monte Carlo*, pp. 3–48. Chapman and Hall CRC, New York.

Gilks, W., Thomas, A., and Spiegelhalter, D. (1994). A language and program for complex Bayesian modelling. *American Statistician*, 43, 169–177.

Goodman S. (1999). Toward evidence-based medical statistics. 2: The Bayes factor. *Annals of Internal Medicine*, 130, 1005–1017.

Hastings, W. (1970). Monte Carlo sampling methods using Markov chains and their applications. *Biometrika*, 57, 97–109.

Heck, D. (2019). A caveat on the Savage–Dickey density ratio: the case of computing Bayes factors for regression parameters. *British Journal of Mathematical and Statistical Psychology*, 72, 316–333.

Held, L. and Ott, M. (2018). On *p*-values and Bayes factors. *Annual Review of Statistics and Its Application*, 5, 393–419.

Hobbs, N. and Hooten, T. (2015). *Bayesian Models: A Statistical Primer for Ecologists*. Princeton University Press, Princeton.

Hooten, M. and Hefley, T. (2019). *Bringing Bayesian Models to Life*. Routledge Taylor & Francis Group, New York.

Jeffreys, H. (1939). *Theory of Probability*. Oxford University Press, Oxford.

Jeffreys, H. (1946). Probability and scientific method. *Proceedings of the Royal Society A*, 146, 9–16.

Kass, T. and Raftery, A. (1995). Bayes' factor. *Journal of American Statistical Association*, 90, 773–796.

Kéry, M. and Schaub, M. (2012). *Bayesian Population Analysis using WinBUGS: A Hierarchical Perspective*. Academic Press, New York.

Kroese, D., Brereton, T., Taimre, T. et al. (2014). Why the Monte Carlo method is so important today. *WIREs Computational Statistics*, 6, 383–393.

Kruschke, J. (2015). *Doing Bayesian Data Analysis: A Tutorial with R, JAGS and Stan*, 2nd ed. Academic Press, New York.

Kruschke, J. and Liddell, D. (2017). The Bayesian new statistics: Hypothesis testing, estimation, meta-analysis, and power analysis from a Bayesian perspectives. *Psychonomic Bulletin Review*, 15, 1–28.

Lambert, B. (2018) *A Student's Guide to Bayesian Statistics*. Sage Publishers, Los Angeles.

Laplace, P. (1814). *Essai philosophique sur les probabilités*. Translated by F. Truscott and F. Emory. Cosimo Classics, New York.

Lunn, D., Jackson, C., Best, N., et al. (2012). *The BUGS Book: A Practical Introduction to Bayesian Analysis*. CRC Press / Chapman & Hall, New York.

Ly, A., Stefan, A., van Doorn, J., et al. (2020). Harold Jeffreys' default Bayes factor hypothesis tests: Explanation, extension, and application in psychology. *Computational Brain & Behavior*, 3, 153–161.

Metropolis, N. (1987). The beginning of the Monte Carlo method. *Los Alamos Science*, 15, 125–130.

Metropolis, N., Rosenbluth, A., Rosenbluth, A., et al. (1953). Equation of state calculations by fast computing machines. *Journal of Chemical Physics*, 21, 1087–1092.

Millar, R. (2011). *Maximum Likelihood Estimation and Inference with Examples in R, SAS and ADMB*. John Wiley & Sons, New York.

Monnahan, C., Thorson, J., and Branch, T. (2017). Faster estimation of Bayesian models in ecology using Hamiltonian Monte Carlo. *Methods in Ecology and Evolution*, 8, 339–348.

Morey, R., Romeijn, J., and Rouder, J. (2016). The philosophy of Bayes factors and the quantification of statistical evidence. *Journal of Mathematical Psychology*, 72, 6–18.

Neal, R. (2011). MCMC using Hamiltonian dynamics. In S. Brooks et al. (eds.), *Handbook of Markov Chain Monte Carlo*, pp. 113–152. CRC Press Chapman & Hall, New York.

Neyman, J. and Pearson, E. (1933). IX. On the problem of the most efficient tests of statistical hypotheses. *Philosophical Transactions of the Royal Society A*, 231, 289–337.

Norris, J. R. (1997). *Markov Chains*. Cambridge University Press, Cambridge.

Pawitan, J. (2001) *In All Likelihood: Statistical Modelling and Inference Using Likelihood*. Oxford University Press, Oxford.

Plummer, J. (2003). JAGS: A program for analysis of Bayesian graphical models using Gibbs sampling. In K. Hornik, F. Leisch, and A. Zeileis (eds), *Proceedings of the 3rd International Workshop on Distributed Statistical Computing (DSC 2003)*, Vienna, Austria.

Raifa, H. and Schlaifer, R. (1961). *Applied Statistical Decision Theory*. Harvard Business School, Harvard.

Reid, N. (2010). Likelihood inference. *WIREs Computational Statistics*, 2, 517–525.

Rue, H., Riebler, A., Sørbye, S. et al. (2017). Bayesian computing with INLA: A review. *Annual Review of Statistics and Applications*, 4, 18–45.

Sellke, T. Bayarri, M., and Berger, J. (2001). Calibration of p values for testing precise null hypotheses. *American Statistician*, 55, 62–71.

Vickers A., Cronin A., Björk, T., et al. (2010). Prostate specific antigen concentration at age 60 and death or metastasis from prostate cancer, case-control study. *British Medical Journal*, 341, c4521.

Wagenmakers, E., Lodewyckx, T., Kuriyal, H. et al. (2010). Bayesian hypothesis testing for psychologists: A tutorial on the Savage–Dickey method. *Cognitive Psychology*, 60, 158–189.

Wakenfield, J. (2013). *Bayesian and Frequentist Regression Methods*. Springer-Verlag, New York.

Wilks, S. (1938). The large-sample distribution of the likelihood ratio for testing composite hypotheses. *Annals of Mathematical Statistics*, 9, 60–62.

PART II

Applying the Generalized Linear Model to Varied Data Types

In the following chapters of the book we are going to be fitting statistical models to data. The script or set of R commands necessary for their analyses, data sets in CSV (comma-separated values) format, and the R workspace containing the R objects created for each chapter can be found on the companion website. The R packages needed to fit the models in each chapter are indicated at the start of each chapter and can be loaded with the specified script.

It is worth giving a general outline of the process of fitting the statistical models, which can be decomposed into the following steps:

1. Import and examine the data:
   - Transform the CSV data file into an R data frame.
   - Verify the scales (factor, numeric) of the imported variables.
2. Exploratory data analysis:
   - Assess which probability distribution should be used to model the response variable.
   - Examine the relations between the response and explanatory variable(s) to suggest whether it is reasonable to consider interactions between variables. In confirmatory analysis, the consideration of interactions stems from the hypothesis being tested.
   - Verify the degree of independence of numerical explanatory variables to avoid including strongly correlated variables in the analyses.
   - Characterize the data set by calculating descriptive statistics, eventually per groups.
3. Fitting the statistical model:
   - Frequentist fitting: Obtain and interpret the estimate of all model parameters and their standard errors and 95% confidence intervals. Interpret metrics of model fit such as the adjusted R^2 and the percentage of deviance explained. Use the AIC and Wilks likelihood test in the context of model selection. Carry out a posteriori comparisons when adequate, and interpret their results.
   - Bayesian fitting: Obtain and interpret the posterior means of model parameters and their standard errors and 95% credible intervals. Visually assess the convergence

of the Markov chain Monte Carlo algorithm, and the independence of sampled parameter estimates. When appropriate, use the WAIC in the context of model selection. Obtain and interpret the Bayesian R^2, the posterior predictive distribution, and the posterior predictive checks. Regarding priors, display the prior distributions and make assessments of their influence using prior sensitivity analysis and/or prior predictive distributions.

4. Residual analysis:
   - Assess the quality of fit of statistical models to data through the graphic analysis of plots of residuals vs. fitted, residuals vs. explanatory variables, Q–Q plots of residuals, and the influence of each data point in the model fit (Cook distances in the frequentist framework, the Pareto k statistic in the Bayesian framework).
   - Obtain and understand the different types of residuals needed to make valid assessments of model fit for different types of data.
5. Communicating the model output:
   - Obtain and interpret summary figures of model parameter estimates and their 95% CIs.
   - Obtain and interpret the marginal plots depicting the predicted relations for all explanatory variables and interactions in a model.

At this point, the previous five points are nothing more than a list of as yet undefined and non-explained items. You should not worry, as everything is progressively covered and explained in full detail in the upcoming chapters. All that matters at this point is that you have a general overview of how the analyses of each data set will proceed.

When pertinent, each chapter contains a set of problems that can be solved in either statistical framework or, better, in both. The data set for each problem can be found on the companion website. We suggest following the above list of five points to solve these problems.

CHAPTER 4

The General Linear Model I

Numerical explanatory variables

R packages needed in this chapter:

```
packages.needed<-c("ggplot2","bayesplot","fitdistrplus","reshape2",
"qqplotr","gridExtra","ggeffects", "GGally","broom","qqplotr","brms",
"bayesplot","doBy")
lapply(packages.needed, FUN = require, character.only = T)
```
loads these packages in the working session.

4.1 Introduction

This chapter introduces the general linear model to analyze real-valued response variables with a normal distribution. The continuous and/or categorical explanatory variables in the general linear model are linearly related to, and hence aim to explain the variation of, the mean of the response variable. The mean and the variance of a variable with normal distribution are independent parameters, the latter of which accounts for the variation of the response variable around its mean.

The basic assumptions of the general linear model are that the response variable be normally distributed, that the samples be randomly taken from their statistical populations, and that (unless indicated otherwise) the variance of the response variable is homogeneous across the explanatory variables. The homogeneity of variance (technically known as homoscedasticity) assumption just states how many variances are to be estimated: either one overall variance of the response variable, or several variances for different values of the explanatory variables. In the latter case, both the mean and the variance of the response variable change with the explanatory variables.

The main reason for starting with the general linear model is historical. The normal distribution enjoys a special status among probability distributions due to its frequent use in modeling continuous response variables. The prevalence of the normal distribution is basically due to the central limit theorem (first proved by Laplace in 1813) that states that the result of additive composition of the effects of many (independent, identically distributed) random variables results in a response variable with a normal distribution. Many real-valued variables would seem to be the approximate result of such a mechanism of additive composition of effects.

The general linear model has been the workhorse of many introductory statistics courses and textbooks for most of the 20th century until the formulation of the generalized linear models in 1972 (Chapter 8). It includes as special cases a host of the most frequently used tests and methods of analysis employed by scientists. These include simple and

Statistical Modeling With R. Pablo Inchausti, Oxford University Press. © Pablo Inchausti (2023).
DOI: 10.1093/oso/9780192859013.003.0004

multiple regression, the *t*-test for the difference between the means of two groups, every conceivable form of analysis of variance (one-way, two-way, factorial, etc.) involved in the design of common experiments, and the analysis of covariance. Many textbooks and introductory courses cover these tests or methods in separate chapters and classes, leaving the impression of their being unconnected, or at best loosely related, topics. This is a real pity. In this chapter and the next, we will present this array of tests and methods as just special cases of the general linear model (a fact known since at least Jennings 1967 and Cohen 1968) just depending on the number and types of the explanatory variables involved.

4.2 The lognormal distribution and its relation to the general linear model

Additive composition is by no means the only way in which the effects of many intervening random variables may combined to render a real-valued response variable. A frequent alternative is the multiplication of composing effects of intervening variables. The lognormal distribution results from the product of many positive and independent random variables. In economics, this multiplicative composition is known as Gibrat's law (Sutton 1997) of proportional growth, and a similar case can be made for proportional growth in many other research areas, including ecology, demography, and more. This multiplicative composition of proportional growth events leading to a lognormal distribution of the response variable can be justified by invoking Laplace's central limit theorem in the log domain.

The lognormal distribution has an asymmetric shape with a long upper tail (Fig. 4.1) such that there are many small and a few large values. The basic properties of the lognormal distribution were described long ago (see Limpert et al. 2001 for a historical overview). Its formula is

$$f(Y) = \frac{1}{Y\sqrt{2\pi}\sigma_Y}\exp\left(\frac{-\left(\log(Y)-\mu\right)^2}{2\sigma_Y^2}\right), \tag{4.1}$$

where Y is a strictly positive real-valued random variable, and μ_Y and σ_Y^2 are the mean and the variance of Y measured in log scale. The mean of a variable with lognormal distribution is $\exp(\mu_Y + \sigma_Y^2/2)$ and its variance is $\exp(\sigma_Y^2 - 1)\exp(2\mu_Y + \sigma_Y^2)$. Therefore, the variance and the mean of a random variable with lognormal distribution are not independent: a higher mean means a higher variance (Fig. 4.1). Many strictly positive response variables related to sizes, weights, energies, concentrations, incomes, ages, and the time until an event in many fields of science follow lognormal distributions (Limpert et al. 2001).

When a response variable Y follows a lognormal distribution (Fig. 4.1), $\log(Y)$ has a normal distribution. Likewise, if Y has a normal distribution, then $\exp(Y)$ has a lognormal distribution. Therefore, by applying a logarithmic transformation, we can analyze $\log(Y)$ using the general linear model. Unlike the normal distribution, the lognormal is not part of the exponential family of distributions that gives rise to the generalized linear models (see Chapter 8). This implies that there is no simple way of analyzing data with a lognormal distribution in the frequentist framework (there is no such problem for the Bayesian framework). Given that a logarithmic transformation of the response variable turns the problem of analyzing lognormal data into a well-trodden case, there has been no reason to create separate methods to analyze response variables with a lognormal distribution.

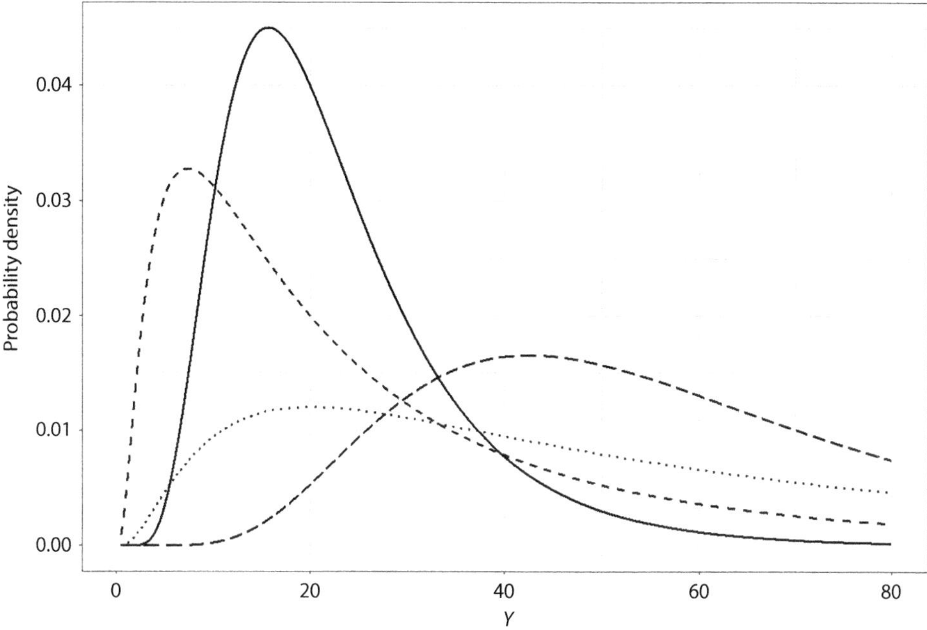

Fig. 4.1 Lognormal probability distribution for four combinations of mean and standard deviation (sd): mean = 3, sd = 0.5 (solid), mean = 4, sd = 0.5 (long dash), mean = 4, sd = 0.5 (short dash), mean = 4, sd = 1 (dotted). The data for the four combinations of parameter values was generated with `lognormal = data.frame(Y = seq(from = 0.5, to = 80, by = 0.01))` and for the four sets of probability densities by putting the appropriate values into `dlnorm(lognormal$Y, meanlog =, sdlog = )`.

4.3 Simple linear regression: One continuous explanatory variable

The goal of the simple linear regression is to establish a linear relationship between a response variable with a normal distribution (Y ~ Normal) and a single numerical explanatory variable. Or, equivalently, to find the best straight line that can be fitted to the data. The statistical model of a simple linear regression is $Y \sim \text{Normal}(\mu_Y, \sigma_Y^2)$, with $\mu_Y = \beta_0 + \beta_1 X$, where β_0 is the intercept or ordinate at the origin that is the predicted value of μ_Y when $X = 0$, and β_1 is the slope of the straight line that denotes the rate of change of μ_Y (and hence of Y) per unit of change of the explanatory variable X. The notion of "best-fit straight line to the data" corresponds to the line that has the smallest overall distance from the line to the entire set of data points. The earliest applications of linear regression date to Legendre in 1805 and Gauss in 1806 (Stigler 1981) in the context of predicting the orbital position of celestial bodies based on astronomical observations (Aldrich 2005). Both Legendre and Gauss used the method of least squares (Chapter 3) to obtain the expressions yielding the values of slope and intercept from the data, and their early work was considerably extended and recast in its modern version by Fisher (1922).

THE DATA IN CONTEXT: Liu et al. (2012) studied plant tolerance to cadmium (Cd), a metal known to have toxic effects on plant and human health. Specifically, they grew a perennial grass in five pots with different concentrations of Cd in a greenhouse, and after some time measured the metal concentration in shoots and roots. If the grass was a bioaccumulator of this metal and hence useful in bioremediation, the concentration of Cd in the plant should be higher than in the soil or, equivalently, the regression slope of (Cd in shoot) vs. (Cd in soil) should be greater than 1.

EXPLORATORY DATA ANALYSIS: Let's first import and examine the data.

```
> DF=read.csv("Ch 04 Cadmium.csv",header=T)
```

The response variable (shoot) has the following descriptive statistics:

```
> summary(DF)
      soil           shoot             root       
 Min.   : 60   Min.   : 16.20   Min.   : 104.6  
 1st Qu.:120   1st Qu.: 52.65   1st Qu.: 245.1  
 Median :180   Median :123.70   Median : 465.0  
 Mean   :180   Mean   :117.02   Mean   : 502.8  
 3rd Qu.:240   3rd Qu.:171.18   3rd Qu.: 664.6  
 Max.   :300   Max.   :217.70   Max.   :1067.1  
```

This is a problem of just two variables: (Cd in soil) vs. (Cd in shoot). Are they related? The response and explanatory variables indeed seem to be linearly related (Fig. 4.2), which justifies the fitting of a linear regression.

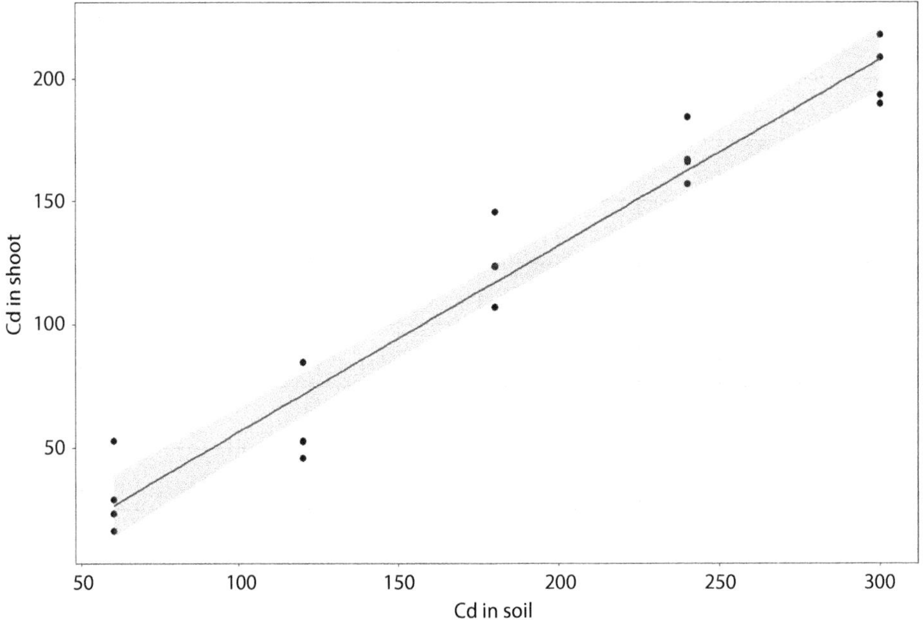

Fig. 4.2 Relation between the response and explanatory variables, cadmium concentration in shoots and soil, respectively. The fitted line is merely indicative.

Now we need to determine the probability distribution that best fits the response variable that is necessary to define the likelihood function to be used in the fitting of the statistical model. Often, this assessment involves eyeballing whether the histogram resembles some probability distribution. Making a histogram requires discretizing the response variable to form intervals and counting the number of data points in each. However, there is no single or optimal criterion for discretization that can be applied to all cases. The potential distortion of judgment caused by discretization is greatest for small sample sizes. Given that the response variable (Cd in shoot) takes positive, real values, we are going to consider the normal and lognormal distributions as plausible candidates. We will use the package `fitdistrplus` to visually assess their fit; this yields the cumulative distribution function (CDF) and the quantile–quantile plot (Q–Q plot) of the data, and compares them to the expected curves if the response variable followed a given probability distribution. We first fit the two candidate distributions as:

```
Norm=fitdist(DF$shoot,"norm")
Lognorm=fitdist(DF$shoot,"lnorm")
```

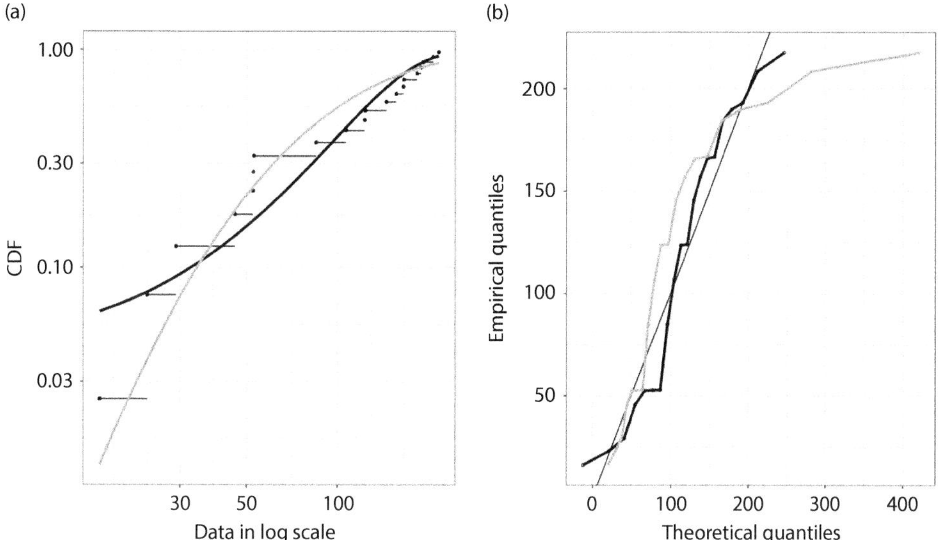

Fig. 4.3 Goodness of fit assessment of the response variable (Cd in shoots) showing the cumulative distribution function (CDF; A) and the quantile–quantile plot obtained using the package `fitdistrplus`.

The CDF (Fig. 4.3a) is the probability density that the response variable is smaller than or equal to any given value ($\Pr(Y \leq y)$), and is shown in the horizontal axis; both axes are shown in log scale for no other reason than to improve the visualization. The CDF curves of normal and lognormal distributions are generated by estimating their respective parameter values from the data, and then calculating the cumulative distributions. The Q–Q plot (Fig. 4.3b) is obtained as follows. Each data point in a distribution represents a quantile of a sample of k values. A useful choice for calculating the quantile for the kth value of the variable Y set in increasing order is $((1, \ldots, k) - 0.5)/k$. A Q–Q plot is the set of quantiles of the response variable in the vertical axis against the theoretical quantiles expected if this variable followed a given probability distribution.

The comparison of the goodness of fit of the response variable to the normal and log-normal distributions is going to be visual and qualitative, and it will not involve any significance test. While neither distribution appears to perfectly fit the data, the normal distribution appears to be closer to the line of perfect fit in the Q–Q plot and only deviates slightly in the lower tail of the CDF (Fig. 4.3a). Hence, as a first approximation, we may consider that the response variable has a normal distribution.

The statistical model: $Y = shoot \sim \text{Normal}(\mu_Y, \sigma_Y^2)$ with $\mu_Y = \beta_0 + \beta_1 soil$. This model can be expressed equivalently as $shoot = \beta_0 + \beta_1 soil + \varepsilon$, where ε represents the residuals or random deviations, which follow a normal distribution with mean = 0 and variance σ^2 (this is abbreviated as $\varepsilon \sim \text{Normal}(0, \sigma^2)$). It also says that the variance σ_Y^2 of the Cd in the shoot is an independent parameter describing the scatter of the expected values of Y that are linearly changing depending on the explanatory variable.

4.4 Simple linear regression: Frequentist fitting

At the start of every R session, the packages `base`, `graphics`, `grDevices`, `stats`, `datasets`, `methods` and `utils` are loaded by default. The command `lm` to fit a general linear model is part of the package `stats`. The fitting of a simple linear regression in R could not be simpler: `m1 = lm(shoot ~ soil, data = DF)`. This is our first encounter with a statistical model in R, so we must understand the syntax. The tilde `~` in R separates the response and explanatory variable(s). Because `soil` is a quantitative explanatory variable, `shoot ~ soil` is fitting a linear model (`lm`) defined by a slope and an intercept. Should we want to be unnecessarily explicit, we could write `shoot ~ 1 + soil`, where the `1` denotes the intercept, and `shoot ~ 0 + soil` if we want a regression through the origin. The basic R syntax for linear statistical models is shown in Table 4.1. We need to indicate where the response and explanatory variables are, and this is done with `data = DF`. We fit the linear model and store the results in the object `m1` (which is a list), from which we will extract many features using the operator `summary`:

```
> m1=lm(shoot~soil, data=DF)
> summary(m1)

Call:
lm(formula = shoot ~ soil, data = DF)

Residuals:
    Min      1Q  Median      3Q     Max
-25.970 -11.220   1.730   7.655  28.680

Coefficients:
             Estimate Std. Error t value Pr(>|t|)
(Intercept) -19.03000    8.35547  -2.278   0.0352 *
soil          0.75583    0.04199  18.001 5.88e-13 ***
---
Signif. codes:  0 '***' 0.001 '**' 0.01 '*' 0.05 '.' 0.1 '

Residual standard error: 15.93 on 18 degrees of freedom
Multiple R-squared:  0.9474,    Adjusted R-squared:  0.9445
F-statistic:   324 on 1 and 18 DF,  p-value: 5.882e-13
```

Table 4.1 R syntax for the effects in linear and generalized linear models. `X` `W` and `Z` are either numerical or categorical explanatory variables affecting the mean of the response variable.

Notation	Meaning of the effects
`X + W`	Main (or single) effects of `X` and `W`
`X : W`	Only interaction between `X` and `W` (but not main effects)
`X * W`	Main effects and interactions of `X` and `W`; equivalent to `X + W + X : W`
`(X + W + Z)^2`	Main effects and two-way interactions for all pairs of variables; equivalent to `X + W + Z + X : W + X : Z + W : Z`
`0 + X`	Main effect for `X` but suppressing the intercept; equivalent to `-1 + X`

In order of importance, we obtain the maximum likelihood estimates of the intercept and the slope, their standard errors, the t-values, i.e., the ratio between a parameter estimate and its standard error, and the p-value, i.e., the probability of obtaining a t-value equal to or more extreme than the one obtained if the null hypothesis $\beta_1 = 0$ were true. The estimated slope means that the shoot Cd concentration increases 0.755 units per increase of one unit of the soil Cd concentration. Is this finding important? The p-value of the slope (5.88×10^{-13}) is *not* a measure of the importance of the finding. Rather, it measures the (very little) compatibility of the data if the null hypothesis of *slope* = 0 were true. To decide upon the importance or relevance of the finding, we must interpret the magnitude of the slope and its 95% confidence interval in the context of the research that motivated the analysis. Only then may we interpret whether the absorption of 75.5% (and the leaving of 24.5%) of the soil Cd by the plants is surprising, relevant, or even interesting. It is the research context what provides the meaning and interpretation to statistical findings. The estimated intercept makes no sense as it implies an average shoot concentration of −19.03 units when there is Cd = 0 in the soil. We will see in Section 4.8 that the standardization of the explanatory variable(s) changes, and gives a meaningful interpretation to, the intercept in linear models.

At the top of the summary table are descriptive statistics of the residuals of the model that are of no particular relevance. At the bottom of the summary table, the residual standard error is the square root of the estimated variance of the residuals or the unexplained variation around the mean in the statistical model. We also obtained the multiple and the adjusted R^2 values that are the proportion of the variation of the response variable explained by the fitted linear regression model. There is an algebraic relationship between the two that corrects the biased nature of multiple R^2: $R^2_{\text{adjusted}} = 1 - \left(1 - R^2_{\text{multiple}}\right) \times \frac{(n-1)}{(k-1)}$, where n is the number of data points and k is the number of explanatory variables in the model. While the higher the R^2, the better the fit of a model to the data, this statistic of goodness of fit should be mostly used in a comparative sense. Hardly anybody is tall or ugly in an absolute sense but only when compared to others. Likewise, R^2 as a measure of goodness of fit of a model should be mostly used by comparing alternative models fitted to the same data. At the bottom of the summary we can see the F-statistic that will be explained in Chapter 5.

We now have estimated the three parameters of the model $Y \sim \text{Normal}(\mu_Y(X) = \beta_0 + \beta_1 X, \sigma_Y^2)$ as $Y \sim \text{Normal}(\mu_Y(X) = -19.03 + 0.75X, 15.93^2)$. For any value of the explanatory variable, we now have a probability function that describes the chances of observing values of the response variable. Therefore, we can view our parameter estimates as a generative model capable of predicting plausible values of the response variable.

It is very simple to obtain the confidence intervals of the parameters of a statistical model fitted in the frequentist framework. Many of us have wrestled with the interpretation of confidence intervals (Chapter 3). The confidence interval is just a range of equally plausible values of a parameter that could be potentially obtained if we repeatedly obtain samples of the same size from the same population. As such, the confidence interval is a thought experiment about potential samples that could be obtained through the repeated sampling of the same population. Ninety-five percent confidence does *not* mean that there is a 95% chance that the interval contains the true but unknown value of the parameter we are estimating; this would be conveyed by the credible interval obtained in the Bayesian framework (see Chapter 3). The true value of a parameter is either included or not in the confidence interval. The 95% confidence interval for the slope of [0.668, 0.844], obtained with the `confint(m1)` command, depicts our uncertainty about the value of the parameter whose best estimate is 0.755: the wider the interval, the higher the uncertainty. Whether the precision of the confidence interval is "largely adequate" or "woefully inadequate" to make decisions would, as always, depend on the research context. Numbers, including parameter estimates and intervals, have no meaning or interest outside a context.

```
> confint(m1)
              2.5 % 97.5 %
(Intercept) -36.584 -1.476
soil          0.668  0.844
```

4.5. Tools for model validation in frequentist statistics

Model validation is an important task of model building aiming to confirm that the fitted model is an acceptable description of the data in hand. A main component of model validation is residual analysis. In its simplest definition, the residuals are the differences between the observed and predicted values: $Y_{obs} - Y_{pred}$. If the model fits the data well, the residuals would be small, approximately random errors without any trends when plotted against the explanatory variable(s). In the case of the general linear model, when the model fits the data well the residuals would have a normal distribution as evaluated by a Q–Q plot. Cook's distance (Cook and Weisberg 1982) is the final component of model validation. It amounts to a sensitivity analysis of the impact that each data point has on the parameter estimates: $D_i = \frac{(b_{-i}-b)^T X^T X(b_{-i}-b)}{(k+1)\sigma^2}$, where b_{-i} and b are the parameter estimates without the ith data point and with the entire data set respectively, X^TX is the variance–covariance matrix of explanatory variables, k is the number of explanatory variables in the model, and σ^2 is the estimated variance of the response variable. Data points with a large Cook distance are likely to have an unduly large influence on the magnitude of parameter estimates. There does not seem to be a clear threshold to define a Cook distance as "large"; several criteria have been proposed (Cook and Weisberg 1982). As with the other components of residual analysis, it seems best to visually inspect the Cook distances and make a qualitative personal judgment.

The simplest way to make a residual analysis in R is to use the command `plot(model)`, which will generate a sequence of five graphs in the Plot pane of RStudio, not all of which are of direct interest. Let's comment on Fig. 4.4. If the model fits the data well, the top left plot (residuals vs. fitted values) and the bottom right (residuals vs. explanatory variable) should show a random scatter in the plot. That is, they should not exhibit a discernible pattern to the eye, nor should their scatter around zero be larger or smaller for the span of values of the explanatory variables. The absence of a discernible trend would imply that the variation of the residuals is homogeneous for the data set, which is an informal assessment of homoscedasticity (i.e., homogeneity of variance, one of the assumptions of the general linear model). We use a Q–Q plot to evaluate whether the residuals are normally distributed (Fig. 4.4, top right) as indicated by their relatively tight scatter of points around the line of perfect fit. The plot of Cook distances (Fig. 4.4, bottom left) only suggests that the third row in our data frame would have a somewhat larger distance than the other values. Should we worry? Experience will progressively teach us that we should be concerned when a Cook distance is vastly larger than the rest of the distances only for small to moderate-sized data sets. At this point, we may distrust this seemingly groundless advice, and examine whether the parameter estimates substantially change when we delete the third row of the data frame (the one having the largest Cook distance). This can be achieved as follows:

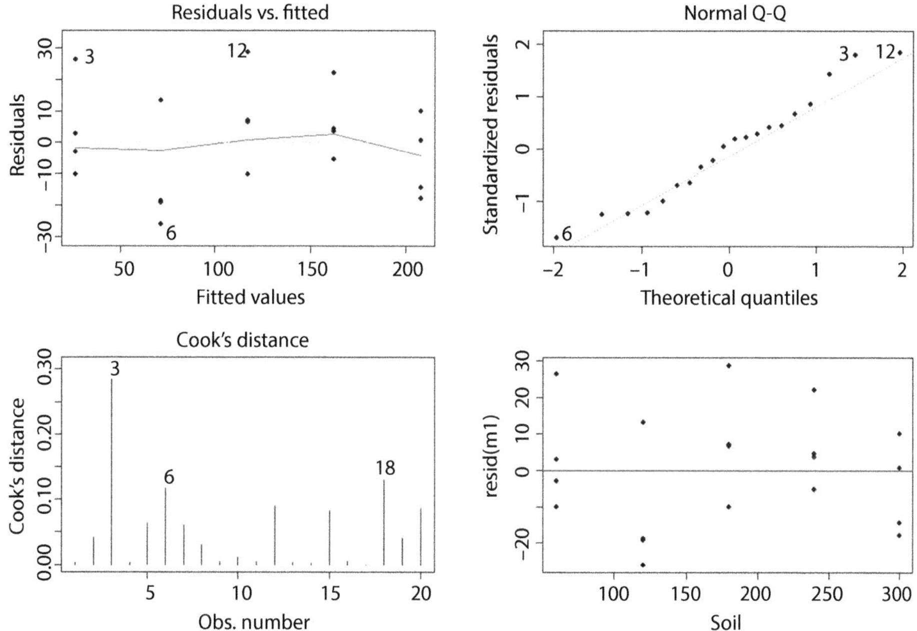

Fig. 4.4 Validation of the frequentist model `m1` by residual analysis, showing the plots of residuals vs. fitted, residuals vs. explanatory variable, Q–Q plot of residuals, and Cook distances. These plots were obtained using the package `stats`.

```
> m2=lm(shoot~soil, data=DF[-3,])
> summary(m2)

Call:
lm(formula = shoot ~ soil, data = DF[-3, ])

Residuals:
   Min     1Q Median     3Q    Max
-22.87 -12.12   1.53   8.23  30.23

Coefficients:
            Estimate Std. Error t value Pr(>|t|)
(Intercept) -25.2371     8.4297   -2.99   0.0082 **
soil          0.7817     0.0414   18.89  7.6e-13 ***
---
Signif. codes:  0 '***' 0.001 '**' 0.01 '*' 0.05 '.' 0.1 '

Residual standard error: 14.9 on 17 degrees of freedom
Multiple R-squared:  0.955,     Adjusted R-squared:  0.952
F-statistic:  357 on 1 and 17 DF,  p-value: 7.6e-13
```

Deleting the third row in the data frame changed the intercept from −19.03 to −25.23 (32% change) and the slope from 0.755 to 0.788 (a 4% change). But the sign, the statistical significance of the slope, and the R^2 barely changed. Are we willing to throw away 1 of the 20 data points (5% of the evidence) when the results are barely sensitive to the third row of the data frame? As a matter of principle, we should keep the entire data set and at most perhaps comment in passing that one of the values might have a rather small influence on the parameters fitted. The factual evidence is sacred and we should never accept the Faustian bargain of deleting data to achieve a better model fit, a higher R^2, or a smaller statistical significance. As a matter of principle again, whenever one or more data points have large Cook distances and the refitting of the statistical model without these data points reveals large quantitative changes, we should report the results with both the entire and the reduced data sets, and aim to openly disclose the differences and seek an explanation in terms of the scientific problem posed.

With some more work, we can generate ggplot-type figures of the residual analysis. We will use the latter type throughout the book. We first make use of the package `broom` that has three very handy functions applicable to a wide array of statistical models; we'll consider them one by one.

```
> tidy(m1, conf.int = T)
# A tibble: 2 x 7
  term        estimate std.error statistic  p.value conf.low conf.high
  <chr>          <dbl>     <dbl>     <dbl>    <dbl>    <dbl>     <dbl>
1 (Intercept) -19.0       8.36       -2.28 3.52e- 2  -36.6       -1.48
2 soil          0.756     0.0420     18.0  5.88e-13    0.668      0.844
```

Now we have a table (actually, a tibble, which is a more sophisticated type of data frame) that can be easily exported as a file or copied into a word processor.

```
> glance(m1)
# A tibble: 1 × 12
  r.squared adj.r.squared sigma statistic  p.value    df logLik   AIC   BIC
      <dbl>         <dbl> <dbl>     <dbl>    <dbl> <dbl>  <dbl> <dbl> <dbl>
1     0.947         0.944  15.9      324. 5.88e-13     1  -82.7  171.  174.
```

The `glance` function collects metrics related to the quality of fit of the statistical model that were scattered in the model summary. The AIC, BIC, and deviance are metrics that denote the degree of empirical support of a model. Their differences are used in the comparison of statistical models in the context of model selection, as explained in Chapter 7.

```
> augment(m1)
# A tibble: 20 x 8
  shoot  soil .fitted  .resid  .hat .sigma  .cooksd .std.resid
  <dbl> <int>   <dbl>   <dbl> <dbl>  <dbl>    <dbl>      <dbl>
1  23.2    60    26.3  -3.12  0.15    16.4 0.00398      -0.212
2  16.2    60    26.3 -10.1   0.15    16.2 0.0419       -0.689
3  52.7    60    26.3  26.4   0.15    14.9 0.285         1.80
4  29.1    60    26.3   2.78  0.15    16.4 0.00316       0.189
5  52.5   120    71.7 -19.2   0.075   15.7 0.0634       -1.25
```

The last command creates a data frame (here we only see the first five rows) with the response and explanatory variable(s), the fitted values with the predicted values of the response variable and their standard error, the raw and standardized residuals, the Cook distances, and the diagonal values of the hat matrix and sigma (which are of limited interest here).

We start by storing the results of `augment(m1)` as `res.m1 = augment(m1)`, and then obtain the same plots of the residual analysis of model `m1` now in the ggplot mode that we will use throughout the book. Figure 4.5 is just a prettier version of Fig. 4.4, and hence the same remarks given above apply to Fig. 4.5.

Generating graphical output of the fitted statistical model is a very important part of communicating the results of the analysis that cannot be overemphasized. We are going to use conditional curves generated by the package `ggeffects`. What `ggeffects` does is first to produce a "pretty data frame" of predicted values for the response variable and its confidence interval, and then generate a ggplot that can be edited at will: `pred.m1 = ggpredict(m1, terms = c("soil"))`. Let's first look at the object `pred.m1`:

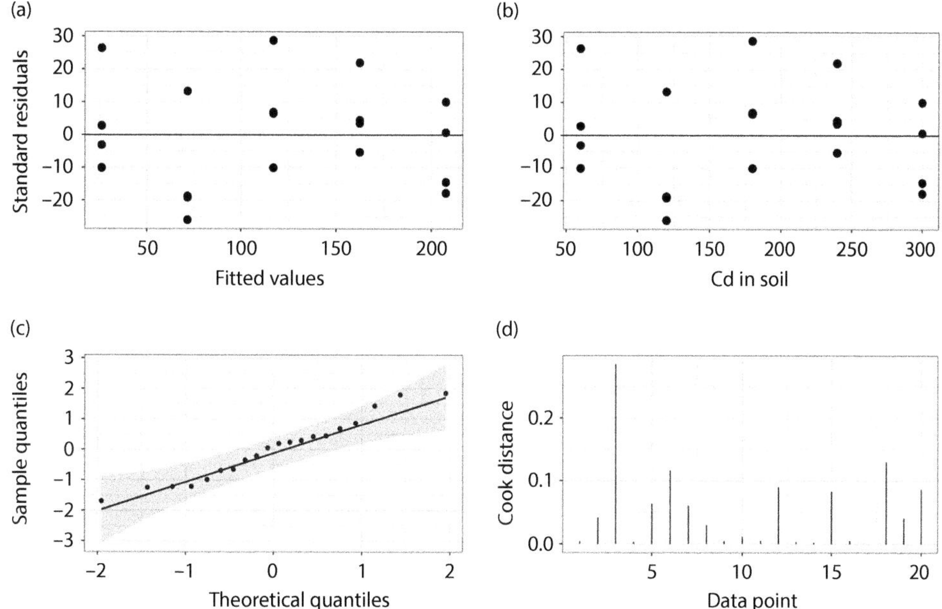

Fig. 4.5 Residual analysis of model `m1` with ggplot-type graphs similar to Fig. 4.4.

```
> ggpredict(m1, terms = c("soil"))
# Predicted values of shoot

soil | Predicted |            95% CI
-----------------------------------
  60 |     26.32 | [ 14.23,  38.41]
 120 |     71.67 | [ 63.12,  80.22]
 180 |    117.02 | [110.04, 124.00]
 240 |    162.37 | [153.82, 170.92]
 300 |    207.72 | [195.63, 219.81]
```

These predicted values allow us to create Fig. 4.6.

4.6 Simple linear regression: Bayesian fitting

Our statistical model, *shoot* $\sim$ Normal(μ_Y, σ_Y^2), with $\mu_Y = \beta_0 + \beta_1 soil$, has three parameters: β_0, β_1, and σ^2. As we discussed in Chapter 2, the Bayesian fitting of a statistical model requires defining the joint likelihood function of the model parameters given the data, Likelihood(data | β_0, β_1, σ^2), and the joint prior probability distribution Pr(β_0, β_1, σ^2) of the three model parameters. More often than not, the joint prior distribution is defined by making the simplistic assumption that the values of the three parameters are independent of one another (this is the rationale behind Gibbs sampling; cf. Chapter 3). That is, Pr(β_0, β_1, σ^2) = Pr(β_0)Pr(β_1)Pr(σ^2). The likelihood and the prior distributions are combined by Bayes' rule (see Chapter 3) to generate the posterior probability distribution Pr(β_0, β_1, σ^2 | data) that will summarize our knowledge about the model parameters.

The likelihood function incorporates the information contained in the data into the parameter estimation to change or update the knowledge on their plausible parameter

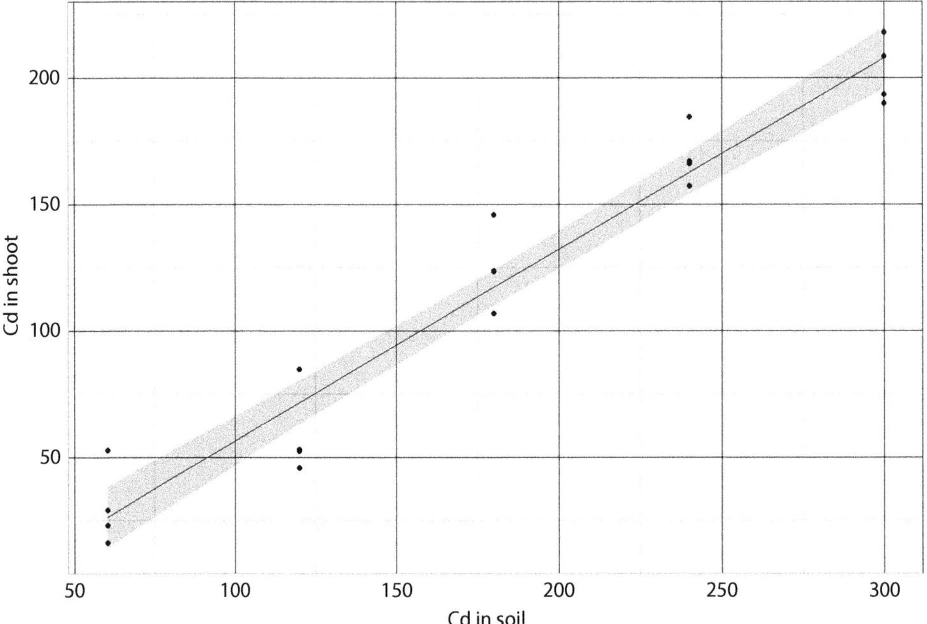

Fig. 4.6 Conditional plot and its 95% confidence interval from model `m1` generated with the `ggeffects` package and the raw data.

values that we had before looking at the current data. It is defined by the probability distribution that best describes the observed variability of the response variable given plausible values of the model parameters. We previously explored whether the shoot concentration of Cd was best modeled by either a normal or a lognormal distribution.

The prior distributions of the model parameters need to be defined on a case-by-case analysis of each model parameter. Defining priors is the single most controversial issue in Bayesian statistics. In this, our first, Bayesian model, we will proceed in a very simple-minded manner. The priors summarize our knowledge of the average value of Cd concentration in the soil (β_0), the relationship between soil and shoot Cd concentration (β_1), and the variability of Cd in the soil (σ^2). We could use web searches to gather information on Cd for our perennial grass in the same or similar environments, or for other perennial or annual grasses, or, failing all of these, perhaps use information for other heavy metals having effects on plants similar to those of Cd. The review of prior evidence could range from almost journalistic, to interviewing experts to elicit information, to a full-fledged meta-analysis, depending on our sophistication and on the information available. Failing all of these, and not being an expert on either Cd or on perennial grasses, we could proceed by defining "weakly informative" priors on the model parameters. "Weakly informative" is a fuzzy way of saying that whenever there is a reasonably large amount of data, the prior would be relatively unimportant so that the posterior distribution would be dominated by the likelihood (i.e., by the information contained in the data). However, if the data were weak in information content, the likelihood should be relatively flat (i.e., a wider range of parameter values could possibly have generated the data), and a "weakly informative prior" would strongly influence the posterior distribution. The phrase "weakly informative" is implicitly in comparison to a default flat prior (that is implicitly in the frequentist framework; Chapter 2), and only has meaning with respect to a specific data set and statistical model (Gelman et al. 2006).

We are going to use the package `brms` to carry out the Bayesian fitting with the Hamiltonian Monte Carlo algorithm (see Chapter 3). What we are going to do with `brms` is to pass the model code, the data, and any additional arguments to another R package, `rstan` (Stan Development Team 2020). In turn, `rstan` will translate the code into C++, compile it, and fit the model using the Hamiltonian Monte Carlo algorithm of the independent program Stan (Hoffman and Gelman 2014). The results of the fitted model are returned to `brms` for some post-processing and visualization in R. Stan is a standalone program that contains a full-fledged probabilistic programming language to carry out Bayesian model fitting (Hoffman and Gelman 2014). We can view the Stan code with the `brms` command `stancode(model name)`. We could, of course, have interacted directly with Stan, or indirectly through `rstan` (or with interfaces to several other platforms including Phyton, Matlab, Julia, and Stata), to write our code. However, this would have required us to learn the Stan syntax. While the Stan syntax is not too difficult and worth learning to fit non-standard models (Stan Development Team 2017), using the package `brms` sidesteps the issue for the set of statistical models considered in this book.

Using the `brms` package requires having a C++ compiler installed and configured. The first step is to check whether there is already a C++ compiler installed by typing in R: `pkgbuild::has_build_tools(debug = TRUE)`. If the answer were `TRUE`, you already have the compiler installed and you only have to install the R libraries required. If, however, the answer were `FALSE`, you have to install the C++ compiler and link it to RStudio. Bürkner (2017) provides instructions for installing the C++ compiler for Windows and Mac operating systems, and users of Linux can find equivalent instructions at https://github.com/stan-dev/rstan/wiki/Installing-Rstan-on-Linux.

The main command of the package `brms` is `brm`, which takes the following arguments: `brm(formula, data, family, prior, warmup, chains, iter)`; `formula` is where we specify the statistical model, `data` is the data frame containing its variables, `family` is where we specify the probability distribution that will be used in the likelihood (here we will also later specify the link function; see Chapter 8), and `prior` is where we specify the prior distributions of the model parameters. In `warmup`, `chains`, and `iter` we indicate the number of iterations of the warm-up period of the algorithm, the number of Markov chains to be used, and the number of iterations of each chain. Because more complex Bayesian models with large data sets are computationally demanding, it is good practice to make the multiple processors of your computer work in parallel to speed up the calculations (Matloff 2016). A simple way of parallelizing the calculations is by installing the package `future` with the command `install.packages("future")`, and then use the option `future = TRUE` inside the `brm` command.

Let's define the prior distributions. `brms` has a very useful command to help the user display the priors that need to be defined for a model:

```
> get_prior(formula=shoot~soil, data=DF,family=gaussian)
                 prior     class coef group resp dpar nlpar bound
1                              b
2                              b soil
3 student_t(3, 124, 101) Intercept
4   student_t(3, 0, 101)     sigma
```

`get_prior` indicates that the model has two parameters of "type b" (coefficients of a linear model): the first one, that is blank under the heading `group`, corresponds to the intercept, and the second one is the slope associated with the explanatory variable `soil`. These two parameters of type b determine the mean of the normal distribution associated with the response variable for each value of the explanatory variable soil. The third parameter, of "type sigma," corresponds to the standard deviation not of the variance but of the normal distribution of the response variable. Lines 3 and 4 show the default weakly informative priors for each parameter predefined in `brms` that could in principle be used to fit the model. Nevertheless, renouncing defining the priors amounts to relinquishing an essential part of the Bayesian statistical analysis and, as a matter of principle, should not be an option when fitting a Bayesian statistical model. Let's define the priors of our model `m1`:

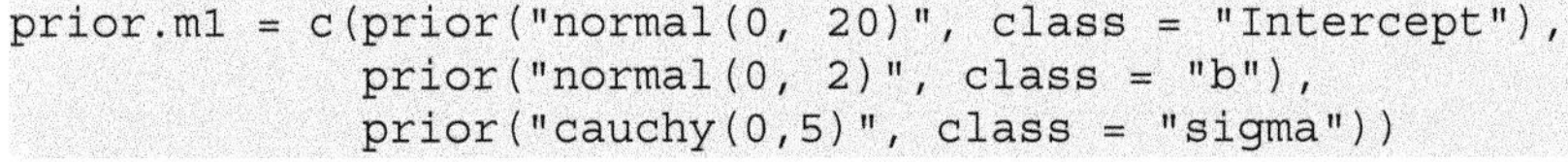

```
prior.m1 = c(prior("normal(0, 20)", class = "Intercept"),
             prior("normal(0, 2)", class = "b"),
             prior("cauchy(0,5)", class = "sigma"))
```

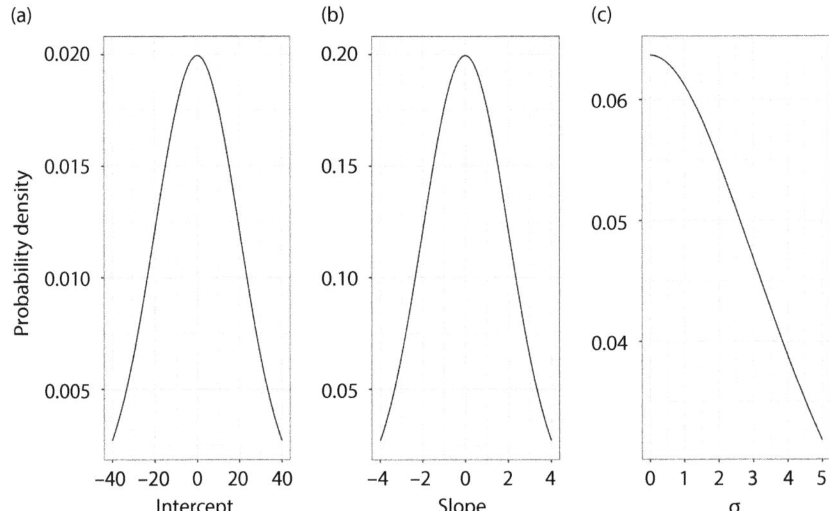

Fig. 4.7 Prior distributions of the three parameters of the Bayesian linear regression model m1.brms.

We recommend plotting the priors to visually gauge and interpret their meaning. What these prior distributions (Fig. 4.7) are stating is that we have grounds to think that the previous evidence for the intercept suggests it should be in [−40, 40] with a more likely value of zero (a similar argument can be made for the slope), and that there will be a small variability of the Cd in the shoots as values of sd > 5 mg are deemed unlikely. Note that in each case there are two parameters defining the shape and spread of the values of the prior distributions. These probability distribution parameters involved in defining the priors are known as hyperpriors in the Bayesian literature (Chapter 3). In particular, the standard deviation of the normal distribution and the scale of the Cauchy distribution control the vagueness of the respective prior distributions: the higher their values, the higher (and the more flattened) the spread of the prior distributions. When the Cauchy distribution is used as a prior for a strictly positive parameter such as the standard deviation, `brms` considers only the positive part, i.e., a half-Cauchy distribution.

Let's finally fit the model using the above-mentioned priors stored in `prior.m1`:

```
m1.brms=brm(formula=shoot~soil, data=DF, family=gaussian, prior = prior.m1,
            warmup = 1000, future=TRUE, chains=3, iter=2000)
```

While the model is being fitted we see several messages indicating the progress for each of the chains; these can be safely ignored (and turned off if wished by adding `silent = T` inside the command). The main features of the output are:

```
> summary(m1.brms)
 Family: gaussian
  Links: mu = identity; sigma = identity
Formula: shoot ~ soil
   Data: DF (Number of observations: 20)
Samples: 3 chains, each with iter = 2000; warmup = 1000; thin = 1;
         total post-warmup samples = 3000

Population-Level Effects:
          Estimate Est.Error l-95% CI u-95% CI Rhat Bulk_ESS Tail_ESS
Intercept   -23.06      8.93   -41.72    -5.99 1.00     1855     1616
soil          0.76      0.04     0.67     0.84 1.00     1994     1566

Family Specific Parameters:
      Estimate Est.Error l-95% CI u-95% CI Rhat Bulk_ESS Tail_ESS
sigma    16.64      3.08    11.88    23.78 1.00     1674     1375
```

Let's dissect the summary of our first fitted Bayesian model. First, `Links: mu = identity; sigma = identity` indicates the link functions used for each of the two parameters (mean `mu`, standard deviation `sigma`) of the normal distribution used in the likelihood part of the Bayesian model fitting. We will fully explain the link functions in Chapter 8 in the context of generalized linear models; for now, suffice it to say that the identity link is just a function that leaves its argument unchanged, much like multiplying a number by one. Second, we have information on the number of chains and their lengths, on the number of warm-up iterations, and the thin rate (here we used the default value) used for sampling parameter values of each of the three chains.

Second, the result of the Bayesian model fitting is a multidimensional distribution of $k + 1$ dimensions, with k being the number of parameters in the statistical model, and the "+ 1" being the joint posterior probability of a triplet [intercept, slope, sigma] of parameter estimates. The summary is saying that we have 3,000 estimates of each of the three model parameters that are the marginal distributions[1] of the multidimensional joint posterior distribution. The 3,000 estimates are the product of the number of chains (3) and the difference between the specified lengths of `iter` (2,000) and `warm-up` (1,000).

Third, under the headings "Population-Level Effects" and "Family Specific Parameters" we can see the values for `Estimate`, `Est.Error`, `l-95% CI`, and `u-95% Ci`, which are the mean, the standard deviation, and the limits of the 95% credibility interval (see Chapter 3 for the interpretation) of each model parameter. These values are calculated from the 3,000 estimates of the marginal distributions generated from the joint posterior distribution. Finally, we also have three metrics related to the model fitting for each model parameter: `Rhat`, `Bulk_ESS`, and `Tail_ESS`. `Rhat` was defined in Chapter 3, and it is a crude measure that the three chains have converged to the same stationary distribution. The closer `Rhat` is to 1 the better, with values smaller than 1.1 considered as "acceptable" in common practice and tradition. The other metrics (`Bulk_ESS`, `Tail_ESS`) are estimates

[1] The marginal distributions would be the projection of the four-dimensional (intercept, slope, standard deviation, and probability density) posterior distribution; see Fig. 4.8.

of the effective sample size (i.e., the number of statistically independent estimates of each parameter; see Chapter 3) of the three chains obtained by Stan. Obviously, the higher and closer these values are to 3,000 the better (see https://mc-stan.org/misc/warnings.html for details). Should `Rhat` be greater than the "acceptable threshold" and/or `Bulk_ESS` worryingly smaller than (in our case) 3,000 draws, we should run longer chains and perhaps consider changing the starting values of the chains to check if the default random starting values of each chain had an undue influence on the posterior distributions.

Let's translate `summary(m1.brms)` into words. First, recall that our prior for the slope (Fig. 4.7) indicated that we expected or had evidence of positive slope values (i.e., plants absorb Cd from soil) being as likely as negative values (the opposite). Now, the compounded influence of our overly vague prior for the slope and the information contained in the data (conveyed by the likelihood) in `m1.brms` suggests otherwise (Fig. 4.8). Our single best estimate of the slope is 0.75 (the mean of its posterior distribution) and there is a probability (not confidence!) of 0.95 that the slope lies in [0.66, 0.95]. Given that we have a (marginal) posterior probability distribution for the slope, we can also indicate that not all values in the credible interval are equally likely.

Given that `m1.brms` is our first Bayesian model, we are going to start by visualizing and interpreting the model output, and then we'll verify whether the MCMC algorithm has converged to a stationary distribution. This is obviously backwards as it is pointless visualizing and interpreting an inadequately fitted Bayesian model, but allow us the mischief this time just for didactic reasons. While `m1.brms` has only three parameters whose means and credible intervals shown in `summary(m1.brms)` can be put in table form, it

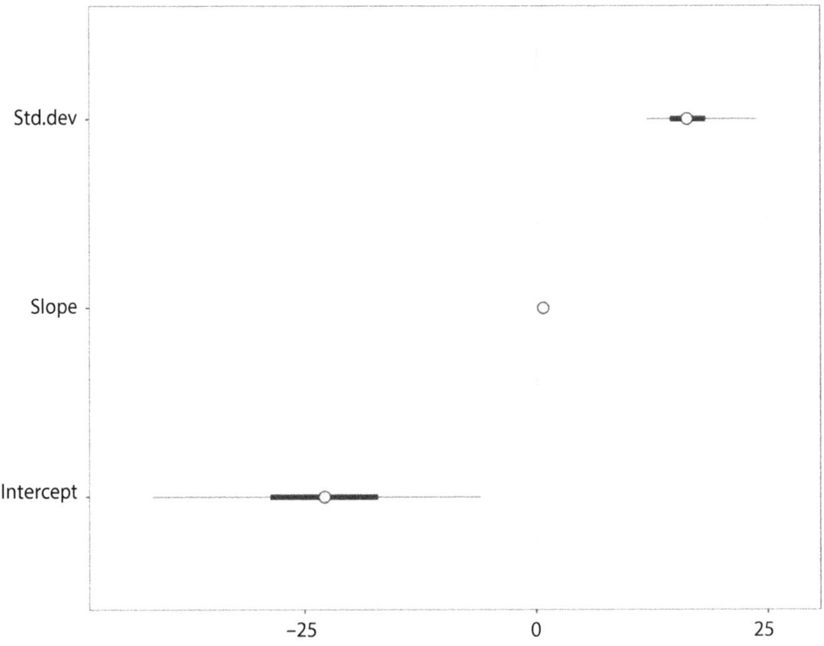

Fig. 4.8 Mean (dot) and 50% and 95% credible intervals of the parameters of the Bayesian linear regression model `m1.brms`. The slope does have a 95% credible interval [0.66, 0.95] that is swamped by the scale needed to show the credible intervals of the other parameters.

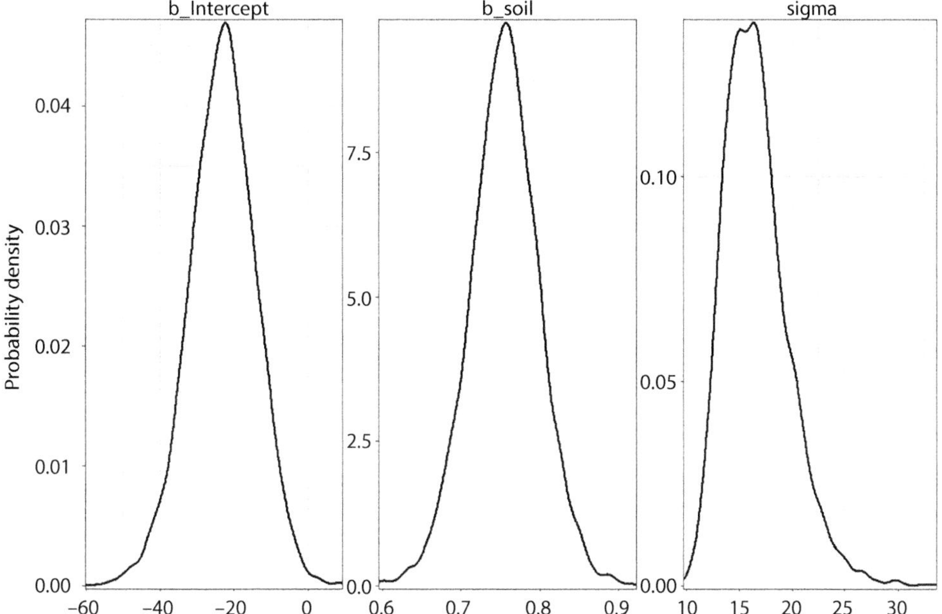

Fig. 4.9 Posterior distributions of the parameters of the Bayesian linear regression model `m1.brms`.

is useful to learn how to depict the model output in a graphical form by visualizing the marginal posterior probabilities of `m1.brms` parameters.

The marginal posterior distributions of Fig. 4.9 depict the triplets of parameters estimated contained in the three chains used to fit `m1.brms`. Had the data not contained any relevant empirical information on the relation between the response and the explanatory variables, the posterior distributions of each model parameter (Fig. 4.9) would have resembled the respective prior distributions (Fig. 4.7).

We should verify that these three chains actually converged to the same marginal stationary distribution for each model parameter using the package `bayesplot`. It is crucial to verify that the three chains converged to a stationary posterior distribution (see Chapter 3), for which we are going to use several visual checks.

The three chains shown in Fig. 4.10 need not be, and indeed will never be, identical or have the same values of descriptive statistics. What they need to be is consistent in the degree of support that they provide to sets of model parameter values. We can see that Fig. 4.10 is consistent with the convergence metric `Rhat` discussed above.

We can also examine the parameter values sampled in the three chains and compare descriptive statistics to further verify their convergence to a common stationary distribution. The summary statistics can be obtained with:

```
> summaryBy(b_Intercept+b_soil+sigma~chain, FUN=c(mean, var), data=m1.brms.s)
  chain b_Intercept.mean b_soil.mean sigma.mean b_Intercept.var b_soil.var sigma.var
1     1            -23.7       0.757       16.8            87.8    0.00205      9.97
2     2            -23.1       0.756       16.6            75.0    0.00180      9.54
3     3            -22.4       0.753       16.5            75.6    0.00195      9.02
```

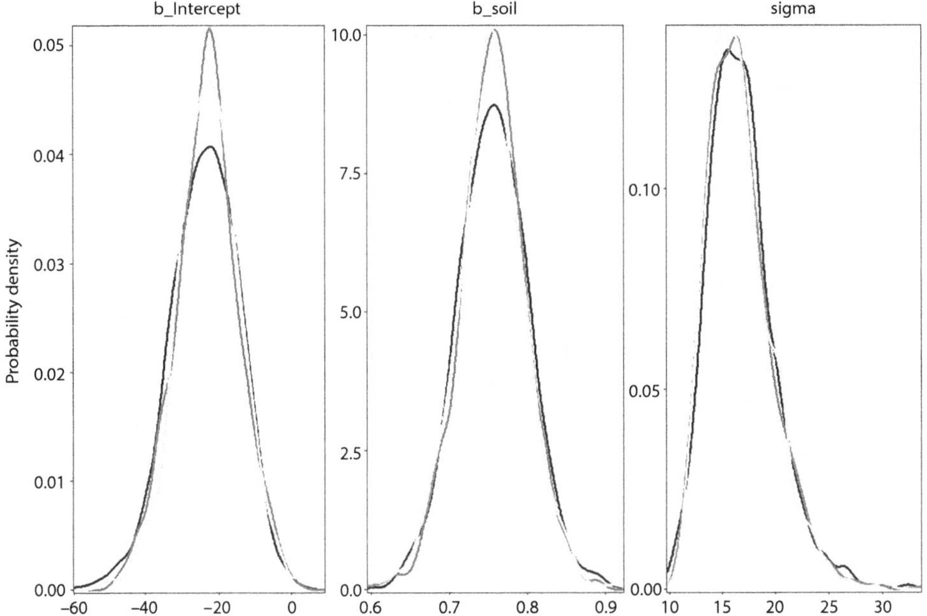

Fig. 4.10 Posterior distributions of the intercept, slope, and standard deviation (from left to to right) showing the convergence of the three chains of the Bayesian linear regression model `m1.brms` to a common stationary posterior distribution.

We can see that the means and variances of the three model parameters are very similar among the chains, thus corroborating their convergence to a common stationary distribution (cf. Fig. 4.9).

The final check of the adequacy of the parameter values drawn is the degree of statistical independence of the sampled parameter values (Fig. 4.11). Intuitively, several perfectly correlated values are the same value and contain less information than several different values that by necessity would have a less than perfect correlation. However, strong autocorrelation of sampled parameter estimates is not by itself a signal of lack of convergence of the MCMC algorithm, it may only indicate some misbehavior of a chain or suggest that one should run longer chains to achieve convergence.

Figure 4.11 shows that the sampled parameter values of model `m1.brms` have satisfactorily low levels of autocorrelation. So, what can we safely conclude thus far? The model `m1.brms` has converged to a stationary distribution (i.e., the three chains can be considered as equivalent) that is the posterior distribution. This implies that we can combine and use the three chains to make a statistical inference, as was done by the command `summary(m1.brms)`.

We had stated that the statistical model to be fitted was *shoot* $\sim$ Normal(μ_Y, σ_Y^2) with $\mu_Y = \beta_0 + \beta_1 soil$. Using the means of the posterior distributions of `m1.brms` we can now state that *shoot* ~ Normal$(\mu_Y = -22.77 + 0.75 soil,\ \sigma_Y^2 = 16.72^2)$, just as we discussed for the frequentist fitting of the same linear regression model. However, there is an important and crucial difference between the outputs of the frequentist (`m1`) and the Bayesian (`m1.brms`) models that bears emphasizing. While the former comprises the maximum likelihood parameter estimates and their standard errors and associated confidence intervals, the latter consists of three full-fledged marginal posterior probability functions of the model

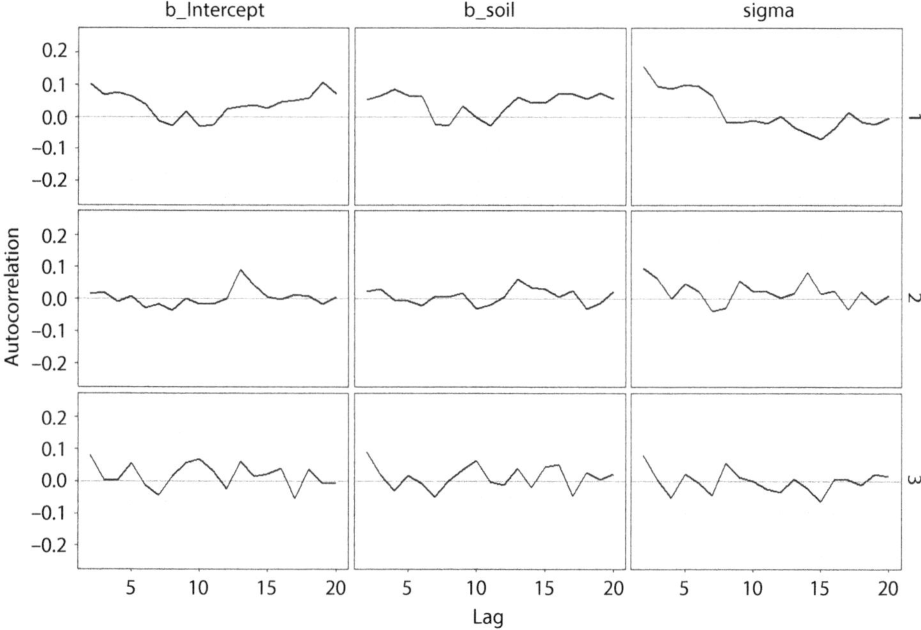

Fig. 4.11 Autocorrelation of the time series of the drawn values of each model parameter (from top to bottom: intercept, slope, and standard deviation) of the Bayesian model linear regression model `m1.brms` of each chain. The scale limit on the x-axis was set to avoid showing the autocorrelation of lag = 0 (the correlation of a value with itself) that is equal to one by definition.

parameters from which we calculate their means, standard deviation (or standard error), and credible intervals. These posterior distributions (Fig. 4.9) are based on 3,000 triplets of parameter values sampled in the three chains that can be obtained with:

```
> posterior_samples(m1.brms,pars=pars.m1.brms, add_chain = F)
  b_Intercept b_soil sigma
1      -17.06  0.738  14.4
2      -32.29  0.786  16.7
3      -36.19  0.817  16.4
4      -15.54  0.680  17.2
5      -34.71  0.824  17.3
```

Each of these 3,000 (here we only show the first five) triplets of parameter values defines a different probability distribution describing the chances of obtaining a value of the response variable Y for *any* given value of the explanatory variable X. Based on the posterior distributions of model parameters, we can calculate the associated goodness-of-fit metrics such as R^2, and the residuals for the 20 observed values of Y given an X. Therefore, we will have a posterior distribution of the 3,000 estimates of R^2 and of the 20 residuals.

We can also obtain the proportion of the variation of the response variable explained by the fitted model:

```
> bayes_R2(m1.brms)
   Estimate Est.Error  Q2.5 Q97.5
R2    0.944   0.00923 0.919  0.95
```

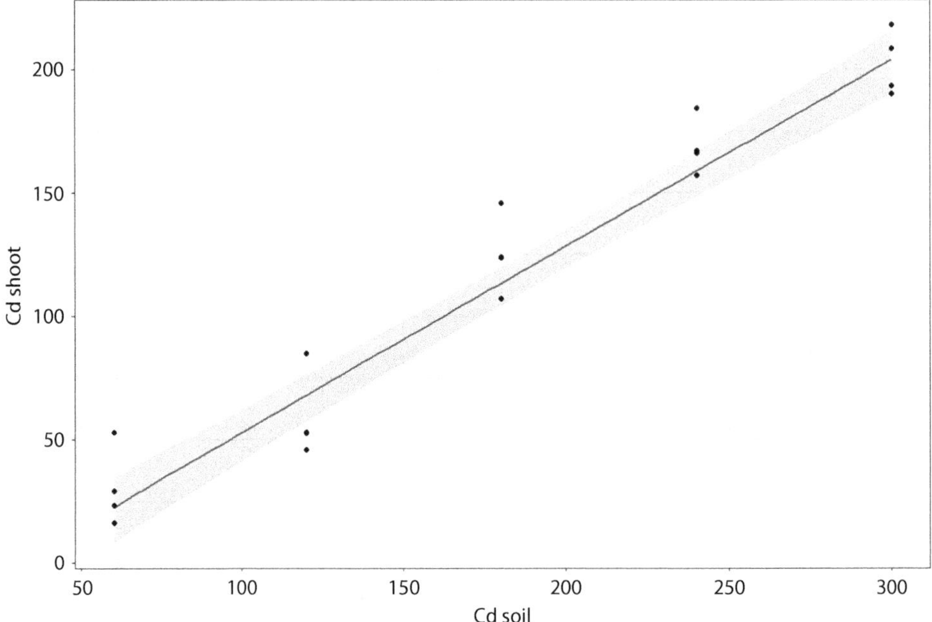

Fig. 4.12 Conditional plot showing the relation between the response (Cd in shoot) and explanatory variable (Cd in soil) predicted by the Bayesian linear regression model `m1.brms` and its 95% credible interval, along with the data.

Gelman et al.'s (2018) measure of R^2 is defined as $R^2 = \frac{\text{Variance}(Y_{\text{fitted}})}{\text{Variance}(Y_{\text{fitted}})+\text{Variance}(Y_{\text{residuals}})}$, where Y_{fitted} are the predicted values of the response variable Y obtained by applying the equation $Y = \beta_0 + \beta_1 X$ using the posterior distributions of the intercept (β_0) and the slope (β_1) in the three chains, and $Y_{\text{residuals}}$ are the raw residuals defined as ($Y_{\text{fitted}} - Y_{\text{observed}}$) for each value of the explanatory variable X. The actual values of the posterior distribution of R^2 could be seen by using `bayes_R2(m1.brms, summary = F)`; this posterior distribution was summarized above by its mean, standard deviation, and its 2.5 and 97.5 quantiles that define its credible interval.

We can also plot the fitted model and visually examine how it fits the available data, just as we did for the frequentist linear regression. We use the `brms` command `conditional_effects` to generate the predicted curve with its 95% credible interval, to which we will add the observed data (Fig. 4.12).

4.7. Tools for model validation in Bayesian statistics

Verifying that the MCMC algorithm has converged to a stationary distribution does not imply that the fitted model provides a good description of the empirical evidence. A badly fitting model may also attain algorithm convergence to a stationary distribution. Convergence to a stationary distribution is a quality-control step of the MCMC algorithm involved in Bayesian model fitting. Just like we did for the corresponding frequentist model, the validation of our first Bayesian model will be based on the analysis of residuals.

We need first to obtain the residuals. Here we will use the Pearson residuals,Pearson residuals = $\frac{Y_{\text{fitted}} - Y_{\text{observed}}}{\text{std deviation}(Y_{\text{observed}})}$, a kind of standardized raw residuals. The command `residuals(m1.brms, type = "pearson", nsamples = 1000, summary = F)` generates a matrix of 1,000 Pearson residuals for each of the 20 values of the response variable Y. Recall that in Bayesian statistics we never obtain a single best point estimate of any quantity (parameter values, residuals, R^2, etc.), but a posterior distribution that needs to be summarized using descriptive statistics. Now, if instead we execute `resid.m1.brms = as.data.frame(residuals(m1.brms, method = "pearson", nsamples = 1000, summary = T))` we get a data frame with summary statistics (mean, standard deviation, and 2.5 and 97% quantiles) for each of the 20 Pearson residuals corresponding to the 20 values of the response variable Y:

```
> head(res.m1.brms)
  Estimate Est.Error   Q2.5  Q97.5
1   0.0346     0.352 -0.603  0.757
2  -0.3614     0.366 -1.024  0.390
3   1.6425     0.351  1.006  2.364
4   0.3650     0.360 -0.287  1.104
5  -0.8519     0.259 -1.330 -0.260
6  -1.2677     0.267 -1.759 -0.658
```

We can now put these summary statistics of the Pearson residuals of the 20 data points in the original data frame `DF`:

```
DF$resid=res.m1.brms$Estimate
DF$resid.Q2.5=res.m1.brms$Q2.5
DF$resid.Q97.5=res.m1.brms$Q97.5
```

We can similarly obtain a posterior distribution of the fitted or predicted values by `fit.m1.brms = as.data.frame(fitted(m1.brms, scale = "linear", summary = T, nsamples = 1000))` look at its summary statistics by:

```
> head(res.m1.brms)
  Estimate Est.Error   Q2.5  Q97.5
1   0.0346     0.352 -0.603  0.757
2  -0.3614     0.366 -1.024  0.390
3   1.6425     0.351  1.006  2.364
4   0.3650     0.360 -0.287  1.104
5  -0.8519     0.259 -1.330 -0.260
6  -1.2677     0.267 -1.759 -0.658
```

and then put the 20 mean fitted values and their 95% credible intervals into the data frame `DF`:

```
DF$fit=fit.m1.brms$Estimate
DF$fit.Q2.5=fit.m1.brms$Q2.5
DF$fit.Q97.5=fit.m1.brms$Q97.5.
```

We are now ready to make the residuals vs. fitted and residuals vs. explanatory variable plots, along with the Q–Q plot of the residuals. Figure 4.13 shows that the residuals appear to have a random scatter without any discernible trends when plotted against either the fitted values or the explanatory variable, and that the Pearson residuals are normally distributed, as they should be for a correctly fitting statistical model (Faraway 2016). We can then safely conclude that the model `m1.brms` was satisfactorily validated by the residual analysis. There is an analogue of Cook distances in the Bayesian framework that will be explained in Chapter 7 and used thereafter.

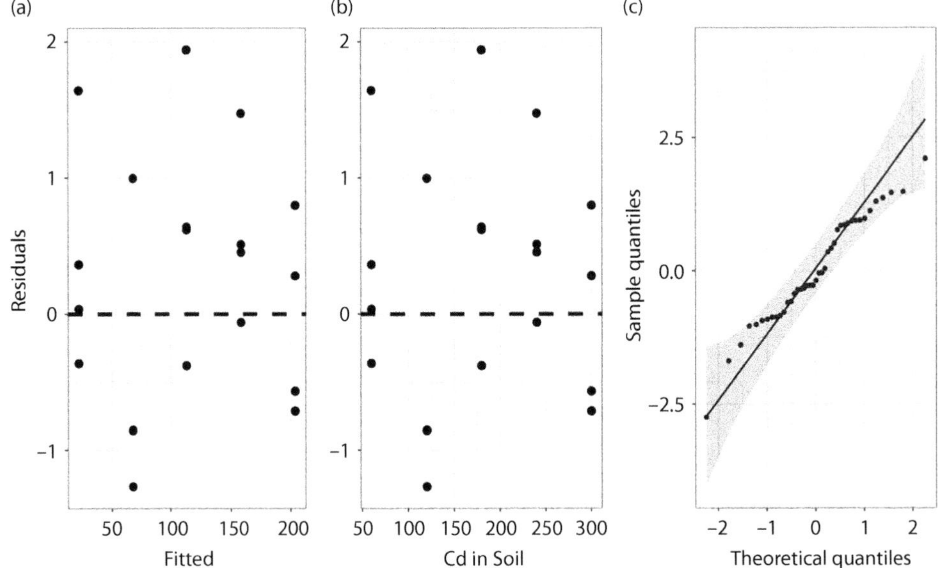

Fig. 4.13 Residual analysis of the Bayesian linear regression model `m1.brms` showing the mean of the Pearson residuals vs. the mean of the fitted value (A), the mean of the Pearson residuals vs. the explanatory variable in the model (B), and a quantile–quantile plot of the mean of the Pearson residuals (C).

4.8 Multiple linear regression: More than one numerical explanatory variable

The goal of multiple linear regression is to establish a linear relationship between a response variable with a normal distribution (Y ~ Normal) and a set of quantitative explanatory variables. Multiple linear regression is a generalization of simple multiple regression, and the latter is of course a special case of the former.

The statistical model of a multiple linear regression is $Y \sim \text{Normal}(\mu_Y, \sigma_Y^2)$, with $\mu_Y = \beta_0 + \beta_1 X_1 + \beta_2 X_2 + \cdots + \beta_k X_k$, where β_0 is the intercept or the predicted value of μ_Y when $X = 0$, and the β_i (with $i = 1$ to k, the number of numerical explanatory variables) are the partial slopes that denote the rate of change of μ_Y per unit of change of the explanatory variable X_i while holding all other explanatory variables constant. Thus, the partial slope has the same meaning as a partial derivative of a continuous function. In geometric terms, the multiple linear regression aims to find the equation of the (hyper)plane that best fits the data, with "best" being defined in terms of the smallest overall deviations from the data points.

Given that multiple regression involves more than one numerical explanatory variable, at least two new questions arise that were not relevant for the simple linear regression. First, how many and which explanatory variables should be initially included in the model? From the statistical standpoint there cannot be an answer to this fundamental question, for the very good reason that it is a scientific, not a statistical, question. If we carrying out a confirmatory analysis (Chapter 1), we should only consider the explanatory variables specifically involved in the scientific hypotheses that motivated the work.

But if ours is an exploratory analysis (Chapter 1) without specific hypotheses, most people tend to follow the approach of "everything but the kitchen sink" and incorporate all explanatory variables to see what comes out of the fishing expedition for results. The second question is whether the explanatory variables are to be included only as simple effects, or should we also evaluate their interactions (i.e., X_iX_j)? By interactions we mean that the magnitude of the effect of the explanatory variable X_i on the (mean of the) response variable Y changes or varies depending on the value of another variable X_j. The product X_iX_j is nothing more than a crude and simple way of expressing an effect above and beyond the linear relations that we are fitting. Reasons for considering (or not) interactions between explanatory variables include whether we have specific hypotheses suggesting the interactions, and the results of careful graphical exploratory analysis of the data.

THE DATA IN CONTEXT: Sokal and Rohlf (1995) compiled public data on atmospheric pollution from 41 US cities and wished to relate it to several numerical explanatory variables in order to formulate a simple predictive model. The response variable was the average annual SO_2 concentration in mg per m^3, and the explanatory variables were the average annual temperature in degrees Celsius, the number of industrial companies with more than 20 employees (`ManufEnter`), the population size (1970 census) in thousands, the average annual average wind speed in miles per hour, the average annual precipitation in cm^3, and the average number of rainy days per year.

EXPLORATORY DATA ANALYSIS: Let's first import and examine the data:

```
> DF1= read.csv("Ch 04 regr multiple.csv", header=T)
```

It is useful to obtain basic descriptive statistics of the data frame just created to get a first idea of the central tendency (means and medians) and the ranges of the response and explanatory variables:

```
> summary(DF1)
          City         SO2            Temp         ManufEnter      Population1970  AvgWindSpeed
 Albany     : 1   Min.   :  8   Min.   :43.5   Min.   :  35   Min.   :  71   Min.   : 6.0
 Albuquerque: 1   1st Qu.: 13   1st Qu.:50.6   1st Qu.: 181   1st Qu.: 299   1st Qu.: 8.8
 Atlanta    : 1   Median : 26   Median :54.6   Median : 347   Median : 515   Median : 9.4
 Baltimore  : 1   Mean   : 30   Mean   :55.8   Mean   : 463   Mean   : 609   Mean   :12.9
 Buffalo    : 1   3rd Qu.: 35   3rd Qu.:59.3   3rd Qu.: 462   3rd Qu.: 717   3rd Qu.:10.6
 Charleston : 1   Max.   :110   Max.   :75.5   Max.   :3344   Max.   :3369   Max.   :92.0
 (Other)    :35
   AvgPrecip      AvgRainyDays
 Min.   : 7.0   Min.   : 36
 1st Qu.:31.0   1st Qu.:103
 Median :38.7   Median :115
 Mean   :36.8   Mean   :114
 3rd Qu.:43.1   3rd Qu.:128
 Max.   :59.8   Max.   :166
```

The next step is to decide a plausible probability density function describing the variability of the response variable, for which we will use the package `fitdistrplus` as in Section 4.3. Again, we start by fitting the two distributions that we are to compare:

```
norm.m2=fitdist(DF1$SO2,"norm")
lognorm.m2=fitdist(DF1$SO2,"lnorm")
```

Figure 4.14 clearly shows that the response variable `SO2` is closer to having a lognormal than a normal distribution. This finding implies that we should logarithmically transform the response variable to make $\log(SO_2)$ normally distributed (see Section 4.2) so that we

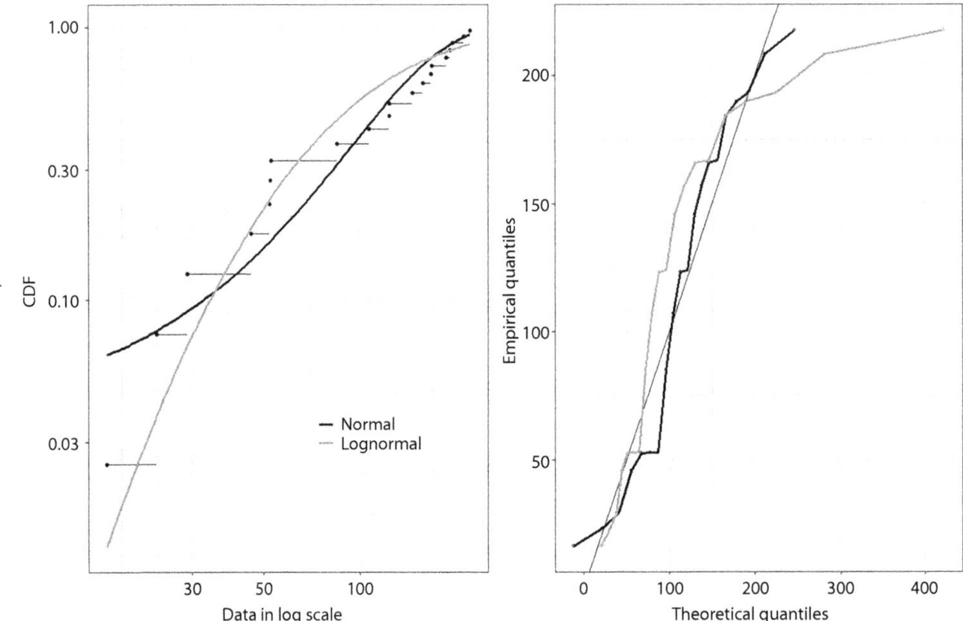

Fig. 4.14 Goodness of fit assessment of the response variable (`SO2`) showing the cumulative distribution function (CDF; left) and the quantile-quantile plot obtained using the package `fitdistrplus`.

can make use of all the inferential theory and estimation machinery of the general linear model. To make the other exploratory analysis plots with the variable to be modeled, we create the new variable `DF1$logSO2 = log(DF1$SO2)`.

We now need to examine whether the response variable `logSO2` is linearly related to the numerical explanatory variables, and to assess whether the latter are strongly correlated to each other. Whenever two explanatory variables are perfectly correlated (i.e., collinear), the model cannot be fitted. However, because perfect correlation (regardless of the sign) never occurs in actual data, we need to have a working criterion to call two explanatory variables strongly correlated. Even when the correlation between pairs of explanatory variables is too strong but less than perfect, the statistical model is deemed to be "numerically unstable," meaning that its parameter estimates and standard errors are unreliable. Above and beyond numerical problems, whenever two explanatory variables are strongly correlated, there would be a redundancy (in colloquial terms, we would be almost adding the same information twice) in their effects on the response variable since one of them would not add any extra explanatory capacity to the statistical model. Dormann et al. (2013) carried out extensive simulations involving many of the statistical models covered in this book, and suggested an heuristic (which means that it appears to be reasonable for most of the cases considered) criterion to only include numerical explanatory variables whose correlation is smaller than 0.7 in absolute value. We will use this heuristic criterion to select and include explanatory variables in the statistical models across this book. We use the package `GGally` to make the exploratory graphic analysis.

The bottom row in Fig. 4.15 contains the plots of `logSO2` vs. each of the explanatory variables; we see that there are grounds to think that temperature and `ManufEnter` may be

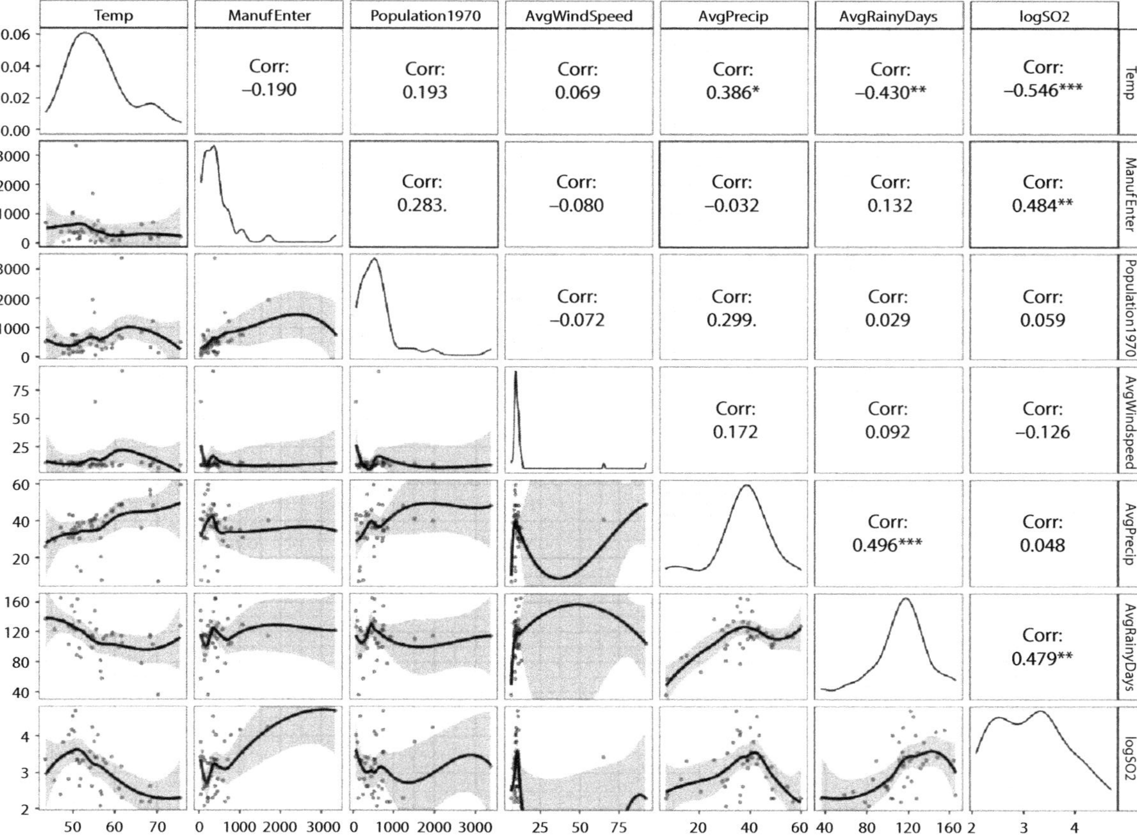

Fig. 4.15 Pairwise plots of the response (`logSO2`) and explanatory variables, showing the smoothed lines (below diagonal), the density plots of each variable (diagonal), and the pairwise Pearson correlation between pairs of variables. The figure was obtained using the package `GGally`.

linearly related, and that `AvgPrecip` and `AvgRainyDays` may be quadratically related, to `logSO2`. Just for the sake of simplicity at this point we are not going to consider quadratic effects; they will be considered in Section 4.10. Incidentally, the width of the gray band (the standard error around the smoothed lines) reflects the amount of data available in certain ranges of the explanatory variables: as there are a few cities with strong `AvgWindSpeed` and large `ManufEnter` there is a large uncertainty in these portions of the respective plots (center of the bottom row of Fig. 4.15). The highest correlation between explanatory variables is 0.496, and thus we could include all of them in the following statistical model.

4.9 Multiple linear regression: Frequentist fitting

Based on the exploratory data analysis, we will fit the model `m2 = lm(logSO2 ~ Temp + ManufEnter + Population1970 + AvgWindSpeed + AvgPrecip+AvgRainyDays, data = DF1)` whose main outputs are the `summary` and the `anova` table.

```
> summary(m2)

Call:
lm(formula = logSO2 ~ Temp + ManufEnter + Population1970 + AvgWindSpeed +
    AvgPrecip + AvgRainyDays, data = DF1)

Residuals:
    Min      1Q  Median      3Q     Max
-1.2956 -0.2940 -0.0751  0.4223  1.0257

Coefficients:
               Estimate Std. Error t value Pr(>|t|)
(Intercept)     4.89e+00   1.31e+00    3.74  0.00067 ***
Temp           -4.85e-02   1.96e-02   -2.47  0.01886 *
ManufEnter      4.76e-04   1.58e-04    3.01  0.00487 **
Population1970 -3.93e-05   1.61e-04   -0.24  0.80826
AvgWindSpeed   -5.14e-03   5.47e-03   -0.94  0.35377
AvgPrecip       1.35e-02   1.30e-02    1.04  0.30682
AvgRainyDays    2.99e-03   5.72e-03    0.52  0.60510
---
Signif. codes:  0 ‘***’ 0.001 ‘**’ 0.01 ‘*’ 0.05 ‘.’ 0.1 ‘ ’ 1

Residual standard error: 0.52 on 34 degrees of freedom
Multiple R-squared:  0.535,     Adjusted R-squared:  0.453
F-statistic: 6.51 on 6 and 34 DF,  p-value: 0.000121
```

Under the heading “Coefficients” in the summary we have the intercept (the mean of `logSO2` when all explanatory variables are zero) and the partial slopes (the rate of change of `logSO2` per unit change of each explanatory variable), their standard errors, and the ratio between each coefficient and its standard error (t-value) that is used to assess whether the intercept is statistically different from zero, and the p-values. These p-values test the actual effect of each explanatory variable on the mean of `logSO2` when all other variables are excluded from the model. Using the much-decried criterion of statistical significance, only `Temp` and `ManufEnter` would seem to have statistically significant effects. The current model explains 45.3% of the variation of the response variable `logSO2`.

A second main output of the statistical model is its `anova` table:

```
> anova(m2)
Analysis of Variance Table

Response: logSO2
                Df Sum Sq Mean Sq F value  Pr(>F)
Temp             1   5.89    5.89   21.82 4.6e-05 ***
ManufEnter       1   2.96    2.96   10.98  0.0022 **
Population1970   1   0.03    0.03    0.12  0.7286
AvgWindSpeed     1   0.07    0.07    0.26  0.6135
AvgPrecip        1   1.52    1.52    5.63  0.0235 *
AvgRainyDays     1   0.07    0.07    0.27  0.6051
Residuals       34   9.18    0.27
```

This table shows a set of sequential tests of the effects of the explanatory variables that basically assess whether an explanatory variable has a statistically significant effect on the mean of the response variable `logSO2` given the other variables already in the model. In order to understand what the sequential tests are, let's first realize that the variables are entered in the model in the same (arbitrary) order in which we wrote the model `m2`. (We will explain the sum of squares (`Sum Sq`), the mean squares (`Mean Sq`), and the *F*-test in detail in Chapter 5.) For now, let's focus on the *p*-values of each explanatory variable. The *p*-value for `Temp` indicates whether the addition of `Temp` to a model that only had the intercept significantly improved the explanatory power of the model. The *p*-value of `ManufEnter` evaluates whether including this explanatory variable in a model that contained the intercept and `Temp` improved its explanatory power, and so on.

Let us note that these sequential tests are conceptually different from the tests involved in `summary(m2)`. Hence, there could be differences between the statistical significance of variables in the `summary` and the `anova` table of a model. By default, the user of R controls the order of entry of explanatory variables in the statistical model, a point that some statisticians and statistical software (i.e., SAS and Stata) consider an anathema but that may be of importance when used wisely. The choice of entry of variables into a model can be used for different purposes. For instance, we could first enter a confounding variable (i.e., a variable known to be related to the response variable but that is not part of our hypothesis) to account for its effect, and then test the effects of all other explanatory variables. Of course, labeling an explanatory variable a "confounding variable" (Chapter 14) is contingent on having preliminary knowledge or evidence of its effect (otherwise we would not have measured it!), and on the statistical hypotheses being tested. We could also choose to enter the explanatory variables in decreasing order of their correlation with the response variable, or to include first those explanatory variables which we think have stronger mechanistic reasons to be related to the response variable.

Before validating the fitted statistical model by residual analysis and depicting its results in terms of predicted relations and figures, it is important to note that there are two main difficulties when interpreting the estimated parameters in terms of the problem. First, the intercept of `summary(m2)` is the mean value of `logSO2` for a city with no inhabitants, no rain or rainy days, etc. Obviously, such a city cannot cannot possibly exist. Second, we cannot compare the values of partial slopes because the explanatory variables were measured in different scales. Standardization of the explanatory variables can resolve both problems and provide a different meaning to the fitted regression parameters.

4.10 The importance of standardizing explanatory variables

Standardization of the explanatory variables involves the centering of the variables by subtracting each value from each variable's mean, and dividing the centered variable by its standard deviation. The result of standardization of the explanatory variables is that they will all have a mean of zero and a unit standard deviation. Schielzeth (2012) gave a simple and lucid explanation of the reasons for, and the benefits of, standardizing the explanatory variables when fitting statistical models. The standardization is accomplished by `DF1s = scale(DF1[,3:8], center = T, scale = T)`. This command selects the numerical explanatory variables from the data frame `DF1` and stores the result in the new data frame `DF1s`. We now need to place the response variable in the new data frame with `DF1s = as.data.frame(cbind(logSO2 = DF1$logSO2, DF1s))`. The command `cbind` attaches the vector `DF1$logSO2` and the data frame `DF1s` by columns provided that they have the same number of rows (and they do), and generates a matrix that is converted into a data frame and written over `DF1s`. We now have a new data frame with the response variable and the standardized explanatory variables that can be used for fitting the model: `m2s = lm(logSO2 ~ Temp + ManufEnter + Population1970 + AvgWindSpeed + AvgPrecip + AvgRainyDays, data = DF1s)`. The summary of the fit is:

```
> summary(m2s)

Call:
lm(formula = logSO2 ~ Temp + ManufEnter + Population1970 + AvgWindSpeed +
    AvgPrecip + AvgRainyDays, data = DF1s)

Residuals:
    Min      1Q  Median      3Q     Max
-1.2956 -0.2940 -0.0751  0.4223  1.0257

Coefficients:
               Estimate Std. Error t value Pr(>|t|)
(Intercept)      3.1530     0.0811   38.86   <2e-16 ***
Temp            -0.3502     0.1420   -2.47   0.0189 *
ManufEnter       0.2684     0.0891    3.01   0.0049 **
Population1970  -0.0228     0.0931   -0.24   0.8083
AvgWindSpeed    -0.0791     0.0842   -0.94   0.3538
AvgPrecip        0.1589     0.1531    1.04   0.3068
AvgRainyDays     0.0792     0.1517    0.52   0.6051
---
Signif. codes:  0 '***' 0.001 '**' 0.01 '*' 0.05 '.' 0.1 ' ' 1

Residual standard error: 0.52 on 34 degrees of freedom
Multiple R-squared:  0.535,     Adjusted R-squared:  0.453
F-statistic: 6.51 on 6 and 34 DF,  p-value: 0.000121
```

There are similarities and differences between the summaries of models `m2` and `m2s`. The signs, t-values, and p-values are the same in both model summaries, and so is the percentage of variation explained by the models. The actual differences are the magnitudes of the standard errors of the estimated parameters and their interpretation. First, the intercept of the model `m2s` with standardized explanatory variables is the expected value or mean of `logSO2` for the mean values of the explanatory variables. These mean values are:

```
> summary(DF1[,3:8])[4,]
       Temp          ManufEnter     Population1970      AvgWindSpeed           AvgPrecip       AvgRainyDays
"Mean   :55.8  " "Mean   : 463  " "Mean   : 609  " "Mean   :12.9  " "Mean   :36.8  "  "Mean   :114  "
```

and they correspond to the "average or typical city" in the data set. Second, the partial slopes are the changes in the expected value of `logSO2` when an explanatory variable changes a standard deviation. Thus, if `Temp` changes, `sd(DF1$Temp)[1] = 7.2` degrees Fahrenheit, the mean value of `logSO2` would decrease by 0.3501. Let's also note that the magnitudes of the partial slopes of `m2s` are more similar to each other than those of `m2`. Because the explanatory variables are now all in the same relative scale, we can compare the relative magnitudes of the partial slopes. Let's take the two explanatory variables having statistically significant effects, `Temp` and `ManufEnter`. We can now say that the magnitude of the effect of `Temp` on `logSO2` is 31% (1.31 = 0.3501 / 0.2684) stronger than that of `ManufEnter`. Therefore, making a simple ratio of partial slopes of a model fitted with standardized explanatory variables yields a much richer interpretation of the relative importance of these variables in the context of the problem.

We can now validate the model `m2s` using residual analysis. We start by putting all the metrics involved in the residual analysis in a single data frame with `res.m2s = augment(m2s)` from the package `broom`. The residual analysis will involve the following nine ggplots: residuals vs. fitted values, Q–Q plot of residuals, Cook distances for each data point, and residuals vs. each of the six explanatory variables.

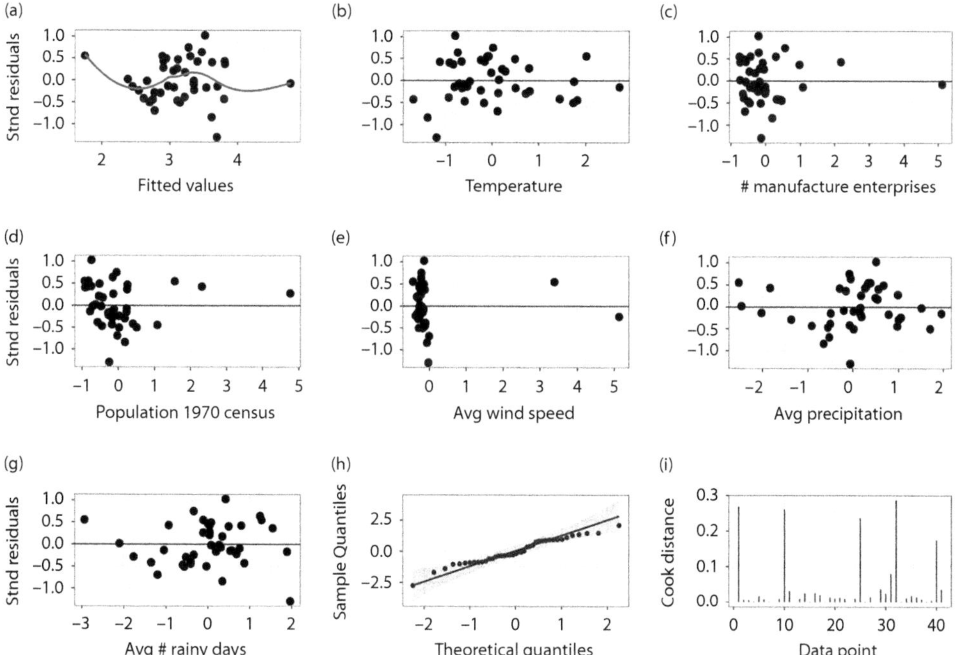

Fig. 4.16 Residual analysis of the frequentist multiple regression model `m2s` showing the residuals vs. the fitted values (a), the residuals vs. each explanatory variable (b to g), the quantile–quantile plot of the residuals (h), and the Cook distances of each data point (i).

The residuals vs. fitted values plot (Fig. 4.16a) shows similar numbers of positive and negative residuals, and also that the points in the range [2, 4] have a slightly smaller variation in the lower than in the upper end of this range. This might suggest that the residuals do not have a homogeneous variation, which would be a mild violation of the model assumptions. The Q–Q plot (Fig. 4.16h) shows that the residuals are normally distributed, as they should be. The Cook distances (Fig. 4.16i) suggest that, depending on the threshold used, there might be four or five points with stronger influence on the parameter estimates. We might wish to examine these points and perhaps consider refitting the model without them to check whether there are important qualitative differences (i.e., changes in sign or in statistical significance of the partial slopes, large changes in the model R^2) between the two model fits. The remaining plots of Fig. 4.16 show a reasonably random scatter of residuals, in the sense that one could hardly imagine fitting a relationship in any of these clouds of points. Of course, the latter statement is very qualitative and is based on the experience of fitting many similar models. What we are really looking for is large deviations from the expected patterns, not small glitches that are bound to exist in just about every data set. It is very important not to over-interpret the plots of residuals vs. number of manufacturers, 1970 population census, and average wind speed (Fig. 4.16 c, d and e). The right-hand sides of these plots have very few data points simply because most cities had a small number of enterprises, were of small-to-medium population size, and had moderate wind speed. However, if we just focus on the left-hand sides of these plots where the data are, we can conclude that the residuals have a reasonably random scatter in these small ranges of the explanatory variables.

On balance, the only one of the nine plots of Fig. 4.16 that contains a mild suggestion of model lack-of-fit. What could we do about it? First, we could add more explanatory variables or more data if we have them (we don't, otherwise we would have included them in the analysis!). Second, we could change the likelihood component of the model, and consider alternatives to the lognormal distribution (such as the gamma distribution, see Chapter 12), and compare their respective residual analyses to see if there is any improvement. Third, we could change the model structure and contemplate the explicit modeling of the heterogeneity of the variation of the response variable; we show one such example in Chapter 15. What would be the consequences for our inference if we keep `m2s` as it is now? The main impact of the mild feature detected in the residual analysis of Fig. 4.16 is likely to be a small bias in the model parameters and/or in their standard errors, but the main conclusions thus far extracted will probably remain unchanged.

We should now extract the main model outputs to be communicated: a plot of the partial slopes and their confidence intervals that replaces `summary(m2s)` (Fig. 4.17) and the set of conditional plots for each explanatory variable (Fig. 4.17). The model `m2s` has six explanatory variables. Thus, it needs values for all these variables in order to predict values of the response variable `logSO2` and the plotting of the predictions would entail a seven-dimensional plot that cannot be visualized. What we could do instead is to make simpler conditional plots considering one or two explanatory variables at a time, while holding the other explanatory variables at some fixed values. We use the package `ggeffects`. This is exactly what `pred.m2s = ggpredict(m2s, terms = c("Temp[all]", "ManufEnter"))` does: it generates a data frame (`pred.m2s`) storing the predicted values of `logSO2` for all values of `Temp` and for three values of `ManufEnter` (its mean and mean ± SD) while holding at zero all the other explanatory variables not included in `terms`. Because the explanatory variables in `m2s` were standardized, holding their values at zero in `ggpredict` amounts to setting them to their mean values in the original data set (Fig. 4.18).

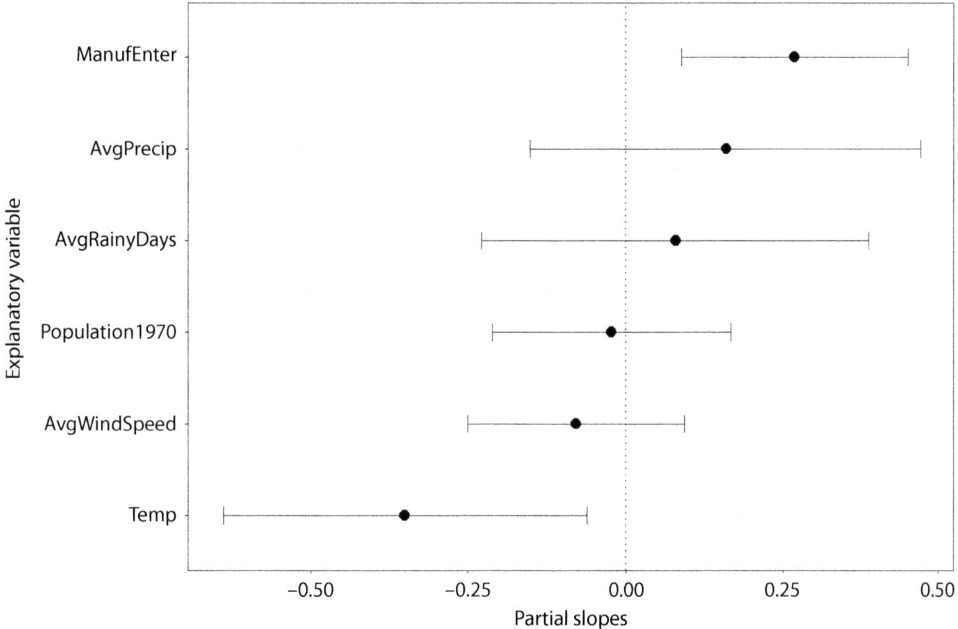

Fig. 4.17 Estimated partial slopes and their 95% confidence intervals from the frequentist multiple regression model `m2s`.

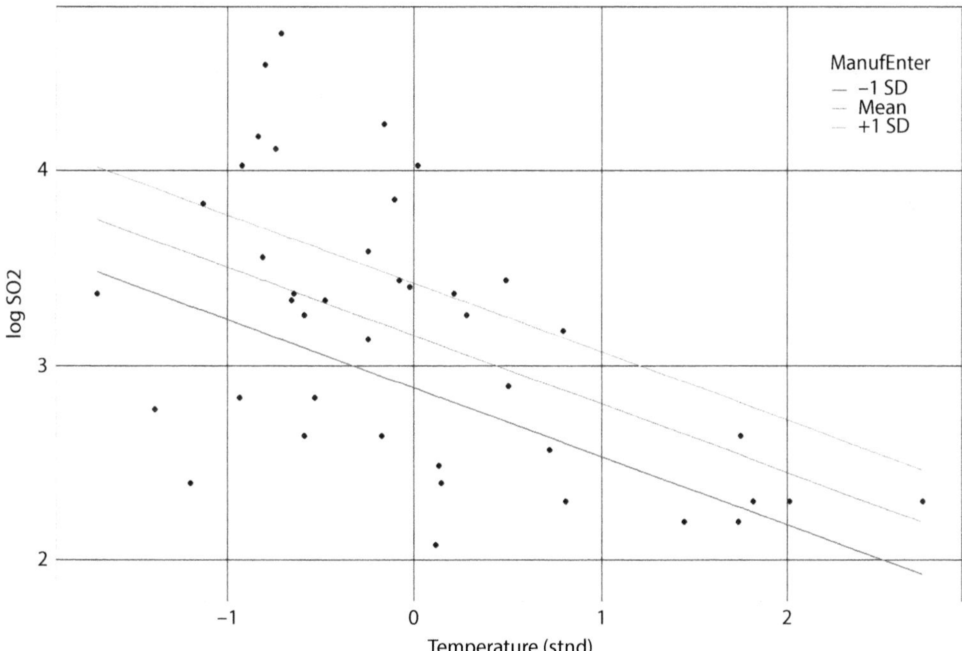

Fig. 4.18 Conditional curves depicting the predicted values of the response variable `logSO2` by the frequentist multiple regression model `m2s` for different values of temperature (standardized to have zero mean and unit standard deviation) and for three values of the number of manufacturing enterprises in a city (mean and mean ± SD), while holding the other standardized explanatory variables at zero. The points are the actual values of the response variable `logSO2` and the explanatory variable `Temp` in the standardized scale.

We could, of course, generate other plots of the predicted values of the response variable for other combinations or selected sets of values of the explanatory variables using the package `ggeffects`. Here we have just focused on the two explanatory variables having statistically significant effects in model `m2s`.

4.11 Polynomial regression

Standardizing the explanatory variables is also essential in polynomial regression (Schielzeth 2012). A polynomial regression is just a multiple regression in which the explanatory variable is fitted in powers of a polynomial. In terms of model fitting, the polynomial regression is a linear problem, despite the resulting curve being obviously nonlinear. The statistical model of a polynomial regression with one explanatory variable is $Y \sim \text{Normal}(\mu_Y, \sigma_Y^2)$, with $\mu_Y = \beta_0 + \beta_1 X + \beta_2 X^2 + \cdots + \beta_k X^k$. Based on the suggested relations of Fig. 4.15, let's fit a quadratic relation between `logSO2` and `AvgPrecip` with the original data in data frame `DF2` and with the standardized explanatory variable in data frame `DF2s`:

```
> m3=lm (logSO2 ~ AvgPrecip  + I(AvgPrecip^2), data=DF1)
> m3s=lm (logSO2 ~ AvgPrecip + I(AvgPrecip^2), data=DF1s).
```

The syntax `I(AvgPrecip^2)` uses the R command I meaning "as is" to interpret `AvgPrecip^2` as AvgPrecip^2. The model summaries are:

```
> summary(m3, cor=T)

Call:
lm(formula = logSO2 ~ AvgPrecip + I(AvgPrecip^2), data = DF:

Residuals:
    Min      1Q  Median      3Q     Max
-1.3080 -0.4502  0.0614  0.2610  1.3009

Coefficients:
                Estimate Std. Error t value Pr(>|t|)
(Intercept)     1.503884   0.555524    2.71   0.0101 *
AvgPrecip       0.113496   0.033337    3.40   0.0016 **
I(AvgPrecip^2) -0.001697   0.000495   -3.43   0.0015 **
---
Signif. codes:  0 '***' 0.001 '**' 0.01 '*' 0.05 '.' 0.1 '

Residual standard error: 0.629 on 38 degrees of freedom
Multiple R-squared:  0.238,     Adjusted R-squared:  0.198
F-statistic: 5.94 on 2 and 38 DF,  p-value: 0.00568

Correlation of Coefficients:
               (Intercept) AvgPrecip
AvgPrecip      -0.93
I(AvgPrecip^2)  0.81        -0.97
```

and:

```
> summary(m3s, cor=T)

Call:
lm(formula = logSO2 ~ AvgPrecip + I(AvgPrecip^2), data = DI

Residuals:
    Min      1Q  Median      3Q     Max 
-1.3080 -0.4502  0.0614  0.2610  1.3009 

Coefficients:
               Estimate Std. Error t value Pr(>|t|)    
(Intercept)      3.3825     0.1188   28.47   <2e-16 ***
AvgPrecip       -0.1332     0.1107   -1.20   0.2364    
I(AvgPrecip^2)  -0.2352     0.0686   -3.43   0.0015 ** 
---
Signif. codes:  0 '***' 0.001 '**' 0.01 '*' 0.05 '.' 0.1 '

Residual standard error: 0.629 on 38 degrees of freedom
Multiple R-squared:  0.238,	Adjusted R-squared:  0.198 
F-statistic: 5.94 on 2 and 38 DF,  p-value: 0.00568

Correlation of Coefficients:
               (Intercept) AvgPrecip
AvgPrecip      -0.25                
I(AvgPrecip^2) -0.56        0.44    
```

The residual standard error (i.e., the square root of the estimate of σ_Y^2), the adjusted R^2, and the F-statistics are the same in both models, but the estimated intercepts, the coefficients for the linear and quadratic effects, their standard errors, and their statistical significance strongly differed between models `m3` and `m3s`. We also see that the estimated coefficients were very highly correlated in `m3` but markedly less so for `m3s`. The interpretation of the estimated parameters also differs between models `m3` and `m3s` (Schielzeth 2012). When the explanatory variable was in its original scale, we could not interpret the linear effect because the quadratic effect (a kind of "self-interaction") was statistically significant, but the latter lacks a clear interpretation. However, when the explanatory variable was standardized in `m3s` the two coefficients are independent and now have different and interesting interpretations. The coefficient for the linear effect can be interpreted as the effect of the change in `AvgPrecip` in one standard deviation on the mean of the response variable. The coefficient for the quadratic effect reflects the impact that a change of one standard deviation in the explanatory variable would have on both tails of the distribution of values of the response variable `logSO2`. As always, model `m3s` should be validated by residual analysis and the results extracted, but in the interests of brevity we will not pursue these points here.

4.12 Multiple linear regression: Bayesian fitting

We are going to fit exactly the same statistical model, `logSO2 ~ Temp + ManufEnter + Population1970 + AvgWindSpeed + AvgPrecip + AvgRainyDays` with standardized data using Bayesian methods.

The command

```
> get_prior(formula=logSO2~Temp+ManufEnter+Population1970+AvgWindSpeed+AvgPrecip+AvgRainyDays,
+           data=DF1s,family=gaussian)
                prior     class           coef group resp dpar nlpar bound       source
               (flat)         b                                                 default
               (flat)         b      AvgPrecip                             (vectorized)
               (flat)         b   AvgRainyDays                             (vectorized)
               (flat)         b   AvgWindSpeed                             (vectorized)
               (flat)         b     ManufEnter                             (vectorized)
               (flat)         b Population1970                             (vectorized)
               (flat)         b           Temp                             (vectorized)
 student_t(3, 3.3, 2.5) Intercept                                               default
   student_t(3, 0, 2.5)     sigma                                               default
```

clarifies the prior distributions that need be defined to fit this model: the intercept, the six partial slopes, and the standard deviation of the response variable `logSO2`. Recalling that the intercept is the average of `logSO2` for the mean values of the standardized explanatory variables, we could use the average and standard deviation of `logSO2` in other data sets in similar US or European cities to define the hyperparameters of the normal prior distribution for the intercept. Hyperparameters define the prior distributions of other model parameters. We could similarly use previous data sets to showing the relation between the response and explanatory variables to define the hyperpriors for the partial slopes. We are going to use weakly informative prior distributions for these parameters:

```
prior.m2 = c(set_prior("normal(0, 4)", class = "Intercept"),
             set_prior("normal(0, 2)", class = "b"),
             set_prior("cauchy(0,5)",  class = "sigma"))
```

These distributions can be visualized (Fig. 4.19) and interpreted beforehand to verify that they are compatible with what is known about the problem of SO_2 contamination.

The prior distribution for the intercept says that on the average $\log(SO_2) = 0$ (i.e., the mean SO_2 in a city would be 1 mg/m^3), and that we expect that 95% of values would lie in $[\exp(-4) = 0.0018, \exp(4) = 54.6]$, a very wide distribution. The prior distributions for the partial slopes indicate that there is as much evidence of negative relationships as

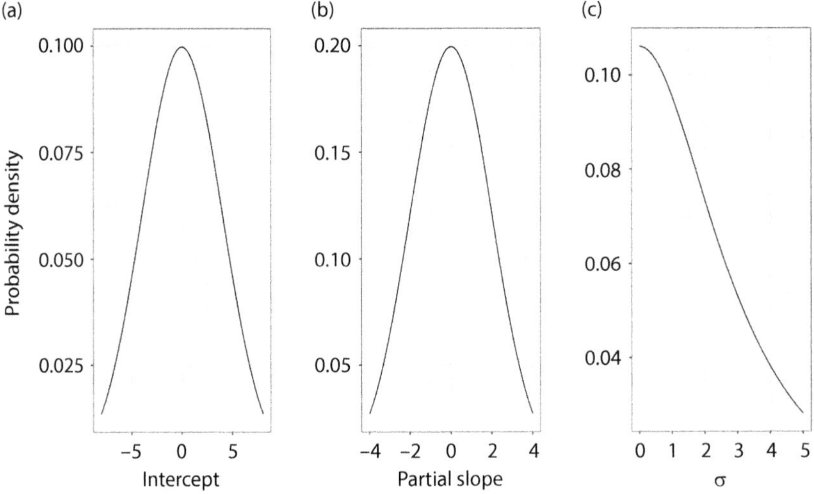

Fig. 4.19 Prior distributions of the intercept, partial slopes, and the standard deviation of the response variable `logSO2` of the Bayesian multiple regression model `m2.brms`.

there is for positive between the explanatory variables and `logSO2` (otherwise we would have used an asymmetric prior distribution), and that 95% of the slope values would lie in [–4, 4]. To some extent, it may seem easier to define weakly informative priors for slopes of standardized variables than for those in the original scales, precisely because we are not strictly required to think about the absolute magnitudes of their effects on the response variable. On the other hand, it is probably harder to elicit strongly informative priors for the slopes from experts or from the literature because we also need to inquire and think about the standard deviation of each explanatory variable. As for the standard deviation of the response variable, its prior distribution states that we believe that smaller values are vastly more plausible than large values of this parameter.

The Bayesian multiple regression model is `m2.brms = brm(formula = logSO2 ~ Temp + ManufEnter + Population1970 + AvgWindSpeed + AvgPrecip + AvgRainyDays, data = DF1s, family = gaussian, prior = prior.m2, warmup = 1000, future = TRUE, chains = 3, iter = 2000, thin = 3)`. Setting `thin = 3` means that although each of the three chains will sample 2,000 values, at most only one of every three sampled values (2000 / 3 ≈ 668) will be retained as a parameter estimate. Thinning the sampled values of a chain would decrease their autocorrelation. The model summary is:

```
> summary(m2.brms)
 Family: gaussian
  Links: mu = identity; sigma = identity
Formula: logSO2 ~ Temp + ManufEnter + Population1970 + AvgWindSpeed + AvgPrecip + AvgRain
yDays
   Data: DF1s (Number of observations: 41)
Samples: 3 chains, each with iter = 2000; warmup = 1000; thin = 3;
         total post-warmup samples = 1000

Population-Level Effects:
               Estimate Est.Error l-95% CI u-95% CI Rhat Bulk_ESS Tail_ESS
Intercept          3.15      0.08     2.99     3.31 1.00      787      957
Temp              -0.35      0.15    -0.64    -0.05 1.00     1051      886
ManufEnter         0.27      0.09     0.09     0.45 1.00      938      958
Population1970    -0.02      0.10    -0.21     0.18 1.00     1067     1011
AvgWindSpeed      -0.08      0.09    -0.24     0.10 1.00     1031      910
AvgPrecip          0.16      0.16    -0.15     0.47 1.00      871      851
AvgRainyDays       0.08      0.16    -0.22     0.40 1.00      927      779

Family Specific Parameters:
      Estimate Est.Error l-95% CI u-95% CI Rhat Bulk_ESS Tail_ESS
sigma     0.54      0.07     0.42     0.68 1.00     1036      986
```

Although we should first assess model convergence, let's interpret these results. We have the mean, standard deviation, and 95% credible intervals calculated from the marginal posterior distributions of all model parameters. We will shortly visualize these posterior distributions. All the `Rhat` values were equal to one, suggesting convergence of the chains, and the high effective sample sizes are a crude measure that the three chains have converged to the same stationary posterior distribution. We see that only the 95% credible intervals of the partial slopes for `Temp` and `ManufEnter` did not contain the value of zero, a result that is coincident with the frequentist model `m2s`. Our Bayesian multiple regression model explained a proportion of the variation of the response variable similar to its frequentist analogue (0.54; see Section 4.6):

```
> bayes_R2(m2.brms)
   Estimate Est.Error  Q2.5 Q97.5
R2     0.52    0.0739 0.338  0.63
```

We can first assess whether the posterior distributions of each chain are similar or convergent for each model parameter using the package `bayesplot`. Figure 4.20 shows that the three chains converged to the same marginal distribution for all model parameters, in agreement with the `Rhat` shown in `summary(m2.brms)`. Let's evaluate the autocorrelation of the sampled parameter values in each chain. The autocorrelation profiles (Fig. 4.21) for all model parameters and chains are smaller than 0.15 in absolute value, thus indicating that the parameter estimates can be considered as largely statistically independent. This corroborates the large effective sample sizes shown in `summary(m2.brms)`.

Having verified that the model `m2.brms` converged to a stationary distribution, we can finally visualize the marginal posterior distributions of its main parameters (Fig. 4.22). We can also produce plots of the conditional (or marginal) effects, depicting the effect of each response variable on the response variable `logSO2` predicted by the model `m2.brms`. (Fig. 4.23).

It is interesting to see that the marginal predicted relations for `ManufEnter Population1970` and `AvgWindSpeed` (Fig. 4.23) seem to be driven by a few data points that are very different from the bulk of the data. This is not to say that these points are outliers, but their values should be double-checked as they seem to have a strong impact on the fitted relationships.

We should finally validate the model `m2.brms` by residual analysis. Just as we did in Section 4.6, we must first obtain the Pearson residuals and the fitted values related to this statistical model:

```
res.m2.brms=as.data.frame(residuals(m2.brms, type="pearson", nsamples=1000,
                          summary=T))
fit.m2.brms=as.data.frame(fitted(m2.brms, scale="linear", summary=T,
                          nsamples=1000))
```

and then put them into the data frame containing the explanatory variables:

```
DF1s$resid=res.m2.brms$Estimate # means of Pearson residuals for each data point
DF1s$fit=fit.m2.brms$Estimate # means of fitted values for each data point.
```

In analyzing Fig. 4.24, we can almost repeat what was written for the residual analysis of the frequentist multiple regression: this Bayesian model has some mild problems of lack of fit related to possible heterogeneity of variation of the residuals (Fig. 4.26, top left), and we should be cautious to avoid over-interpreting the ranges of plots having few points. Overall, we would say that the model `m2.brms` fits the data reasonably well.

This has been a very long chapter. We fitted our first frequentist and Bayesian models, which required a fair amount of detailed explanation to understand the model outputs and their validation by residual analysis. This was not wasted effort, for we are going to use and reuse the same code with minor variations across this book. We insisted on detailed interpretation of the residual analysis based on qualitative overall judgment that can only be progressively developed.

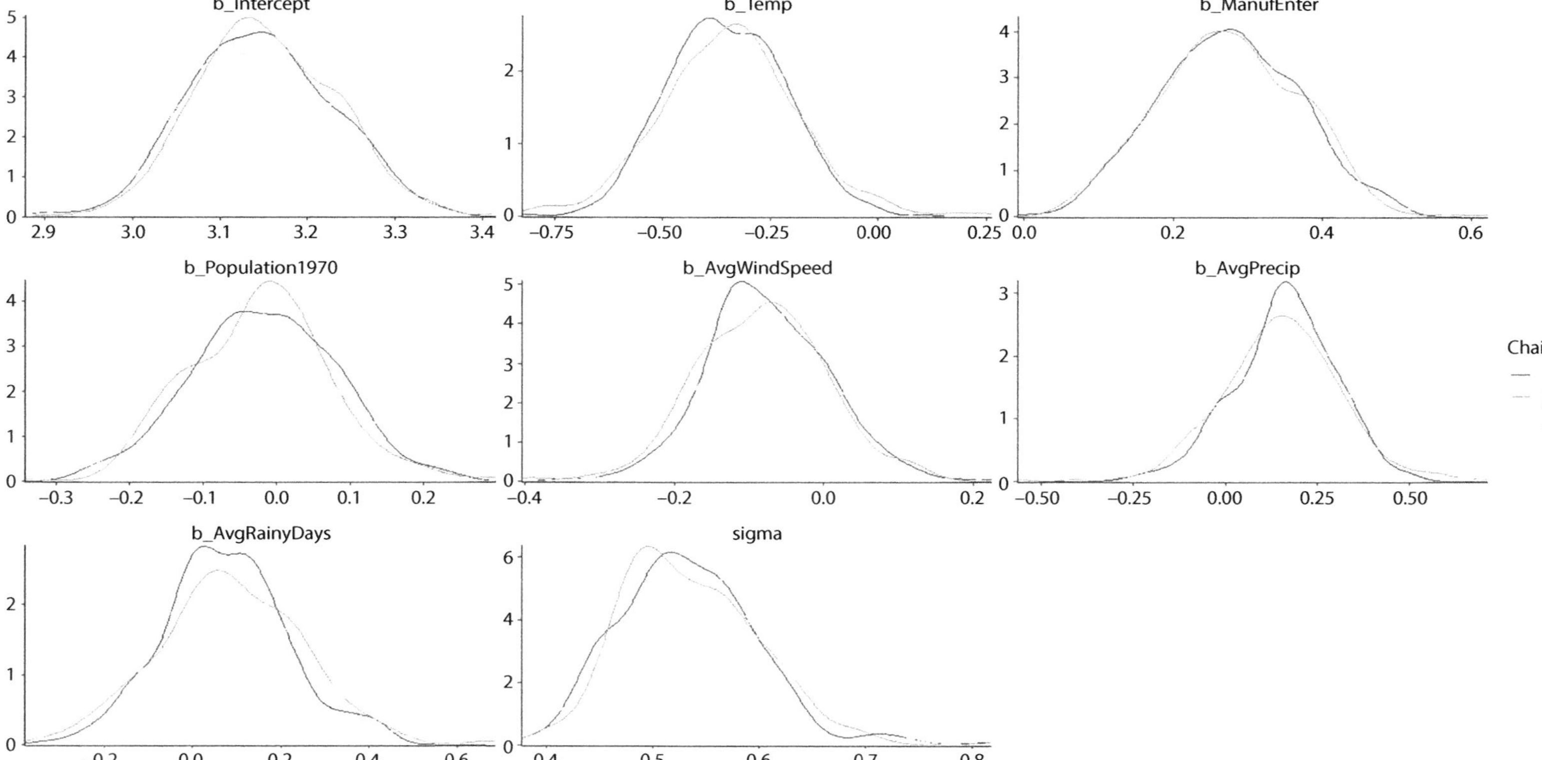

Fig. 4.20 Posterior distributions of the intercept, partial slopes, and standard deviation showing the convergence of the three chains of the Bayesian multiple regression model `m2.brms` to a common stationary posterior distribution.

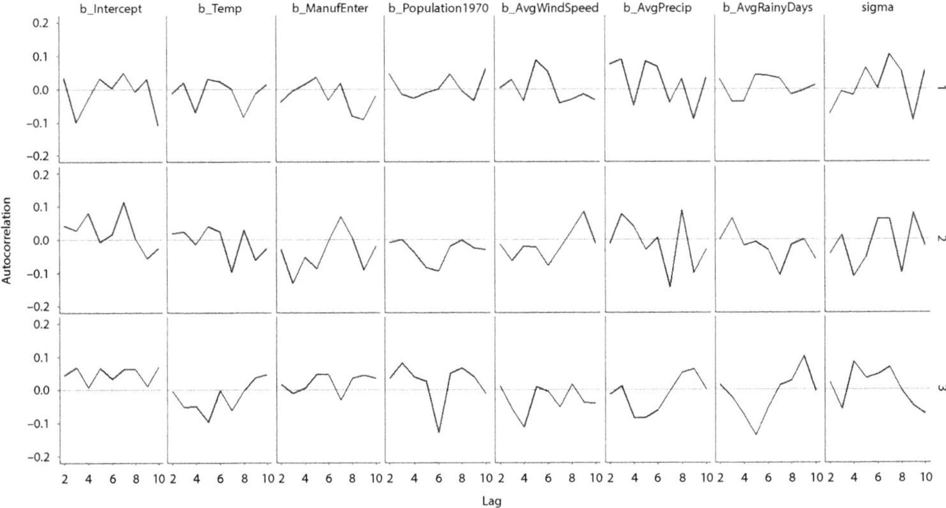

Fig. 4.21 Autocorrelation of the time series of the sampled values of each model parameter of each chain of the Bayesian multiple regression model `m2.brms`. The scale limit on the *x*-axis was set to avoid showing the autocorrelation of lag = 0 (the correlation of a value with itself) that is equal to one by definition.

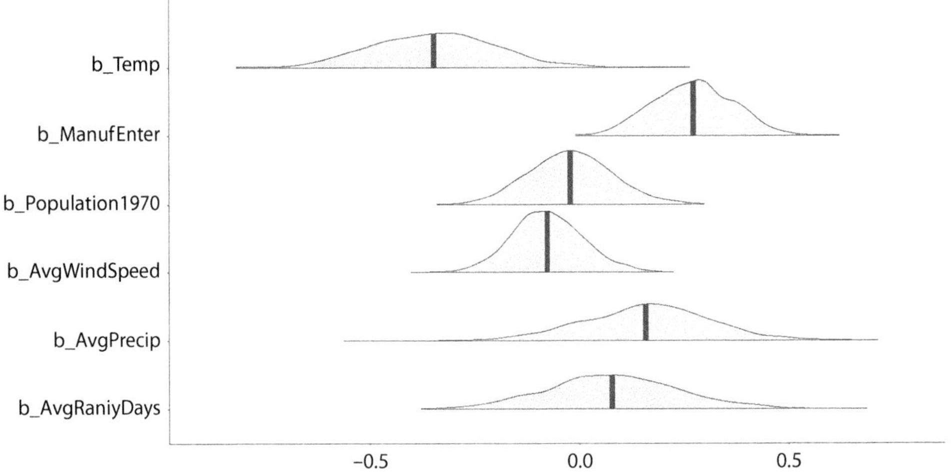

Fig. 4.22 Marginal posterior distributions of the partial slopes of the Bayesian multiple regression model `m2.brms`. The black line is the mean and the shaded area the 95% credible interval of each parameter.

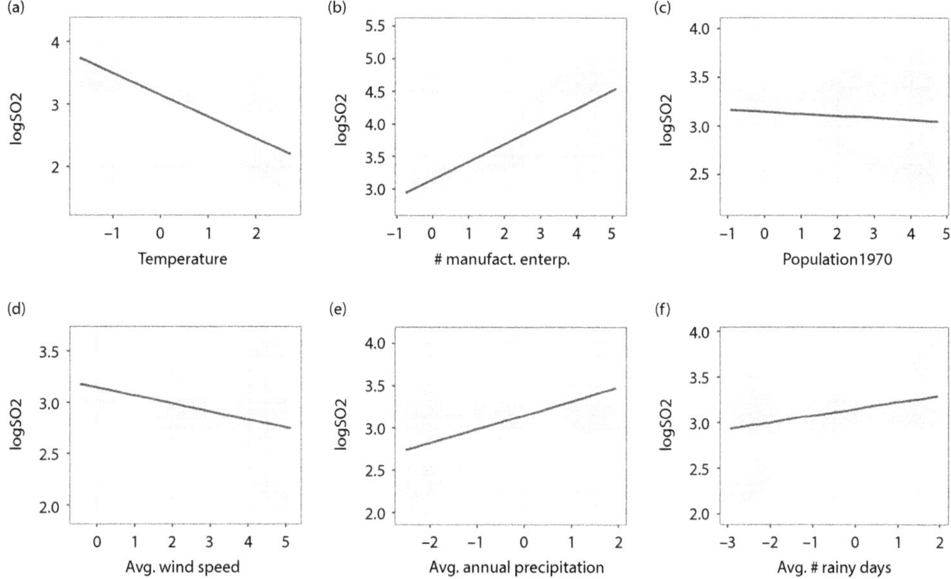

Fig. 4.23 Conditional plots for each explanatory variable predicted by the Bayesian multiple regression model `m2.brms` while holding all other explanatory variables at zero. The points are the observed values, the line is the marginal prediction, and the gray ribbon is its 95% credible interval.

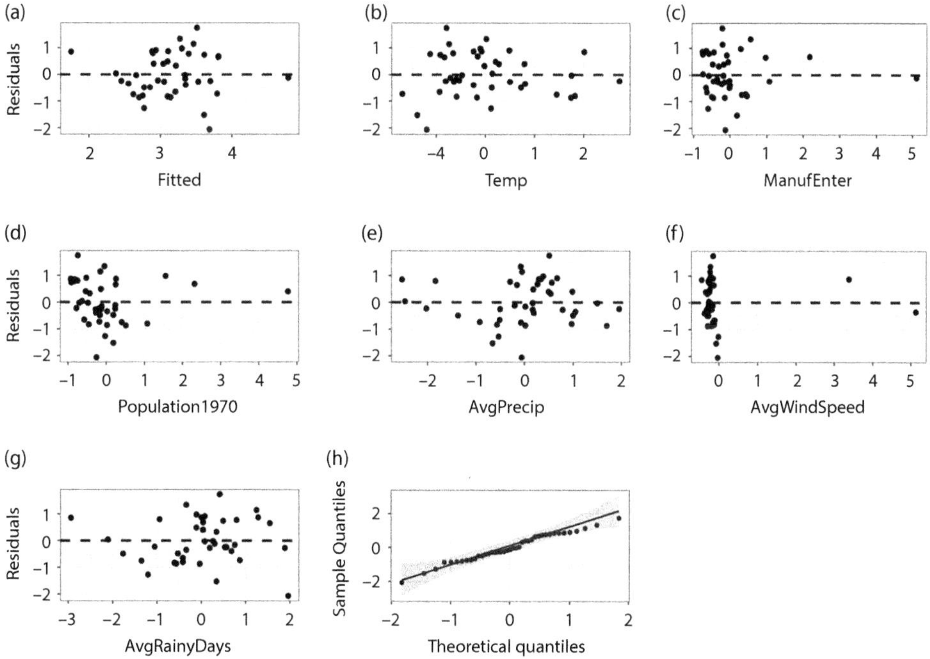

Fig. 4.24 Validation of the Bayesian multiple regression model `m2.brms` by residual analysis, showing the mean of the Pearson residuals vs. the fitted values (A), the mean of the Pearson residuals vs. each standardized explanatory variable (B to G), and the quantile–quantile plot of the mean of the Pearson residuals (H).

4.13 Problems

Visit the companion website at www.oup.com/companion/InchaustiSMWR to obtain the data sets for these problems.

4.1 Huel et al. (2008)'s goal was to determine the relationship between Hg exposure (Hg_hair, μg/g) in mothers and the erythrocyte Ca pump activity (act_Ca, nmol/mg/hr) in the umbilical cord at delivery of their newborns. File: Pr4-1.csv

4.2 Loyn (1987) wanted to know if the average number of forest birds (abundance, calculated from several independent 20 min counting sessions) in forest patches in southeastern Australia was related to the landscape structure. He quantified the following explanatory variables: patch area (patch.area, ha), distance to the nearest forest patch (dist.nearest, km), distance to the nearest forest patch that is larger than the current patch (dist.larger, km), altitude (m), time since fragmentation (yrs.isolation). File: Pr4-2.csv

References

Aldrich, J. (2005). Fisher and regression. *Statistical Science*, 20, 401–417.

Bürkner, P.-C. (2017). brms: Bayesian regression models using Stan. R package. *Journal of Statistical Software*, 80, 1–28.

Cohen, J. (1968). Multiple regression as a data-analytic system. *Psychological Bulletin*, 70, 426–443.

Cook, R. and Weisberg, S. (1982). *Residuals and Influence in Regression*. Chapman and Hall, New York.

Dormann, C., Elith, J., Bacher, S., et al. (2013). Collinearity, a review of methods to deal with it and a simulation study evaluating their performance. *Ecography*, 36, 27–46.

Faraway, J. (2016). *Extending the Linear Model with R: Generalized Linear, Mixed Effects and Nonparametric Regression Models*. CRC Press / Chapman and Hall, New York.

Fisher, R. (1922). On the mathematical foundations of theoretical statistics. *Proceedings of the Royal Society A*, 222, 309–368.

Gelman, A., Goodrich, B., Gabry, J. et al. (2018). R-squared for Bayesian regression models. *American Statistician*, 73, 307–309.

Gelman, A. Jakulin, A., Pittau, M. et al. (2006). A weakly informative default prior distribution for logistic and other regression models. *Annals of Statistics*, 2, 1360–1383.

Hoffman, M. and Gelman, A. (2014). The No-U-Turn sampler: Adaptively setting path lengths in Hamiltonian Monte Carlo. *Journal of Machine Learning Research*, 15, 1351–1381.

Huel, G., Sahuquillo, J., Debotte, G. et al. (2008). Hair mercury negatively correlates with calcium pump activity in human term newborns and their mothers at deliver. *Environmental Health Perspectives*, 118, 263–267.

Jennings, E. (1967). Fixed effect analysis of variance by regression analysis. *Multivariate Behavioral Research*, 2, 95–108.

Limpert, E. Stahel, W., and Abbt, M. (2001). Lognormal distributions across the sciences: Keys and clues. *Bioscience*, 51, 341–352.

Liu, Y., Kai Wang, K., Peixian, X., et al. (2012). Physiological responses and tolerance threshold to cadmium contamination in Eremochloa ophiuroides. *International Journal of Phytoremediation*, 14, 467–480.

Loyn, R. (1987). Effects of patch area and habitat on bird abundances, species numbers and tree health in fragmented Victorian forests. In D. Saunders, G. Arnold, A. Burbridge, et al. (eds), *Nature Conservation: The Role of Remnants of Native Vegetation*, pp. 65–77. Surrey Beatty and Sons, Chipping Norton, Australia.

Matloff, N. (2016). *Parallel Computing for Data Science with Examples in R, C++ and CUDA*. CRC Press / Chapman and Hall, New York.

Schielzeth, H. (2012). Simple means to improve the interpretability of regression coefficients. *Methods in Ecology and Evolution*, 1, 103–113.
Sokal, R. and Rohlf, F. (1995). *Biometry*, 3rd edn. W. H. Freeman, New York.
Stan Development Team (2017). The Stan C++ library, version 2.17.0. Available at: https://mc-stan.org/.
Stan Development Team (2020). RStan, the R interface to Stan. R package version 2.21.2.
Stigler, S. (1981). Gauss and the invention of least squares. *Annals of Statistics*, 9, 465–474.
Sutton, J. (1997). Gibrat's legacy. *Journal of Economic Literature*, 35, 40–59.

CHAPTER 5

The General Linear Model II

Categorical explanatory variables

Packages needed in this chapter:

```
packages.needed <- c("ggplot2","bayesplot","fitdistrplus","reshape2",
"qqplotr","ggeffects","GGally","broom","qqplotr","brms",
"bayesplot","doBy","ggmcmc","multcomp")
lapply(packages.needed, FUN = require, character.only = T) # loads these
packages in the working session.
```

5.1 Introduction

This chapter will extend the introduction to the general linear model to include categorical explanatory variables. These categorical variables or factors divide the data into several non-overlapping parts. By comparing the parameters (means, variances, etc.) among the groups defined by these variables, we would account for, and in a sense explain, part of the variation of the response variable.

Contrasting the parameter estimates between groups or categories denoting different conditions is one of the fundamental aspects of experimental design. The groups being compared can be defined by just one categorical variable, or by combinations of two or more. The general aim is that by estimating sets of differences between group means, we can ascertain the importance of categorical explanatory variables (and their interactions).

After covering the simplest possible case of a categorical explanatory variable of two levels (the *t*-test), we proceed to its generalization for more than two levels (one-way analysis of variance). We finish with the important topic of multiple comparisons. The even more general cases of having two (or more) categorical explanatory variables (factorial analysis of variance) and both numerical and categorical explanatory variables (analysis of covariance) will be covered in Chapter 6.

5.2. Student's *t*-test: One categorical explanatory variable with two groups

At this level, we all know how to test whether the means of two groups are different. The *t*-test to compare two means requires that the response variable be normally distributed,

Statistical Modeling With R. Pablo Inchausti, Oxford University Press. © Pablo Inchausti (2023).
DOI: 10.1093/oso/9780192859013.003.0005

$Y \sim \text{Normal}(\mu_Y, \sigma_Y^2)$. Let's recall that we can center a normally distributed variable Y at zero by subtracting its mean μ_Y, such that the difference $\varepsilon = (Y - \mu_Y)$ would be ~ Normal(0, σ_Y^2).

The statistical model of the t-test is

$$Y = \mu_{Y,j} + \varepsilon, \tag{5.1}$$

where $\mu_{Y,j}$ is the mean of the response variable for the group j (j = 1,2). $\varepsilon = (Y - \mu_Y)$ denotes the random and normally distributed differences between each value of the response variable and its mean, whose scatter is determined by the variance of the response variable, σ_Y^2. Equation (5.1) is saying that the mean, but not the variance, of the response variable may differ between the two groups. We could also view it as the mean of the response variable μ_Y being a function of a categorical explanatory variable μ_Y (factor) with two levels. Equation (5.1) is often rewritten as

$$Y = (\mu_Y + \alpha_j) + \varepsilon, \tag{5.2}$$

where μ_Y is the overall mean of the response variable and $\alpha_j = \mu_{Y,j} - \mu_Y$ denotes the difference between the mean of the jth group and the overall mean of the response variable. The α_j are the main parameters of interest as they reflect the magnitude of the differences between the group means and the overall mean. Given that there are two groups, the statistical model of Eq. (5.2) would a priori have four parameters: μ_Y, α_1, α_2, and σ_Y^2. However, the latter is not actually true, and it is best understood through a fictitious and simple example with a few data points.

Let's say that we have a total of six samples of a response variable, three of which belong to one of two groups defined by a categorical explanatory variable (say, sex) that is denoted as 0 and 1, as in Table 5.1. Using the sample means of each group to estimate their population means, and given that $\alpha_j = \mu_{Y,j} - \mu_Y$, then $\alpha_1 = 3.6 - 5.8 = -2.2$ and $\alpha_2 = 8.0 - 5.8 = +2.2$. Therefore, $-\alpha_1 = \alpha_2$ or $\sum \alpha_j = 0$, and α_1 and α_2 are not independent parameters. The statistical model for the t-test, Eq. (5.1), then has only three independent parameters: μ_Y, α_1 (or α_2), and σ_Y^2. This seemingly trivial feature will become important in understanding how R solves the t-test as a single linear regression, and to interpret the R output of a general linear model involving categorical explanatory variables.

Table 5.1 Example data samples for a response variable, three of which belong to one of two groups defined by a categorical explanatory variable that is denoted as 0 and 1.

Explanatory variable	Y	Means
0	3.4	
0	4.1	3.6
0	3.2	
1	6.8	
1	7.2	8.0
1	9.9	
Overall mean		5.8

We can rewrite Eq. (5.2) yet again in the form Y = Xβ + ε:

$$\begin{pmatrix} 3.4 \\ 4.1 \\ 3.2 \\ 6.8 \\ 7.2 \\ 9.9 \end{pmatrix} = \begin{pmatrix} 1 & 1 & 0 \\ 1 & 1 & 0 \\ 1 & 1 & 0 \\ 1 & 0 & 1 \\ 1 & 0 & 1 \\ 1 & 0 & 1 \end{pmatrix} \begin{pmatrix} \mu_Y \\ \alpha_1 \\ \alpha_2 \end{pmatrix} + \begin{pmatrix} \varepsilon_1 \\ \varepsilon_2 \\ \varepsilon_3 \\ \varepsilon_4 \\ \varepsilon_5 \\ \varepsilon_6 \end{pmatrix}, \text{ or } \begin{aligned} 3.4 &= \mu_Y + \alpha_1 \quad + \varepsilon_1 \\ 4.1 &= \mu_Y + \alpha_1 \quad + \varepsilon_2 \\ 3.2 &= \mu_Y + \alpha_1 \quad + \varepsilon_3 \\ 6.8 &= \mu_Y + \quad \alpha_2 + \varepsilon_4 \\ 7.2 &= \mu_Y + \quad \alpha_2 + \varepsilon_5 \\ 9.9 &= \mu_Y + \quad \alpha_2 + \varepsilon_6 \end{aligned},$$

$$\text{which leads to: } \begin{aligned} 3.4 &= \mu_{Y,0} + \varepsilon_1 \\ 4.1 &= \mu_{Y,0} + \varepsilon_2 \\ 3.2 &= \mu_{Y,0} + \varepsilon_3 \\ 6.8 &= \mu_{Y,1} + \varepsilon_4 \\ 7.2 &= \mu_{Y,1} + \varepsilon_5 \\ 9.9 &= \mu_{Y,1} + \varepsilon_6 \end{aligned}$$

In these equations X is called the design matrix. This binary matrix automatically written by R denotes the potential influence of each level of the explanatory categorical variable on the mean of the response variable. Each data point is associated with one of the two levels of the categorical explanatory variable, as indicated in the second and third columns of X. The effect of this association is reflected in the mean of the response variable $\mu_{Y,\,j}$ for each group.

The design matrix X has a repetitive structure, with each row being repeated as many times as there are samples for each group (three in the above example). We could summarize the design matrix and the vector of parameters as $\begin{pmatrix} 1 & 1 & 0 \\ 1 & 0 & 1 \end{pmatrix} \begin{pmatrix} \mu_Y \\ \alpha_1 \\ \alpha_2 \end{pmatrix}$. Now we have two equations and three unknowns (parameters) to estimate, an over-parameterized linear system with more equations than parameters that has infinite solutions. The alert reader might have noticed that the first column of the design matrix X is actually the sum (i.e., a linear combination) of the other columns. Thus, the "summarized design matrix" cannot be inverted to yield a unique solution in terms of the parameter vector β. Solving this problem requires constraining this over-parameterized system by setting one of the model parameters to zero, thereby deleting the corresponding column of the design matrix.

A useful way to constrain the over-parameterized design matrix involves setting $\alpha_1 = 0$ (we could have set $\alpha_2 = 0$ with identical results) and consequently deleting the second column of this design matrix, and redefining α_2. Now, with $\alpha_1 = 0$, $\mu_{Y,1} = \mu_{Y,\text{REF}}$ is now called the reference group or level of the categorical explanatory variable. The newly defined term $\alpha_2^* = \mu_{Y,2} - \mu_{Y,\text{REF}}$ corresponds to the magnitude of the difference between the means of groups 1 and 2. This is precisely what we are aiming to estimate, and Eq. (5.2) now becomes

$$Y = (\mu_{Y,\text{REF}} + \alpha_2^*) + \varepsilon. \tag{5.3}$$

Equation (5.3) now has three parameters: $\mu_{Y,\text{REF}}$, α_2^* (with which we can obtain both $\mu_{Y,1}$ and μ_Y), and σ_Y^2. This way of constraining the design matrix is the default option used in R and it is known as "treatment contrasts." There are other ways of constraining the design matrix that will not be discussed here. Table 4.1 show the R syntax for linear statistical models.

THE DATA IN CONTEXT: Young et al. (2004) investigated whether the length of the horn of the horned lizard differed between those that were killed by a bird predator (the loggerhead shrike) and those that were still alive. They took advantage of the gruesome but convenient predator behavior that skewers its victims on thorns or barbed wire for later eating. They measured the horns in the remains of 30 lizards killed by this avian predator, and in 155 living lizards.

EXPLORATORY DATA ANALYSIS: Let's first import and examine the data.

```
> DF=read.csv("Ch 05 horned.lizards.csv",header=T)
> str(DF)
'data.frame':   185 obs. of  2 variables:
 $ HornLength: num  25.2 26.9 26.6 25.6 25.7 25.9 27.3 25.1 30.3 25.6 ...
 $ Survival  : Factor w/ 2 levels "killed","living": 2 2 2 2 2 2 2 2 2 2 ...
```

The summary of the imported data gives a first overview:

```
> summary(DF)
   HornLength      Survival
 Min.   :13.1   killed: 30
 1st Qu.:22.4   living:155
 Median :24.1
 Mean   :23.9
 3rd Qu.:25.7
 Max.   :30.3
 NA's   :1
```

We will exclude the single missing value of `HornLength` by `DF = na.omit(DF)`. This command will delete all rows from a data frame having at least one missing (`NA`) value for any variable. This is a problem of just two variables: `HornLength` and `Survival`. Are they related?

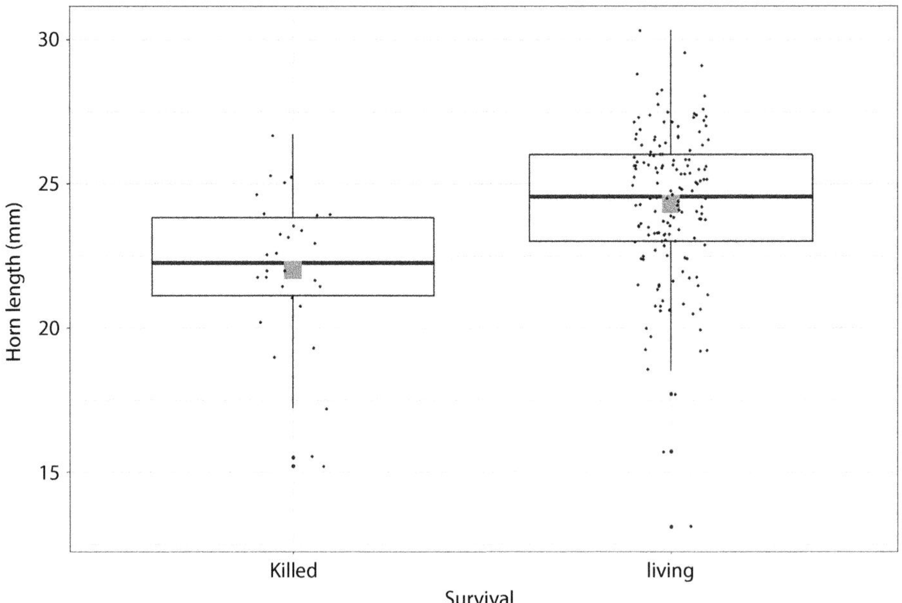

Fig. 5.1 Boxplot showing the relationship between horn length and lizard survival. The gray squares indicate the means for each group.

Yes, they are (Fig. 5.1). We see that both the median and mean horn length of living lizards were higher than those of killed ones, and that their variability (represented by the interquartile distance shown as the box widths) were similar. It is useful to overlay the data points to provide an idea of both the number of points and the actual variation in each group.

Given the categorical explanatory variable `Survival`, the exploratory analysis should include the descriptive statistics for each group using the package `doBy` to apply the function `desc.stats`:

```
> desc.stats=function(x){c(mean=mean(x), median=median(x),sd=sd(x), n=length(x))}
> summaryBy(HornLength~Survival, data=DF, FUN=desc.stats)
  Survival HornLength.mean HornLength.median HornLength.sd HornLength.n
1   killed            22.0              22.2          2.71           30
2   living            24.3              24.6          2.63          154
```

There is a 10.5% = [100 × (24.3 – 22.0) / 22.0] difference in the average horn length between killed and living lizards (Fig. 5.1).

We need to determine the probability distribution that best fits the response variable in order to define the likelihood function to be used in fitting the statistical model. We evaluate the goodness of fit of the normal and lognormal distributions using the package `fitdistrplus`:

```
Norm=fitdist(DF$HornLength, "norm")
Lognorm=fitdist(DF$HornLength,"lnorm")
```

to obtain the cumulative distribution function and quantile–quantile plots (the explanation of the CDF and Q–Q plots was given in Chapter 4). While neither probability distribution fitted the tails of the distribution of the response variable particularly well, the normal distribution seems to have a slightly better fit (Fig. 5.2).

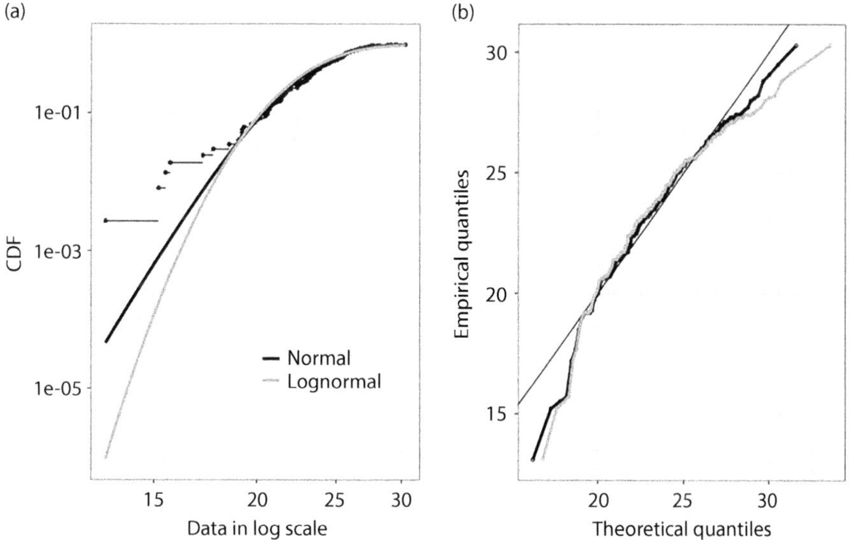

Fig. 5.2 Goodness of fit assessment of the response variable horn length showing (a) the cumulative distribution function and (b) the quantile–quantile plot obtained using the package `fitdistrplus`.

We can visualize (a shortened version of) the model matrix that will be used to fit the statistical model:

```
> unique model.matrix(DF$HornLength~DF$Survival)
    (Intercept) DF$Survivalliving
              1                 0
              1                 1
```

We can also see the label `DF$Survivalliving`, indicating that the level `killed` was used as the reference group whose mean will appear under the heading `(Intercept)`. What we are saying is that α_2^* in Eq. (5.2) will correspond to the difference between the mean horn length of the "killed" group (the reference group) and the "living." Why was "killed" chosen as the reference group? It is just because R by default uses an alphabetical criterion to define the reference group; this could be changed with `relevel(DF$Survival, ref = "living")`.

5.3. The *t*-test: Frequentist fitting

Let's finally fit the statistical model `m1 = lm(HornLength ~ Survival, data = DF)` and look at its summary:

```
> summary(m1)

Call:
lm(formula = HornLength ~ Survival, data = DF)

Residuals:
    Min      1Q  Median      3Q     Max
-11.181  -1.281   0.269   1.719   6.019

Coefficients:
               Estimate Std. Error t value Pr(>|t|)
(Intercept)      21.987      0.483   45.56  < 2e-16 ***
Survivalliving    2.295      0.528    4.35  2.3e-05 ***
---
Signif. codes:  0 '***' 0.001 '**' 0.01 '*' 0.05 '.' 0.1 ' '

Residual standard error: 2.64 on 182 degrees of freedom
Multiple R-squared:  0.0942,	Adjusted R-squared:  0.0892
F-statistic: 18.9 on 1 and 182 DF,  p-value: 2.27e-05
```

The estimated average of the response variable for the reference group (killed) shown as `(Intercept)` is 21.987 mm with standard error 0.483 mm, and the `t value` is the ratio estimate/SE, which is hugely different from zero. The latter is completely irrelevant in the context of the problem being solved, which is to find out whether the mean horn length differs between killed and living lizards. The second parameter, `Survivalliving`, is the α_2^* in Eq. (5.3) that corresponds to the difference between the means of the reference (killed) and the living groups. The average horn length of the living group was 2.295 mm longer than that of the reference group, 0.528 is the standard error of the difference of means, and the `t value` is again their ratio (4.35), which is statistically different from zero. The *t*-statistic $T = \frac{(\overline{Y}_{\text{killed}} - \overline{Y}_{\text{living}})}{\text{SE}(\overline{Y}_{\text{killed}} - \overline{Y}_{\text{living}})}$ is precisely the 4.35 just obtained. Is the difference in means of 2.295 / 21.987 ~ 10% relevant, interesting, or surprising? This should not be decided on the basis of the associated *p*-value (2.27×10^{-5}), but by reinterpreting this magnitude in the research context that motivated the gathering of data. The mean horn lengths of the killed and living groups were 21.987 and 21.987 + 2.295 = 24.3 mm. The squared residual standard error ($2.35^2 = 5.52$ mm^2) is the estimate of the variance of the response variable. The statistical model fitted explains 8.92% of the variation of horn length. This is a very small adjusted R^2, but what could we realistically expect in a statistical model with just a binary explanatory variable?

What R has done is precisely fit a single linear regression of `HornLength` on the binary explanatory variable `Survival`. In so doing, it transformed the *t*-test for comparing two means of independent samples into a simpler problem of simple linear regression that had previously been solved. Older textbooks call it "regression on a dummy variable." The estimates of the parameters of the regression on a dummy variable by maximum likelihood have a one-to-one correspondence with the parameters of the constrained statistical model, Eq. (5.3).

We can obtain a tidy table of the parameter estimates with the package `broom`,

```
> tidy(m1, conf.int = T)
# A tibble: 2 x 7
  term            estimate std.error statistic   p.value conf.low conf.high
  <chr>              <dbl>     <dbl>     <dbl>     <dbl>    <dbl>     <dbl>
1 (Intercept)        22.0      0.483     45.6  1.90e-101     21.0      22.9
2 Survivalliving      2.29     0.528      4.35 2.27e-  5      1.25      3.34
```

and also collate in `res.m1` what we need for the residual analysis: `res.m1 = augment(m1)`.

The fit of the statistical model to the data is not very good. The range of the positive residuals (two units) was smaller than that of the negative residuals (four units), whereas they should have been similar had the model fitted well (Fig. 5.3a). The Q–Q plot (Fig. 5.3c) showed that a fair number of points were not contained in the 95% envelope of the perfect fit at the lower tail. The typical (median) residuals were similar for both groups, and the range of residuals was larger for the living group (Fig. 5.3c). And finally, the Cook distances (Fig. 5.3d) showed five points (5 / 184 $\approx$ 2%) with a potentially large influence on the estimated parameters.

The points with large (arbitrarily using 0.04 as the threshold) Cook distances were:

```
> which(res.m1$.cooksd> 0.04)
[1] 142 166 175 179 180
```

and those having "potentially problematic residuals" (using –2 as a threshold) were:

```
> which(res.m1$.std.resid < -2)
[1] 103 142 153 154 175 180
```

Unsurprisingly, there are three coincident data points (142, 175, and 180) between these two sets. This is *NOT* to say that these three points are outliers that could or should be thrown away for the sake of attaining a "better model fit." Let's refit model `m1` without these three points to assess whether they have a detectable influence on the estimated parameters:

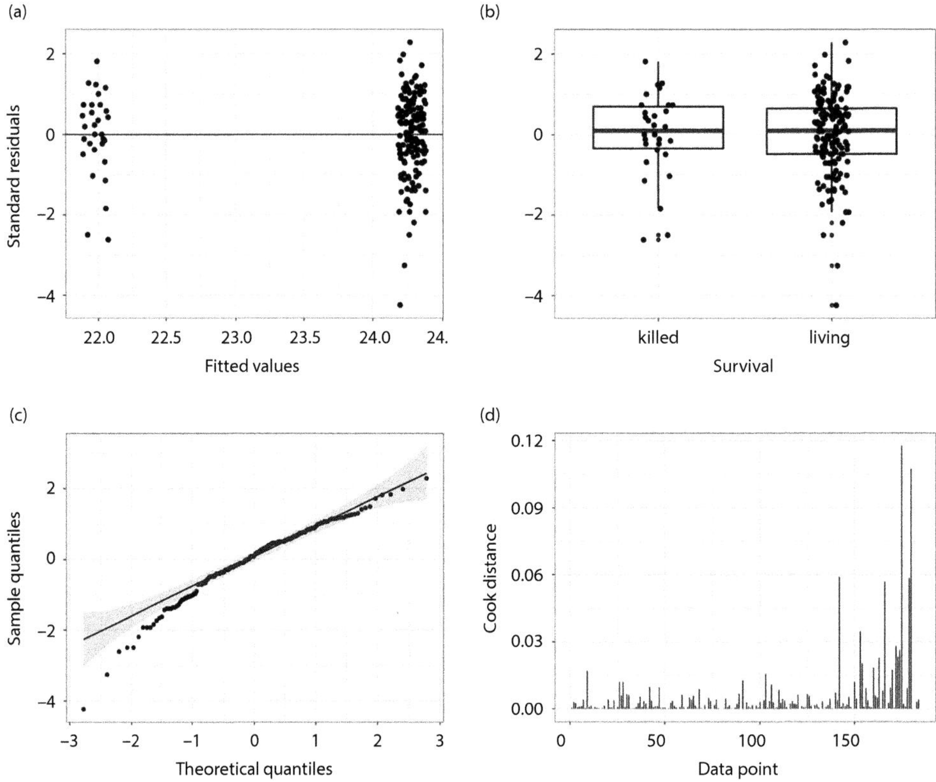

Fig. 5.3 Residual analysis of model `m1`, showing the residuals vs. fitted values (a), a boxplot of the residuals vs. the categorical explanatory variable (b), the quantile–quantile plot of residuals (c), and the Cook distances (d).

```
> m1.1=lm(HornLength~Survival, data=DF[-c(142,175,180),])
> summary(m1.1)

Call:
lm(formula = HornLength ~ Survival, data = DF[-c(142, 175, 1
    ])

Residuals:
   Min     1Q Median     3Q    Max
-8.654 -1.354  0.146  1.546  5.946

Coefficients:
                Estimate Std. Error t value Pr(>|t|)
(Intercept)       22.461      0.458    49.0   <2e-16 ***
Survivalliving     1.894      0.498     3.8    2e-04 ***
---
Signif. codes:  0 '***' 0.001 '**' 0.01 '*' 0.05 '.' 0.1 ' '

Residual standard error: 2.42 on 179 degrees of freedom
Multiple R-squared:  0.0747,    Adjusted R-squared:  0.0695
F-statistic: 14.4 on 1 and 179 DF,  p-value: 0.000198
```

There were no important qualitative changes between the two models: the parameters did not change their signs or their statistical significance, so the main conclusions remain the same. However, there were changes in the magnitude of the mean horned length of the "killed" group (21.987 → 22.416: +2.1%), in both the magnitude of the difference between the means (2.295 → 1.894: −17.8%) and in its precision (0.528 → 0.498: +5.6%), and in the residual standard error (2.42 → 2.64: +9.1%). We should definitely double-check the accuracy and veracity of these values that could induce an overestimation of the effect size. When we cannot make such a verification (as in the current case), we should explain the outcome of the residual analysis that led to model `m1.1` and, at the very least, also highlight that the magnitude of the differences in mean horn length between killed and living lizards could be overestimated (+17.8%) because of the undue influence of 3 / 154 ≈ 2% of the available data. (The residual analysis of model `m1.1` was, of course, slightly better than that of model `m1`, but nothing to write home about.)

What are the consequences of using the not-so-well-validated `m1` model for inference? The general linear model is known to be relatively robust to minor violations of its main assumptions, particularly for large sample sizes and nearly balanced data sets (e.g., Winer et al. 1991). That is, we can still trust the main findings (i.e., that there is a significant difference between the means of living and killed lizards) in spite of the problems revealed in the residual analysis. However, the estimates of model parameters and their standard errors might be biased. The only alternatives would be to fit a different statistical model, perhaps using another probability distribution rather than normal, to include other explanatory variables if available, and/or to account for the slight heterogeneity of variance between groups that is suggested in Fig. 5.3.

The conditional plot of model `m1` (or, if preferred, `m1.1`) can be obtained with the package `ggeffects` to produce the object `pred.m1` that can be visualized as in Fig. 5.4.

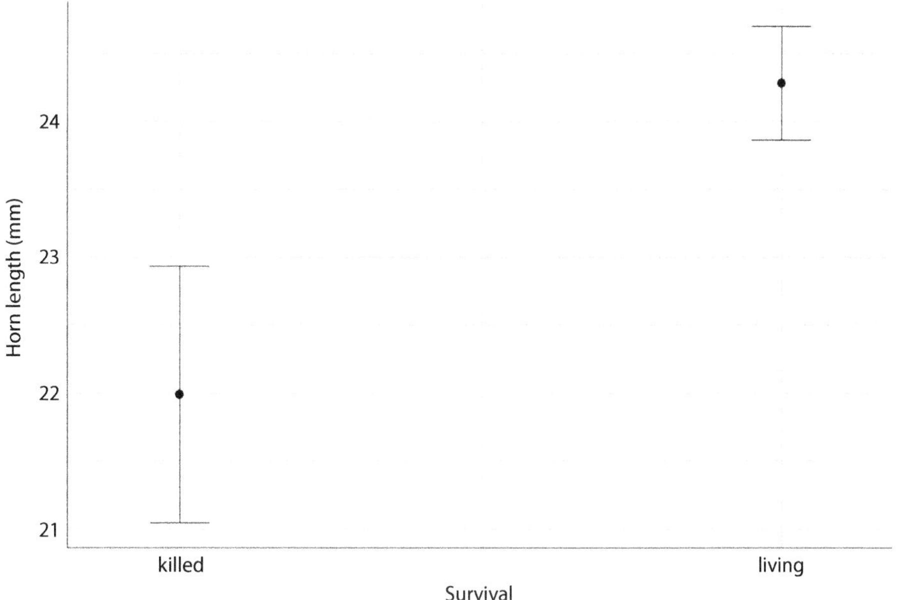

Fig. 5.4 Conditional plot showing the predicted means of the response variable and its 95% confidence interval for model `m1` generated with the `ggeffects` package.

```
> pred.m1
# Predicted values of HornLength

Survival | Predicted |           95% CI
---------------------------------------
killed   |     21.99 | [21.04, 22.93]
living   |     24.28 | [23.86, 24.70]
```

5.4. The *t*-test: Bayesian fitting

We use the package `brms` to fit the statistical model for the Bayesian *t*-test. We start by defining the prior distributions for the three parameters of Eq. (5.3). Let's see the default priors:

```
> get_prior(formula=HornLength~Survival, data=DF,family=gaussian)
                prior     class coef          group resp dpar nlpar bound
1                             b
2                             b Survivalliving
3 student_t(3, 24, 10) Intercept
4  student_t(3, 0, 10)     sigma
```

We need to define a prior for the mean `HornLength` of the killed group (`Intercept`) for the magnitude of the difference of means (`Survivalliving`), and for the standard deviation of the response variable (`sigma`). Before seeing the fit, we might anticipate that the mean `HornLength` could range between 10 and 30 mm, and thus a normal distribution with mean = 20 and SD = 5 would lead to an interval [10, 30] enclosing 95% of the possible values of `HornLength` for the reference group (killed). For the magnitude of the difference between means, a normal prior with mean = 0 and SD = 4 would define a 95% interval of [–8, 8], such that the mean of the living group would lie in [12, 28]. And for the standard deviation of `HornLength`, we might choose a weakly informative half-Cauchy distribution. It is always very helpful to plot the prior distributions to verify whether the degree of support of these probability distributions is consistent with our previous knowledge of each parameter (see Fig. 5.5).

There is nothing "canonical" or "fundamental" about using the normal and half-Cauchy distributions for the priors. Any probability distribution would do the job, as long as it prevents impossible parameter values such as negative mean `HornLength` or negative standard deviations. We use the normal distribution (Fig. 5.2) as a reasonable approximation for the likelihood of the Bayesian fitting equation.

The priors of the Bayesian *t*-test are set by:

```
> prior.m2 = c(set_prior("normal(20, 5)", class = "Intercept"),
+              set_prior("normal(0, 4)", class = "b"),
+              set_prior("cauchy(0,5)", class = "sigma"))
```

and the statistical model is run with:

```
m2.brms=brm(formula=HornLength~Survival,  data=DF,  family=gaussian,  prior  =
  prior.m2, warmup = 1000, future=TRUE, chains=3, iter=2000, thin=3)
```

(we refer you to the explanations of these options given in Section 4.5). The model summary is:

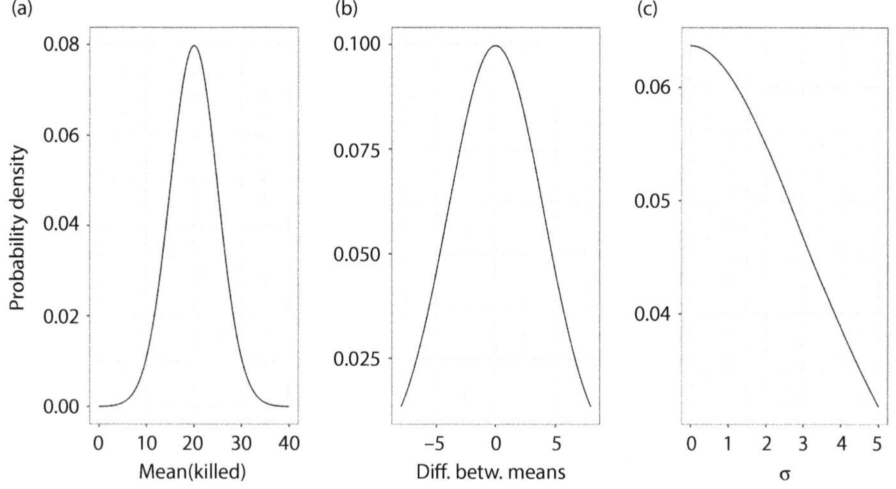

Fig. 5.5 Prior distributions for the three parameters of the Bayesian *t*-test `HornLength ~ Survival`: mean of the reference group (a), magnitude of the mean difference (b), and standard deviation of the response variable (c).

```
> summary(m2.brms)
 Family: gaussian
  Links: mu = identity; sigma = identity
Formula: HornLength ~ Survival
   Data: DF (Number of observations: 184)
Samples: 3 chains, each with iter = 2000; warmup = 1000; thin = 3;
         total post-warmup samples = 1000

Population-Level Effects:
               Estimate Est.Error l-95% CI u-95% CI Rhat Bulk_ESS Tail_ESS
Intercept         22.03      0.48    21.06    22.92 1.00      913      917
Survivalliving     2.24      0.51     1.28     3.27 1.00      987      821

Family Specific Parameters:
      Estimate Est.Error l-95% CI u-95% CI Rhat Bulk_ESS Tail_ESS
sigma     2.66      0.14     2.40     2.96 1.00      901      994
```

The model summary gives the mean, the standard deviation, and the 2.5% and 97.5% quantiles of the posterior distributions of the three model parameters under the headings `Estimate`, `Est.Error`, `l-95% CI`, and `u-95% CI`, respectively. The `l-95% CI` and `u-95% CI` values are the limits of the 95% credible intervals of each model parameter. There are also three metrics (`Rhat`, `Bulk_ESS`, and `Tail_ESS`) related to the quality of the numerical model fitting. `Rhat` being close to one indicates that the three chains have converged to the same stationary distribution. The other metrics (`Bulk_ESS` and `Tail_ESS`) are estimates of the number of statistically independent estimates of each parameter in the three chains. The three chains were 2,000 iterations long and we chose `thin = 3` to minimize the autocorrelation of the sampled chain values (see Chapter 4), and hence we will shortly see that we have 2,000 / 3 $\approx$ 668 parameter estimates per chain. Putting together the 668 parameter estimates of the three chains (which would do if the changes actually converged to the same stationary distribution), we would then have 2,000 estimates per parameter. The `Bulk_ESS` and `Tail_ESS` metrics show how many of these pooled 2,000 estimates can be considered statistically independent. While the closer to 2,000 the better, as a rule of thumb `Bulk_ESS` and `Tail_ESS` should not be too far from 1,000.

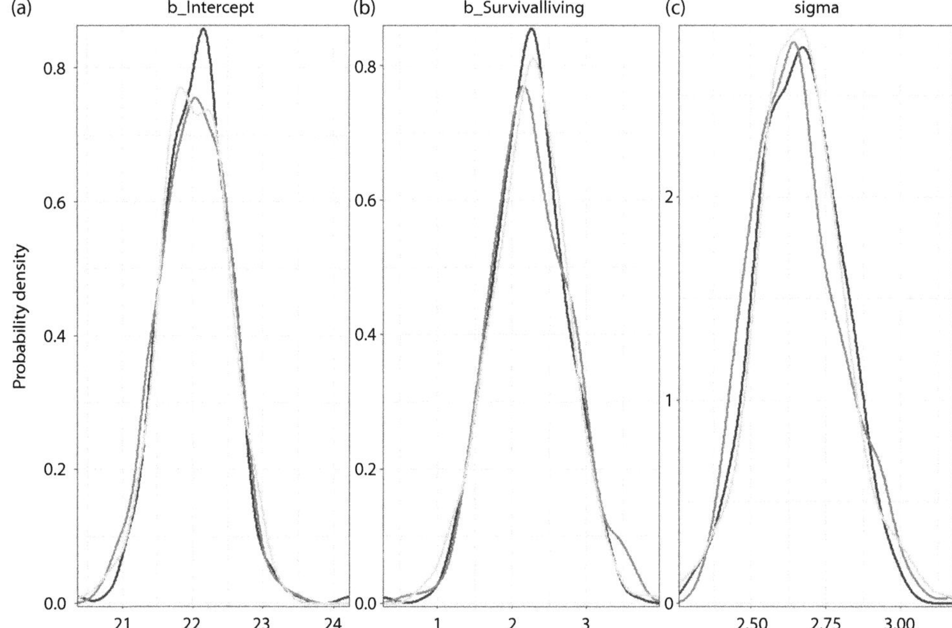

Fig. 5.6 Marginal posterior distributions of the model parameters for the three chains of the Bayesian *t*-test `m2.brms`: mean of the reference group (a), magnitude of the mean difference (b), and standard deviation of the response variable (c).

We can plot and compare the marginal posterior distribution of model parameter across chains, as in Fig. 5.6. The conditional posterior distributions of all parameters for the three chains were indeed very similar. The difference between the prior (Fig. 5.5) and the posterior distributions (Fig. 5.6) is due to the information contained in the data that entered into the estimation via the likelihood.

The autocorrelation plots (Fig. 5.7) show that the successive parameter estimates of each chain can be considered to be statistically independent. So far so good with model `m2.brms`.

After validating the numerical fit of the model `m2.brms`, we can display (Fig. 5.8) the fitted parameters and their credible intervals obtained by merging the three chains that attained a common posterior stationary distribution. Rather than making a single plot, in this case we need to make two separate plots because of the disparate magnitudes of the parameters (see Fig. 5.6).

The fitted model `m2.brms` explained a modest amount of the variation of the response variable using Gelman et al. (2018)'s goodness of fit metric (see Chapter 4), which is similar to the adjusted R^2 obtained in the frequentist fitting of the same model.

```
> bayes_R2(m2.brms)
   Estimate Est.Error   Q2.5 Q97.5
R2   0.0926    0.0364 0.0305 0.172
```

We now need to assess the goodness of fit of the statistical model `m2.brms` to the data through residual analysis. We first need to calculate and store the Pearson residuals (see Chapter 4) with `res.m2.brms = as.data.frame(residuals(m2.brms, type = "pearson", nsamples = 1000, summary = T))`. Then `res.m2.brms` will contain

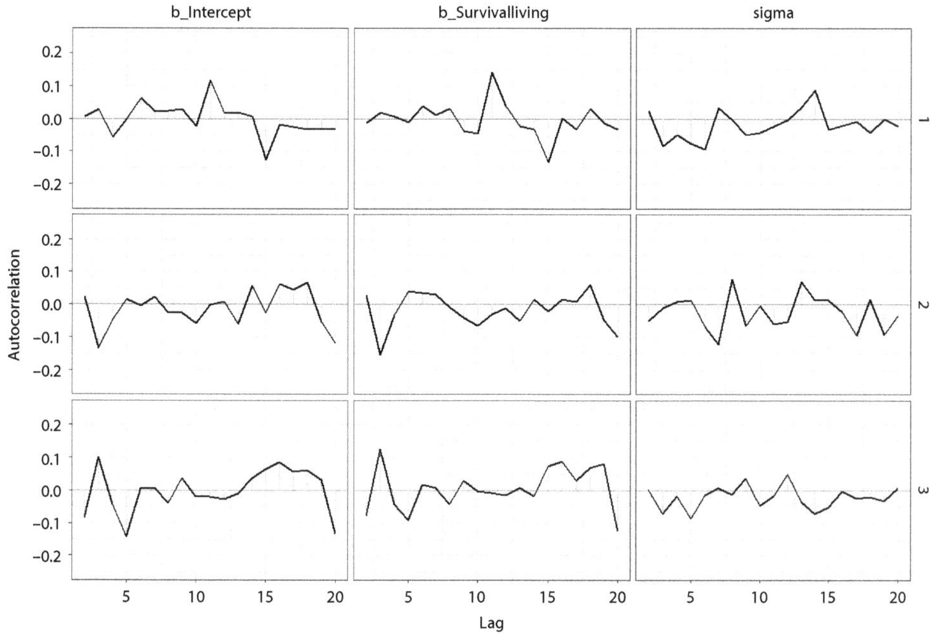

Fig. 5.7 Autocorrelation plots of the model `m2.brms` for the three chains. The scale limit on the *x*-axis was set to avoid showing the autocorrelation of lag = 0 (the correlation of a value with itself) that is equal to one by definition.

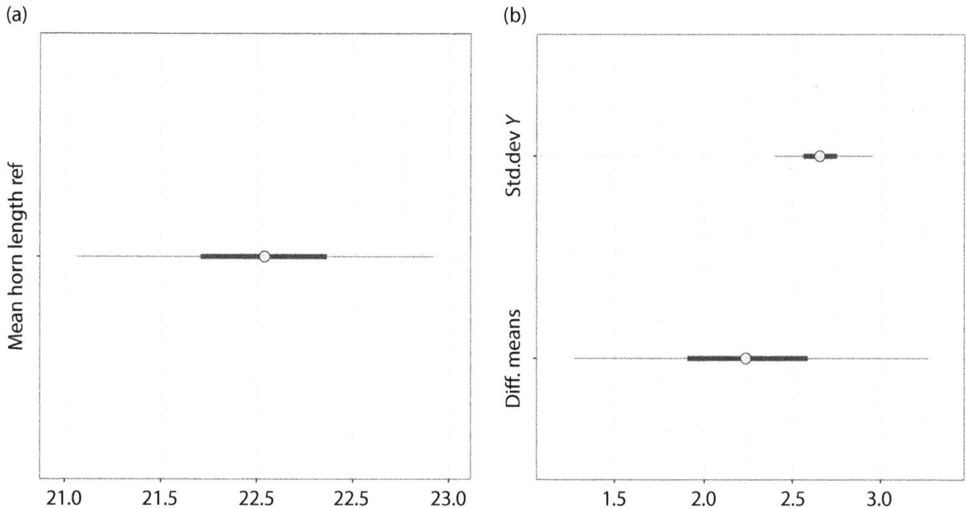

Fig. 5.8 Mean and 80% (thick) and 95% (thin) credible intervals of the three model parameters of model `m2.brms`: mean horn length of the reference group (killed; a); the difference of means between the reference group and living, and the standard deviation of the response variable horn length (b).

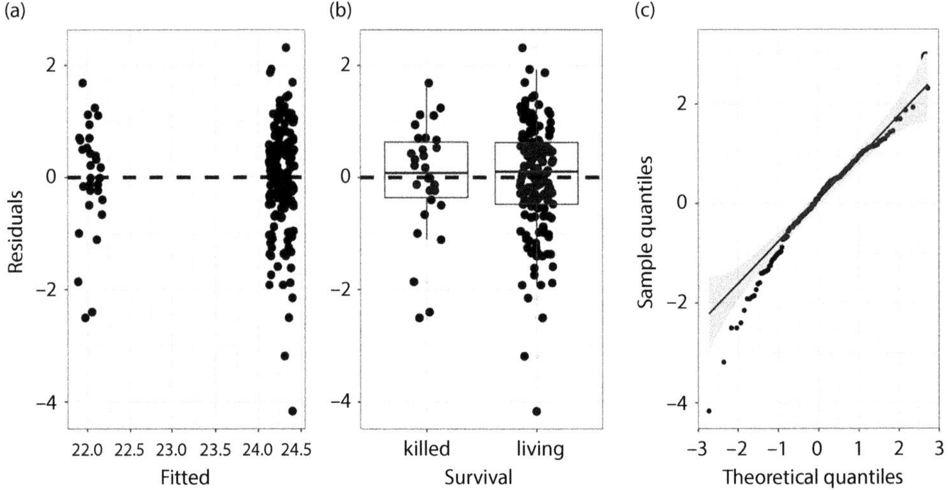

Fig. 5.9 Residual analysis of model `m2.brms`, showing the mean of the Pearson residuals vs. mean of the fitted value (a), mean of the Pearson residuals vs. the explanatory variable in the model (b), and a quantile–quantile plot of the mean of the Pearson residuals (c).

the mean, standard error, and the lower and upper limits for the 95% credible intervals of the Pearson residuals. Likewise, we can obtain the same statistics for the fitted or predicted values of the response variable of model `m2.brms` with `fit.m2.brms = as.data.frame(fitted(m2.brms, scale = "linear", summary = T, nsamples = 1000))`. For ease of manipulation when producing the residual analysis plots, we put the means of the Pearson residuals and of the fitted value in the original data frame: `DF$fit = fit.m2.brms$Estimate` and `DF$resid = res.m2.brms$Estimate`.

The fit of the model `m2.brms` (Fig. 5.9) was almost identical to the one obtained from the frequentist fitting (Fig. 5.3). Given that the very same deficiencies and comments apply, they will not be repeated here. After validating (or at least provisionally accepting) the model `m2.brms`, we can display its main result (Fig. 5.10).

Doing all this work to make a simple comparison of two means is akin to killing a mosquito with a flamethrower. But the point here was to highlight the statistical model underlying the *t*-test as a means of introducing the analysis of variance.

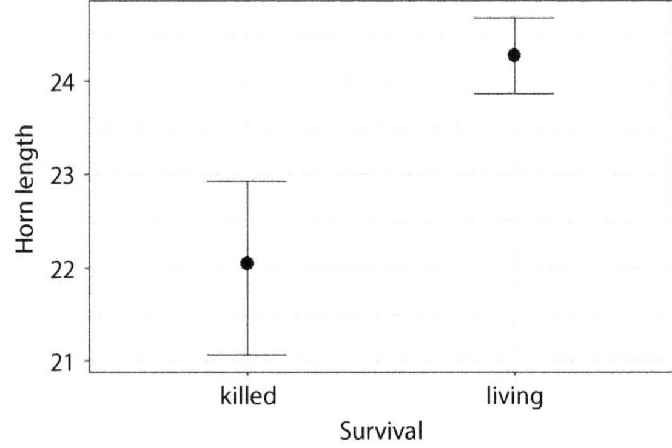

Fig. 5.10 Predicted mean horn length and their 95% credible intervals from model `m2.brms`.

5.5. Viewing one-way analysis of variance as a multiple regression

One-way or single-factor analysis of variance was introduced by Fisher (1921) with the mundane example of comparing the mean potato yields between different applications of manure. One-way analysis of variance allows assessing whether the means of two or more groups defined by a single categorical variable differ from each other. We can view the *t*-test as a special case of one-way analysis of variance, or the latter as a generalization of the former. One-way analysis of variance is another special case of the general linear model, and hence it entails the same assumptions discussed in Section 4.1.

The statistical model of one-way analysis of variance is the same as for the *t*-test (Eq. (5.1)), $Y = (\mu_Y + \alpha_j) + \varepsilon$, with the only difference that the explanatory categorical variable might have two or more groups. Just as for the *t*-test (Section 5.2), it is useful to understand how the linear model is fitted in the case of one-way analysis of variance because it will help us to understand and interpret the outputs of the fitted models.

Let's say that we have the smallest possible data set for a one-way analysis of variance consisting of six samples of a normally distributed response variable (say, the length of something), two of which belong to each of three groups, denoted as A, B, and C, defined by a single categorical explanatory variable (say, Site), as in Table 5.2.

Our basic $Y = (\mu_Y + \alpha j) + \varepsilon$ model would now have the following parameters: μ_Y, α_1, α_2, α_3, and σ_Y^2, where $\alpha_j = \mu_{Y,\,j} - \mu_Y$ as before with j = 1,2,3 denoting the sites A, B, and C. Writing Eq. (5.1) as a linear equation of the form $Y = X\beta + \varepsilon$ for this data set gives

$$\begin{pmatrix} 3.4 \\ 4.1 \\ 3.2 \\ 6.8 \\ 7.2 \\ 9.9 \end{pmatrix} = \begin{pmatrix} 1 & 1 & 0 & 0 \\ 1 & 1 & 0 & 0 \\ 1 & 0 & 1 & 0 \\ 1 & 0 & 1 & 0 \\ 1 & 0 & 0 & 1 \\ 1 & 0 & 0 & 1 \end{pmatrix} \begin{pmatrix} \mu_Y \\ \alpha_1 \\ \alpha_2 \\ \alpha_3 \end{pmatrix} + \begin{pmatrix} \varepsilon_1 \\ \varepsilon_2 \\ \varepsilon_3 \\ \varepsilon_4 \\ \varepsilon_5 \\ \varepsilon_6 \end{pmatrix} \quad \text{or} \quad \begin{aligned} 3.4 &= \mu_Y + \alpha_1 + \varepsilon_1 \\ 4.1 &= \mu_Y + \alpha_1 + \varepsilon_2 \\ 3.2 &= \mu_Y + \alpha_2 + \varepsilon_3 \\ 6.8 &= \mu_Y + \alpha_2 + \varepsilon_4 \\ 7.2 &= \mu_Y + \alpha_3 + \varepsilon_5 \\ 9.9 &= \mu_Y + \alpha_3 + \varepsilon_6 \end{aligned}$$

which leads to

$$\begin{aligned} 3.4 &= \mu_{Y,1} + \varepsilon_1 \\ 4.1 &= \mu_{Y,1} + \varepsilon_2 \\ 3.2 &= \mu_{Y,2} + \varepsilon_3 \\ 6.8 &= \mu_{Y,2} + \varepsilon_4 \\ 7.2 &= \mu_{Y,3} + \varepsilon_5 \\ 9.9 &= \mu_{Y,3} + \varepsilon_6 \end{aligned}$$

Table 5.2 Example data samples for a response variable, three of which belong to one of two groups defined by the categorical explanatory variable Site that is denoted as A, B, and C.

Site	*Y*	Means
A	3.4	3.8
A	4.1	
B	3.2	5.0
B	6.8	
C	7.2	8.4
C	9.9	
Overall mean		5.7

where X is again the design matrix. Two things can be noticed from the design matrix X. First, it has a repetitive structure with each row being repeated as many times as there are samples from each of the three groups. Second, its first column is the sum of the other matrix columns (i.e., a linear combination). It is again an over-parameterized model since only two of the α_j are independent of each other and the third one is set to a fixed value. The overall mean of Y is $\mu_Y = (\mu_{Y,A} + \mu_{Y,B} + \mu_{Y,C}) / 3$, and replacing $\mu_{Y,j} = \mu_Y + \alpha j$ in the previous equation gives $\mu_Y = (\mu_Y + \alpha_A + \mu_Y + \alpha_B + \mu_Y + \alpha_C) / 3$, from which we have $\alpha_1 + \alpha_2 + \alpha_3 = 0$. You can check this yourself (up to rounding errors): (5.7 – 3.8) + (5.7 – 5.0) + (5.6 – 8.4) = 2.0 + 0.7 + (–2.7) = 0.

We need to perform the same trick of constraining the design matrix by setting one of the α_j to zero. Setting $\alpha_1 = 0$, and by necessity deleting the second column of the design matrix, requires redefining the model parameters as we did before for the *t*-test (Section 5.2). Now, α_j^* is defined as the difference between the means of the *j*th group and that of the reference level: $\alpha_j^* = \mu_{Y,\text{REF}} - \mu_j$.

The constrained design matrix can also be formulated by defining two binary indicator variables, say X_1 and X_2, whose combination allows assigning each data item to each of the three groups defined by the explanatory variable site. Let's then define $X_1 = 0$ if from Site A or 1 if from other Sites, and $X_2 = 0$ if from Site B or 1 if from other Sites. We require $k - 1$ indicator variables to differentiate data from k groups. This approach will allow the estimation of the parameters of Eq. (5.2) by making a multiple regression on the binary (or dummy) variables X_1 and X_2.

One-way analysis of variance allows the partitioning of the variation of the response variable into two non-overlapping fractions. In doing so, it analyzes the relative contributions of identifiable sources of variation to the total variation of the response variable. These sources of variation are: the part associated with the mean of the response variable μ_Y being a linear function of the categorical explanatory variables, and the residual or unexplained variation that corresponds to the variance of the response variable, σ_Y^2.

A few lines of high-school level algebra will show this point. The overall or total squared variation of the six values of the response variable can be written as $\sum\left(Y - \overline{\overline{Y}}\right)^2$, where $\overline{\overline{Y}}$ is the overall or grand mean. An equation cannot be altered by adding and subtracting the same term in one side: $\sum\left(Y - \overline{\overline{Y}}\right)^2 = \sum\left(Y - \overline{\overline{Y}} + \overline{Y}_j - \overline{Y}_j\right)^2$. We can rearrange the right-hand side as $\sum\left(Y - \overline{\overline{Y}}\right)^2 = \sum\left((Y - \overline{Y}_j) + \left(\overline{Y}_j - \overline{\overline{Y}}\right)\right)^2$, and then obtain $\sum\left(Y - \overline{\overline{Y}}\right)^2 = \sum(Y - \overline{Y}_j)^2 + 2\sum(Y - \overline{Y}_j)\left(\overline{Y}_j - \overline{\overline{Y}}\right) + \sum\left(\overline{Y}_j - \overline{\overline{Y}}\right)^2$. The middle term, $2\sum(Y - \overline{Y}_j)\left(\overline{Y}_j - \overline{\overline{Y}}\right)$, is the sum of the products of deviations of each value from its group mean and the group means from the overall mean, either of which is equal to zero. Therefore, we end up with:

$$\sum\left(Y - \overline{\overline{Y}}\right)^2 = \sum(Y - \overline{Y}_j)^2 + \sum\left(\overline{Y}_j - \overline{\overline{Y}}\right)^2. \tag{5.4}$$

Equation (5.4) states that the overall variation (left-hand side) equals the sum of the within-group variation and the variation between the group means from the overall mean (Fig. 5.11). This is the basis of the partitioning of the explained variation underlying the analysis of variance and of every special case of the general linear model. This is what it is expressed as the adjusted R^2 in the summary of the fitted models. The partitioning of the variation for the one-way analysis of variance can be depicted in graphical terms as in Fig. 5.11.

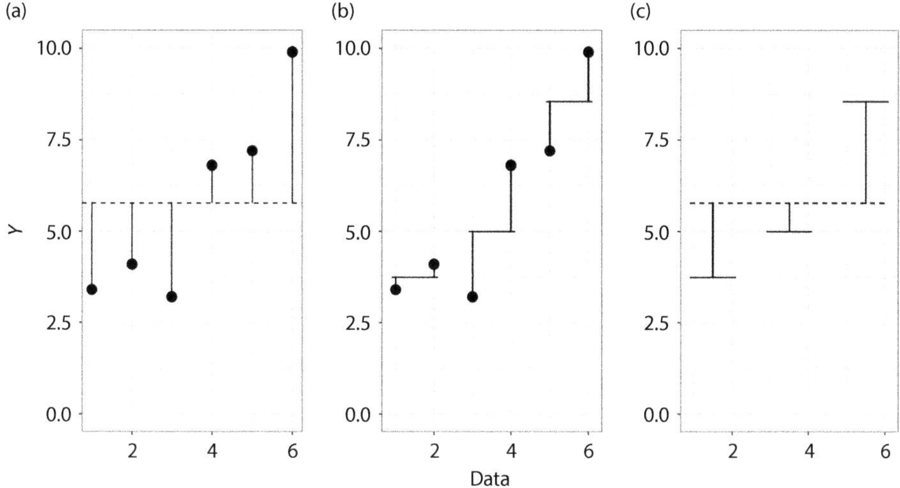

Fig. 5.11 Partitioning of the total variation (difference between each data point at the overall mean; a) of the response variable in the one-way analysis of variance into the within-group variation (difference between each data point at its group mean; b), and the between-group variation (difference between each group mean and the overall mean; c) shown in Eq. (5.3). Each vertical line shows the deviation of each of six data points from the mean in each of the three sources of variation involved in the one-way analysis of variance.

We can view one-way analysis of variance as a comparison of the variation of "batches" or sets of coefficients (i.e., the α_j^* in Eq. (5.3)) (*sensu* Gelman 2005) reflecting the differences between the means of groups defined by the categorical explanatory variable (Fig. 5.11a) in relation to the unexplained variation in the data (Fig. 5.11b). But we cannot directly compare among sums of squares. We need to obtain true estimates of variances by dividing the within-group variation $\sum\left(Y-\overline{Y_j}\right)^2$ and the between-group variation $\sum\left(\overline{Y_j}-\overline{\overline{Y}}\right)^2$ by the corresponding degrees of freedom, yielding two estimators of variance. Fisher (1925) found the expected values of these two mean-square estimators (and here we show only the case when the groups have equal sample sizes; when sample sizes differ the expressions are uglier but the intuition remains the same) to be

$$\mathrm{E}\left(MS_{\text{within}}\right)=\mathrm{E}\left(\frac{\sum\left(Y-\overline{Y_j}\right)^2}{k(n-1)}\right)=\sigma_Y^2$$

$$\mathrm{E}\left(MS_{\text{between}}\right)=\mathrm{E}\left(\frac{\sum\left(\overline{Y_j}-\overline{\overline{Y}}\right)^2}{k-1}\right)=\sigma_Y^2+k\frac{\sum\left(\overline{Y_j}-\overline{\overline{Y}}\right)^2}{k-1},\tag{5.5}$$

where E is the mathematical expectation (or expected value), and k is the number of groups. The well-known statistical test of one-way analysis of variance is the ratio between these two independent estimators of variance, $F_0=\frac{\mathrm{E}(MS_{\text{between}})}{\mathrm{E}(MS_{\text{within}})}=\frac{\sigma_Y^2+k\frac{\sum\left(\overline{Y_j}-\overline{\overline{Y}}\right)^2}{k-1}}{\sigma_Y^2}$. We can see that whenever there are large differences between the group means, we must expect the test statistic F_0 to be >1. F_0 is compared with a quantile from the F-distribution with the appropriate degrees of freedom in the numerator and denominator at a given significance level.

This has been a rather quick, informal, and rough explanation of one-way analysis of variance. You may find more formal and exhaustive presentations of the same material in Winer et al. (1991) and Aho (2014).

The main statistical test in one-way analysis of variance is a so-called "omnibus test" that simultaneously compares more than two conditions or features that define a compound null hypothesis. Here, the compound null hypothesis is that none of the possible pairs of group means defined by the categorical explanatory variable differ from zero. After rejecting a compound null hypothesis, we would need to carry out post hoc or a posteriori tests (Section 5.5) to determine which specific pairs of group means are actually significantly different.

THE DATA IN CONTEXT: Savage and West (2007) investigated the sleep time patterns of several species of mammals in relation to their body weight and other explanatory variables. Their original data set was updated by including information on each species' diet (carnivore, herbivore, insectivore, omnivore) and stored as the data frame `msleep` in the package `ggplot2`. We are going to assess whether the total daily sleep time of different mammalian species varies depending on their diet. Savage and West's (2007) original analyses were more complicated as they also used body weight as an explanatory variable, and accounted for the non-independence of the data points by including their degree of closeness in a phylogeny.

EXPLORATORY DATA ANALYSIS: Let's first import and examine the data:

```
> DF2=as.data.frame(ggplot2::msleep)
```

We start by converting the explanatory variable `vore` from the tibble `msleep` from character to factor form: `DF2$vore = as.factor(DF2$vore)`.

The summary of the relevant variables for the problem at hand in the data frame gives:

```
> summary(DF2[,c("vore", "sleep_total")])
      vore      sleep_total
 carni  :19   Min.   : 1.90
 herbi  :32   1st Qu.: 7.85
 insecti: 5   Median :10.10
 omni   :20   Mean   :10.43
 NA's   : 7   3rd Qu.:13.75
              Max.   :19.90
```

We have seven missing values for the explanatory variable and a much smaller number of insectivore species compared to the other diets. We are going to clean up the data frame by deleting the missing values (NA) with `DF2 = na.omit(DF2)`.

A comparison of the descriptive statistics and a boxplot will allow us to assess whether `vore` and `sleep_total` are related:

```
> desc.stats=function(x){c(mean=mean(x),median=median(x),sd=sd(x),n=length(x))}
```

```
> summaryBy(sleep_total~vore, data=DF2, FUN=desc.stats)
     vore sleep_total.mean sleep_total.median sleep_total.sd sleep_total.n
1   carni            13.33               12.5           3.72             3
2   herbi             9.52               10.9           4.38            10
3 insecti            14.05               14.1           7.99             2
4    omni            12.24               10.1           4.02             5
```

(we used the same code in Section 5.3). The summary of descriptive statistics per diet can be visualized in a boxplot (Fig. 5.12). Both the table of summary statistics and Fig. 5.12 show that the total duration of sleep differs depending on diet, with insectivores sleeping on average more than the other groups. However, we should not miss that there were just two insectivore species.

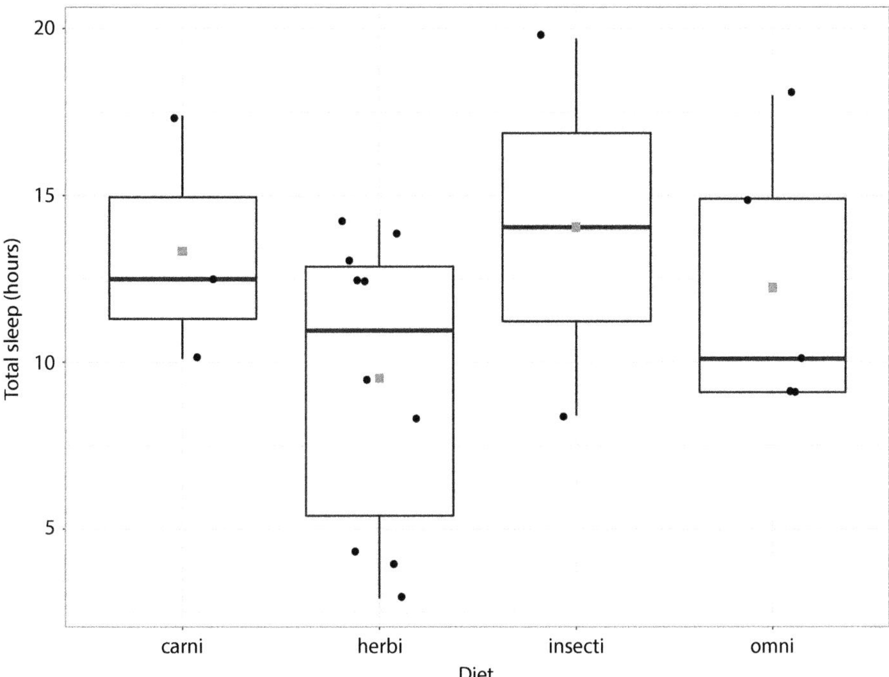

Fig. 5.12 Boxplot showing the relationship between `diet` and `sleep_total`. The gray squares indicate the means for each group.

We again use the package `fitdistrplus` to assess whether a normal or lognormal distribution should be used to model the response variable. We start with:

```
norm.sleep=fitdist(DF2$sleep_total, "norm")
lognorm.sleep=fitdist(DF2$sleep_total,"lnorm")
```

It seems clear from Fig. 5.13 that `total_sleep` is closer to a normal distribution.

The command `model.matrix(sleep_total ~ vore, data = DF2)` obtains the model matrix with as many rows (83) as data points (species) are in the data frame. We can obtain a shortened version of the model matrix as:

```
> unique(model.matrix(sleep_total~vore, data=DF2))
   (Intercept) voreherbi voreinsecti voreomni
4            1         0           0        1
5            1         1           0        0
9            1         0           0        0
22           1         0           1        0
```

What has happened?. The explanatory variable `vore` had four levels or categories:

```
> unique(DF2$vore)
[1] carni   omni    herbi   insecti
Levels: carni herbi insecti omni
```

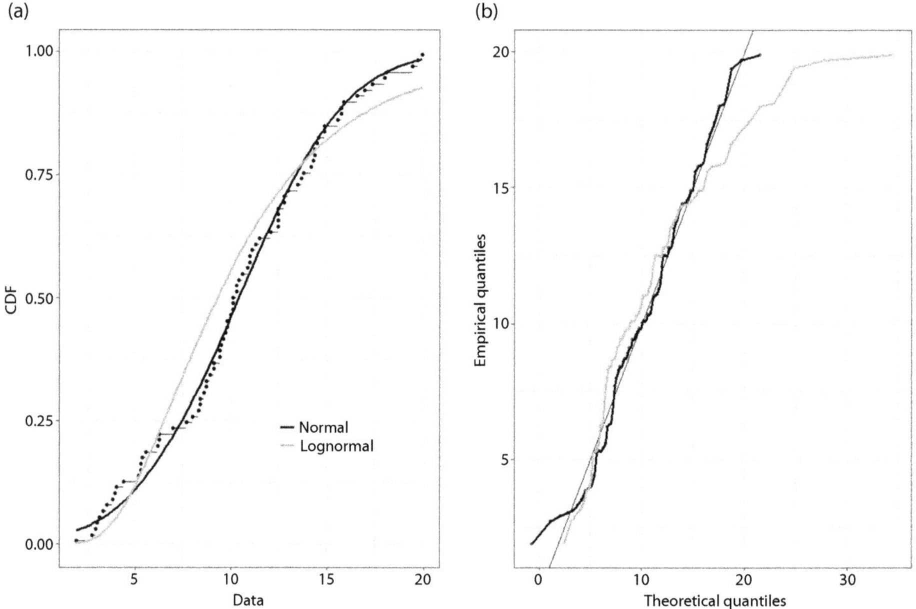

Fig. 5.13 Goodness of fit assessment of the response variable `sleep_total` showing the cumulative distribution function (a) and the quantile–quantile plot obtained using the package `fitdistrplus` (b).

By default, R used the alphabetic criterion to define `carni` as the reference group (shown as `(Intercept)`), and defined three binary indicator variables whose combination allows the unique assignment of each data point to one of the levels of the categorical explanatory variable `vore`. That is, R has constrained the original over-parameterized one-way analysis of variance design matrix by deleting one column and eliminating one of the α_j^* of Eq (5.3). The other redefined α_j correspond to the differences between the mean of the jth group and the group defined as the reference group (the one whose α was deleted from the original linear model).

5.6 One-way analysis of variance: Frequentist fitting

Let's finally fit the statistical model, `m3 = lm(sleep_total ~ vore, data = DF2)` (see Table 4.1 for the R syntax), and then examine the omnibus F-test:

```
> anova(m3)
Analysis of Variance Table

Response: sleep_total
          Df Sum Sq Mean Sq F value Pr(>F)
vore       3    134    44.6    2.24  0.091
Residuals 72   1436    19.9
```

This table suggests that the explanatory variable `vore` would not be very relevant to explain the mean of the response variable `sleep_total` because the p-value associated with the F-statistic is greater than the cut-off of 0.05 traditionally employed to declare statistical significance. Therefore, the mean sleep time would not greatly differ among the groups defined by diet.

The summary yields the parameter estimates of the statistical model:

```
> summary(m3)

Call:
lm(formula = sleep_total ~ vore, data = DF2)

Residuals:
   Min     1Q Median     3Q    Max
-7.679 -3.427 -0.429  4.012  9.021

Coefficients:
            Estimate Std. Error t value Pr(>|t|)
(Intercept)   10.379      1.024   10.13  1.7e-15 ***
voreherbi     -0.870      1.293   -0.67    0.504
voreinsecti    4.561      2.244    2.03    0.046 *
voreomni       0.546      1.431    0.38    0.704
---
Signif. codes:  0 '***' 0.001 '**' 0.01 '*' 0.05 '.' 0.1 ' '

Residual standard error: 4.47 on 72 degrees of freedom
  (7 observations deleted due to missingness)
Multiple R-squared:  0.0852,    Adjusted R-squared:  0.0471
F-statistic: 2.24 on 3 and 72 DF,  p-value: 0.0914
```

`(Intercept)` is the mean of `sleep_total` for the reference group (`carni`) not appearing in the rest of the summary table. The standard error (1.0245) and the p-value (1.7×10^{-15}) are of no interest in the context of the problem being solved: comparing the means, not verifying whether one of them differs from zero. The other coefficients of the summary are the magnitude of the differences in the means of `sleep_total` between each group (the α_j^* of Eq. (5.3)) and the reference group (`carni`). Herbivore species sleep on average fewer, and omnivore and insectivore species more, hours per day than carnivores. Each of these differences between means, their standard errors, t-values, and p-values are simply specific t-tests for differences in the mean of the response variable between each group and the reference group. We can roughly see that the magnitudes of the differences in average `sleep_total` are (–0.86 / 10.37 =) –8.3%, +44.0%, and +5.3% for herbivores, insectivores, and omnivores with respect to carnivores. The summary only shows us three specific contrasts that are just a fraction of the $k(k - 1) / 2 = 4(4 - 1) / 2 = 6$ possible pairwise comparisons for the categorical explanatory variable `vore` with four levels (see Section 5.5). We also have the estimated standard deviation of the response variable (4.47) and the proportion of its variation explained by the model (0.047). Finally, the F-statistic at the bottom of the summary table is the same as previously shown in the `anova` table. There might seem to be a contradiction between the p-value of the omnibus F-test (0.09143) and that of the contrast between carnivore and insectivore species (0.0458). Rest assured, there is no contradiction. Nonetheless, we have to defer the explanation to Section 5.5.

We can obtain a table of the parameter estimates and their confidence intervals with the package `broom`:

```
> tidy(m3, conf.int = T)
# A tibble: 4 x 7
  term        estimate std.error statistic  p.value conf.low conf.high
  <chr>          <dbl>     <dbl>     <dbl>    <dbl>    <dbl>     <dbl>
1 (Intercept)   10.4        1.02    10.1   1.70e-15   8.34       12.4
2 voreherbi     -0.870      1.29    -0.672 5.04e- 1  -3.45        1.71
3 voreinsecti    4.56       2.24     2.03  4.58e- 2   0.0868      9.04
4 voreomni       0.546      1.43     0.382 7.04e- 1  -2.31        3.40
```

We notice that the difference of the average `sleep_total` between carnivores (the reference group shown as the intercept) and insectivores is very imprecise, since its 95% confidence interval ranges between 0.08 (= 3 min) and 9.04 hours. We recall that all values in this very wide 95% confidence interval must be considered equally plausible. The worthless estimate of the mean differences in average sleep time between carnivores and insectivores and its wide confidence interval is due to the large variation in sleep time of the very small (n = 2) sample size for insectivores (Fig. 5.12).

We again collate all we need to carry out the residual analysis: `res.m3 = augment(m3)`. The plots of the residual analysis of model `m3` are shown in Fig 5.14, where the Q–Q plot has an inadequate fit in both tails, the residuals appear to have heterogeneous variation among diets, with the one of four points of the herbivore data being very different from the others, and two data points have large Cook distances, thus suggesting their potential influence on the parameter estimates. We can identify these two influential points by:

```
> DF2[which(res.m3$.cooksd> 0.10),c("name", "vore", "sleep_total")]
               name    vore sleep_total
62 Giant armadillo insecti        18.1
68       Cotton rat   herbi        11.3
```

Given the poor quality of the model fit to the data, we should at the very least examine its sensitivity by refitting the same model `m3` without these two influential data points:

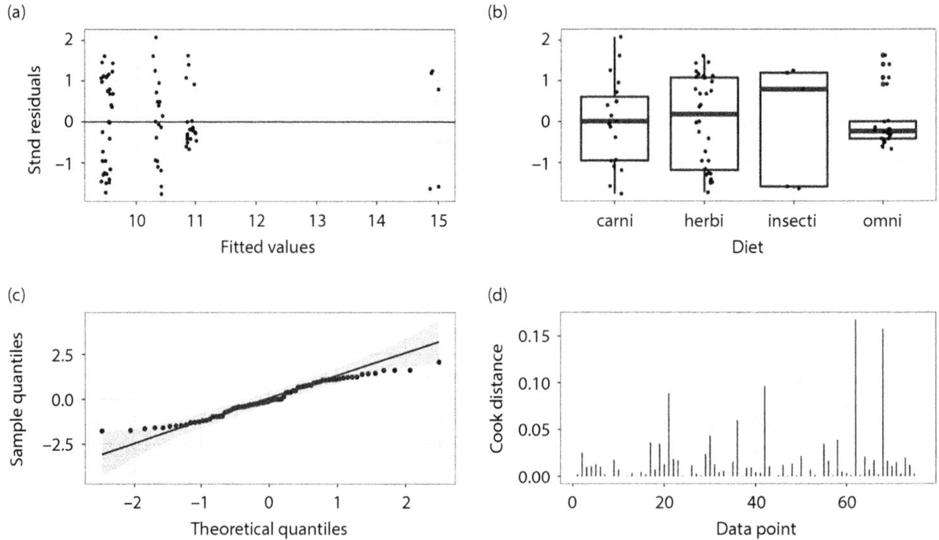

Fig. 5.14 Residual analysis of model `m3`, showing the residuals vs. fitted values (a), a boxplot of the residuals vs. the categorical explanatory variable (b), the quantile–quantile plot of residuals (c), and the Cook distances (d).

```
> summary(m3.1)

Call:
lm(formula = sleep_total ~ vore, data = DF2[-which(res.m3$.co
    0.1), c("vore", "sleep_total")])

Residuals:
   Min     1Q Median     3Q    Max
-7.679 -3.847 -0.602  4.085  9.021

Coefficients:
            Estimate Std. Error t value Pr(>|t|)
(Intercept)   10.379      1.033   10.04  3.3e-15 ***
voreherbi     -0.927      1.312   -0.71     0.48
voreinsecti    3.771      2.478    1.52     0.13
voreomni       0.546      1.443    0.38     0.71
---
Signif. codes:  0 '***' 0.001 '**' 0.01 '*' 0.05 '.' 0.1 ' '

Residual standard error: 4.5 on 70 degrees of freedom
  (7 observations deleted due to missingness)
Multiple R-squared:  0.0592,    Adjusted R-squared:  0.0189
F-statistic: 1.47 on 3 and 70 DF,  p-value: 0.23
```

We can observe two main qualitative differences when comparing models `m` and `m3.1`: the contrast in the mean `sleep_total` between carnivore and insectivore species is no longer statistically significant, and the already small adjusted R^2 of `m3` decreases even further. We should therefore be very wary before ascribing much relevance to the statistical significance found in model `m3` before we can check the accuracy of these two data points and until we have a model that fits these data adequately.

Nevertheless, and just for completeness, we show in Fig. 5.15 a conditional plot of model `m3` obtained with the command `ggpredict` of package `ggeffects`, giving the means and their 95% confidence intervals.

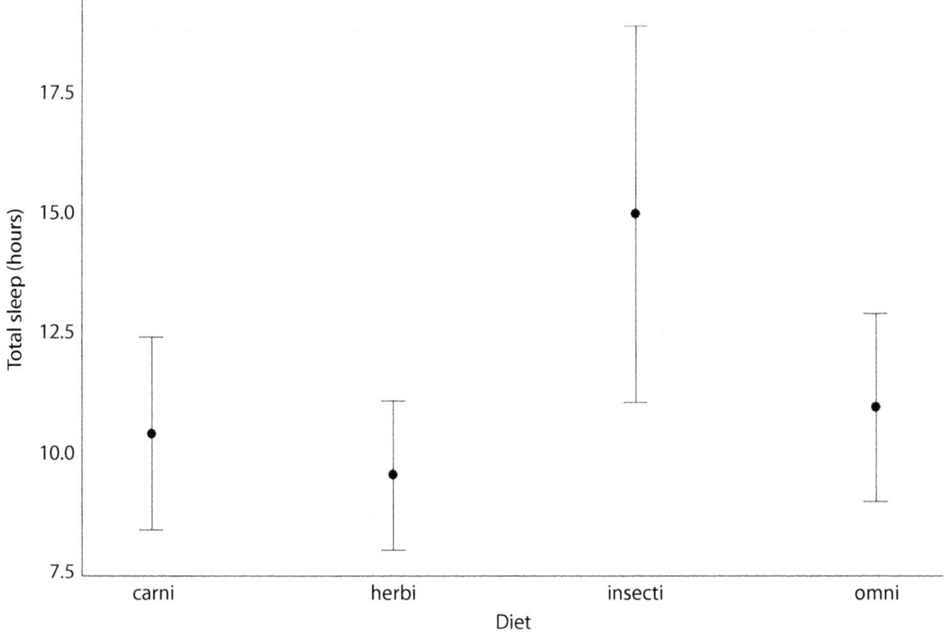

Fig. 5.15 Conditional plots predicting the average total sleep time according to diet and their 95% confidence intervals from model `m3` generated with the `ggeffects` package.

5.7 One-way analysis of variance: Bayesian fitting

We again use the package `brms` to fit the Bayesian one-way analysis of variance using the constrained statistical model of Eq. (5.3). Let's look at the default `brms` priors:

```
> get_prior(formula=sleep_total~vore, data=DF2,family=gaussian)
                 prior     class        coef group resp dpar nlpar lb ub       source
                (flat)         b                                              default
                (flat)         b   voreherbi                             (vectorized)
                (flat)         b voreinsecti                             (vectorized)
                (flat)         b    voreomni                             (vectorized)
 student_t(3, 11.3, 4) Intercept                                              default
    student_t(3, 0, 4)     sigma                                   0          default
```

As before, the carnivore diet is (by default) the reference group:

```
> unique(model.matrix(sleep_total~vore, data=DF2))
    (Intercept) voreherbi voreinsecti voreomni
1             1         0           0        0
2             1         0           0        1
3             1         1           0        0
22            1         0           1        0
```

We must define the priors for the mean of `sleep_total` for carnivores (the reference group, indicated as `Intercept`), the magnitude of the differences between the means of the reference group and the other groups (the α_j^* of Eq. (5.3), indicated as `voreherbi`, `voreinsecti`, and `voreomni`), and the standard deviation of the response variable (indicated as `sigma`).

In the context of the current problem, a principled way of proceeding would be to gather information on the most plausible ranges of the model parameters from published scientific literature. Or we could consult with experts on the subject matter (sleep in mammals) to shed light on our ignorance. What do we know about plausible values of the model parameters? Let's say that we gathered information that carnivores sleep on average 9 hours per day and sleeping time would range between 4 and 15 hours. Our prior for this parameter is a probability distribution that accommodates these sketchy empirical features. One, but by no means the only, suitable probability distribution would be Lognormal($\mu_{\log Y}$ = log(9), $\sigma_{\log Y}$) to support strictly positive real values. What value for $\sigma_{\log Y}$ we should use is less clear from the sketchy prior information, but we could choose (by trial and error) its value to accommodate the range [4, 15] just reported. If we ignore the prior magnitude of mean differences that could be expected between diets, a range of ±40% difference from the average of 9 hours for carnivores is plausible. This would imply a range from 9 × 0.6 = 5.4 hours to 9 × 1.4 = 12.6 hours, with a mean of zero. A normal distribution with mean = 0 and sd = (max − min) / 4, i.e., (12.6 − 5.4) / 4 = 1.8 hours, seems reasonable. Guessestimating the standard deviation from a range is a rough approach often used in meta-analysis (Wan et al. 2014). The prior for the standard deviation of the daily hours of sleep is harder to guess, and perhaps here we should use a vague prior such as Lognormal (μ = log(6), σ = 0.5). It is useful to plot the priors before analysis (Fig. 5.16) to make sure that they conform to our previous knowledge of the model parameters and do not contain inadmissible parameter values such as negative variances or negative average daily sleep.

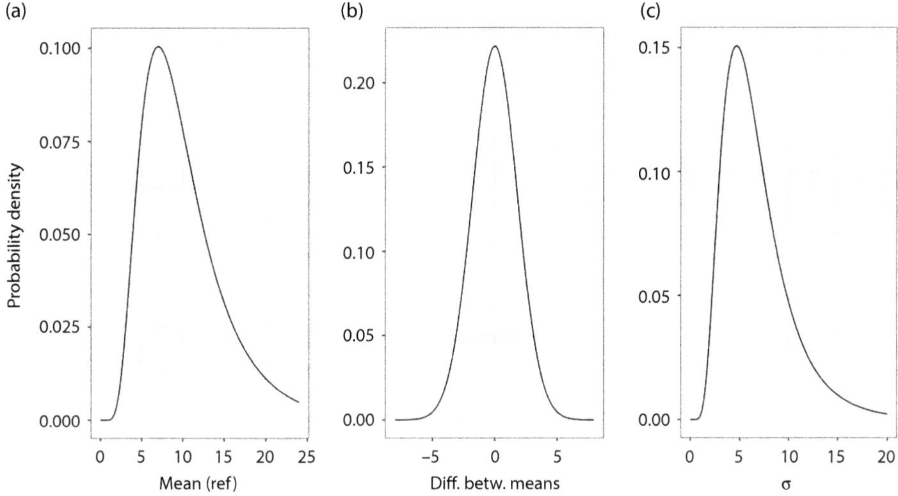

Fig. 5.16 Prior distributions for the three parameters: the mean of the reference group (carnivores; a), the magnitude of the difference between each mean and the reference group (b), and standard deviation of the response variable (c) of the Bayesian one-way analysis of variation `sleep_total ~ vore`.

These priors were put into a single object:

```
> prior.m3 = c(set_prior("lognormal(log(9), 0.5)", class = "Intercept"),
+              set_prior("normal(0, 1.8)", class = "b"),
+              set_prior("lognormal(log(6), 0.5)", class = "sigma"))
```

that was used in the fitting of the Bayesian one-way analysis of variance:

```
m3.brms=brm(formula=sleep_total~vore, data=DF2, family=gaussian, prior =
            prior.m3, warmup = 1000, future=TRUE, chains=3, iter=2000, thin=3)
```

```
> summary(m3.brms)
 Family: gaussian
  Links: mu = identity; sigma = identity
Formula: sleep_total ~ vore
   Data: DF2 (Number of observations: 76)
Samples: 3 chains, each with iter = 2000; warmup = 1000; thin = 3;
         total post-warmup samples = 1000

Population-Level Effects:
            Estimate Est.Error l-95% CI u-95% CI Rhat Bulk_ESS Tail_ESS
Intercept      10.70      0.81     9.13    12.26 1.00     1014      845
voreherbi      -0.98      0.99    -2.86     0.93 1.00      956      910
voreinsecti     1.86      1.37    -0.90     4.52 1.00     1090      956
voreomni        0.14      1.05    -1.91     2.11 1.01      972      910

Family Specific Parameters:
      Estimate Est.Error l-95% CI u-95% CI Rhat Bulk_ESS Tail_ESS
sigma     4.55      0.38     3.87     5.41 1.00     1095      856
```

The summary has the mean, standard deviation, and the 2.5% and 97.5% percentiles (the 95% credible interval) of the marginal posterior distributions of the five model parameters. The interpretation of the metrics `Rhat`, `Bulk_ESS`, and `Tail_ESS` was given in Section 5.3. `Rhat` being close to one and `Bulk_ESS` and `Tail_ESS` being not too far from 1,000 statistically independent parameter estimates suggest adequate model fitting that can be further evaluated with the autocorrelation plots. Fig. 5.17 illustrates that

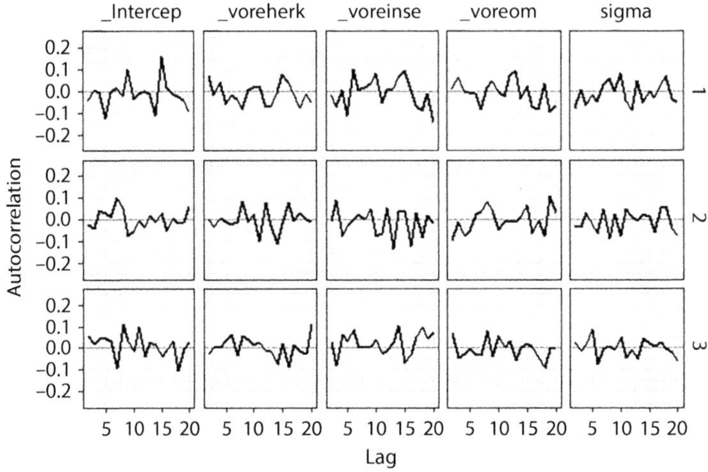

Fig. 5.17 Autocorrelation plots of each parameter of model `m3.brms` for the three chains. The parameter names are those shown in Fig. 5.18. The scale limit on the *x*-axis was set to avoid showing the autocorrelation of lag = 0 (the correlation of a value with itself) that is equal to one by definition.

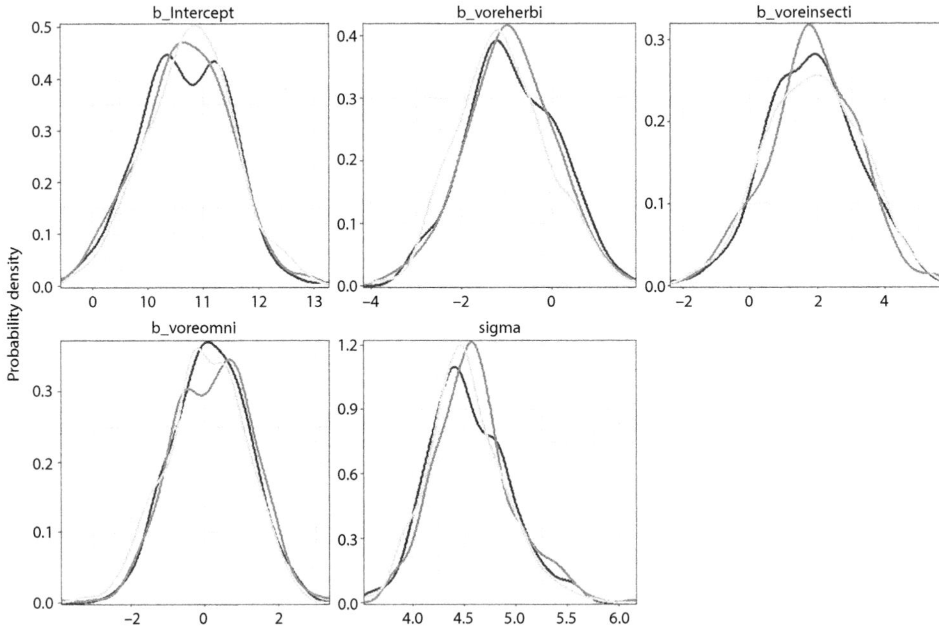

Fig. 5.18 Marginal posterior distributions of the parameters of model `m3.brms` for the three chains.

the three chains showed the parameter estimates sampled to have a very low degree of autocorrelation.

The marginal posterior distributions of the model parameters (Fig. 5.18) show that the three chains converged to a posterior stationary distribution. These three chains can then be merged into a single posterior distribution to yield what we saw in `summary(m3.brms)`, as graphically depicted in Fig. 5.19.

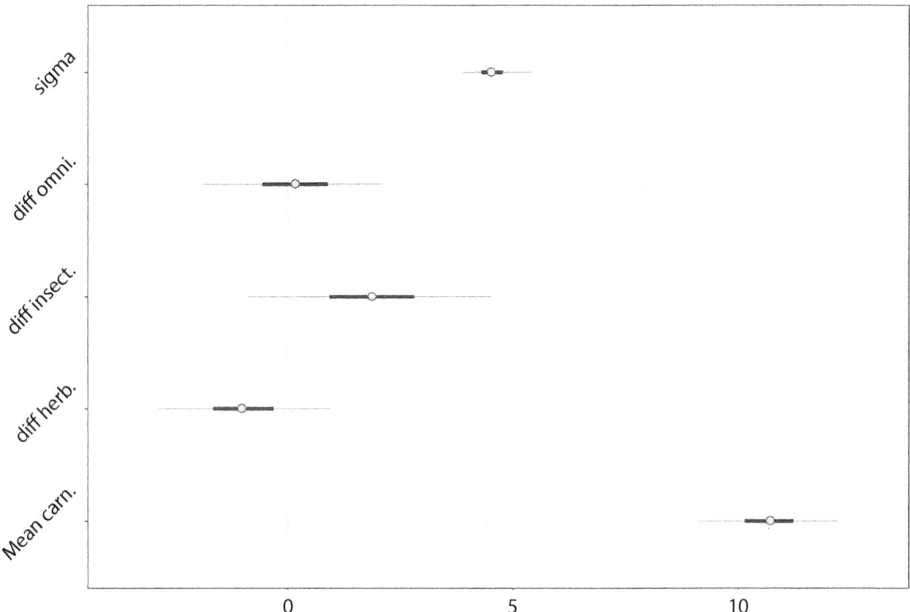

Fig. 5.19 Means (dots) and 50% (thick) and 95% (thin) credible intervals of the parameters of model `m3.brms`. The parameter names are those shown in Fig. 5.18.

Because the 95% credible intervals of the differences of the mean `sleep_total` with the reference group (carnivores) include zero, we can conclude that, within the precision of these estimates, the group means hardly differ from the mean `sleep_total` of carnivores.

Similar to its frequentist analogue, model `m3.brms` explained a small percentage of the variation of the response variable:

```
> bayes_R2(m3.brms)
   Estimate Est.Error    Q2.5 Q97.5
R2   0.0488    0.0324 0.00461 0.125
```

We can now compare the prior (Fig. 5.17) with the marginal posterior (Fig. 5.18) distributions of the model parameters. If the empirical data contained a limited amount of information, the two sets of distributions should exhibit a substantial overlap. The difference in scales between the prior and posterior distributions is not relevant to judging their degree of overlap; what matters for the comparison are changes in the ranges of parameter values supported by the prior and posterior distributions of each parameter. We can see the amount of information of the empirical data only narrowed the range of plausible values for the mean of the reference group (Fig. 5.20a) and the standard deviation of the response variable (Fig. 5.20c), but did not substantially change the prior distribution for the most important model parameters, namely the magnitude of differences between the means between groups (Fig. 5.20b). This is consistent with the small differences among

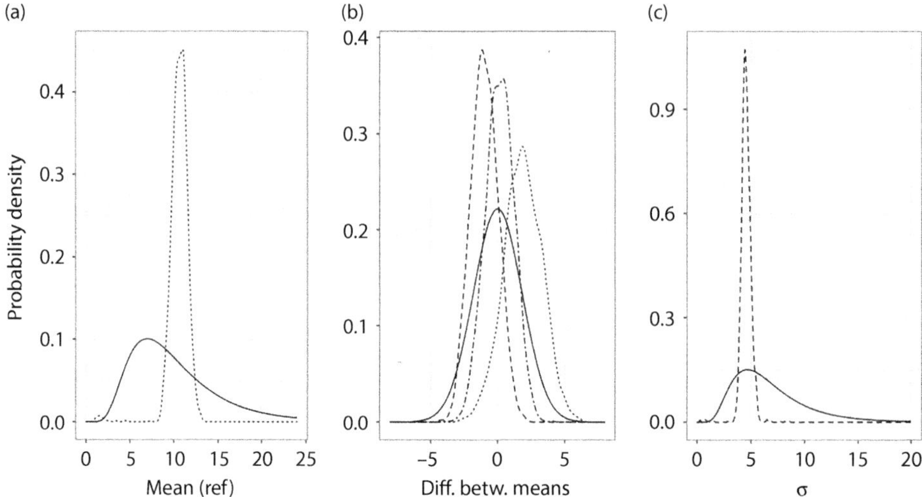

Fig. 5.20 Overlap between the prior (continuous lines) and posterior distributions for the three parameters of the Bayesian one-way analysis of variance (model `m3.brms`). The three lines in the central plot correspond to the mean differences between herbivores (long dash), insectivores (short dash), and omnivores (dotted) and the reference group (carnivores).

the group means, and their 95% credible intervals are likely to contain zero difference between means (Fig. 5.19), as also shown in `summary(m3.brms)`.

We now evaluate the goodness of fit of the model `m3.brms` with residual analysis (see Chapter 4). We start by simulating the Pearson residuals with `res.m3.brms = as.data.frame(residuals(m3.brms, type = "pearson", nsamples = 1000, summary = T))` and the predicted values for the response variable with `fit.m3.brms = as.data.frame(fitted(m3.brms, scale = "linear", summary = T, nsamples = 1000))`. Then we store the means of the Pearson residuals and of the fitted values in the data frame `DF2`. But now a small problem arises: there were seven NA values for the variable `vore`, and hence the statistical models were fitted for 76 rather than the 83 data points of the data frame `DF2`. Accordingly, we will have 1,000 simulations for the 76 actual values rather than the 83 data points in `DF2` and we will have a problem when trying to incorporate the average residuals and fitted values into the data frame `DF2`. The simplest solution is generate another data frame without rows from `DF2` having NA in the variable `vore`, `DF2c = DF2[!is.na(DF2$vore),]`, and then write `DF2c$fit = fit.m3.brms$Estimate` and `DF2c$resid = res.m3.brms$Estimate`.

Just as its frequentist analogue was, model `m3.brms` has a mediocre fit to the data: the Pearson residuals are not normally distributed (Fig. 5.21a) and their variation is not homogeneous across diets (Fig. 5.21b). We cannot be satisfied with this statistical model and should be wary of trusting it too much for inferring differences in the daily sleep patterns across diets. What can we do now? One option is to introduce other explanatory variables (and we have several such candidates such as body weight) to account for the heterogeneity of the data. Another possibility for improving the model would be to include the degree of non-independence of species, since all mammals are to some degree related through their phylogeny (see, for instance, Paradis 2006). It is always possible that we did not detect important differences in the average daily sleep duration among diets simply because we are using an incomplete or plainly incorrect model.

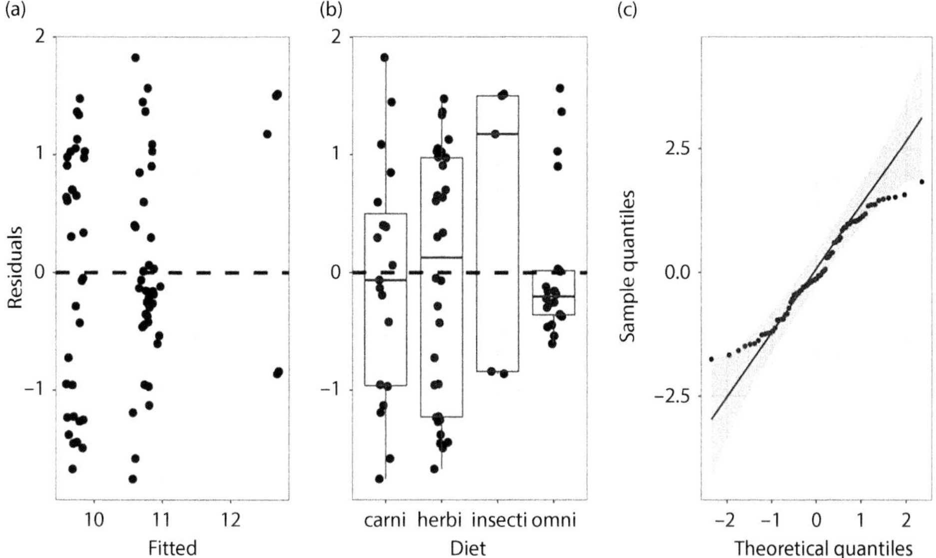

Fig. 5.21 Residual analysis of model `m3.brms`, showing the mean of the Pearson residuals vs. the mean of the fitted value (a), the mean of the Pearson residuals vs. the explanatory variable in the model (b), and the quantile–quantile plot of the mean of the Pearson residuals (c).

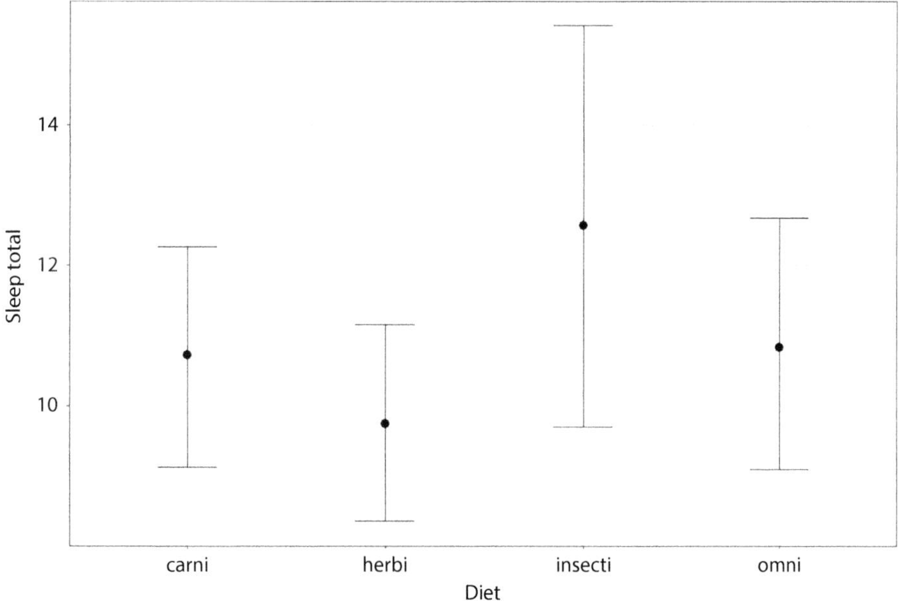

Fig. 5.22 Conditional plot of the mean of the response variable and the 95% credible intervals of `sleep_total` per diet from model `m3.brms`.

Nonetheless, we can produce a conditional plot (Fig. 5.22) summarizing the fitted mean values of the response variable and their associated 95% credible intervals. It should be no surprise that the 95% credible intervals of the mean `sleep_total` among diets have a high degree of overlap, suggesting that their means are barely different from each other.

5.8. A posteriori tests in frequentist models

The null statistical hypothesis in the one-way analysis of variance is that all means μ_j of the j groups defined by the categorical explanatory variable were identical (Winer et al. 1991). In the frequentist framework, the omnibus F-test simultaneously evaluates the composite hypothesis that there is no (statistically significant) difference between any set of means of the k groups. Whenever this null statistical hypothesis is rejected, we just learn that there is some set of group means that are significantly different from each other, and we must strive to identify this set.

The simplest and most commonly used approach consists of finding the set of pairs of group means that differ significantly from each other. Given that there are $k = 4$ diets (carnivore, herbivore, insectivore, and omnivore), there are $k(k - 1) / 2 = 6$ pairwise comparisons. The composite statistical hypothesis in one-way analysis of variance is subject to Boole's (also called Bonferroni's) inequality (Blitzstein and Hwang 2012). This states that the probability of at least one of the pairwise comparisons being equal is no greater than the sum of the probabilities of the individual comparisons or contrasts:

$$\Pr\left[(\mu_{\text{carni}} = \mu_{\text{herbi}}) \cup \cdots \cup (\mu_{\text{insecti}} = \mu_{\text{omni}})\right] \leq \Pr(\mu_{\text{carni}} = \mu_{\text{herbi}}) + \cdots + \Pr(\mu_{\text{insecti}} = \mu_{\text{omni}}) . \tag{5.6}$$

The left-hand side of Eq. (5.6) is called the family-wise error rate (FWER) for the set of comparisons, and each pairwise comparison of means on the right-hand side is of course a t-test. Equation (5.6) shows that the probability of finding a significant difference increases with the number of comparisons. This is known as the "multiplicity problem": the more comparisons we make, the higher the chances of finding statistically significant results, both true and false positives. The goal would be to correct the α of each comparison so as to keep FWER = 0.05, and thus avoid fooling oneself by wrongly rejecting the null hypotheses of no differences.

While most statistics textbooks recommend the Bonferroni-type corrections for the general linear model, they are hardly ever used for the generalized linear model (Chapter 9 onward). This is very wrong and most unfortunate. The eventual need to adjust the significance level in a posteriori tests is entirely independent of the probability distribution used in the likelihood to model the response variable in the frequentist framework.

Stemming from Eq. (5.6), Bonferroni (1936) proposed the first solution (and one of most popular) to the multiplicity problem: divide α by the number of comparisons k to be made. Bonferroni's correction does not make any assumption about the degree of dependence of a set of comparisons. Assuming for now that the individual comparisons are mutually independent, the probability of not making any false positive errors in k comparisons is $(1 - \alpha)^k$. Therefore, the FWER, the probability of making at least one false positive judgment in the set of pairwise comparisons, is $1 - (1 - \alpha)^k$. If we use $\alpha = 0.05$ for $k = 6$ comparisons, FWER = 0.26. Therefore, at least one Type I error will be committed for a set of six tests 26% of the time. If we want FWER = 0.05 for k comparisons, the significance level for each comparison should be $\alpha = 1 - (1 - \text{FWER})^{1/k}$. With $k = 6$, $1 - (1 - 0.05)^{1/6} = 0.0085$, i.e., we should use a significance level 5.8 ($\approx 0.05 / 0.0085$) times smaller for all pairwise comparisons than the customary 5%. A pairwise comparison with $\alpha = 0.0085$ is called "conservative" because it rarely finds significant differences unless the magnitude of the difference between means is very large. Decreasing the significance level α will automatically increase the probability of detecting false negatives (Aho 2014) and inevitably decrease statistical power (i.e., the probability of detecting true differences between means). Therefore, the more severe the Bonferroni-type of correction, the higher the chances of not detecting a true difference in any individual comparison.

Applying a slightest decrease of the significance level of individual comparisons to keep FWER = 0.05 is an old problem for all statistical models in the frequentist framework (see, e.g., Bretz et al. 2011, Sauder and Demars 2019). A bewildering number of methods have been proposed: Fisher, Sidak et al., Student, Newman and Keuls, Dunnet, Tukey's HSD, Waller–Duncan, Dunnet's C, and many more (Bretz et al. 2011, Aho 2014). This is an active research topic in the area of genomics where often many thousands of comparisons are made in the search for associations between features of gene sequences and the occurrence of diseases (e.g., Noble 2009).

Statisticians and scientists both remain divided on the need to adjust the significance levels in multiple comparisons (e.g., Perneger 1995, Moran 2003, Nakagawa 2004, Aho 2014, Bretz et al. 2011, Greenland 2006, Rothman 1990). The consensus seems to be that such adjustments are essential in confirmatory (those whose goal is the definite proof of a pre-defined set of hypothesis to be tested) but not in exploratory (those examining data without a pre-specified key set of hypotheses) studies (Wakenfield 2013, Tong 2019). In practice, the overwhelming majority of scientific studies lie somewhere between these two clear-cut and principled extreme cases. Lacking a prior statement of the hypotheses to be tested and the degree of relationship between them, it seems unclear whether one should adjust the significance levels of multiple comparisons. And the strength of the adjustment depends on the number of comparisons and on their degree of independence, none of which may be trivial to assess. Should we correct for the actual, or for the potential, number of comparisons (In this paper? In the researcher's career?) carried out (Sjolander and Vansteeladt 2019; Greenland and Hofman 2020)? Should we correct only for those comparisons related to the previously stated hypothesis and forgo the post hoc comparisons merely suggested by the data analysis? Let's be clear here: all researchers are competing for ever scarcer journal space in the best journals in the "publish or perish" race. Over-correcting the significance levels decreases the statistical power of the multiple comparisons and conspires against attaining the statistical significance needed to publish under current statistical practices. This should not be interpreted to mean that adjustment of the significance levels in multiple comparisons can be waived (*contra* Moran 2003, Rothman 1990, Nakagawa 2004), but that statistics still lacks a clear and principled way of proceeding that puts all research on the same footing regarding the need to adjust the significance levels, and the manner of doing so, in multiple comparisons.

Methods to adjust the significance levels in multiple comparisons can make either simultaneous or sequential adjustments to α (Aho 2014, Bretz et al. 2011). The adjustment method $\alpha = 1 - (1 - \text{FWER})^{1/k}$ was simultaneous, and it led to having the same adjusted α in all comparisons. Sequential procedures employ stepwise, progressive adjustments of α and they are less conservative than those based on simultaneous adjustments (Aho 2014). Without the assurance of being universally appropriate for all cases, we will use Benjamini and Hochberg's (1995) false discovery rate (FDR) throughout this book to adjust the significance level in a posteriori comparisons of all statistical models. In a nutshell, the FDR sorts the p-values of individual comparisons and sequentially attempts to reject the hypotheses starting from the one having the smallest p-value, and making slight, progressive adjustments of the significance levels. We use the package `multcomp`, whose main command is:

```
m3.post=glht(model=m3, linfct=mcp(vore="Tukey"),test=adjusted ("fdr"))
```

The command `glht` (meaning general linear hypothesis) takes model `m3`, `linfct` specifies the linear hypotheses to be tested involving the multiple comparisons of the categorical explanatory variable `vore`, `"Tukey"` means that we will make all pairwise comparisons (we will shortly see other possibilities), and finally we specify the chosen method (`"fdr"`) for adjusting the significance levels. The results are:

```
> summary(m3.post)

         Simultaneous Tests for General Linear Hypotheses

Multiple Comparisons of Means: Tukey Contrasts


Fit: lm(formula = sleep_total ~ vore, data = DF2)

Linear Hypotheses:
                      Estimate Std. Error t value Pr(>|t|)
herbi - carni == 0      -0.870      1.293   -0.67     0.90
insecti - carni == 0     4.561      2.244    2.03     0.18
omni - carni == 0        0.546      1.431    0.38     0.98
insecti - herbi == 0     5.431      2.147    2.53     0.06 .
omni - herbi == 0        1.416      1.273    1.11     0.67
omni - insecti == 0     -4.015      2.233   -1.80     0.27
```

The summary gives the magnitudes of the differences between all pairs of means and their adjusted significance levels. We can see that all the adjusted p-values of the individual comparisons were greater than the customary significance level of 0.05. Hence, we conclude that the mean `sleep_total` did not significantly differ between any pair of diets. Comparing `summary(m3.post)` with `summary(m3)` above, we can now see that the "barely statistically significant" p-value of 0.0458 between carnivores and insectivores (that disappeared when we deleted two data points in model `m3.1`) was simply because `summary(m3)` contains a series of contrasts with the reference group without correcting the significance level of pairwise comparisons. We can also obtain the FDR-adjusted confidence limits of the pairwise differences of means, as shown in Fig. 5.23. All 95% confidence intervals (also obtained after correcting the significance level) contain zero difference between pairs of means. We also see that all 95% confidence intervals of pairwise differences involving insectivores are very wide (about 10 hours) and of hardly any interest. This is because there were only two insectivore species.

We just performed a posteriori tests comparing all possible pairs. This approach may be called a "brute force approach." On one hand, it may have obtained more pairwise tests than we wanted and needed (and hence applied a needlessly stronger correction of the significance levels). On the other, it may not have yielded other comparisons that were interesting for our research goals. We can be wiser and should do better. Let's assume that all we really want to know may be summarized in the following three comparisons: carnivores vs. herbivores, omnivores vs. the average of carnivores and herbivores, and omnivores vs. the average of the other three diets. These specific comparisons should have been stated in advance as part of the research hypothesis. To formulate these three contrasts, we need to know the order in which the levels of the factor `vore` were entered:

```
> unique(DF2$vore)
[1] carni   omni    herbi  insecti
Levels: carni herbi insecti omni
```

We may change the order of the factor levels at will (e.g., putting omnivores as the first level) by writing `DF2$vore = factor(DF2$vore, levels = c("omni", "carni",`

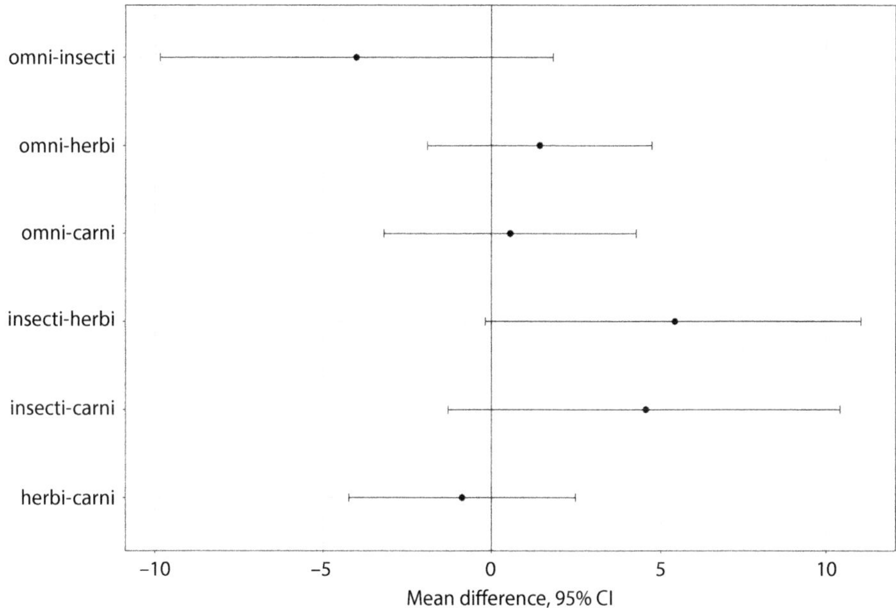

Fig. 5.23 A posteriori tests of all pairwise differences of means and adjusted 95% confidence intervals of `sleep_total` per diet from model `m3`.

`"herbi", "insecti")`. We are going to write a set of three vectors whose coefficients denote the levels of the factor `vore` to be compared in each contrast. In more general terms, these are known as "orthogonal contrasts" (Aho 2014, Bretz et al. 2011), and they are characterized by their row algebraic sum being equal to zero. Bearing in mind the vector DF2$vore just ordered, the first vector of coefficients, `d1 = c(1,-1,0,0)`, will evaluate whether $(\overline{Y}_{\text{carni}} - \overline{Y}_{\text{herbi}}) = 0$; the second, `d2 = c(-0.5,0,-0.5,1)`, denotes $\overline{Y}_{\text{omni}} - \frac{\overline{Y}_{\text{carni}}+\overline{Y}_{\text{herbi}}}{2} = 0$; and the third, `d3 = c(-0.33,-0.33,-0.33,1)`, refers to $\overline{Y}_{\text{omni}} - \frac{\overline{Y}_{\text{carni}}+\overline{Y}_{\text{herbi}}+\overline{Y}_{\text{insecti}}}{3} = 0$. We can now put the three vectors in a matrix,

```
> contr.mat=rbind("carni vs herbi"=d1,
+                 "omni vs avg of carni and insecti"=d2,
+                 "omni vs avg of diets"=d3)
```

which is used to obtain the desired contrasts while correcting their significance levels: `m3.post2 = glht(m3, linfct = mcp(vore = contr.mat), test = adjusted("fdr"))`. The summary is shown below, and the corresponding plot in Fig. 5.24:

```
> summary(m3.post2)

         Simultaneous Tests for General Linear Hypotheses

Multiple Comparisons of Means: User-defined Contrasts


Fit: lm(formula = sleep_total ~ vore, data = DF2)

Linear Hypotheses:
                                      Estimate Std. Error t value Pr(>|t|)
carni vs herbi == 0                      0.870      1.293    0.67     0.78
omni vs avg of carni and insecti == 0   -1.734      1.502   -1.15     0.46
omni vs avg of diets == 0               -0.672      1.273   -0.53     0.86
(Adjusted p values reported -- single-step method)
```

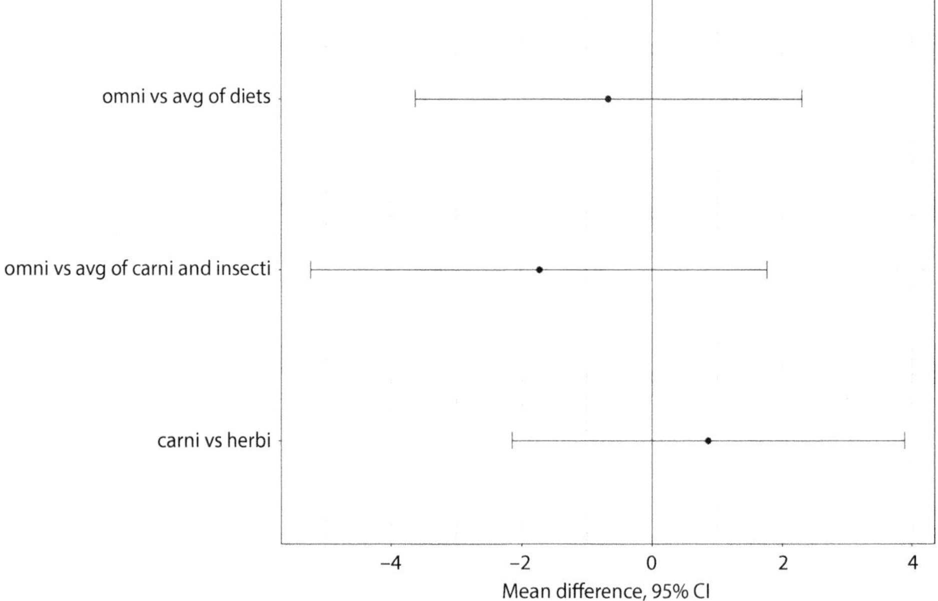

Fig. 5.24 Pairwise difference of means and adjusted 95% confidence intervals of `sleep_total` for a specific set of a posteriori comparisons from model `m3`.

It is obviously better to make fewer, more informative, a posteriori comparisons related to specific hypotheses than to use the "brute force approach," and then make up an overarching explanation. The difference between these two ways of proceeding becomes more blatant as the number of a posteriori comparisons becomes very large.

5.9 A posteriori tests in Bayesian models?

On the Bayesian side of the statistical world, there is much less overall concern about the multiplicity problem. In general, the adjustment of the significance levels in multiple comparisons is viewed as unnecessary. There are several reasons for this.

First and foremost, the flow of time and information in frequentist and Bayesian statistics operate in opposite directions. In frequentist statistics, the significance level α is the probability of obtaining results as extreme as those observed, or more so, if the null hypothesis were true: Pr(current or more extreme data | H_0 is considered as true). In frequentist statistics, the statement "H_0 is true" is considered as fixed, and the phrase "current or more extreme possible data" refers to possible data that could be potentially sampled in the future from the same statistical population. The significance level calculated from the current data refers to the sampling distribution of a test statistic (see Chapter 3). The adjustment of the significance level in multiple comparisons aims to correct for the false-positive extreme results (leading to the rejection of no differences) that could occur because the same data could now, or in the future, be used several times in several a posteriori comparisons.

In contrast, Bayesian statistics does not rely on sampling distributions, and makes no explicit reference either to statistical significance or to potential data to be gathered in the future. The posterior distribution in Bayesian statistics reads Pr(parameters | the current data). The statement "parameters" refers to the values of model parameters that are, of course, what we hypothesize about. And "current data" considers it as a given and fixed body of evidence that enters only once in the analysis. Should we gather more data, we would incorporate them in a second analysis that could possibly use as priors the posterior distributions of the previous analyses (Lindley 1972). In Bayesian statistics, only the current data enters into the statistical inference through the likelihood function. The posterior distribution portrays the degree of plausibility of every plausible combination of parameter values, given the data and the joint prior distribution. The posterior distribution then contains all possible dependencies and correlations between the parameters stemming from both the likelihood (i.e., from the data) and the prior distribution. There is just one joint posterior distribution whose marginal projections can be looked at and combined in any number of ways. The marginal posterior distributions of each α_j^* of Eq. (5.3) can be used in any number of arithmetic operations on model parameters as those involved in a posteriori comparisons without actually having to adjust the posterior distribution.

Second, let's recall that we had set the priors $\alpha_j^* \sim \text{Normal}(\mu_{\alpha_j} = 0, \sigma^2_{\alpha_j} = 1.4)$ after arguing for the plausible magnitude of the differences in the average sleep duration. More generally, we could have left the hyperparameters $\mu_{\alpha j}$ and $\sigma^2_{\alpha j}$ undefined so that they could be estimated from the data. In the latter case, the model $Y \sim \text{Normal}(\mu_{Y,\text{REF}} + \alpha_j^*, \sigma_Y^2)$ would have become a two-level or hierarchical model where some high-level parameters (α_j^*) depended on the values of low-level parameters ($\mu_{\alpha j}$ and $\sigma^2_{\alpha j}$). Hierarchical models are fully addressed in Chapter 13. For now, suffice it to say that hierarchical models allow a more objective approach to inference by estimating the parameters of prior distributions from data rather than requiring them to be specified using external information (Gelman 2006) as we did in Section 3.4. The dependency between higher and lower parameters in hierarchical models induces a phenomenon called "shrinkage" (also known as "partial pooling") whereby the values of the lower-level parameters are pulled closer together than they would be if they were not a higher-level distribution (Kruschke 2011). In the context of our one-way analysis of variance for the mammalian sleep times depending on diet, shrinkage occurs because the data from all species influence the high-level distribution of α_j^*, which in turn influence the estimates of each species. Again, we fully explain this important feature of hierarchical models in Chapter 13. It turns out that by pulling low-level parameter estimates toward the mean at the higher level, the shrinkage effect holds the key to minimizing the risk of false alarms in Bayesian statistics (e.g., Gelman et al. 2012, Berry and Hochberg 1999, Kruschke 2011).

While Bayesians would seem to be at an extraordinary advantage over frequentists regarding multiple comparisons, there is never such a thing as a free lunch. Bayesian statistics mitigate the risk of false alarms involved in multiple testing by incorporating knowledge into the prior distributions of model parameters (Gelman et al. 2012, Kruschke 2010, 2011). The prior distributions may include structural knowledge whereby data from different groups mutually inform and constrain each other's estimates to minimize the risk of false alarms in multiple comparisons (e.g., Gelman et al. 2012, Kruschke 2010, 2011). In some sense, prior distributions prevent the likelihood from "learning too much from the actual data" by constraining the acceptance of extreme combinations of parameter values that would lead to the detection of spurious relationships or contrasts.

If setting the appropriate priors holds the key to minimizing the risk of false alarms in Bayesian multiple comparisons, then we should have a clear and principled way of summarizing, eliciting, and setting the priors. What would be an "appropriate prior" in the context of minimizing the risk of false alarms involved in multiple a posteriori tests? The consensus is that "weakly informative priors" are likely to do the job most of the time (Gelman et al. 2014, 2017, Lemoine 2019, Banner et al. 2019, and references therein). Weakly informative priors not only reflect a diluted amount of knowledge on plausible ranges of parameter values, but also, and more importantly, constrain the parameter space by providing low support to impossible and or implausible parameter values (Banner et al. 2019, Gabry et al. 2019, Gelman et al. 2014). Weakly informative priors also serve as a method of statistical regularization (Hobbs and Hooten 2015; McElreath 2019) that help Bayesian models obtain more accurate parameter estimates (Gelman et al. 2017, Hobbs and Hooten 2015) and an adequately conservative assessment of multiple comparisons with an acceptable risk of false alarms (Gelman et al. 2012, Lemoine 2019).

How can we know whether a prior distribution is acceptably "weakly informative?" While there are suggestions of weakly informative priors for parameters of specific models (see http://github.com/stan-dev/stan/wiki/Prior-Choice-Recommendations; Gelman 2006, Gelman et al. 2008, Banner et al. 2019), the short and general answer is that the degree of informativeness of prior distributions can only be assessed in the context of the likelihood for specific data (Gelman et al. 2017). The latter implies that there are *NOT* foolproof, iron-cast proposals of weakly informative priors equally valid for all data. Weakly informative priors aim to generate data with a span larger than the observed evidence while incorporating as much previous knowledge on parameter values and their interdependence (Gabry et al. 2019). In Chapter 7 we use the method of the prior predictive distribution to that effect. And, of course, we can always perform a prior sensitivity analysis on the uncertain and perennially debatable choice of priors in Bayesian modeling (Depaoli and van de Shoot 2017).

Are we certain that Bayesian models remain safe from the risk of false alarms involved in multiple comparisons? While Gelman et al. (2012) showed that the shrinkage effect involved in hierarchical models would largely eliminate the need for Bonferroni-type corrections, Ogle et al.'s (2019) results give a more nuanced view. While the shrinkage effect involved in hierarchical models always helps in attaining a lower FWER in multiple comparisons, their effectiveness depends on the degree of pooling, which in turn results from the total sample size, the number of groups being compared and their degree of sampling balance, and the overall variance of the response variable (Ogle et al. 2019). That is, it is the magnitude of the shrinkage effect that matters in keeping the risk of false alarms at acceptable levels, not its mere presence due to the (often hierarchical) nature of Bayesian statistical models. In closing, while the jury is still out, the balance of the evidence implies that Bayesian models are on much safer ground in multiple comparisons than equivalent frequentist models fitted to the same data.

5.10 Problems

Visit the companion website at www.oup.com/companion/InchaustiSMWR to obtain the data sets for these problems.

5.1 Naya et al. (2008) investigated the impact of lactation (the most energetically demanding period in the life cycle of female mammals) on the resting metabolic rate (in joules per ml of CO_2) of small female caviomorph rodents under laboratory conditions. File: Pr5-1.csv

5.2 García-Alonso et al. (2008) compared the total amounts of fatty acids in polychaetes under three feeding regimes. Polychaetes are a key prey species for economically important fish. Aquaculture produces huge amounts of sludge, which might be recycled by feeding to polychaetes. They measured the amount of fatty acids (fatty.acids; μg per mg of dry weight) and the categorical explanatory variable food, with three levels: field or natural food, fish food, and eel sludge from aquaculture farms. File: Pr5-2.csv

References

Aho, K. (2014). *Foundational and Applied Statistics for Biologists Using R.* CRC Press / Chapman and Hall, New York.

Banner, K., Irvinde, K. and Rodhouse, T. (2019). The use of Bayesian priors in ecology: The good, the bad and the not great. *Methods in Ecology and Evolution*, 11, 882–889.

Benjamini, Y. and Hochberg, M. (1995). Controlling the false discovery rate, a practical and powerful approach to multiple testing. *Journal of the Royal Statistical Society* B, 57, 289–300.

Berry, D. and Hochberg, M. (1999). Bayesian perspectives on multiple comparisons. *Journal of Statistical Planning and Inference*, 82, 215–227.

Blitzstein, J. and Hwang, J. (2012). *Introduction to Probability.* CRC Press / Chapman and Hall, New York.

Bonferroni, C. (1936). *Teoria statistica delle classi e calcolo delle probabilità.* Pubblicazioni del Istituto Superiore di Scienze Economiche e Commerciali di Firenze, Florence.

Bretz, F., Hothorn, T., and Westfall, P. (2011). *Multiple Comparisons Using R.* CRC Press / Chapman and Hall, New York.

Depaoli, S. and van de Shoot, R. (2017). Improving transparency and replication in Bayesian statistics: The WAMBS-Checklist. *Psychological Methods*, 22, 240–261.

Fisher, R. (1921). Studies in crop variation. I. An examination of the yield of dressed grain from Broadbalk. *The Journal of Agricultural Science*, 11, 107–135.

Fisher, R. (1925). *Statistical Methods for Research Workers.* Oliver and Boyd, Edinburgh.

Gabry, J., Simpson, D., Vehtari, A., et al. (2019). Visualization in Bayesian workflow. *Journal of the Royal Statistical Society A*, 182, 389–402.

García-Alonso, J., Müller, C., and Hardege J. (2008). Influence of food regimes and seasonality on fatty acid composition in the ragworm. *Aquatic Biology*, 4, 7–13.

Gelman, A. (2005). Analysis of variance: Why it is more important than ever. *Annals of Statistics*, 33, 1–33.

Gelman, A. (2006). Prior distributions for variance parameters in hierarchical models. *Bayesian Analysis*, 1, 515–533.

Gelman, A., Carlin, J. Stern, H. et al. (2014). *Bayesian Data Analysis*, 3rd. edn. Chapman and Hall / CRC Press, New York.

Gelman, A., Goodrich, B., Gabry, J. et al. (2018). R-squared for Bayesian regression models. *American Statistician*, 73, 307–309.

Gelman, A., Hill, J. and Yajima, M. (2012). Why we (usually) don't have to worry about multiple comparisons. *Journal of Research on Educational Effectiveness*, 5, 189–211.

Gelman, A., Jakulin, A., Pittau, M. et al. (2008) A weakly informative default prior distribution for logistic and other regression models. *Annals of Statistics*, 2, 1360–1383.

Gelman, A., Simpson, D., and Betancourt M. (2017). The prior can often only be understood in the context of the likelihood. *Entropy*, 19, 555–568.

Greenland, S. (2006). Multiple comparisons and association selection in general epidemiology. *International Journal of Epidemiology*, 37, 430–434.

Greenland, S. and Hofman, A. (2020). Multiple comparisons controversies are about context and costs, not frequentism versus Bayesianism. *European Journal of Epidemiology*, 34, 801–808.

Hobbs, N. and Hooten, T. (2015). *Bayesian Models: A Statistical Primer for Ecologists.* Princeton University Press, Princeton.

Kruschke, J. (2010). Bayesian data analysis. *Cognitive Science*, 1, 658–676.

Kruschke, J. (2011). *Doing Bayesian Data Analysis: A Tutorial with R and BUGS.* Academic Press, New York.

Lemoine, N. (2019). Moving beyond non-informative priors: Why and how to choose weakly informative priors in Bayesian analyses. *Oikos*, 128, 912–928.

Lindley, D. (1972). *Bayesian Statistics: A review*. SIAM, Philadelphia.
McElreath, R. (2019). *Statistical Rethinking: A Bayesian Course with Examples in R and Stan*, 2nd edn. CRC Press / Chapman and Hall, New York.
Moran, M. (2003). Arguments for rejecting the sequential Bonferroni in ecological studies. *Oikos*, 100, 403–405.
Nakagawa, S. (2004). A farewell to Bonferroni: The problems of low statistical power and publication bias. *Behavioral Ecology*, 15, 1044–1045.
Naya, D., Ebesperger, L., Sabat, P. et al. (2008). Digestive and metabolic flexibility allows fe-males degus to cope with lactation costs. *Physiological and Biochemical Zoology*, 81, 186–194.
Noble, W. (2008). How does multiple correction work? *Nature Genetics*, 27, 1135–1137.
Ogle, K., Peltier, E., Felt, L. et al. (2019). Should we be concerned about multiple comparisons in hierarchical Bayesian models? *Methods in Ecology and Evolution*, 10, 553–564.
Paradis, E. (2006). *Analysis of Phylogenetics and Evolution with R*. Springer, New York.
Perneger, T. (1995). What's wrong with Bonferroni adjustments. *British Medical Journal*, 316, 1236–1239.
Rothman, K. (1990). No adjustments are needed for multiple comparisons. *Epidemiology*, 1, 43–46.
Sauder, D. and Demars, C. (2019). An updated recommendation for multiple comparisons. *Advances in Methods and Practices in Psychological Science*, 2, 26–44.
Savage, V. and West, G. (2007). A quantitative, theoretical framework for understanding mammalian sleep. *Proceedings of the National Academy of Sciences USA*, 104, 1051–1056.
Sjolander, A. and Vansteeladt, S. (2019). Frequentist versus Bayesian approaches to multiple testing. *European Journal of Epidemiology*, 34, 809–821.
Tong, C. (2019). Statistical inference enables bad science; statistical thinking enables good science. *The American Statistician*, 73, 246–261.
Wakenfield, J. (2013). *Bayesian and Frequentist Regression Methods*. Springer, New York.
Wan, X., Wang, W., Liu, J. et al. (2014). Estimating the sample mean and standard deviation from the sample size, median, range and/or interquartile range. *BMC Medical Research Methodology*, 14, 135–147.
Winer, B., Brown, D., and Michels, K. (1991). *Statistical Principles in Experimental Design*, 3rd edn. McGraw and Hill, New York.
Young, K., Brodie, E., and Brodie, E. (2004). How the horned lizard got its horns. *Science*, 305, 1909–1910.

CHAPTER 6

The General Linear Model III

Interactions between explanatory variables

Packages needed in this chapter:

```
packages<-c("ggplot2","bayesplot","fitdistrplus","reshape2",
"gridExtra","qqplotr","ggeffects","GGally","broom","qqplotr","brms",
"bayesplot")
lapply(packages.needed,FUN=require,character.only=T)#
```

loads these packages in the working session.

6.1 Introduction

This chapter builds on the introduction to the general linear model of Chapters 4 and 5, and extends it to include interactions between explanatory variables. Understanding the material in the previous two chapters is essential for grasping this one. Whenever there is more than one explanatory variable, the effect or influence of these variables on the mean of the response variable can be either independent or interactive. The former means the effects of each explanatory variable are additive, while the latter implies that the impact of two levels of one of the explanatory variables (measured as the difference of their means) varies or changes with at least one level of another explanatory variable. Whenever there is a significant/important interaction between explanatory variables, the impact of the interacting variables must be interpreted together, as the effect of each variable in isolation on the mean of the response variable is of little interest.

In Chapter 5, after covering the simplest possible case of a categorical explanatory variable of two levels (i.e., the *t*-test), we proceeded to its generalization for more than two levels (i.e., one-way analysis of variance). Now we are taking one step further to consider two categorical explanatory variables (factorial analysis of variance). Finally, we consider a yet more general case of having both numerical and categorical explanatory variables (analysis of covariance). The main ideas involved in the interactions between explanatory variables are of course valid for all the statistical models in Part II.

6.2 Factorial analysis of variance

This involves an instance of the general linear model of a response variable with a normal distribution and (at least) two categorical explanatory variables. The combinations of levels of these categorical explanatory variables define the groups whose means are being compared.

Statistical Modeling With R. Pablo Inchausti, Oxford University Press. © Pablo Inchausti (2023).
DOI: 10.1093/oso/9780192859013.003.0006

We are going to linit the discussion of factorial analysis of variance to two categorical explanatory variables, on the understanding that it can be extended to any number of factors. Thus, factorial analysis of variance has the following statistical definition: the response variable Y has a normal distribution (μ_Y, σ_Y^2), and its statistical model is (see Section 5.3):

$$Y = (\mu_Y + \alpha_i + \beta_j + \alpha\beta_{ij}) + \varepsilon, \tag{6.1}$$

where μ_Y is the overall mean of the response variable, α_i and β_j denote the impact of each explanatory variable whose number of levels are indexed by i and j respectively, $\alpha\beta_{ij}$ denotes the interactive effects of the two factors and, as in Eq. (5.1), ε denotes the random and normally distributed differences between each value of the response variable and its group mean now defined by the combinations of two categorical explanatory variables. According to the model, the mean of the response variable will reflect the effect of the categorical explanatory variables and their interaction (the part between parentheses in Eq. (6.1)), and $\varepsilon \sim \text{Normal}(0, \sigma_Y^2)$ such that the random variation about the mean would have an average of zero, and its scatter would be determined by the variance of the response variable, σ_Y^2. The underlying assumptions of Eq. (6.1) are the same for any instance of the general linear model and need not be reiterated here (see Section 4.1).

Table 6.1 Example of the simplest possible data set leading to factorial analysis of variance for a response variable Y with a normal distribution. The mean of Y reflects the effect of the categorical explanatory variables A and B and their interaction (Eq. (6.1)). The data are shown shaded in gray. There are four groups (A_1B_1, A_1B_2, A_2B_1, A_2B_2) defined by the combinations of the two levels of each categorical explanatory variable, and their means are shown in italics. The marginal means and the overall mean (in bold, at the lower right) are also shown.

	B_1		B_2		Marginal means of A_i
A_1	16.2	*15.8*	13.4	*12.1*	14.0
	15.4		10.8		
A_2	12.6	*12.8*	15.6	*15.8*	14.3
	12.9		15.9		
Marginal means of B_j	14.3		14.0		**14.2**

The simplest way to explain and interpret the terms of Eq. (6.1) is to introduce a small data set (see Table 6.1). Starting with $Y = Y$, we can add and subtracting the same terms to the right-hand side to maintain the equality: $Y = Y + (\overline{Y_i} - \overline{Y_i}) + (\overline{Y_j} - \overline{Y_j}) + 3\left(\overline{\overline{Y}} - \overline{\overline{Y}}\right) + (\overline{Y_{ij}} - \overline{Y_{ij}})$, where $\overline{Y_{ij}}$, $\overline{Y_i}$, $\overline{Y_j}$, and $\overline{\overline{Y}}$ are the means of Y in each of the four combinations A_1B_1, A_1B_2, A_2B_1, and A_2B_2, the marginal means for each level of the factors A and B, and the overall mean, respectively. Rearranging the terms in the latest equation will allow matching with the statistical model of Eq. (5.6) on a term-by-term basis:

$$\begin{aligned} Y &= \overline{\overline{Y}} + \left(\overline{Y_i} - \overline{\overline{Y}}\right) + \left(\overline{Y_j} - \overline{\overline{Y}}\right) + \left[\left(\overline{Y_{ij}} - \overline{\overline{Y}}\right) - \left(\overline{Y_j} - \overline{\overline{Y}}\right) - \left(\overline{Y_i} - \overline{\overline{Y}}\right)\right] + (Y - \overline{Y_{ij}}), \\ Y &= \mu_Y + \alpha_i + \beta_j + \alpha\beta_{ij} + \varepsilon. \end{aligned} \tag{6.2}$$

The α_i denote the single or pure effect of factor A; they are α_1 = 14.0 – 14.2 = –0.2 and α_2 = 14.3 – 14.2 = +0.2 (Table 6.1). Similarly, the β_j for factor B are 14.3 – 14.2 = +0.1 and 14.0 – 14.2 = –0.2 for β_1 and β_2, respectively. What we have done thus far is nothing more than a simple extension of the one-way analysis of variance statistical model, Eq. (5.4), to include the effect of a second categorical explanatory variable. The alert reader would not have missed that (up to rounding errors) $\sum \alpha_i + \sum \beta_j = 0$, just as we explained in Section 5.4.

The interaction term $\alpha\beta_{ij}$ in Eq. (6.2) quantifies the extent to which the difference between the cell means and the overall mean $\left(\overline{Y}_{ij} - \overline{\overline{Y}}\right)$ differs from the simple effects of each factor denoted by $\alpha_i = \left(\overline{Y}_i - \overline{\overline{Y}}\right)$ and $\beta_j = \left(\overline{Y}_j - \overline{\overline{Y}}\right)$. Whenever the latter difference is large, the joint effect of both factors on the response variable is above and beyond the simple effect of each explanatory variable in isolation. The magnitude of the difference $(\overline{Y}_{A_1} - \overline{Y}_{A_2})$ is not constant but varies with, or depends on, the level of the factor B considered: +3.0 for B_1 and −3.7 for B_2 (and likewise for $(\overline{Y}_{B_1} - \overline{Y}_{B_2})$ for the factor A: +3.7 for A_1 and −3.0 for A_2 (Table 6.1)). Again, $\sum \alpha\beta_{ij} = 0$. The last term ε of Eq. (6.2) is the variation of each data point from each of the four group means that, as before, reflects the variation of the response variable unexplained by the effects of the two categorical explanatory variables and their interaction. In the absence of replication for all combinations of the explanatory variables, the terms $\alpha\beta_{ij}$ and ε in Eq. (5.5) are compounded into the latter error term.

Just as we did in Section 5.4, we can rewrite Eq. (6.2) as a linear equation of the form $Y = X\beta + \varepsilon$ for this data set to present it as a multiple linear regression:

$$\begin{pmatrix} 16.2 \\ 15.4 \\ 12.6 \\ 12.9 \\ 13.4 \\ 10.8 \\ 15.6 \\ 15.9 \end{pmatrix} = \begin{pmatrix} 1 & 1 & 0 & 1 & 0 & 1 & 0 & 0 & 0 \\ 1 & 1 & 0 & 1 & 0 & 1 & 0 & 0 & 0 \\ 1 & 0 & 1 & 0 & 1 & 0 & 1 & 0 & 0 \\ 1 & 0 & 1 & 0 & 1 & 0 & 1 & 0 & 0 \\ 1 & 1 & 0 & 1 & 0 & 0 & 0 & 1 & 0 \\ 1 & 1 & 0 & 1 & 0 & 0 & 0 & 1 & 0 \\ 1 & 0 & 1 & 0 & 1 & 0 & 0 & 0 & 1 \\ 1 & 0 & 1 & 0 & 1 & 0 & 0 & 0 & 1 \end{pmatrix} \begin{pmatrix} \mu_Y \\ \alpha_1 \\ \alpha_2 \\ \beta_1 \\ \beta_2 \\ \alpha\beta_{11} \\ \alpha\beta_{12} \\ \alpha\beta_{21} \\ \alpha\beta_{22} \end{pmatrix} + \begin{pmatrix} \varepsilon_1 \\ \varepsilon_2 \\ \varepsilon_3 \\ \varepsilon_4 \\ \varepsilon_5 \\ \varepsilon_6 \\ \varepsilon_7 \\ \varepsilon_8 \end{pmatrix}, \tag{6.3}$$

where X is again the design matrix. This matrix has as many rows as data, and as many columns as the model has parameters affecting the mean of the response variable. The design matrix X in Eq. (6.3) is created using binary indicator variables denoting the presence/absence of each level of the single effects of the categorical explanatory variables and their interaction. The value for the first cell in the data table is $16.2 = (1)\mu_Y + (1)\alpha_1 + (0)\alpha_2 + (1)\beta_1 + (0)\beta_2 + (1)\alpha\beta_{11} + (0)\alpha\beta_{12} + (1)\alpha\beta_{21} + (1)\alpha\beta_{22} + \varepsilon_1$, which means that $16.2 = \mu_Y + \alpha_1 + \beta_1 + \alpha\beta_{11} + \varepsilon_1$. In words, the first value of the top-left cell of the table 16.2) is the sum of the influences of the first levels of factors A and B (α_1 and β_1), their interaction ($\alpha\beta_{11}$), and the residual (ε_1) or variation unexplained by the simple effects of these two categorical variables and their interaction.

Just as for the *t*-test (Section 5.2) and one-way analysis of variance (Section 5.4), the design matrix in Eq. (6.3) is over-parameterized (i.e., it has infinite solutions). It has to be constrained in order to estimate the model parameters. As in Chapter 5, the constraining requires setting to zero one parameter of each of the three effects (α_i, β_j, and $\alpha\beta_{ij}$). To do so, we define one of the levels of each explanatory variable as the "reference group" (Fox 2015). Let's then define A_1 and B_1 as the reference groups for each of these two factors, such that the group defined by the combination (A_1,B_1) in the top-left of thedata table is now the "reference group" ($\mu_{11,\text{REF}}$). This method of constraining the design matrix is the default in R and it has two implications in Eq (6.2). First, we set not only $\alpha_1 = \beta_1 = 0$, but also every other parameter involving the first level of any of these groups: $\alpha\beta_{11} = \alpha\beta_{12} = \alpha\beta_{21} = 0$. We now need $(a - 1)$ parameters for factor A, $(b - 1)$ for factor B, and $(a - 1)(b - 1)$ for their interaction, with a and b being the number of levels of factors A and B (Fox 2015). Second, setting these parameters to zero implies re-expressing the remaining parameters (cf. Section 5.4) and deleting the corresponding columns of the design matrix, such that Eq. (6.3) becomes

$$\begin{pmatrix} 16.2 \\ 15.4 \\ 12.6 \\ 12.9 \\ 13.4 \\ 10.8 \\ 15.6 \\ 15.9 \end{pmatrix} = \begin{pmatrix} 1 & 0 & 0 & 0 \\ 1 & 0 & 0 & 0 \\ 1 & 1 & 0 & 0 \\ 1 & 1 & 0 & 0 \\ 1 & 0 & 1 & 0 \\ 1 & 0 & 1 & 0 \\ 1 & 1 & 1 & 1 \\ 1 & 1 & 1 & 1 \end{pmatrix} \begin{pmatrix} \text{Intercept} \\ \alpha_2^* \\ \beta_2^* \\ \alpha\beta_{22}^* \end{pmatrix} + \begin{pmatrix} \varepsilon_1 \\ \varepsilon_2 \\ \varepsilon_3 \\ \varepsilon_4 \\ \varepsilon_5 \\ \varepsilon_6 \\ \varepsilon_7 \\ \varepsilon_8 \end{pmatrix}. \tag{6.4}$$

We can use the constrained design matrix of Eq. (6.4) to express the predicted means of each of four groups (A_1B_1, A_1B_2, A_2B_1, A_2B_2) and of the marginal means of each level of factors A and B in terms of the newly defined parameters. This is shown in Table 6.2, where α_2^* is the differential (that can be positive or negative) between the means of the reference group A_1B_1 and the group A_2B_1 (note that we keep in the reference level of factor B, i.e., B_1). Likewise, β_2^* is the differential between the means of the reference group A_1B_1 and the group A_1B_2 (note that we keep in the reference level of factor A, i.e., A_1). $\alpha_2^* + \alpha\beta_{22}^*$ would then reflect the differences between the means of A_1B_2 and A_2B_2. We could also obtain and interpret the differences between the marginal means for each categorical variable. These are differences between specific group means, which is after all what we are looking to estimate in this analysis.

Table 6.2 Means of the response variable Y for each of the four groups (A_1B_1, A_1B_2, A_2B_1, A_2B_2) defined by the combinations of the levels of the categorical explanatory variables A and B in terms of the parameters of the constrained design matrix (Eq. (6.4)) of the factorial analysis of variance. The first levels of each categorical variable are considered as the "reference levels," such that A_1B_1 is the "reference group."

	B_1	B_2	Marginal means of A_i
A_1	$\mu_{11} = \text{Intercept}$	$\mu_{12} = \text{Intercept} + \beta_2^*$	$\mu_{A1} = \frac{\mu_{11}+\mu_{12}}{2} = \text{Intercept} + \frac{\beta_2^*}{2}$
A_2	$\mu_{21} = \text{Intercept} + \alpha_2^*$	$\mu_{22} = \text{Intercept} + \alpha_2^* + \beta_2^* + \alpha\beta_{22}^*$	$\mu_{A2} = \frac{\mu_{21}+\mu_{22}}{2} = \text{Intercept} + \frac{\beta_2^*}{2} + \alpha_2^* + \frac{\alpha\beta_{22}^*}{2}$
Marginal means of B_j	$\mu_{B1} = \frac{\mu_{11}+\mu_{21}}{2} = \text{Intercept} + \frac{\alpha_2^*}{2}$	$\mu_{B2} = \frac{\mu_{12}+\mu_{22}}{2} = \text{Intercept} + \frac{\alpha_2^*}{2} + \beta_2^* + \frac{\alpha\beta_{22}^*}{2}$	—

Finally, if we move the first term of the right-hand side of Eq. (6.2) to the left, we obtain a linear equation showing the partitioning of the total variation $\left(Y - \overline{\overline{Y}}\right)$ into the parts due to the simple effect of factor A, of factor B, of their interaction, and the unexplained or residual variation. Employing the same approach used in Section 5.4, by squaring each term of Eq. (6.2) and canceling all terms equal to zero we end up with an equation decomposing the total squared variation into the aforementioned components (see Winer et al. 1991). The resulting decomposition of variability is the basis of the omnibus *F*-tests of factorial analysis of variance. The analysis of variance table arising from the decomposition of the variation in Eq. (6.2) results from the sequential testing of ratios of estimates of the variance (i.e., the mean squares) using *F*-tests. Should you wish to find the expected mean squares and their ratios defining the *F*-tests for factorial analysis of variance, you might consult Winer et al. (1991). These tests actually evaluate the compounded effect of "batches of coefficients" (*sensu* Gelman 2005) to reflect the importance of each source of variation in explaining the data. The same approach can be extended to address more complex research questions involving more than two categorical explanatory variables each with any number of levels.

THE DATA IN CONTEXT: We are going to consider an extremely simple data set of an experiment assessing the effect of two components of their diet (source of protein: beef and cereal; amount of protein: low and high) in the weight gain of rats (Hand et al. 2001). In this laboratory experiment, 10 individual rats were randomly assigned to each of the four groups formed by the 2 × 2 combinations of the categorical explanatory variables. The research questions are whether the amount of protein affected the weight gain of rats (g) in the same or in different ways depending on the source of protein of their diets. Therefore, we would like to compare the amount of protein (high vs. low) within each source of protein (beef and cereal), and we would like to compare whether the average gain for high is similar to the the average gain for low, regardless of the source of protein.

EXPLORATORY DATA ANALYSIS: We start by importing the data set:

```
DF3=read.csv("Ch6 rats.csv", header=T)
```

then verify the total number of observations and the number of levels of each categorical explanatory variable, as well as descriptive statistics of the data set:

```
> summary(DF3)
    source     amount        weight
 beef  :20   high:20    Min.   : 51.0
 cereal:20   low :20    1st Qu.: 75.5
                        Median : 88.5
                        Mean   : 87.2
                        3rd Qu.: 98.0
                        Max.   :118.0
```

They can be further examined for each of the four groups using the function `desc.stats` (Section 5.4) and the package `doBy`:

```
> desc.stats=function(x){c(mean=mean(x), median=median(x),sd=sd(x), n=length(x))}
> summaryBy(weight~source+ amount, data=DF3, FUN=desc.stats) # descriptive statistics
  source amount weight.mean weight.median weight.sd weight.n
1   beef   high       100.0         103.0      15.1       10
2   beef    low        79.2          82.0      13.9       10
3 cereal   high        85.9          87.0      15.0       10
4 cereal    low        83.9          84.5      15.7       10
```

We see that the means, medians, and standard deviations of the four groups are very similar, and the magnitude of the difference (high vs. low) in mean weight differs depending on the sources of protein: ~26% increase from low to high for beef vs. ~2% for wheat. This is a suggestion that the effect of the amount of protein depends on, or varies with, the source of protein. Thus, there might be an interaction between these two categorical explanatory variables that should be included in our statistical models.

Given that our four groups are defined by the 2 × 2 combinations of the factors `source` and `amount`, we should create a new categorical variable to depict them in a graph with `DF3$source.amount = as.factor(paste(DF3$source, DF3$amount, sep = "."))` (we used and explained this instruction in Section 3.4). Let's then examine the variation of the response variable `weight` among these four groups as shown graphically in Fig. 6.1. Violin plots (`geom_violin()`) are an alternative to boxplots that show the density distribution of each group at the expense of not showing descriptive statistics. We should not reach any conclusions on any density plot (nor on any boxplot, for that matter) based on just ten points per group.

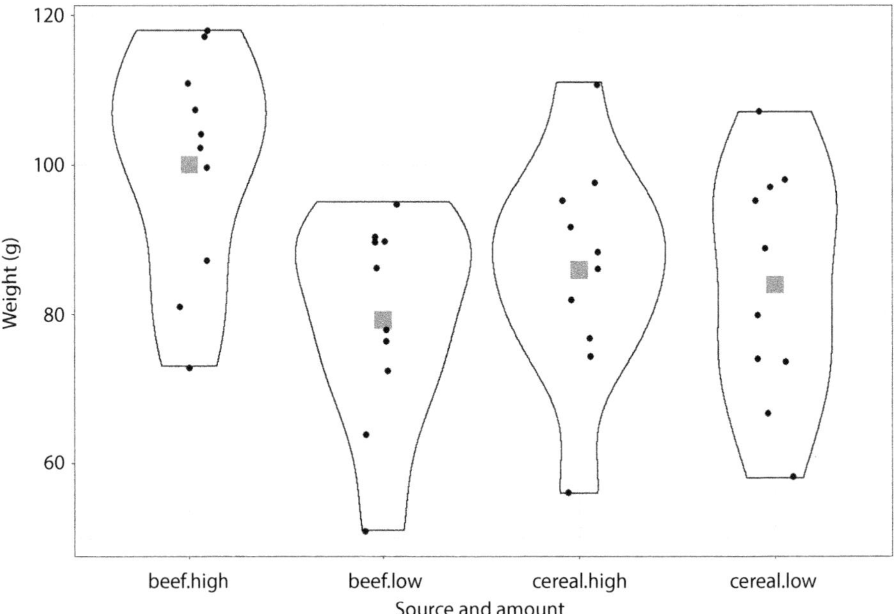

Fig. 6.1 Violin plots showing the variation of the weight of rats among combinations of the explanatory variables source and amount of protein. The group means are represented as gray squares.

We now need to assess which probability distribution can be used in the likelihood part of the statistical model. We use the package `fitdistrplus` to compare the normal and log-normal distributions that we again consider as plausible candidates to model the variation of the response variable:

```
norm.3=fitdist(DF3$weight, "norm")
lognorm.3=fitdist(DF3$weight,"lnorm")
```

Both probability distributions describe equally well the variation of the response variable (Fig. 6.2). We can postulate the statistical model `weight` ~ Normal(μ_Y = `source*amount`,

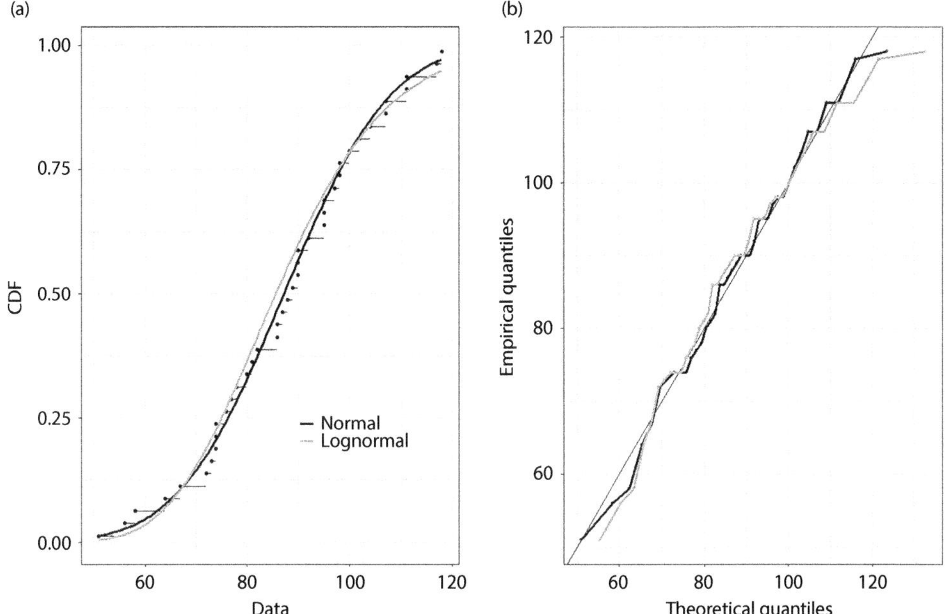

Fig. 6.2 Goodness of fit of the response variable `weight` showing the cumulative distribution function (a) and quantile–quantile plot (b) for two plausible probability distributions obtained with the package `fitdistrplus`.

σ_Y^2). The notation `source*amount` is a shorthand in R to indicate that the mean of the response variable μ_Y reflects the sum of the single effects of each factor and their interaction: `source + amount` + `source:amount` (see Table 4.1).

6.3 Factorial analysis of variance: Frequentist fitting

Let's fit the statistical model `m4 = lm(weight ~ source*amount, data = DF3)`, and examine the omnibus F-tests (see Chapter 5) that summarize the importance of the model terms:

```
> anova(m4)
Analysis of Variance Table

Response: weight
              Df Sum Sq Mean Sq F value Pr(>F)
source         1    221     221    0.99  0.327
amount         1   1300    1300    5.81  0.021 *
source:amount  1    884     884    3.95  0.054 .
Residuals     36   8049     224
```

Whenever a statistical model contains interaction(s) between explanatory variables, the importance of the interaction should be analyzed before any single effect in the model. Should an interaction be deemed important, we cannot disentangle single effects and must interpret them as a "new encompassing variable." In the context of this example, we could interpret the effect of changing the amount of protein on the mean rat weight without knowing the actual source of protein (beef or wheat). On the other hand, whenever an interaction between explanatory variables is not deemed important, we can separate their single effects on the response variable and interpret each of them in isolation. We can see that 0.054 lies not too far from the statistical significance threshold used to judge the relevance of a finding. While some scientists employ "weasel words" to wiggle out of being on the wrong side of the threshold, we are not going to play such a game. If we were to follow the well-ingrained tradition of the frequentist framework, we must say that the interaction between `source` and `amount` of protein was not statistically significant.

Beyond examining the global effect of the explanatory variables, the model summary provides the model parameter estimates:

```
> summary(m4)

Call:
lm(formula = weight ~ source * amount, data = DF3)

Residuals:
   Min     1Q Median     3Q    Max
-29.90  -9.90   2.05  10.85  25.10

Coefficients:
                       Estimate Std. Error t value Pr(>|t|)
(Intercept)              100.00       4.73   21.15   <2e-16
sourcecereal             -14.10       6.69   -2.11   0.0420
amountlow                -20.80       6.69   -3.11   0.0036
sourcecereal:amountlow    18.80       9.46    1.99   0.0545
---
Signif. codes:  0 '***' 0.001 '**' 0.01 '*' 0.05 '.' 0.1 ' '

Residual standard error: 15 on 36 degrees of freedom
Multiple R-squared:  0.23,	Adjusted R-squared:  0.166
F-statistic: 3.58 on 3 and 36 DF,  p-value: 0.023
```

Using an alphabetical criterion, the reference level for the factor `source` is beef, whereas for the factor `amount` it is high (i.e., those that do not appear in the table of coefficients). Therefore, the combination beef–high is the reference group for the analysis (cf. Table 6.3) and it is shown as `(Intercept)`. Knowing that the mean weight for the reference group beef–high equals 100.0 g with a standard error of 4.73 is of little relevance to our goal of estimating the magnitudes of the differences of means between the four groups. For the reference level `amount` = high, `sourcecereal` is the magnitude of the difference between the means of `source` = beef (i.e., the reference level) and `source` = cereal. The latter is 14.1 g (or –14% in this case) lighter than the former, and the standard error of this difference of means is 6.69, which happens to be statistically different from zero. Likewise, for the reference level `source` = beef, `amountlow` (–20.8 g or –20.8% in this case) is the magnitude of the difference between the means of `amount` = high (i.e., the reference level) and `amount` = low, with an interpretation analogous to that just given for `sourcecereal`. Finally, the term `sourcecereal:amountlow` (+18.8 g) denotes the additional differential beyond the single effects of `cereal` and `source` that needs to be added to predict the mean of the last group defined by the combination cereal–low. We can also see that the estimated value of the variance of the response variable `weight` is $15.0^2 = 225$ g^2. The statistical model explains 16.6% of the variation of the response variable `weight`. Finally, the last line of the summary involves an *F*-test of the model. An *F*-statistic of 3.58 is obtained as the ratio of the mean squared of the three sources explaining variation in the model (i.e., `source`, `amount`, and `source:amount`) and the mean squared of the error. The numerator is obtained by dividing the sum of the squares of these sources of variation (221 + 1200 + 884) by the sum of their degrees of freedom (3) to give 768.3, which, divided by 224, gives 3.58 (see, e.g., Winer et al. 1991). The *F*-statistic at the bottom of the summary table is of limited interest and will not be discussed any further. We could obtain such a table with `anova(m4)`.

We can now replace the contents of Table 6.2 with the parameter estimates of the model summary to obtain the predicted means of the response variable for each group (see Table 6.3). And with these estimated means we can plot the implied probability distributions associated with the group means and their estimated standard deviation (Fig. 6.3).

Table 6.3 Means of `weight` for each of the four groups defined by the combinations of the levels of the categorical explanatory variables `amount` (low or high) and `source` (beef or cereal) using the parameter estimates of model `m4`. The group high.beef defined by combination levels of the explanatory variables was used as the "reference group."

		Source	
		Beef	Cereal
Amount	High	100	100 − 14.1 = 85.9
	Low	100 − 20.8 = 79.2	100 − 14.1 − 20.8 + 18.8 = 83.9

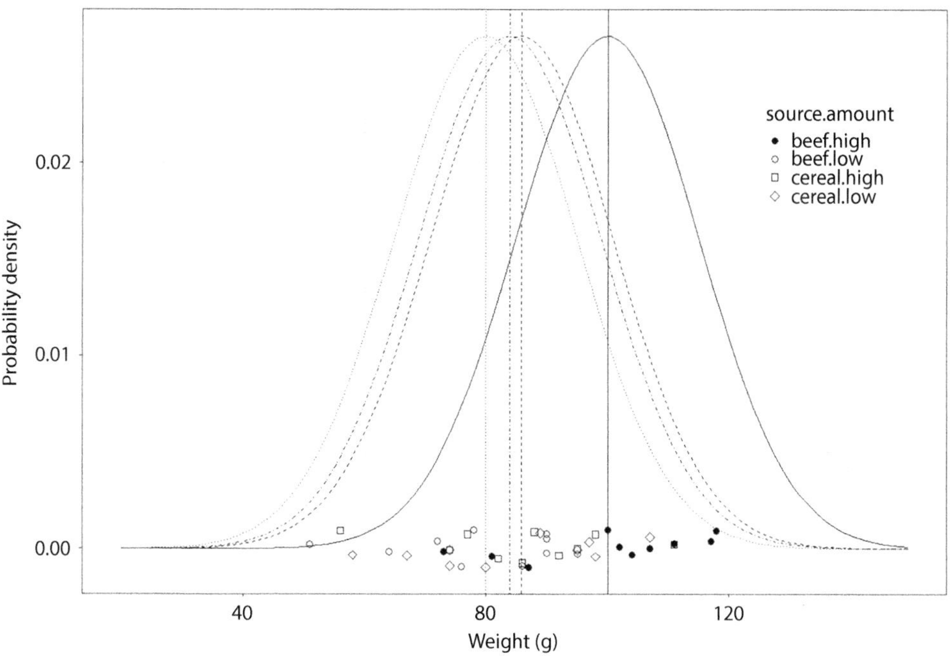

Fig. 6.3 Probability distributions of `weight` for each of the four groups defined by the 2 × 2 combinations of levels of the factors `source` and `amount`. The mean of each distribution and their common standard deviation were obtained from the model summary (see main text). A scatterplot of the actual values of `weight` for each group is show at the bottom.

The model parameters and the confidence intervals can be put in a table using the command `tidy` from the package `broom` :

```
> tidy(m4, conf.int = T)
# A tibble: 4 x 7
  term                   estimate std.error statistic  p.value conf.low conf.high
  <chr>                     <dbl>     <dbl>     <dbl>    <dbl>    <dbl>     <dbl>
1 (Intercept)                100       4.73      21.1  6.84e-22   90.4     110.
2 sourcecereal               -14.1     6.69     -2.11  4.20e- 2  -27.7      -0.538
3 amountlow                  -20.8     6.69     -3.11  3.64e- 3  -34.4      -7.24
4 sourcecereal:amountlow      18.8     9.46      1.99  5.45e- 2   -0.380    38.0
```

and even better summarize the table in a plot (Fig. 6.4).

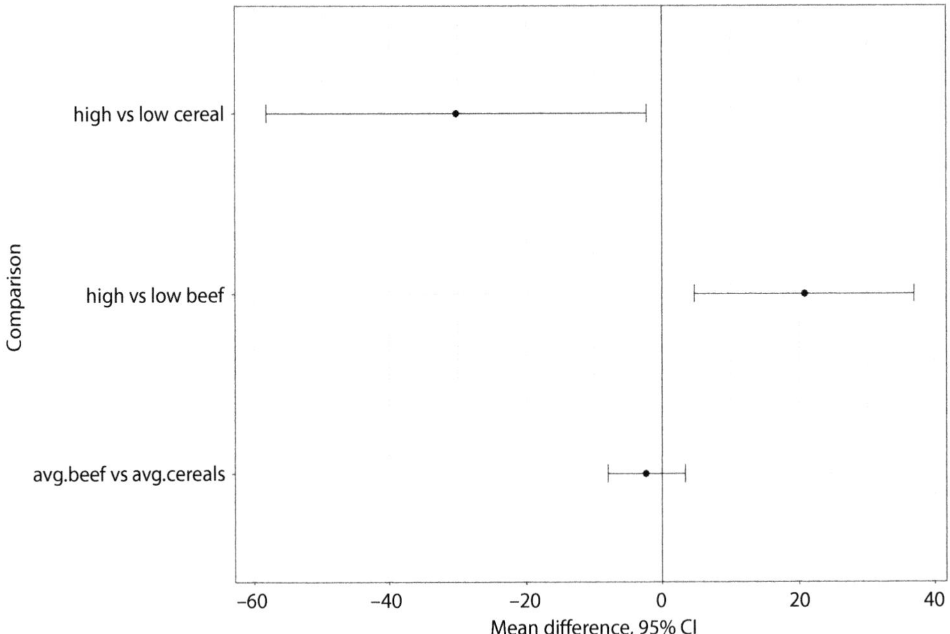

Fig. 6.4 Parameter estimates and their 95% confidence intervals for the model m4.

Now, let's carry out just the three specific a posteriori tests that we pointed out as relevant for the questions we asked: amount of protein (high vs. low) within each source of protein (beef and cereal), and the average gain for high vs. the average gain for low, regardless of the source of protein. We start by finding out the order of the levels:

```
> levels(DF3$source.amount)
[1] "beef.high"   "beef.low"    "cereal.high" "cereal.low"
```

We use this information to define the following contrast vectors (see Section 5.4):

```
> d1=c(1,-1,0,0)
> d2=c(0,0,1,-1)
> d3=c(-0.5,-0.5,0.5,0.5)
> contr.mat=rbind("high vs low beef"=d1,
+                 "high vs low cereal"=d2,
+                 "avg.beef vs avg.cereals"=d3)
```

Note that while that we are using the variable `source.amount` to define these contrasts, the latter was not part of model `m4`. Hence, we need to fit a model with the composite explanatory variable `source.amount` : `m4.1 = lm(weight ~ source.amount, data = DF3)`. We can finally carry out the desired a posteriori tests using the package multcomp: `m4.1post=glht(m4.1, linfct=mcp(source.amount=contr.mat),test=adjusted("fdr"))`

The results are:

```
> summary(m4.1.post)

         Simultaneous Tests for General Linear Hypotheses

Multiple Comparisons of Means: User-defined Contrasts


Fit: lm(formula = weight ~ source.amount, data = DF3)

Linear Hypotheses:
                             Estimate Std. Error t value Pr(>|t|)
high vs low beef == 0           20.80       6.69    3.11   0.0091
high vs low cereal == 0        -30.20      11.58   -2.61   0.0319
avg.beef vs avg.cereals == 0    -2.35       2.36   -0.99   0.5699
```

We can observe that the magnitude of the weight difference of means high vs. low was 50% higher for cereal than for beef, and that both were statistically significant. Just as for Fig. 5.22, we can display the results (point estimates of the differences of means and their confidence intervals) in a figure. Finally, we need to validate the model `m4` with a residual analysis. We will first gather all the variables to be used in a single data frame with `res.m4 = augment(m4)`. We also require the variable `source.amount` (that was not part of model `m4`): `res.m4$source.amount = DF3$source.amount`. The four relevant plots are shown in Fig. 6.5, where we can see that the fit of the model to the data is pretty good. The residuals have a random scatter with similar variability (Fig. 6.5a) that is similar for the four groups (Fig. 6.5b). The residuals appear to have a normal distribution (Fig. 6.5c), and no data point seems to have an inordinate effect on the estimated parameters (similar Cook distances, Fig. 6.5d).

Having validated the model, we should summarize the conditional plots obtained with the predictions, `pred.m4=ggpredict(m4, terms = c("source", "amount"))`, using the package `ggeffects`, to show the main results in a graph. Let's recall that our main goal was to compare the average `weight` of rats among the groups defined by the combination of the factors `source` and `amount`, and, by so doing, we could also assess the global effect of these categorical variables and of their interaction.

The main difference between Fig. 5.16 and Fig. 6.6 is that while the former is just a plot of the model parameters and their 95% confidence intervals that is useful for visualizing effect sizes, the latter displays the means and their 95% confidence intervals for the four groups being compared. Therefore, they should be viewed as complementary, rather than as redundant, displays of the information generated by the statistical model `m4`.

6.4 Factorial analysis of variance: Bayesian fitting

Equation (5.8) contains the constrained version of the statistical model for the factorial analysis of variance. Focusing on the constrained model makes it easier to set the prior distributions of its parameters. We will use the package `brms` to fit the factorial analysis of variance for the rat weight experiment. Let's see the default priors for this model:

```
> get_prior(formula=weight~source*amount, data=DF3,family=gaussian)
                 prior     class                     coef group resp dpar nlpar lb ub       source
                (flat)         b                                                           default
                (flat)         b                amountlow                             (vectorized)
                (flat)         b             sourcecereal                             (vectorized)
                (flat)         b sourcecereal:amountlow                               (vectorized)
 student_t(3, 88.5, 17) Intercept                                                          default
    student_t(3, 0, 17)     sigma                                                0         default
```

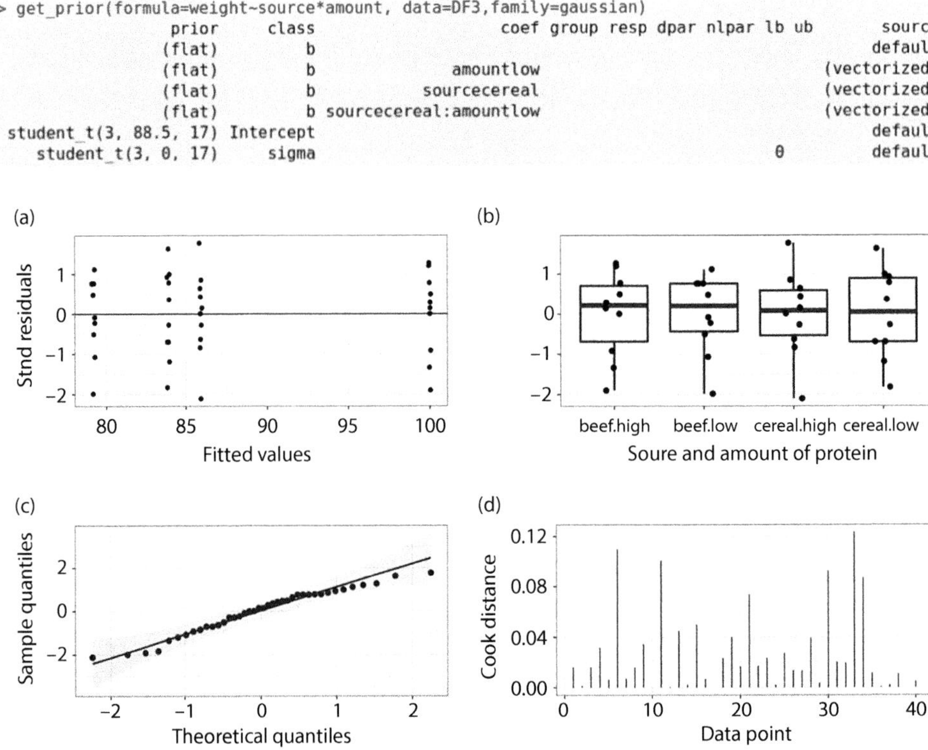

Fig. 6.5 Residual analysis of model m4, showing residuals vs. fitted values (a), a boxplot of the residuals vs. the categorical explanatory variable (b), a quantile–quantile plot of the residuals (c), and the Cook distances (d).

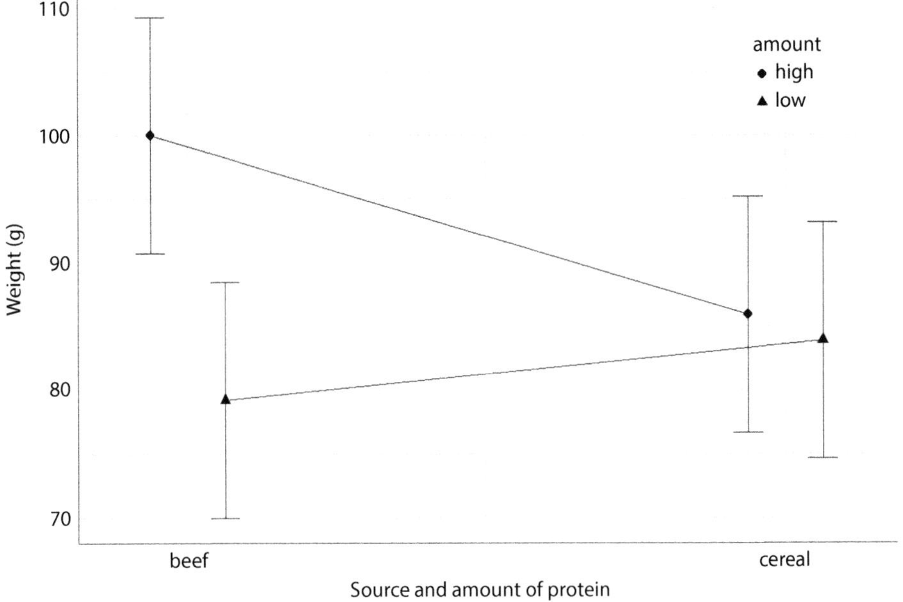

Fig. 6.6 Conditional plot of the predicted means and 95% confidence intervals of model m4 generated with the package ggeffects.

The default priors assigned by `brms` for the parameters are all weakly informative priors.

We now need to set the priors. Let us presume that we have gathered empirical information about the weight average and standard deviation for that rat species or for a similar one. Say that a literature search (or consultation with an expert) provided a published mean and 95% confidence interval for rat weight based on 25 samples under "natural conditions" (whatever that might mean): 80.1 ± 10.4. Based on the well-known expression for the 95% confidence interval for the mean (e.g., Winer et al. 1991), we can compute the standard deviation to be $10.4 \times (25-1)^{0.5} / 2.05 = 24.9$, where 2.05 is the 97.5% quantile of a t-distribution for 25 − 1 degrees of freedom. It would seem hard to set priors for the single effect of `source` (beef vs. cereal) and for `amount` (high vs. low) because these are very loosely defined categories. But perhaps we might sensibly anticipate that any diet could change the average weight of rats by no more than ±25% for any experiment of comparable duration. The average change in weight for rats subject to any diet might then lie in the range [56 g, 93 g], which should help us define the following priors:

- Average weight of reference group: Intercept ~ Lognormal($\mu_{\text{log.weight}}$ = log(74.8), $\sigma_{\text{log.weight}}$ = 0.125). Given the mean, we could have used a normal distribution and still not obtained impossible negative weights. The value of 0.125 was set to obtain the range of plausible mean weight values.
- Generic effect of diet on average weight: These are the other three parameters under the heading "b," and we could set them as Normal(0, 0.25 × 74.8).
- Standard deviation of weight: We could use either a gamma distribution or a half-Cauchy distribution for this strictly positive parameter of the class `sigma`. Let's choose Half-Cauchy(location = 24.9, scale = 5) with a debatable but plausible scale = 5 to permit a range of feasible standard deviations of individual rat weight.

The setting of priors is accomplished with:

```
prior.m4 = c(set_prior("lognormal(log(80.1), 0.125)", class = "Intercept"),
             set_prior("normal(0, 0.25*74.8)", class = "b"),
             set_prior("cauchy(24.9,5)", class = "sigma"))
```

A plot helps verify that priors reflect the plausible ranges; see Fig. 6.7.

We can now fit the model using the same specifications discussed in Section 5.4:

```
m4.brms=brm(formula=weight~source*amount, data=DF3, family=gaussian, prior =
prior.m4, warmup = 1000, future=TRUE, chains=3, iter=3000, thin=3)
```

The summary is:

```
> summary(m4.brms)
 Family: gaussian
  Links: mu = identity; sigma = identity
Formula: weight ~ source * amount
   Data: DF3 (Number of observations: 40)
Samples: 3 chains, each with iter = 3000; warmup = 1000; thin = 3;
         total post-warmup samples = 2000

Population-Level Effects:
                     Estimate Est.Error l-95% CI u-95% CI Rhat Bulk_ESS Tail_ESS
Intercept               96.71      4.92    86.94   106.15 1.00     1841     1851
sourcecereal            -9.95      6.49   -22.57     2.94 1.00     2025     1866
amountlow              -15.94      6.56   -28.43    -3.10 1.00     1911     1830
sourcecereal:amountlow  11.78      8.64    -5.13    28.66 1.00     1970     1816

Family Specific Parameters:
      Estimate Est.Error l-95% CI u-95% CI Rhat Bulk_ESS Tail_ESS
sigma    16.29      2.20    12.67    21.07 1.00     1836     1782
```

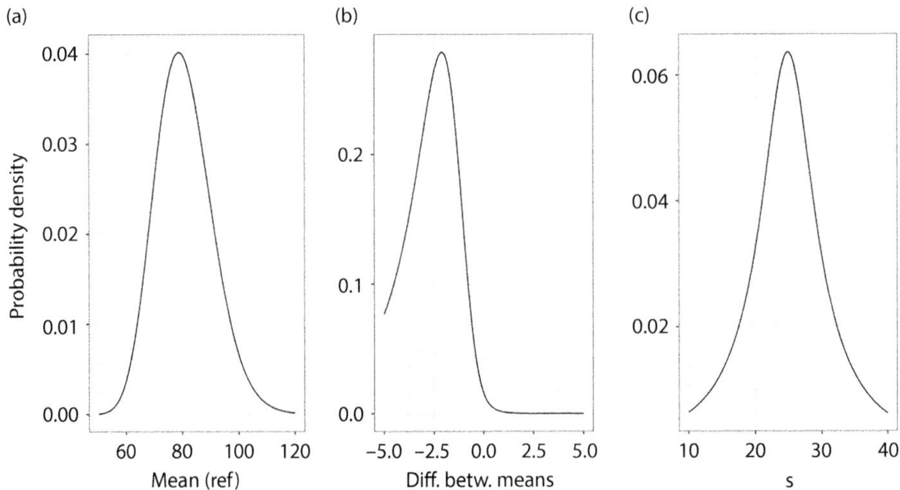

Fig. 6.7 Prior distributions for the parameters (see main text for an explanation of the their meaning) of the Bayesian factorial analysis of variance `weight ~ source*amount` (model `m4.brms`).

We begin by assessing the metrics of model convergence (`Rhat`) and effective sample size (`ESS`) of all the model parameters, and find them both very good: `Rhat` $\approx$ 1 and `ESS` > 1000.

The chains of model `m4.brms` had good mixing properties (the trace plot is not shown, but see the code on the companion web site), and the parameter estimates of each chain have a very low autocorrelation such that they can essentially be considered as statistically independent (Fig. 6.8). The three chains have converged to a common stationary distribution (Fig. 6.9) that can be relied upon for inferential purposes.

Returning to the summary of model `m4.brms`, we have the means, standard errors, and 95% credible intervals for the parameters determining the mean (`Population-Level Effects`) and the standard deviation (`Family Specific Parameters`) of the response variable `weight`. These are summary statistics obtained by combining the three chains to form marginal posterior distributions (Fig. 6.9). The interpretation of these parameters is identical to those discussed in Section 6.3, as their meaning is independent of the framework employed in their estimation. Better than showing a table of numbers is to depict the same information in a figure (Fig. 6.10).

The model explains a modest amount of the variance of the response variable (Gelman et al. 2018):

```
> bayes_R2(m4.brms)
   Estimate Est.Error   Q2.5 Q97.5
R2     0.19    0.0886 0.0321 0.363
```

At this point, we could combine the chains with `chains.m4.brms = data.frame(as.mcmc(m4.brms, pars = m4.pars, combine_chains = T))` in order to plot a single posterior distribution for each parameter against their prior distributions (cf. Fig 5.22). Recalling the fundamental equation of Bayesian statistics, Posterior ~ Likelihood × Prior (Chapter 3), if the data contained a sufficient amount of empirical information, the posterior would differ from the prior distribution.

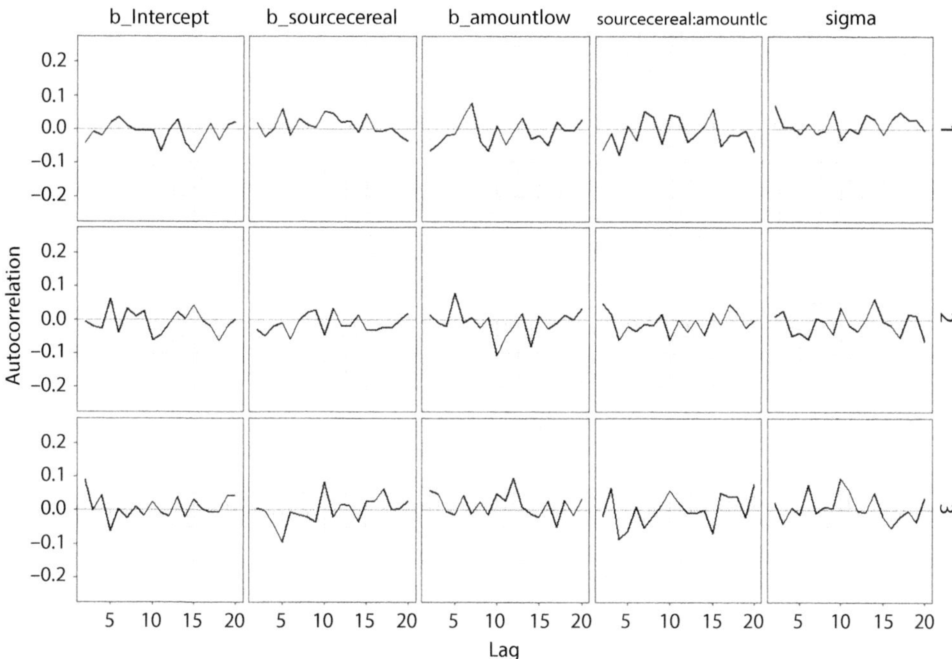

Fig. 6.8 Autocorrelation plots for the three chains of each parameter of model m4.brms.

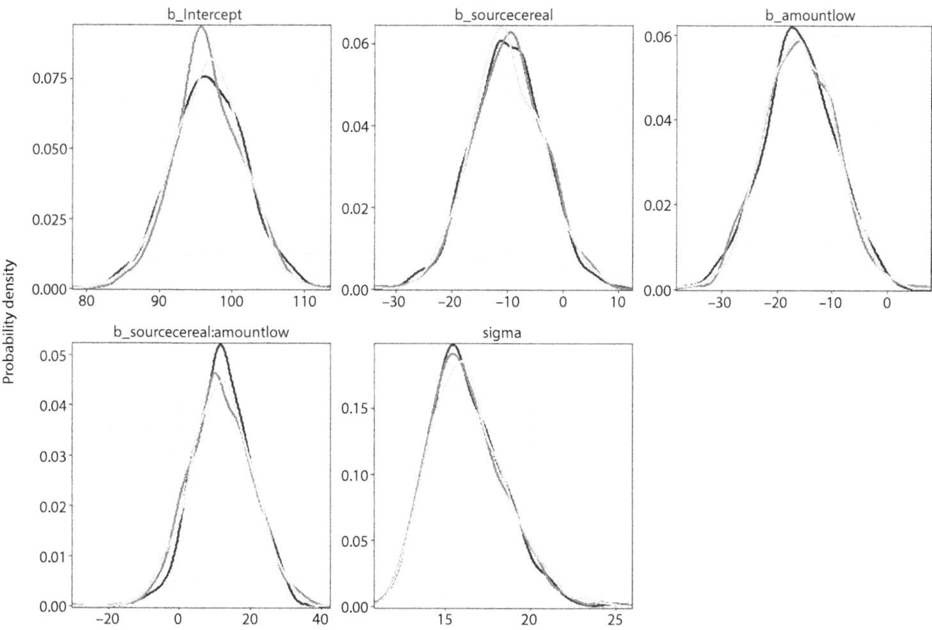

Fig. 6.9 Marginal posterior distributions of the parameters of model `m4.brms` showing the convergence of the three chains to a marginal common stationary distribution.

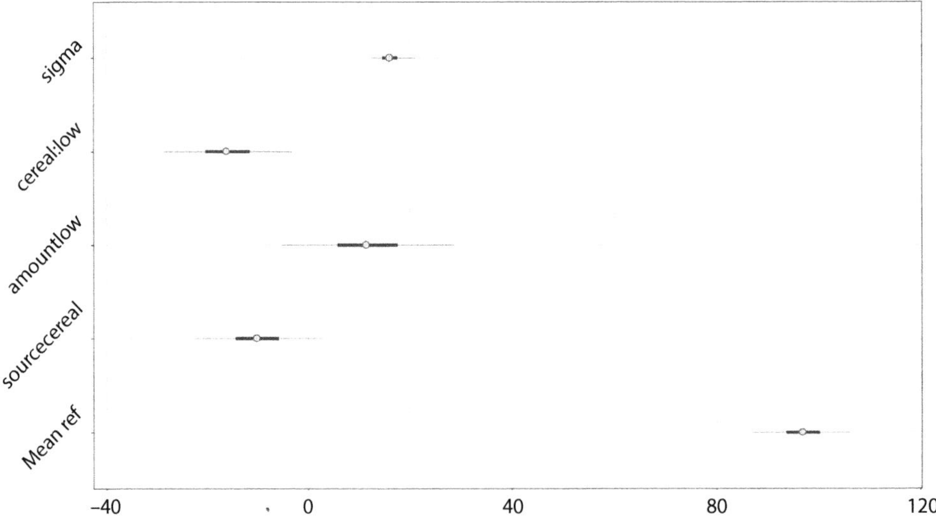

Fig. 6.10 Means (dots) and 50% (thick) and 95% (thin) credible intervals of the parameters of model m4.brms. The parameter names are explained in the main text.

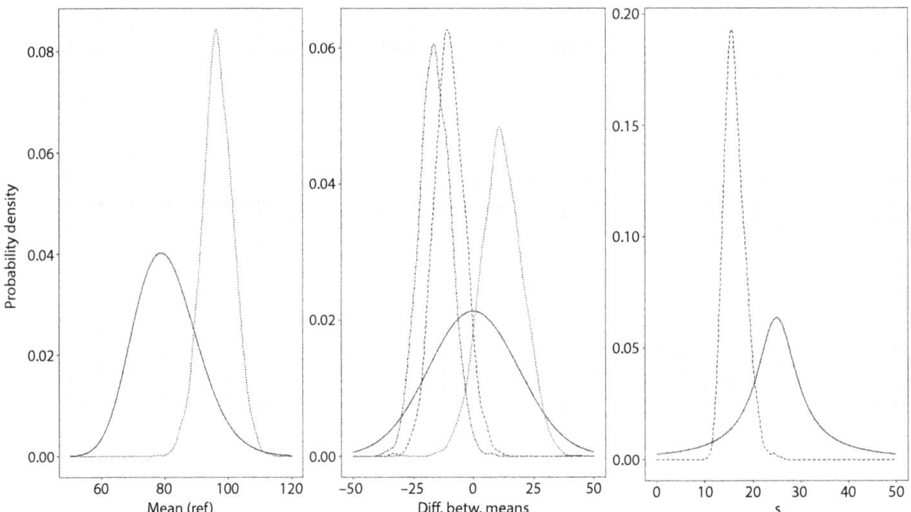

Fig. 6.11 Overlap between the prior (continuous lines) and posterior distributions for the three parameters of the Bayesian factorial analysis of variance (cf. Fig. 5.22). The three lines in the central plot correspond to the mean differences between `source = cereal` and `source = beef` for `amount = low` (long dash), the interaction (long dash, see Table 6.3), between `amount = high` and `amount = low` for `source = cereal`.

We see from Fig. 6.11 that all the posterior distributions were narrower than the priors (i.e., we learned something) and that the specific effects (source and/or amount of protein) were always smaller than the 25% that we had envisaged. By combining the changes, we now have 2,000 estimates of each parameter:

```
> head(chains.m4.brms)
  b_Intercept b_sourcecereal b_sourcecereal.amountlow b_amountlow    sigma
1    92.00322       3.603621                 7.715992  -18.607873 13.86565
2   105.28607     -24.616235                18.172388  -19.216927 15.13031
3    96.94309      -7.101060                13.466076  -21.831456 14.95189
4    88.42101      -1.398223                 4.789305   -7.441631 15.29174
5    93.54611      -2.197092                -3.921079  -12.068296 19.03985
6   102.58724     -14.200233                 4.170646  -18.056882 14.10258
```

We can now obtain 2,000 estimates of the posterior distribution of the means for each group by simply adding the columns of `chains.m4.brms` as we did to obtain Table 6.3. We might also easily generate the posterior distribution of the percentage difference between specific pairs of means.

Finally, we validate the model `m4.brms` with residual analysis. As in Section 6.3, we start by putting the means of the residuals and fitted values for each data point in the same data frame that holds the original data:

```
res.m4.brms=as.data.frame(residuals(m4.brms, type="pearson", nsamples=1000, summary=T))
fit.m4.brms=as.data.frame(fitted(m4.brms, scale="linear", summary=T, nsamples=1000))
DF3$resid=res.m4.brms$Estimate # means of Pearson residuals for each data point
```

and then generate the plots in Fig. 6.12, which suggest that the model m4.brms fits the data reasonably well. The overall assessment of the goodness of fit of m4.brms is essentially the same as for its frequentist counterpart, and it does not need repeating here.

Having validated the model, we must obtain one final figure to visualize the main model findings in terms of the means of the four groups and their 95% credible intervals (Fig. 6.13).

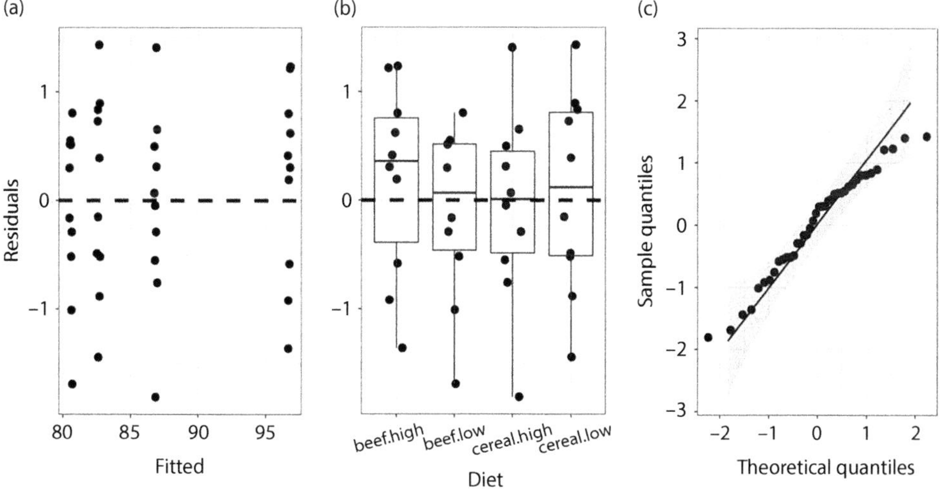

Fig. 6.12 Residual analysis of model `m4.brms`, showing the mean of the Pearson residuals vs. the mean of the fitted value (a), the mean of the Pearson residuals vs. the explanatory variable in the model (b), and a quantile–quantile plot of the mean of the Pearson residuals (c).

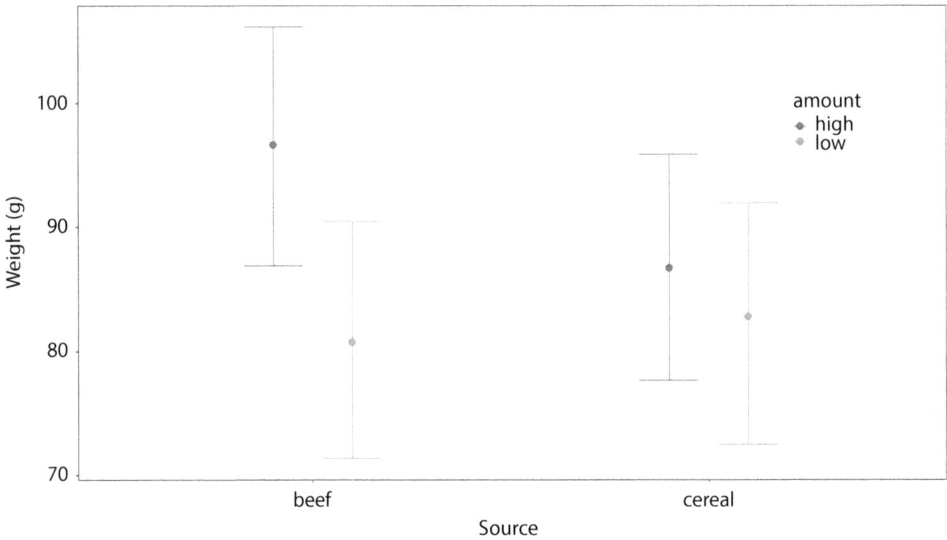

Fig. 6.13 Conditional plots predicting the mean and 95% credible intervals of `weight` depending on the `source` and `amount` of protein in the diet from model `m4.brms`, obtained with the package `ggeffects`.

6.5 Analysis of covariance: Mixing continuous and categorical explanatory variables

Analysis of covariance is yet another variation of the general linear model in which the mean of a response variable with a normal distribution now reflects the influence of numerical and categorical explanatory variables. That is, analysis of covariance combines what we learned in Chapter 4 with what we have been covering thus far in this chapter.

To keep it simple for now, let's assume that we have only one explanatory variable of each type. The linear relationship between the mean of the response variable μ_Y and the numerical explanatory variable would be denoted by a straight line defined by its intercept β_0 and slope β_1 (Chapter 4). The effect of the categorical explanatory variable (say with two levels) would be manifested as a difference in the mean of the response variable μ_Y according to the factor level. But since μ_Y varies with the numerical explanatory variable, it follows that any effect of the categorical explanatory variable would be reflected as differences in the slope and/or intercept according to the levels of the factor.

In this simple example, four cases are logically possible: the intercept (but not the slope) varies with the factor, the slope (but not the intercept) varies with the factor, both the intercept and the slope vary with the factor, and neither the intercept nor the slope change between the factor levels. Expressed in the most general form (i.e., the third case just mentioned), the statistical model of analysis of covariance is

$$Y \sim \text{Normal}\left(\mu_Y = \beta_{0,j} + \beta_{1,j}X, \sigma_Y^2\right), \tag{6.5}$$

where the slope and the intercept may differ depending on the categorical variable $j = 1,2$ and σ_Y^2 is the variance of the response variable around its mean predicted by the categorical and numerical explanatory variables.

Analysis of covariance was devised by Fisher (1935) with the original goal of discounting the effect of a numerical explanatory variable in order to enable consideration of the differences between the levels of a factor in the context of a controlled experiment (see Chapter 14). In this context, the numerical explanatory variable is called a "confounding variable" whose effect on the mean of the response variable is known (or can be anticipated) in advance. The advance knowledge of the potential impact of the confounding variable on the response variable is the very reason we recorded its value.

THE DATA IN CONTEXT: We are going to use a microbiological data set from Daalgard (2006) and aim to show the effect of glucose (a categorical variable with two levels, yes or no) on the diameter of bacteria (the response variable, in microns). This experiment involves the use of cultures in flasks under controlled conditions in which the concentration of bacterial cells per milliliter can never be identical between flasks. While the concentration cannot be tightly controlled and kept constant, it can be measured for it is known in advance that it affects the cell diameter. There were 32 flasks with added glucose and 19 with only a a mineral medium to serve as the control.

EXPLORATORY DATA ANALYSIS: Let's import the data and make a first summary:

```
> DF5=read.csv("Ch6 bacteria.csv", header=T)
> summary(DF5)
 glucose        conc              diameter
 NO :19    Min.   : 11000    Min.   :19.2
 YES:32    1st Qu.: 27500    1st Qu.:21.4
           Median : 69000    Median :23.3
           Mean   :164325    Mean   :23.0
           3rd Qu.:243000    3rd Qu.:24.4
           Max.   :631000    Max.   :26.3
```

How are these variables related? Figure 6.14 shows that the diameter and the concentration have a negative exponential relationship $y = a\exp(-bx)$, as shown by the smoothed lines. Applying logarithms on both sides, we can linearize the relationship between these variables as required in the general linear model. We can also see that the slopes seem similar but the intercepts differ between the two groups defined by `glucose`.

Given that we are going to use `log(diameter)` as the response variable, we should verify which probability distribution would better describe its overall variation, comparing yet again the normal and lognormal distributions using the package `fitdistrplus`:

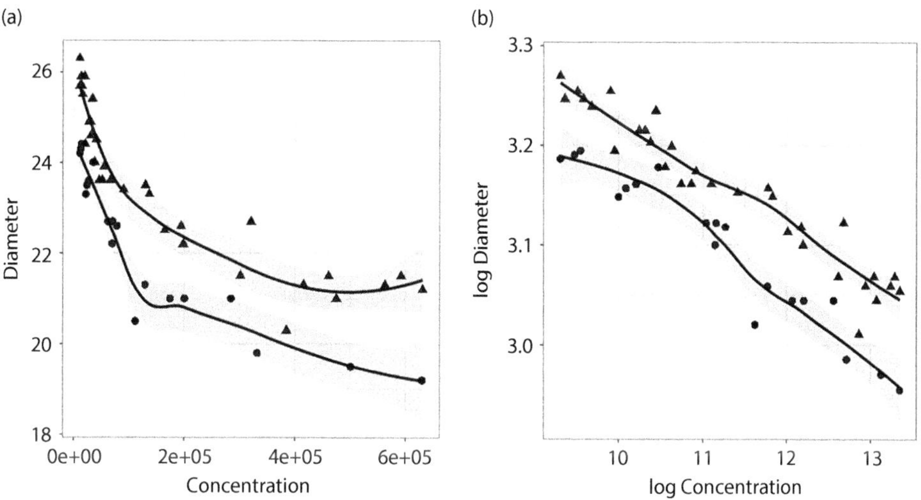

Fig. 6.14 Relation between cell diameter and bacterial concentration depending on the presence of glucose in the original and in a logarithmic scale.

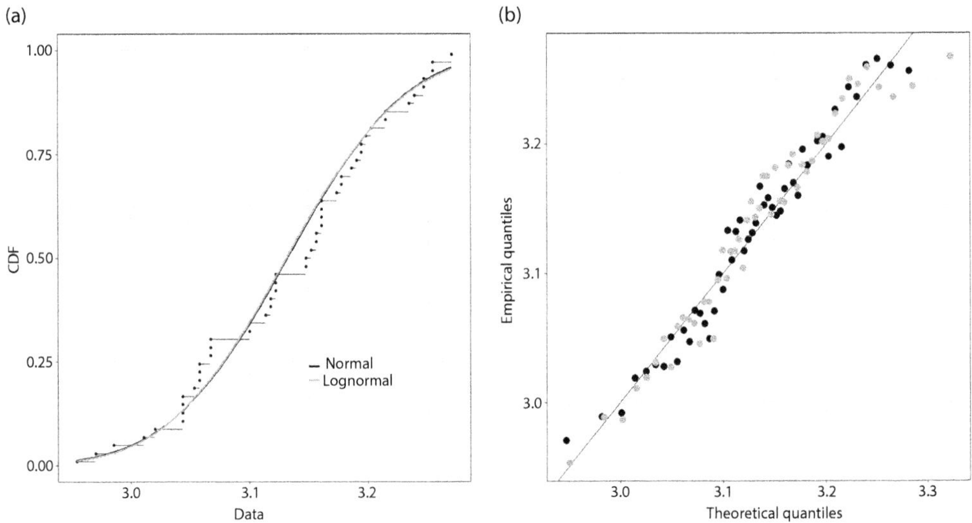

Fig. 6.15 Goodness of fit assessment of the response variable log(conc) showing the cumulative distribution function (a) and a quantile–quantile plot (b) obtained using the package `fitdistrplus`.

```
bac.norm=fitdist(log(DF5$diameter),"norm")
bac.lognorm=fitdist(log(DF5$diameter),"lnorm")
```

Figure 6.15 shows that the normal and lognormal probability density distributions are essentially indistinguishable for the response variable `log(diameter)`. Thus, we will use the former to comply with one of basic assumptions of the general linear model (Chapter 4).

6.6 Analysis of covariance: Frequentist fitting

We fit the following model (see Table 4.1 for the R syntax):

```
> summary(m5)

Call:
lm(formula = log(diameter) ~ log(conc) * glucose, data = DF

Residuals:
     Min       1Q   Median       3Q      Max
-0.06153 -0.01125  0.00013  0.00867  0.04054

Coefficients:
                     Estimate Std. Error t value Pr(>|t|)
(Intercept)           3.76418    0.04422   85.12   <2e-16 *
log(conc)            -0.05968    0.00392  -15.22   <2e-16 *
glucoseYES           -0.00787    0.05456   -0.14     0.89
log(conc):glucoseYES  0.00648    0.00482    1.34     0.19
---
Signif. codes:  0 '***' 0.001 '**' 0.01 '*' 0.05 '.' 0.1 '

Residual standard error: 0.0209 on 47 degrees of freedom
Multiple R-squared:  0.936,     Adjusted R-squared:  0.932
F-statistic:  230 on 3 and 47 DF,  p-value: <2e-16
```

The coefficient associated with the numerical explanatory variable `log(conc)` is obviously the slope (–0.059) that was estimated with great precision (SE = 0.00392), and it turns out to be significantly different from zero. But since there are two groups (`glucose`: YES or NO), what slope is this? The coefficient `glucoseYES` is the difference of intercepts relative to the reference group (`glucoseNO`) that is in `(Intercept)`. Hence, the intercept for `glucoseNO` is 3.76 and that of `glucoseYES` is 3.76 – 0.07 = 3.69. The slope of `log(conc)` is the the slope of the reference group `glucoseNO`. Finally, the term `log(conc):glucoseYES` is the magnitude of the differences between the slopes of `glucoseNO` and `glucoseYES`. The latter is very small (0.00648), and it is not significantly different from zero. Therefore, we can conclude that that the two lines are nearly parallel with a small difference in their intercepts (cf. Fig. 6.15b). The fitted statistical model explains 93.2% of the variation of the response variable `log(diameter)`, whose standard deviation was estimated to be 0.0209. In sum, we have two distributions yielding the probability density of obtaining bacterial cells of certain `log(diameter)` depending on `log(conc)` and the presence of `glucose`:

$$\text{log(diameter)}_{\text{GLUCOSE=NO}} \sim \text{Normal}(\mu_Y = 3.76 - 0.059\text{log(conc)}, \sigma_Y = 0.0209),$$
$$\text{log(diameter)}_{\text{GLUCOSE=YES}} \sim \text{Normal}(\mu_Y = 3.69 - 0.032\text{log(conc)}, \sigma_Y = 0.0209).$$

The omnibus *F*-tests of the analysis of variance table evaluate the sequential addition of explanatory variables into the model (cf. Chapter 4):

```
> anova(m5)
Analysis of Variance Table

Response: log(diameter)
                  Df Sum Sq Mean Sq F value  Pr(>F)
log(conc)          1 0.2486  0.2486  571.44 < 2e-16
glucose            1 0.0503  0.0503  115.70 2.9e-14
log(conc):glucose  1 0.0008  0.0008    1.81    0.19
Residuals         47 0.0204  0.0004
```

The interaction `log(conc):glucose` denotes the strength of the changes in the slope that relates the numerical explanatory variable `log(conc)` to the mean of `log(diameter)` between the two groups defined in the factor `glucose`. Its *F*-value (1.81) does not pass the customary threshold of 0.05, which is consistent with the small difference between these slopes (0.00648; SE = 0.00482) previously mentioned. Therefore, we may consider the effect of `glucose` and of `log(conc)` on the mean of `log(diameter)` to be largely independent of each other. `log(conc)` has the strong effect on the response variable that we saw in Fig. 6.15; `glucose` reflects the difference in intercepts between the two levels of this factor, and it happens to be statistically significant. There might seem to be a contradiction between results of `anova(m5)` and `summary(m5)` for `glucose`. However, let's recall that they correspond to very different things. While `summary(m5)` denotes the effect of each parameter in isolation as if the other parameters were not in the model, `anova(m5)` depicts the effect of "batches of coefficients" conforming sets of sources of variation that are sequentially evaluated, and hence they are conditional on the terms that entered the model previously.

Let's now reflect on the interpretation of an intercept in model `m5`: it would be the `log(diameter)` of a bacteria grown in a culture of `log(conc)` = 0 that has just one bacterium. What we can do to provide a better meaning to the intercepts is to center (subtract the mean) the variable `log(conc)` : `DF5$log.conc =`

`as.vector(scale(log(DF5$conc), scale = F, center = T))`. The analysis of variance table is identical to the one for model m5, and needs no further comment:

```
> m5.1=lm(log(diameter)~log.conc*glucose, data=DF5)
> anova(m5.1)
Analysis of Variance Table

Response: log(diameter)
                 Df Sum Sq Mean Sq F value  Pr(>F)
log.conc          1 0.2486  0.2486  571.44 < 2e-16 ***
glucose           1 0.0503  0.0503  115.70 2.9e-14 ***
log.conc:glucose  1 0.0008  0.0008    1.81    0.19
Residuals        47 0.0204  0.0004
```

The only changes from `summary(m5)` to `summary(m5.1)` are in the values and interpretation of `(Intercept)` and `glucoseYES`. After centering the numerical explanatory variable, the `(Intercept)` is now the mean `log(diameter)` for the condition `glucose` = NO for a culture having the average `log.conc` = 11.3,[1] which is something meaningful and easily interpretable. The difference in the intercepts of 0.0652 is now statistically significant, in agreement with what we found in `anova(m5)` :

```
> summary(m5.1)

Call:
lm(formula = log(diameter) ~ log.conc * glucose, data = DF5

Residuals:
     Min       1Q   Median       3Q      Max
-0.06153 -0.01125  0.00013  0.00867  0.04054

Coefficients:
                    Estimate Std. Error t value Pr(>|t|)
(Intercept)          3.09133    0.00479  645.26  < 2e-16 **
log.conc            -0.05968    0.00392  -15.22  < 2e-16 **
glucoseYES           0.06520    0.00605   10.78  2.7e-14 **
log.conc:glucoseYES  0.00648    0.00482    1.34     0.19
---
Signif. codes:  0 '***' 0.001 '**' 0.01 '*' 0.05 '.' 0.1 '

Residual standard error: 0.0209 on 47 degrees of freedom
Multiple R-squared:  0.936,     Adjusted R-squared:  0.932
F-statistic:  230 on 3 and 47 DF,  p-value: <2e-16
```

We can gather (cf. Section 5.4) the parameter estimates of model `m5.1` and their 95% confidence intervals with `tidy(m5.1, conf.int = T)` and depict this table as a plot using the command `ggcoef` from the package `GGally`. However, the differences of four orders of magnitude in the estimates of `summary(m5.1)` nullify the interest in the plot.

Let's then validate model `m5.1` with residual analysis and start by gathering all the variables to plot in `res.m5.1 = augment(m5.1)` ; the plots are shown in Fig. 6.16. We can see that the quality of model fit was barely acceptable. For a start, the plots of the first row do not show all the residuals forming a "parallel window" around zero, as there were two "extreme" residuals with values smaller than 2. The Q–Q plot shows a few points lying close but definitely outside the 95% confidence band. And finally, the data of the seventh row had a Cook distance much larger than the rest, perhaps suggesting that it

[1] `mean(log(DF5$conc))`
11.3

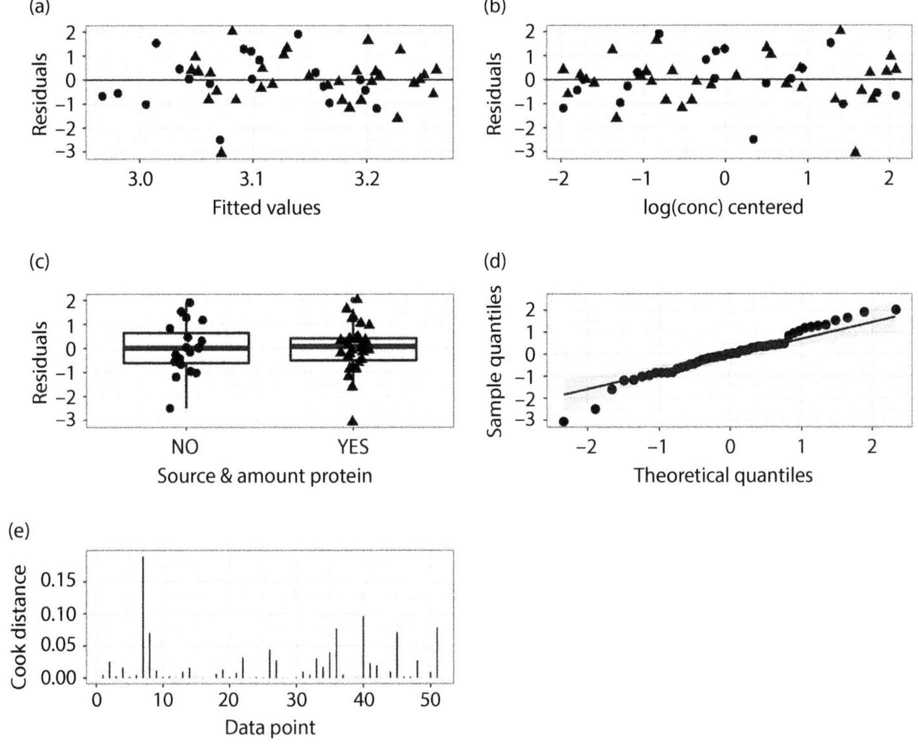

Fig. 6.16 Residual analysis of model m4, showing the residuals vs. the fitted values (a), a boxplot of the residuals vs. the categorical explanatory variable (b), a quantile–quantile plot of the residuals (c), and the Cook distances (d).

had a large influence on the parameter estimates. Let's identify these values with extreme residuals and large Cook distance (0.19):

```
> res.m5.1[res.m5.1$.std.resid < -2,]
# A tibble: 2 x 8
  `log(diameter)` log.conc glucose .fitted .std.resid   .hat .sigma .cooksd
            <dbl>    <dbl> <fct>     <dbl>      <dbl>  <dbl>  <dbl>   <dbl>
1            3.01    1.59  YES        3.07      -3.07 0.0748 0.0189  0.190
2            3.02    0.342 NO         3.07      -2.49 0.0583 0.0196  0.0963
```

If we run `m6 = lm(log(diameter) ~ log.conc*glucose, data = DF5[res.m5.1$.std.resid > 2,])` for all data not having these "extreme" residuals, we find no qualitative differences with model `m5`, and that all the parameter estimates and R^2 differed by less than 1/1000 between the two models (results not shown). While this quick exercise does not assure us that these two "offending" points are either correct or valid, it may suggest that the less-than-perfect model fit would not greatly affect the main conclusions to be extracted from this model.

The final step is to generate a visual depiction of the model findings, which is definitely preferable to tables of numbers. We generate the data to be plotted using the command `ggpredict` from the package `ggeffects`: `pred.m51 = ggpredict(m5.1, terms = c("log.conc", "glucose"))`; `pred.m51` holds the set of values of the response variable predicted by the model `m5.1` using the single effects of above-listed explanatory variables and their interaction. Figure 6.17 shows that the response variable (for no glucose) changes from (approximately) 25 to 19 microns ((25 – 19) / 19 = –31.6%) in the span

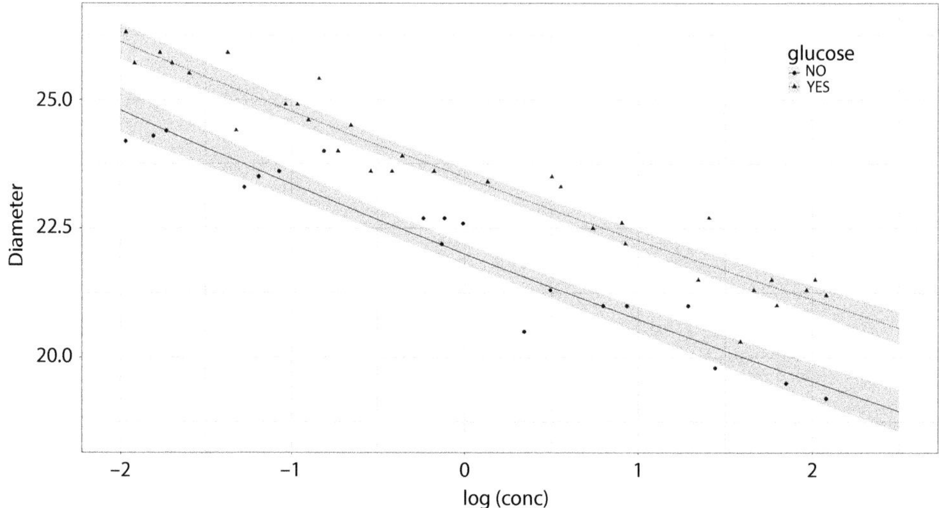

Fig. 6.17 Predicted relationships between the response and explanatory variables from the model m5.1. Note that the response variable was back-transformed to allow the plotting of the data, and that the numerical explanatory variable `log(conc)` is centered and shown in units of standard deviations. The value of zero corresponds to `mean(log(DF5$conc))` = 11.3, and each unit correponds to `sd(log(DF5$conc))` = 1.29, such that −1 corresponds to 11.3 − 1.29 = 12.01 in log scale or 164,391 cells in the original units.

of values of concentration. However, for any concentration (since the curves are almost parallel), the presence of glucose changes the average diameter by exp(0.06) = 1.06, which is 6% bigger. While glucose certainly has a positive impact on cell diameter, being able to carefully control the culture concentration would seem to have more effect on the outcome of the experiment.

6.7 Analysis of covariance: Bayesian fitting

The Bayesian fitting as always starts by setting the priors for the model parameters. We will use Eq. (6.4) as the statistical model. Let's check which are the model parameters whose priors must be defined:

```
> get_prior(formula=log(diameter)~log.conc*glucose, data=DF5,family=gaussian)
                  prior     class                coef group resp dpar nlpar lb ub        source
                 (flat)         b                                                       default
                 (flat)         b          glucoseYES                              (vectorized)
                 (flat)         b            log.conc                              (vectorized)
                 (flat)         b log.conc:glucoseYES                              (vectorized)
 student_t(3, 3.1, 2.5) Intercept                                                       default
   student_t(3, 0, 2.5)     sigma                                           0           default
```

We have four parameters in `class` = `b` that determine the mean of the response variable `log(diameter)`; the other parameter of `class` = `sigma` is the standard deviation of the response variable. In descending order, they are the intercept (or mean of `log(diameter)` for the reference group `glucoseNO`), the difference of intercepts between `glucoseYES` and the reference group, the slope of `log.conc` (i.e., log(conc) centered to have a mean of zero for the reasons explained above) for the reference group, and the difference in slopes between `glucoseYES` and the reference group.

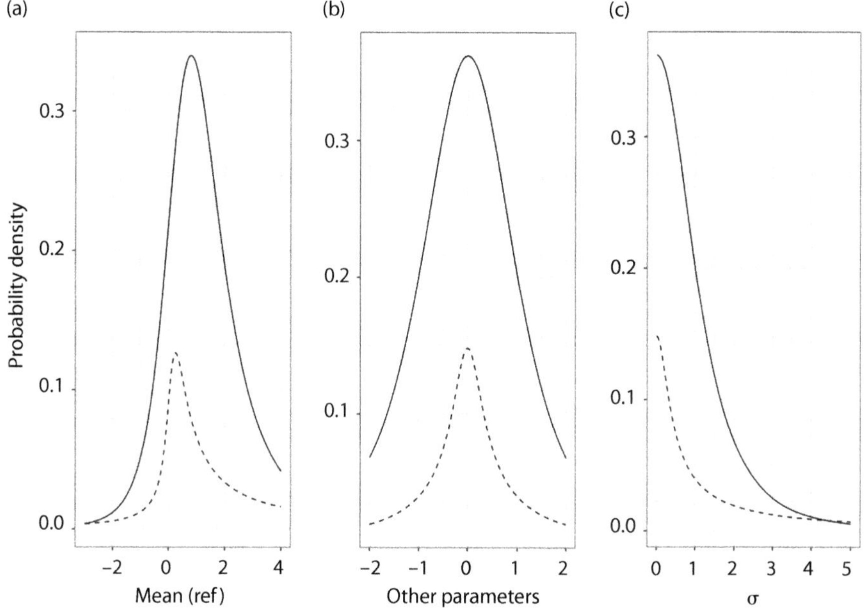

Fig. 6.18 Weakly and strongly (dashed lines) informative priors for parameters of the model m5.brms showing different degrees of support for certain ranges of each parameter.

We are going to slightly adapt these default weakly informative priors in a first analysis, and then change them for strongly informative priors to assess their influence on the posterior distributions.[2] We see in Fig. 6.18 that the main differences between the two types of priors is the concentration of probability for some ranges of values of each parameter.

The weakly informative priors we are going to use are:

```
> weakprior.m5 = c(set_prior("student_t(3, 2.5, 2.5)", class = "Intercept"),
+                  set_prior("student_t(3, 0, 2.5)", class = "b"),
+                  set_prior("student_t(3, 0, 2.5)", class = "sigma"))
```

and the model summary for weakly informative priors is:

```
> summary(m5.brms.w)
 Family: gaussian
  Links: mu = identity; sigma = identity
Formula: log(diameter) ~ log(conc) * glucose
   Data: DF5 (Number of observations: 51)
Samples: 3 chains, each with iter = 3000; warmup = 1000; thin = 3;
         total post-warmup samples = 2000

Population-Level Effects:
                   Estimate Est.Error l-95% CI u-95% CI Rhat Bulk_ESS Tail_ESS
Intercept              3.76      0.04     3.67     3.85 1.00     1522     1319
logconc               -0.06      0.00    -0.07    -0.05 1.00     1525     1375
glucoseYES            -0.01      0.06    -0.12     0.11 1.00     1490     1581
logconc:glucoseYES     0.01      0.00    -0.00     0.02 1.00     1475     1593

Family Specific Parameters:
      Estimate Est.Error l-95% CI u-95% CI Rhat Bulk_ESS Tail_ESS
sigma     0.02      0.00     0.02     0.03 1.00     1530     1784
```

[2] The means and variances of the *t*-distribution used in brms are a somewhat complex function of the two parameters `df` and `ncp`. The interested reader can find details in Evans et al. (1996). The parameterization used in `brms` (and Stan) of the *t*-distribution differs from the one used in the R package `base`.

The strongly informative priors we are going to use are:

```
> strongprior.m5 = c(set_prior("student_t(300, 2.5, 2.5)", class = "Intercept"),
+                    set_prior("student_t(300, 0, 2.5)", class = "b"),
+                    set_prior("student_t(300, 0, 2.5)", class = "sigma"))
```

What changes between the two set of priors is the number of degrees of freedom of the *t*-distribution that essentially determines its degree of flatness: the fewer degrees of freedom, the flatter and the less informative the prior distribution.

```
m5.brms.s=brm(formula=log(diameter)~log(conc)*glucose, data=DF5,family=gaussian,
prior = strongprior.m5, warmup = 1000, future=TRUE, chains=3, iter=3000, thin=3)
```

And the summary is:

```
> summary(m5.brms.s)
 Family: gaussian
  Links: mu = identity; sigma = identity
Formula: log(diameter) ~ log(conc) * glucose
   Data: DF5 (Number of observations: 51)
Samples: 3 chains, each with iter = 3000; warmup = 1000; thin = 3;
         total post-warmup samples = 2000

Population-Level Effects:
                   Estimate Est.Error l-95% CI u-95% CI Rhat Bulk_ESS Tail_ESS
Intercept              3.76      0.05     3.67     3.86 1.00     1380     1621
logconc               -0.06      0.00    -0.07    -0.05 1.00     1321     1578
glucoseYES            -0.01      0.06    -0.12     0.10 1.00     1308     1536
logconc:glucoseYES     0.01      0.01    -0.00     0.02 1.00     1302     1498

Family Specific Parameters:
      Estimate Est.Error l-95% CI u-95% CI Rhat Bulk_ESS Tail_ESS
sigma     0.02      0.00     0.02     0.03 1.00     1614     1625
```

The differences in the descriptive statistics (means, standard errors, and 95% credible intervals) of the posterior distributions between the weakly and strongly informative priors are really tiny. Our prior sensitivity analysis showed that the amount of information contained in the data largely swamped the priors and determined the posterior distribution obtained. This by no means needs to be the case every time, hence our interest in carrying out a prior sensitivity analysis.

Let's quickly assess the convergence of model `m5.brms.s`. All the chains converged to a common stationary distribution (Fig. 6.19) as also suggested by the `Rhat` values of the model summary. The trace plots (not shown, but see the script on the companion web site) also indicate very good mixing of the chains. On the other hand, the autocorrelation plot (again, not shown) shows the sampled values to be essentially independent, in agreement with the large (> 1,000) values of `Bulk_ESS` above.

Having established that the chains converged to a common stationary distribution, we can obtain the marginal probability densities of the slopes and intercepts of the two groups defined by the glucose. Doing so just requires recalling the meaning of the estimated parameters of the linear model that we discussed before:

```
chains.m5.brms.s=data.frame(as.mcmc(m5.brms.s,pars=m5.pars, combine_chains = T))
```

The names of the variables in `chains.m5.brms.s` are:

```
names(chains.m5.brms.s)
[1] "b_Intercept" "b_logconc" "b_logconc.glucoseYES" "b_glucoseYES" "sigma"
```

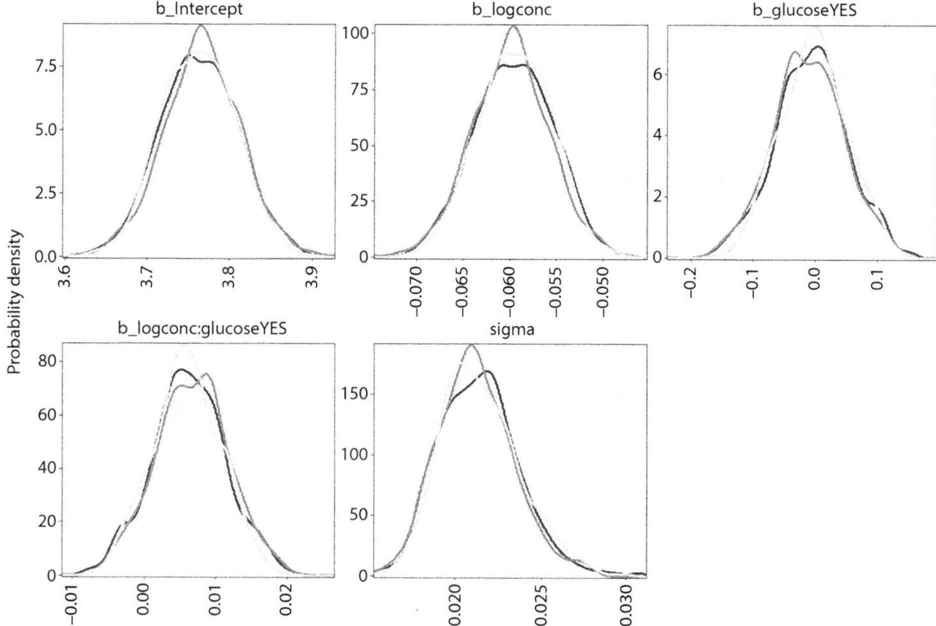

Fig. 6.19 Marginal posterior distributions of the parameters of model m5.brms.s showing the convergence of the three chains to a marginal common stationary distribution.

We can now do the following:

```
params.m5.brms.s=data.frame(int.GlucNO=chains.m5.brms.s$b_Intercept)
params.m5.brms.s$int.GlucYES=chains.m5.brms.s$b_Intercept+
                                    chains.m5.brms.s$b_glucoseYES
params.m5.brms.s$slope.GlucNO=chains.m5.brms.s$b_logconc
params.m5.brms.s$slope.GlucYES=chains.m5.brms.s$b_logconc+
                                    chains.m5.brms.s$b_logconc.glucoseYES
params.m5.brms.s$sigma=chains.m5.brms.s$sigma
```

Now we have 2,001 statistically independent estimates of the slopes and intercepts of the two regression lines estimated from the posterior distribution of `m5.brms.s`. We can compute any statistic we wish and use it to make inferences. For instance, what is the probability of the ratio of slopes being smaller than one?

```
table (params.m5.brms.s$slope.GlucYES/params.m5.brms.s$slope.GlucNO > 1))/2001
FALSE  TRUE
0.897 0.103
```

There is a probability of 0.103 of the slope of glucose = YES being greater than the slope of glucose = NO as calculated from a posterior distribution of ratios based on 2,001 values. These are trivial calculations and are just shown to illustrate the wealth of possibilities available once we have the posterior distribution of a statistical model.

We can also visualize the covariation among the marginal posterior distributions of each model parameter. Figure 6.20 shows what we already know from any simple linear regression: that the values of the slope and the intercept are strongly and negatively correlated. We also see that while the marginal distributions of the slopes and intercepts are very similar to a Gaussian distribution, that of the standard deviation is asymmetric and it has a longer upper tail. This is common for parameters constrained by a hard boundary such as the standard deviation having to be strictly positive.

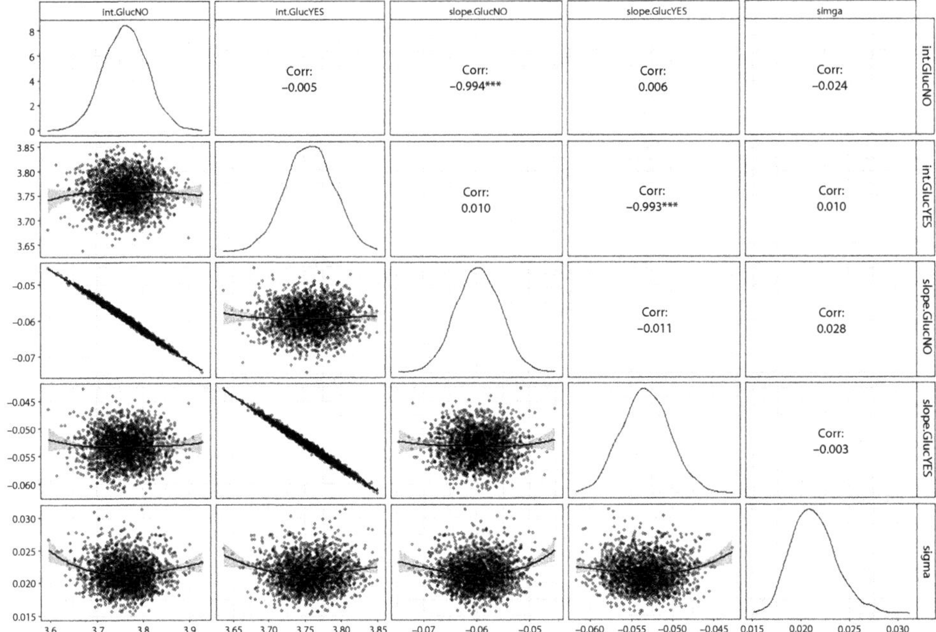

Fig. 6.20 Marginal posterior distributions (diagonal) and covariation (below diagonal) of the parameters of model `m5.brms.s`; the correlations between pairs of parameters are shown above the diagonal. These plots are based on the 2,001 sampled parameter estimates obtain by merging the three chains of the statistical model.

`m5.brms.s` explains a large percentage of the variance of the response variable (Gelman et al. 2018):

```
> bayes_R2(m5.brms.s)
   Estimate Est.Error  Q2.5 Q97.5
R2    0.932   0.00614 0.917 0.939
```

The mean R^2 is very close to the value we obtained in the frequentist fitting.

The validation of the model m5.brms.s proceeds in the same way as for previous Bayesian models. We must first calculate and then store the average values of the residuals and the fitted values of `m5.brms.s` for each point:

```
res.m5.brms.s=as.data.frame(residuals(m5.brms.s, type="pearson",
nsamples=1000, summary=T))
fit.m5.brms.s=as.data.frame(fitted(m5.brms.s, scale="linear", summary=T,
nsamples=1000))
DF5$fit=fit.m5.brms.s$Estimate
DF5$resid=res.m5.brms.s$Estimate
```

Now we can make the usual plots for the model validation. Figure 6.21 shows the same problems that we commented on for the frequentist fitting of the same data (Fig. 6.17): 2 (one for each group) out of 92 (≈2.1%) data points lay outside the parallel window in [–2, 2], and these two points also lie outside the envelope of the Q–Q plot. Should we discard these two "offending" points? No, absolutely not. They are part of the hard-won evidence gained in the experiment, and they by no means compromise either the reliability of the parameter estimates or the main conclusions that can be obtained from an otherwise well-fitting statistical model.

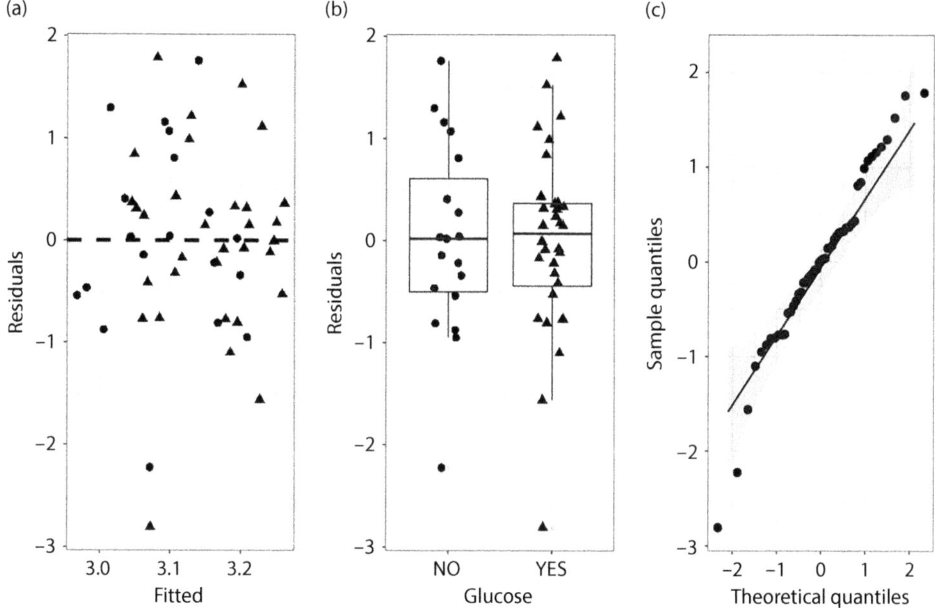

Fig. 6.21 Marginal posterior distributions of the slopes and intercepts of the regressions for the groups defined by the factor `glucose`, and the standard deviation of the response variable `log(diam)`. All distributions were obtained from the posterior distribution of model `m5.brms.s`.

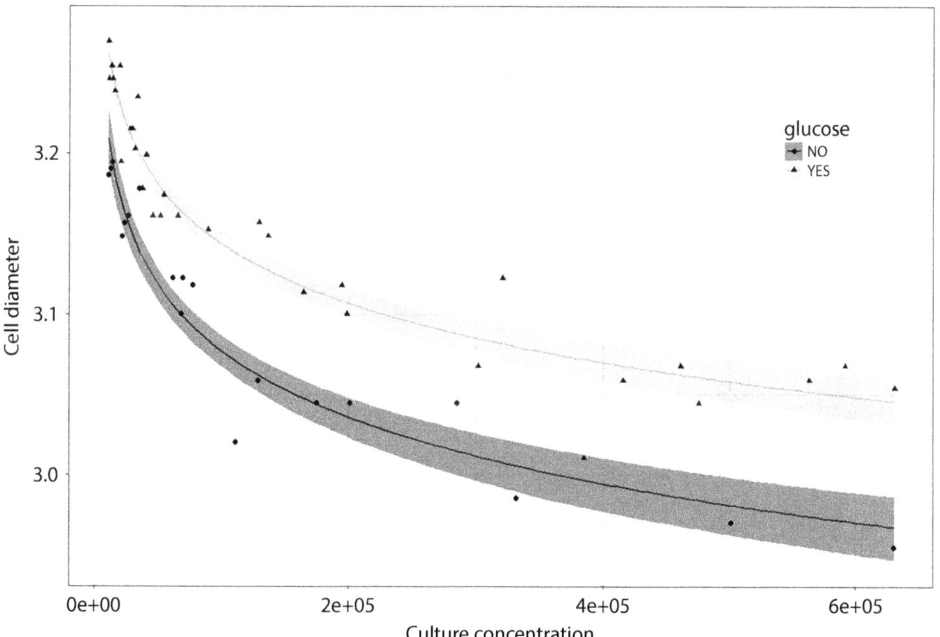

Fig. 6.22 Conditional plot of the relationship between the response and explanatory variables for the model `m5.brms.s`.

As usual, the final figure renders the visual representation of the fitted curves with the estimated parameters and the data. Figure 6.22 shows that the model `m5.brms.s` provided a very good fit to the experimental data. Of course, the two parallel negative exponential curves correspond to the parallel straight lines of Fig. 6.15 obtained with the frequentist fitting.

This chapter concludes the presentation of the general linear model. What we have learned about the fitting of statistical models involving interactions between explanatory variables, and the interpretation of their parameters, can and will be used in the coming chapters.

6.8 Problems

Visit the companion website at http://www.oup.com/companion/InchaustiSMWR to obtain the data sets for these problems.

6.1 Naya and Božinovič (2006) studied the physiological adjustments to changes in food availability and tail loss in the South American lizard in the laboratory. They used changes in food availability as an indicator of competition, and tail autotomy as a proxy of predation pressure because voluntary tail loss occurs when a predator attack occurs or is imminent. They measured two categorical experimental factors: Food (Normal/Low) and Autotomy (Yes/No), and the response variable metabolic.rate (in joules/h). File: Pr6-1.csv

6.2 The same authors carried out another experiment with the same lizard species to investigate whether animals adjusted their internal organs to meet their functional demands. They hypothesized that lizards reared with unrestricted food would have larger digestive tracts (and associated organs) than lizards reared with low food availability, after adjusting for body mass. They measured the following variables: dry weight of the small intestine (small.int; g), body mass (g), and the same categorical variables as in the previous problem. File: Pr6-2.csv

References

Daalgard, P. (2006). *Introductory Statistics with R*. Springer, New York.

Evans, M., Hastings, H., and Peacock, B. (1996). *Statistical Distributions*, 3rd edn. John Wiley & Sons, New York.

Fisher, R. (1935). *The Design of Experiments*, 1st edn. Oliver and Boyd, Edinburgh.

Fox, J. (2015). *Applied Regression Analysis and Generalized Linear Models*, 3rd edn. Sage, New York.

Gelman, A. (2005). Analysis of variance: Why it is more important than ever. *Annals of Statistics*, 33, 1–33.

Gelman, A., Goodrich, B., Gabry, J. et al. (2018). R-squared for Bayesian regression models. *American Statistician*, 73, 307–309.

Hand, D., Daly, R., McConway, F. et al. (eds.) (2001). *A Handbook of Small Data Sets*. Springer, New York.

Naya, D. and Božinovič, F. (2006) The role of ecological interactions on the physiological flexibility of lizards. *Functional Ecology*, 20, 601–608.

Winer, B., Brown, D., and Michels, K. (1991). *Statistical Principles in Experimental Design*, 3rd edn. McGraw Hill, New York.

CHAPTER 7

Model Selection

One, two, and more models fitted to the data

R packages needed in this chapter:

```
packages.needed = c("ggplot2","gridExtra","reshape2","brms")
lapply(packages.needed, FUN = require, character.only = T)
```
loads these packages in the working session.

7.1 Introduction

Scientists may often fit several statistical models to the same data, particularly in exploratory data analysis. These models may include different sets of explanatory variables, interactions between sets of variables, polynomial terms, etc. With $k = 4$ numerical explanatory variables, we could formulate a single model containing all variables, three models with trios ($X_1 + X_2 + X_3$, $X_1 + X_2 + X_4$, $X_1 + X_3 + X_4$), six models with pairs ($X_1 + X_2$, $X_1 + X_3$, $X_1 + X_4$, $X_2 + X_3$, $X_2 + X_4$, $X_3 + X_4$), and four models with just one explanatory variable. That is, 14 candidate statistical models to describe a single data set. But if it just depends on brute force or on vague and loosely defined hypotheses, there are many more possibilities still within the realm of linear models. Arbitrarily trying many models until finding one or more that attain statistical significance is a pervasive misbehavior known as *p*-hacking that plagues the practice of statistics (e.g., Simmons et al. 2011, Gelman and Loken 2014, Simonsohn et al. 2014). The fitting of an arbitrarily large number of models could be called "the squid approach to scientific research," aiming to hide our ignorance and lack of clear thinking behind a cloud of ink under the pretense of "letting the data speak" to find a "truth." Rest assured: data are silent. Clarity and simplicity in a plausible explanation may only emerge after clear thinking and specific analysis stemming from scientific hypotheses.

This chapter discusses the main ideas of model selection in statistics based on the principle of parsimony. This is the conceptual basis of model selection employing criteria based on the theory of information. We discuss the Akaike information criterion (AIC) used in the frequentist framework to arrive at the most parsimonious model or to obtain the "average model." Thereafter, we explain two criteria used in the Bayesian framework, the deviance information criterion (DIC) and the widely applicable information criterion (WAIC). In order to understand the latter, we first need to explain the posterior predictive distribution, and show its connection to posterior predictive checks, an important tool for the validation of Bayesian models. Finally, we explain the prior predictive distribution, a recently developed tool for calibrating the priors of Bayesian models. The chapter will

Statistical Modeling With R. Pablo Inchausti, Oxford University Press.
 DOI: 10.1093/oso/9780192859013.003.0007

be largely conceptual as we defer the application of these tools of model selection and validation to subsequent chapters.

7.2 The problem of model selection: Parsimony in statistics

Whenever we consider more than one statistical model, having a criterion to select the "best" model becomes an issue. Within the realm of general linear models (Chapters 4 to 6), the R^2 can only increase for higher numbers of explanatory variables and terms (reflecting, for example, interactions) in a statistical model. Thus, if we base the selection of the "best" model on R^2, we should always add to it "everything, including the kitchen sink." This is because an explanatory variable containing random numbers would account for some minuscule fraction of the variation of the response variable, and hence increase R^2. If measures of goodness of fit inevitably improve with model complexity, where should one stop? Besides the non-trivial task of finding intelligible interpretations for overly complex models, it is well known that statistical models with too many unjustifiable terms (i.e., variables and interactions) are very prone to overfit the data and produce absurd predictions. Polynomial regressions provide a simple example to illustrate these issues. We simulated 10 points of a response variable with a normal distribution using second-degree polynomial data (`DF = data.frame(x = runif(n = 10, min = 100, max = 500)), DF$y = rnorm(n = 10, mean = -1.2 + 2*DF$x, sd = 100)`), and then fitted linear, quadratic, and ninth-degree polynomials, thus having two, three, and ten parameters (including the intercept).

Figure 7.1 displays the well-known attribute of high-degree polynomials of making horribly wrong predictions for certain ranges of the explanatory variables. We also obtain steadily decelerating increases of R^2 as the model becomes more over-fitted. By over-fitting we mean that a statistical model is overly sensitive to little details and small idiosyncrasies in the data. By being sensitive to minute details in the data probably related to

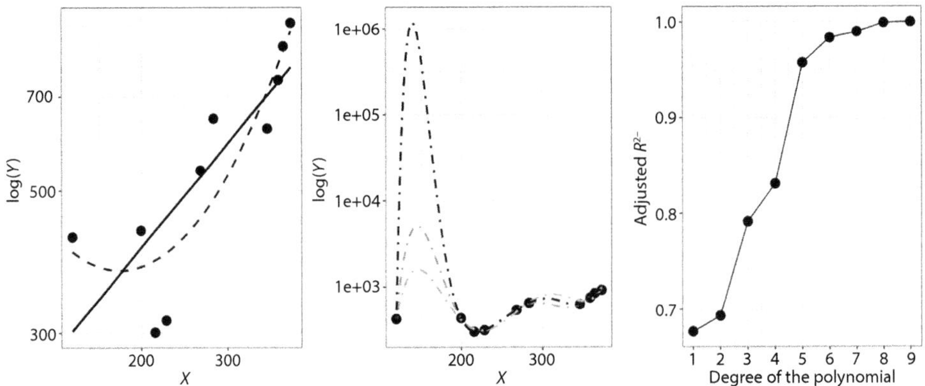

Fig 7.1 Fitting of 10 points with polynomial regressions of increasing degree. The data was generated from $Y \sim \text{Normal}(\mu_Y = -1.2 + 2X, \sigma_Y = 100)$, with the values of X being drawn from a uniform distribution in [100, 500]. The first two panels show the fitting of data by a straight line and a quadratic regression (left), and by polynomial regressions of fifth (gray), seventh (dark gray), and ninth (black) degrees. The increase in the polynomial degree leads to better fitting of the actual data but to worse predictions for non-observed data. The right panel shows the decelerating increase of R^2 with the polynomial degree.

random variation and labeled as "noise," an over-fitting model badly reflects the general trends and its parameters are estimated with small precision (i.e., large standard errors). Under-fitting models behave in the opposite way, tending to fail to identify the effects of relations that are actually supported by the data; they provide biased estimates of model parameters. True, under-fitting models tend to be "robust" or "insensitive" (depending on whether we adopt a positive or negative viewpoint) to changes in a few data points because they only very roughly reflect the regular trends of the data (Burnham and Anderson 2002). In statisticians' parlance, under-fitting models (in our example, polynomials of too low a degree) tend to be biased, and overfitting ones (in our example, polynomials of too high a degree) tend to estimate parameters with poor precision and provide spurious identification of the effects of explanatory variables. Needless to say, over- and under-fitting are the extremes of a gradient. The model selection problem can be posed as striking the "right balance" between under- and over-fitted models with the same data (Burnham and Anderson 2002).

Model selection can also be viewed in terms of hypothesis testing. This is particularly so when models are part of a nested hierarchy such that one model is a special case of another more general one when a given parameter is equal to zero. The formal statistical test to make the comparison between two models using Wilks' test was explained in Chapter 3. More generally, model selection can involve situations in which models cannot be neatly arranged as a nested hierarchy whereby one model is a special case of another. We will find many such situations later in the book in which Wilks' test should not be applied. We thus need other, more general, criteria for model selection.

The principle of parsimony is the conceptual pillar of model selection in statistics. Box and Jenkins (1970, p. 17) suggested that using the principle of parsimony should lead to a model with "the smallest possible number of parameters for adequate representation of the data." The principle of parsimony has a long history in the philosophy of science. It is attributed to the thirteenth-century English Franciscan priest William of Ockham, who allegedly wrote "Entia non sunt multiplicanda preaeter necessitatem" (entities should not be multiplied unnecessarily). This quote is known as "Ockham's razor," and it has an interesting and erratic history. For instance, Ockham's razor is another example of Stigler's (1980) law of epinomy[1] since it can be found in the earlier works of Aristotle, Maimonides, and Thomas Aquinas (Lazar 2010). William of Ockham, a leading medieval philosopher, surely used razors during his tempestuous life, which involved frequent quarrels with his Franciscan superiors, and an excommunication by the pope. However, the above Latin quote cannot be found anywhere in his writings (Lazar 2010); the term "Ockham's razor" was coined many centuries later by the Irish mathematician William Rowan Hamilton in 1852 (Lazar 2010).

Be that as it may, the principle of parsimony in statistics based on Ockham's razor is the basis for choosing simpler over more complex statistical models. In general, more parsimonious models are easier to interpret and explain, less prone to over-fitting, and are easier to generalize to other samples taken under similar conditions (Lazar 2010). The principle of parsimony emphasizes simplicity and elegance of explanation, and it seems to agree with common scientific practice. Given a set of alternative models, the principle of parsimony would not lead us to the true data-generating model (provided that it is in the set of models being compared), but to the simplest model that is consistent with the

[1] This law states that many discoveries are not attributed to the original discoverer. It also applies to itself: the same idea was previously stated by the American sociologist Thomas Merton.

data. George Box (1976) quipped that "All models are wrong, but some are useful." Box's quote underlies the view that statistical models are provisional scaffolds allowing us to grasp tentative truths contingent on the data available. The principle of parsimony is then a heuristic guide that agrees with our perception of how scientific research proceeds in leading to reasonably simple findings that we can translate into words to build intelligible interpretations. The word "principle" in science often refers to a statement that cannot be proven to be true beyond reasonable doubt, but that we believe to be correct and apparently reasonable as a working guide to proceed in our trade.

7.3 Model selection criteria in the frequentist framework: AIC

The *Akaike information criterion* (Akaike 1973) is by far the most frequently used metric in model selection in the frequentist framework. Its conceptual underpinning can be found in information theory (Shannon 1948), which lies at the intersection between statistics, probability theory, statistical mechanics, computer science, and engineering. We need to make a detour to provide a short explanation of information theory before returning to explain the AIC.

Information theory was born out of the secret wartime work of Claude Shannon at Bell Labs. One of Shannon's main interests was quantifying the information content of messages. The etymology of "to inform" (from the Latin *informare*) is to give form or to form an idea, and by so doing to reduce the initial uncertainty about the message. When developing a metric to measure uncertainty, Shannon postulated three intuitive desiderata (it should be continuous, it should increase with the length of the message, and it should be additive) that were uniquely satisfied by the information entropy $H = \sum P \log(P)$ (or $- \int P \log(P)\, dP$ for continuous functions), where P is the probability of an event or an actual letter in the message.[2] Applied to probability distributions, Shannon's uncertainty is E(log(Pr(Y))) where E() is the expected value, which in words means that the information entropy is the average log-probability of Y. Shannon's information entropy is then a measure of the information content in a message or, conversely, a measure of the extent of the reduction of the initial uncertainty upon receiving the message.

Kullback and Leibler (1951) aimed to provide a rigorous definition of information in relation to Fisher's (1922) notion of "sufficient statistics" in the context of the maximum likelihood method (Chapter 3). Kullback and Leibler (1951) derived an information measure that turns out to be the negative of Shannon's information entropy. Their measure is both a fundamental quantity in information theory and the basis of the AIC (Burnham and Anderson 2002). So, what is the Kullback and Leibler (1951) measure? Let's call "truth" the true statistical model generating the data with a given number of parameters; this is a perfect and ideal model, a continuous, smooth function that precisely and only connects all values of Y. Some gods may know the "truth," but sadly we humans do not. Let's call "approx" a statistical model estimated from data that we hope to be a close approximation to the "truth." Kullback and Leibler's (1951) distance measure is

$$D_{\mathrm{KL}} = \int \mathrm{truth}(Y) \times \log\left(\frac{\mathrm{truth}(Y)}{\mathrm{approx}(Y)}\right) dY, \qquad (7.1)$$

[2] It was John von Neumann who, after noticing that Shannon's metric of uncertainty was identical to Boltzmann's entropy, suggested the name "information entropy" in a meeting with Claude Shannon.

i.e., a measure of discrepancy between "approx" and "truth" for every value of the response variable Y; the integral is just a sum of the discrepancy for every value of Y. The lower D_{KL} is, the better the approximation to the true model. Technically speaking, D_{KL} is a "directed distance" because the measure from "approx" to "truth" does not have the same value the other way around. D_{KL} is an extension of Shannon's entropy that measures how much information is lost by approximating the truth (Burnham and Anderson 2002).

Akaike (1973) proposed using D_{KL} as the fundamental basis for model selection. However, we cannot evaluate Eq. (7.1) because we do not know the distribution of truth(Y). Let's first realize that Eq. (7.1) can be written as

$$D_{KL} = \int \text{truth}(Y) \times \log(\text{truth}(Y))\, dY - \int \text{truth}(Y) \log(\text{approx}(Y))\, dY, \tag{7.2}$$

where the first term is an unknown constant C, and the second is a relative expected D_{KL} between truth(Y) and approx(Y) that turns out to be the log-likelihood of approx(Y) (the actual derivation is more involved; see the details in Burnham and Anderson 2002, pp. 43–46). Therefore, Eq. (7.2) becomes

$$D_{KL} \approx C - \int \log(\text{Likelihood}(\text{approx}(Y)))\, dY, \tag{7.3}$$

which can actually be calculated after fitting a statistical model. Akaike (1973) found a relation between the relative expected D_{KL} and the total log-likelihood of a statistical model. This relation became the fundamental breakthrough in model selection. He also found that the second term of Eq. (7.3) had an upward bias that could be corrected by adding the number of parameters k in the statistical model (see Bolker 2008, pp. 277–278 for a derivation). Akaike (1973) defined an information criterion (later named after him) as

$$\text{AIC} = -2 \log(\text{Likelihood}(\text{model})) + 2k. \tag{7.4}$$

The AIC is then the expected, bias-corrected value of the relative Kullback–Leibler distance. The number 2 in the formula appears for "historical reasons" because $-2 \times$ log(Likelihood(model)) is a constituent of the the likelihood ratio test that is used for comparing two statistical models (Chapter 3).

When developing the general theory of generalized linear models, Nelder and Wedderburn (1972; Chapter 8) viewed model fitting as the maximization of the log-likelihood in relation to a perfectly fitting model (called the saturated model) having as many parameters as data points. Fitting saturated models amounts to having a one-to-one map to the world that only makes sense in "thought experiments." At the other extreme, the minimal or simplest model that would only fit the grand mean to the data is often called the "null model." Nelder and Wedderburn (1972) defined the deviance as twice the difference between the log-likelihoods of the saturated and a current, real statistical model. The deviance is a generalization of the residual sum of squares used in the general linear model (Chapters 4 to 6) to cases where model fitting is achieved by maximum likelihood (Chapter 8). We can rewrite Eq. (7.4) as

$$\text{AIC} = \text{Deviance}(\text{model}) + 2k. \tag{7.5}$$

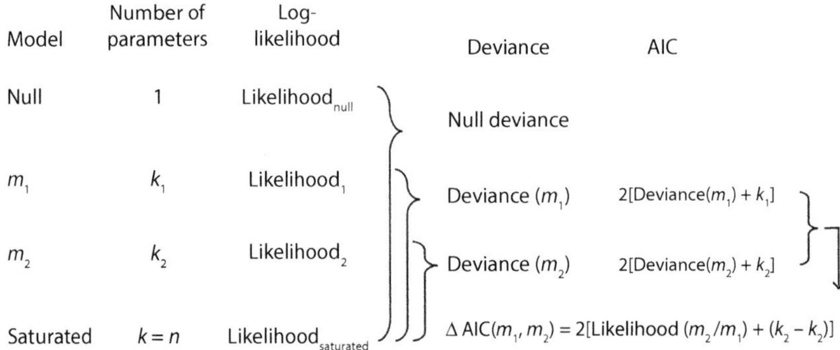

Fig 7.2 Relation between likelihood, deviance, and the AIC. At one extreme, the null model is the simplest (k = one parameter: the overall mean) that can be fitted to data. At the other, the saturated model is the most complex having as many fitted parameters as data points ($k = n$). The null deviance is twice the difference between the log-likelihoods of these models and represents the deviance to be explained by the more realistic models m_1 and m_2 fitted to the same data having k_1 and k_2 parameters ($k_2 > k_1$) and having deviances dev_1 and dev_2, respectively. The AIC, Eq. (7.5), and the differences in AIC between models m_1 and m_2 are also shown.

The deviance is then related to D_{KL} in Eq. (7.3): it reflects the relative distance between the true data-generating model and our current model viewed as our current approximation to the true one. Hence, the higher the deviance of a model, the poorer the fit of our statistical model to the data. Figure 7.2 summarizes the relation between log-likelihood, deviance, and AIC for a set of statistical models.

We can decrease the badness of fit of a model (i.e., its deviance) by adding parameters associated with terms reflecting either explanatory variables or interactions. The two components of AIC in Eq. (7.5) thus change with increasing model complexity: the deviance decreases as the number of parameters (a measure of model complexity) increases. The changes of these two components of AIC in opposite directions eventually reach a balance that prevents, or at least minimizes, the risk of the model over-fitting. The second term in Eq. (7.5), k = the number of parameters, is trivial, innocent, and uncontroversial enough for the statistical models we have covered thus far. But deciding the (effective) number of parameters is neither trivial nor obvious for the mixed models we encounter in Part III. We explain in Part III that the use of AIC with mixed models has some restrictions, and requires some care.

The actual AIC value of a statistical model is irrelevant. A model's AIC value could even be negative, and nobody should care or worry about it. Let's recall that the AIC is the (bias-corrected) relative D_{KL} between a statistical model and the true data-generating model or "saturated model." If we calculate the AICs of a set of candidate models, we could find which of them is the best approximation to the true data-generating model. That is, we would strive to select the model having the smallest AIC as the simplest, most parsimonious model among the set of candidates. The chosen model with the smallest AIC is not necessarily the "truth," it is just the simplest approximation we currently have given the data and the set of models being compared.

To obtain the AIC of a model in R, we simply write `AIC(model name)`. Besides identifying the model with the smallest AIC from a set, we also calculate

$$\Delta\left(\mathrm{AIC}_{\mathrm{model}\ i}\right) = \mathrm{AIC}_{\mathrm{model}\ i} - \min\left(\mathrm{AIC}_{\mathrm{model}\ j \neq i}\right). \tag{7.6}$$

The larger the ΔAIC of a model, the worse its fit, and the lower the empirical support it receives from the data.

The question of model selection using AIC boils down to interpreting ΔAIC to select the best model. Burnham and Anderson (2002, p. 48) wrote in this regard:

> As a rough rule of thumb, models for which $\Delta\text{AIC} \leq 2$ have substantial support and should receive considerations in making inferences. Models having ΔAIC of about 4 to 7 have considerably less support, while models with $\Delta\text{AIC} \geq 10$ have either essentially no support, and might be omitted from further consideration, or at least those models fail to explain some substantial explainable variation in the data.

Please note the careful, almost lawyer-like, language employed in a standard reference on model selection. Since then, many scientists doing model selection have chosen to employ $\Delta\text{AIC} \geq 2$ as the selection criterion. Nobody actually wrote, deduced, or provided a serious or substantial theoretical argument for the criterion $\Delta\text{AIC} \geq 2$, which has become a "folklore theorem" in model selection. Much like the fixing of the significance level at 0.05, it is too late: the dreadful genie of $\Delta\text{AIC} \geq 2$ is now out of the bottle; we cannot now tame it and recall it back into a sealed container holding "the baseless criteria that nobody actually ever stated."

Model selection is most useful in exploratory analysis when we lack a specific statistical hypothesis to test. These exploratory analyses are essentially a "pattern-seeking" exercise aimed at formulating more precise hypotheses that could be tested by designed experiments. But neither life nor scientific practice is as clear-cut as we may wish. We scientists have used, over-used, and abused the unwritten "rule of $\Delta\text{AIC} \geq 2$" to discard models over the last 30 years.

It is important to highlight a few important things. First, model selection is not a statistical test involving significance levels. No model is ever rejected, as they simply possess different degrees of empirical support from the data. It is not a defect of the procedure of model selection using AIC to have parameter estimates that do not happen to be statistically significant. If we just wanted to have a model only with statistically significant parameter estimates, then we might just arbitrarily drop "the offending terms" without dressing up the procedure behind information theory to obtain the desired outcome. This is a baseless, ad hoc method sometimes used because researchers get itchy (and paper reviewers demand explanations) when non-significant terms remain after model selection. Second, we can use AIC to compare any set of models (including those making up a nested hierarchy) provided that they are all fitted to the same data; this includes models fitted using different probability distributions for the response variable (Bolker 2008). Therefore, AIC-based model selection has a wider range of application that the comparison of models with Wilks' likelihood ratio test (Chapter 3). Finally, when applied to general linear models (Chapters 4 to 6), $\text{AIC} = n\log(\sigma^2) + 2k$, where σ^2 is the arithmetic average of the squared residuals (and the maximum likelihood estimate ofσ_Y^2), and n is the number of data points.

Once we have fitted a set of of statistical models and obtained their ΔAICs from Eq. (7.6), we could transform the latter into a [0,1] scale using the Akaike weights (Burnham and Anderson 2002)

$$w_j = \frac{\exp\left(-0.5\ \Delta\text{AIC}_j\right)}{\sum \exp\left(-0.5\ \Delta\text{AIC}_j\right)}. \tag{7.7}$$

Some (e.g., Burnham and Anderson 2004) have interpreted the Akaike weights or model probabilities as the weight of evidence in favor of model j being the actual best model in the candidate set. We could also calculate the ratio of the Akaike weights of two models and obtain the so-called evidence ratio, and conclude for instance that one model enjoys (say) 2.3 times more empirical support than the other. But as we calculate statistics from previous statistics from fitted models we are further and further removed from the actual data. We should observe that both the ΔAICs and the Akaike weights are relative to a specific set of models, in that their values will change if we change the candidate set.

Those not wanting to select the "best" model typically aim to obtain the "average model" by the weighted contributions of a set of candidate models. Believing that the "average model" has any bearing on the true data-generating model requires a great leap of faith and a fertile imagination. Given that any fitted model has its parameter estimates, the parameter estimates of the average model should reflect the weighted contributions of the parameter estimates of each of the models in the set. The parameter estimates of the ith explanatory variable in the "average model" and their standard errors are obtained with

$$\hat{\bar{\beta}}_{j,i} = \frac{\sum I_{j,i} w_j \hat{\beta}_{j,i}}{\sum I_{j,i} w_j} \quad \mathrm{SE}\left(\hat{\bar{\beta}}_i\right) = \sum I_{j,i} w_j \sqrt{\mathrm{var}\left(\hat{\beta}_{i,j}\right) + \left(\hat{\beta}_{i,i} - \hat{\bar{\beta}}_i\right)^2}, \tag{7.8}$$

where w_j are the Akaike weights of each of the models, $\hat{\beta}_{j,i}$are the parameter estimates for the ith explanatory variable of the jth model, and $I_{\mathrm{j},i}$ is an indicator variable denoting whether the ith explanatory variable is in the jth model. The first equation in Eq. (7.8) is nothing more than a weighted average of parameter estimates, and the second one reflects the precision of the parameter estimate $\hat{\beta}_{j,i}$ and how far this estimate is from the weighted average value (Burnham and Anderson 2002). The "average model" is a weighted compromise of parameter estimates of the candidate model set. It is highly recommended to center and standardize all numerical explanatory variables (Chapter 5) prior to fitting the models (Schielzeth and Fortsmeier 2011) to obtain consistent parameter estimates for the "average model." We will not delve any further into model averaging in this book.

Before leaving the AIC, we should mention that it intrinsically assumes that the sample size is much larger than the number of parameters to estimate ($n \gg k$). Because the AIC may perform poorly with overly complex models fitted with limited data, the $\mathrm{AIC_C}$ corrects it (Burnham and Anderson 2002):$\mathrm{AIC_C} = \mathrm{AIC} + \frac{2k(k+1)}{n-k+1}$.

7.4 Model selection criteria in the Bayesian framework: DIC and WAIC

Schwarz (1978) proposed the *Bayesian information criterion* BIC = Deviance(model) + $k \log(n)$whose penalty term for model complexity includes the sample size. The BIC was originally obtained assuming flat, uninformative priors that are typically overwhelmed by the likelihood. Despite its name, the BIC is neither truly based on information theory nor an estimate of the fundamental Kullback–Leibler distance (Burnham and Anderson 2002). The BIC is evaluated at the parameter values that maximize the model likelihood and, as such, it is unrelated to the posterior distribution of model parameters. The BIC is not considered "Bayesian enough" (Hobbs and Hooten 2015), and it is rarely used in practice despite many software programs and R packages still calculating it. It is mentioned here just for the sake of completeness.

Spiegelhalter et al. (2002) introduced the *deviance information criterion* as a Bayesian alternative to the likelihood-based AIC:

$$\text{DIC} = \text{Deviance}\,(\text{model} \mid \text{avg}\,(\text{post - parameters})) + 2p_{\text{D}}. \tag{7.9}$$

You may notice the similarities between the DIC and AIC. We will go slowly to explain Eq. (7.9). For simplicity, let's suppose again that our model is a simple linear regression and, after fitting it with an MCMC algorithm, we have the marginal posterior distributions of the model parameters. We also obtain the posterior distributions of other model summaries. The first term of Eq. (7.9) is the model deviance (= –2 × log(likelihood)) calculated for the average values of the marginal posterior distributions of the model parameters. This should not be surprising since a model deviance is determined by its parameter values, be they the maximum likelihood estimates as in the AIC, Eq. (7.4), or the average as the "typical" value of posterior distributions that often (but not always) have Gaussian shapes as in the DIC, Eq. (7.9). The second term in Eq. (7.9) is the "effective number of parameters" and is a "complexity penalty term" as in the AIC, Eq. (7.4). There are several ways to estimate p_{D} (Spiegelhalter et al. 2002). Gelman et al. (2014) suggested $p_{\text{D}} = \text{Var}\,(\text{posterior deviance})$, which is cryptic enough at first sight. Their estimation of p_{D} is based on the posterior distribution of the deviance. You may wonder: posterior distribution of the deviance? Yes, there is such a thing. Given that the model deviance (= –2 × log(likelihood | parameters)) is determined by its parameter values, we can calculate the deviance for each sampled set of model parameters and later obtain the posterior distribution of the model deviance. The Gelman et al. (2014) measure of the number of effective parameters is the variance of the posterior distribution of the deviance. Gelman et al. (2014, p. 1002) actually proposed another closely related and more numerically stable estimate of p_{D}, but let's leave it as is for now. In an intuitive sense, more complex models have more parameters and a higher posterior uncertainty for each parameter value.

The DIC is a useful information criterion for comparing Bayesian models, including mixed models (see Part III) for which the AIC can only be used with some care. However, the proposed measures of DIC model complexity are all approximations that can sometimes go awry (i.e., a negative effective number of parameters). The various estimates of p_{D} reassuringly converge to the actual integer number of parameters in simple models. However, their interpretation depends on the sample size, and it is far from intuitive for more complex models (Hooten and Hobbs 2015, Vehtari et al. 2017). Also, the DIC should not be used in some cases, such as mixture models (Chapter 10). A further problem is that the DIC assumes the posterior distributions to be a multivariate Gaussian, which is not true for parameters such as variances or correlations that have bounded ranges (Hooten and Hobbs 2015).

Bayesian statisticians have cleverly refused to suggest even rough rules of thumb for model selection, as Burnham and Anderson (2002) surely regret having done for the AIC. In the absence of guidance, scientists have sometimes resorted to using Burnham and Anderson's (2002) rules of thumb when selecting Bayesian models. We can also compute something equivalent to Akaike weights with the DIC using Eq. (7.7). Model averaging using scaled DIC values is, however, on a much less secure theoretical footing (Hooten and Hobbs 2015, p. 18) and should be carried out at your own risk.

The DIC was "the only kid in the block" for Bayesian model selection for some time. Watanabe (2010) more recently proposed another metric for Bayesian model comparison that he called the "widely applicable information criterion" (WAIC), and that others have named the *Watanabe–Akaike information criterion*.

Explaining the WAIC requires us to first explain the posterior predictive distribution. This distribution is the basis for posterior predictive checks, a very useful and important tool for the validation of Bayesian models that has no direct equivalent in the frequentist framework. Thus, we need to take a detour to explain the posterior predictive distribution (Section 7.5), point to its connection with posterior predictive checks, and only then return to finally explain the WAIC.

7.5 The posterior predictive distribution and posterior predictive checks

An important feature of the adequacy of a statistical model is its capacity to predict data not used in its actual fitting. To some extent, evaluating how well the model reproduces the observed data is akin to a high school student stealing the exam in advance, and then achieving a good grade in the test. Chances are that any reasonable model will fit very well the data for which it was tailored. The real "stress test" should be to predict data well that were not used in parameter estimation.

When we have lots of data (i.e., many hundreds or thousands of data points), we could fit the model to a fraction (say, 2/3) of the data set and check how well the model can predict the remaining third. This process is known as cross validation (Stone 1977). Most scientists rarely enjoy the luxury of working with large data sets. A related idea more suitable to modest-sized data sets is "leave-one-out" cross validation, known by the irresistible acronym LOO-CV. In its simplest form, LOO-CV systematically excludes one point at a time in a set of k data points, fits the statistical model with the remaining $k - 1$ points (the "training set"), estimates the predicted value of the response variable for the excluded point, and compares this prediction with the observed value (say, by the squared difference). A good statistical model should be able to accurately predict the values that were not used in its fitting. The problem with LOO-CV is that we need to fit the model k times, thus making it impractical for large data sets. This is why LOO-CV requires some shortcut that minimizes the time spent fitting a large number of statistical models.

The posterior predictive distribution (Rubin 1984) is a probability density distribution for new value(s) of the response variable. The posterior predictive distribution got its name because the prediction of new data is conditional on the data, and on the model's predictive capacity, which in turn depends on the parameter estimated from the same data. To obtain the posterior predictive distribution, let's first recall the example of Section 3.4 concerning the fitting of the parameter μ_Y of a Poisson model (i.e., $Y \sim \text{Poisson}(\mu_Y)$) for plant counts:

$$\Pr(\mu_Y \mid \text{data}) = \frac{\Pr(\text{data} \mid \mu_Y) \times \Pr(\mu_Y)}{\Pr(\text{data})}. \tag{7.10}$$

We now know that the last equation is simplified since we do not need to estimate Pr(data), and it becomes (see Section 3.4)

$$\Pr(\mu_Y \mid \text{data}) \approx \Pr(\text{data} \mid \mu_Y) \times \Pr(\mu_Y). \tag{7.11}$$

In complete analogy to what we did before, we can write the posterior distribution for two unknowns (Y_{NEW} and μ_Y), where Y_{NEW} is a new value of the response variable, as

$$\Pr(Y_{\text{NEW}}, \mu_Y \mid data) \approx \Pr(\text{data} \mid Y_{\text{NEW}}, \mu_Y) \times \Pr(Y_{\text{NEW}}, \mu_Y)\,. \tag{7.12}$$

Note that the likelihood that is the first term of the right-hand side of Eq. (7.12) can be written as $\Pr(\text{data} \mid \mu_Y)$ because the existing data cannot possibly depend on new data to be predicted. The second term is the joint probability of obtaining Y_{NEW}, and μ_Y can be written as $\Pr(Y_{\text{NEW}} \mid \mu_Y) \times \Pr(\mu_Y)$; thus, Eq. (7.12) becomes

$$\Pr(Y_{\text{NEW}}, \mu_Y \mid \text{data}) \approx \Pr(\text{data} \mid \mu_Y) \times \Pr(Y_{\text{NEW}} \mid \mu_Y) \times \Pr(\mu_Y)\,. \tag{7.13}$$

You might notice that the product of the first and third terms on the right-hand side of Eq. (7.12) is the posterior distribution of μ_Y (cf. Eq. (7.11)):

$$\Pr(Y_{\text{NEW}}, \mu_Y \mid \text{data}) \approx \Pr(Y_{\text{NEW}} \mid \mu_Y) \times \Pr(\mu_Y \mid \text{data})\,. \tag{7.14}$$

We are almost there. Given that we are currently only interested in Y_{NEW}, we can obtain its posterior predictive density by marginalizing (or integrating out) the right-hand side of Eq. (7.14):

$$\Pr(Y_{\text{NEW}} \mid \text{data}) \approx \int \Pr(Y_{\text{NEW}} \mid \mu_Y) \times \Pr(\mu_Y \mid \text{data})\, d\mu_Y. \tag{7.15}$$

Let's put Eq. (7.15) in words. First, recall from basic calculus that an integral is basically an infinitesimal sum, and that $d\mu_Y$ denotes the variable across which we are summing for all plausible values of μ_Y in $(0, \infty)$. The posterior predictive distribution of Y_{NEW} given the current data is proportional to the product of two conditional probabilities denoting the uncertainty of predicting Y_{NEW} given the estimated values of the parameter μ_Y, and the uncertainty of obtaining plausible values of μ_Y for the the data at hand. The posterior predictive distribution is then the expected distribution of observing new data given that we have observed certain data and fitted the statistical model. That is, if we were to repeat the same experiment yielding data many times in the future, the posterior predictive distribution is the distribution of values of the response variable that we would expect given what we know (i.e., the current data).

Figure 7.3 illustrates both the making and the meaning of a posterior predictive distribution. First, we fitted the statistical model $Y \sim \text{Poisson}(\mu_Y))$ for plant counts with the R package `brms` (Bürkner 2017). The commands are:

```
plant.counts=brm(formula=plants~1,  data=DF5,family=poisson  (link="log"),
warmup=1000,chains=2, iter=2000, thin=2)
```

Priors do not show up in the fitting of `plant.counts` because we used the default weakly informative priors of `brms` this time. The usual model summary is:

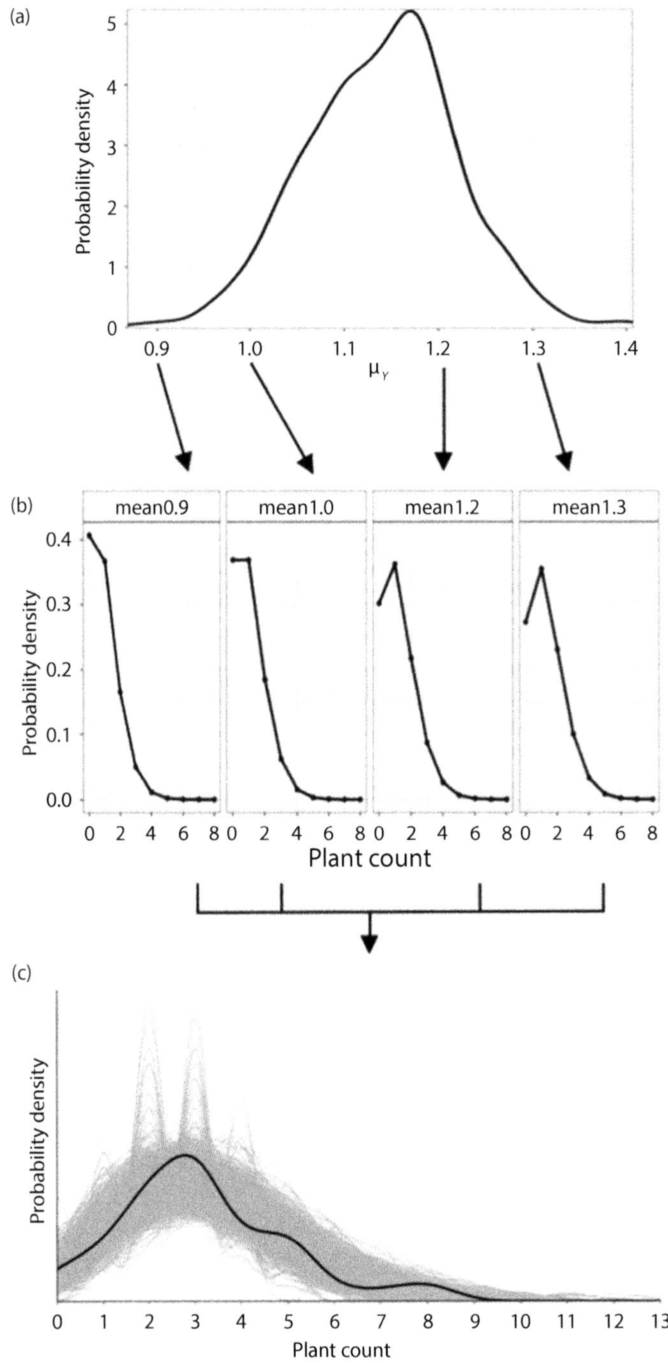

Fig 7.3 Posterior predictive distribution for the plant counts. (a) The posterior distribution of the Poisson parameter μ_Y obtained with the package `brms`. (b) The probability of obtaining Y plant counts for four selected values of the posterior distribution of μ_Y. (c) Ten sampled sets of values of the posterior predictive distribution (gray lines) compared with the density curve of the observed plant counts (black). This figure is taken from McElreath (2015).

```
> summary(plant.counts)
 Family: poisson
  Links: mu = log
Formula: plants ~ 1
   Data: DF5 (Number of observations: 50)
  Draws: 2 chains, each with iter = 2000; warmup = 1000; thin = 2;
         total post-warmup draws = 1000

Population-Level Effects:
          Estimate Est.Error l-95% CI u-95% CI Rhat Bulk_ESS Tail_ESS
Intercept     1.14      0.08     0.98     1.29 1.00      774      700
```

Figure 7.3a shows the posterior distribution of the Poisson parameter μ_Y, which is similar to Fig. 3.9 as obtained with our first hand-made MCMC algorithm. Note that all values of the parameter μ_Y are not equally likely in its posterior distribution. Just for the purpose of illustration, we arbitrarily picked four values of μ_Y and, using `dpois(0:8, lambda = 0.9)`, Figure 7.3b shows the probability of observing plant counts for these four arbitrarily chosen values of μ_Y. Each of these values of μ_Y is differently capable of predicting the plant counts. The posterior predictive distribution, Eq. (7.15), is the weighted sum of (mathematically speaking, infinitely) many values of μ_Y that produce different plots of predicted values of plant counts. Figure 7.3c was obtained with `pp_check(plant.counts, nsamples = 20)` plus a few more commands to make the plot pretty. This bottom graph shows a set of 20 sampled sets of values of plant counts from the posterior predictive distribution that are to be compared with the actually observed plant counts. Just by chance, some of the randomly sampled density profiles from the posterior predictive distribution are similar to the observed density curve of plant counts. If we repeat the same random sampling from the posterior predictive distribution a larger number of times (say, 1,000), we would now have the equivalent of a frequentist sampling distribution. The posterior predictive distribution compounds two sources of uncertainty: the uncertainty associated with the model parameter embodied in its posterior distribution, and the sampling uncertainty of drawing some of the infinitely many possible values of μ_Y to generate the distribution of predicted values of the response variable (Hooten and Hoobs 2015).

We can see from Fig. 7.4 that the data lies more of less in the middle of these 1,000 sampled distributions, which would suggest that the posterior predictive distribution

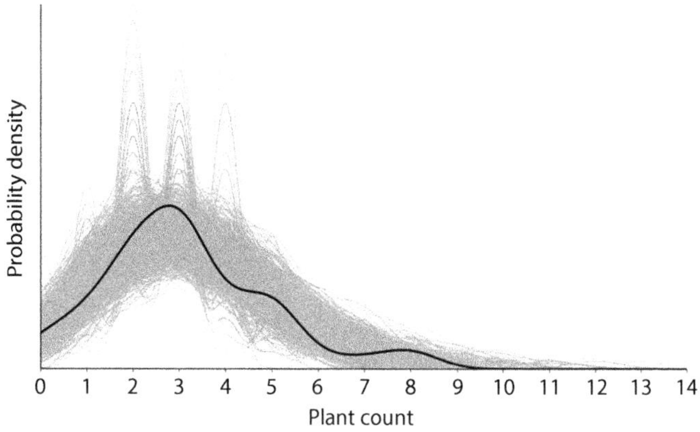

Fig 7.4 One thousand random samples from the posterior predictive distribution obtained after fitting the Poisson parameter for the plant counts (gray lines) and the density curve of the observed plant counts (black).

obtained after the fitted model `plant.counts` is a reasonable generating model capable of reproducing the observed data. Given that we can generate many random samples from the posterior predictive distribution, we could also calculate any actual (e.g., mean, median, the 38th percentile, standard deviation, etc.) or invented (e.g., the proportion of values of $Y \leq 4$) descriptive statistic from these 1,000 random samples and obtain a posterior distribution of its values. The observed values of the same descriptive statistics for the entire data set (or per group whenever the model contains categorical explanatory variable(s)) can now be compared with the actual values of the same statistics in the data. The comparisons between the distributions of a set of descriptive statistics calculated from the posterior predictive distribution and those calculated from the data are known as *posterior predictive checks* (Gelman et al. 2004, 2020; Hobbs and Hooten 2015). Randomly generating many samples from the posterior predictive distribution, computing descriptive statistics from each, and comparing them to the actual data is a Bayesian procedure of model validation that has a distinct frequentist flavor. After all, comparing the statistics calculated from data that is "extreme" in the distribution of posterior predictive checks is analogous to the frequentist procedure of using a sampling distribution of potential data that could arise from the repeated sampling of the same statistical population. Ironically, Bayesian *p*-values were proposed (Bayarri and Berger 2000) and used in posterior predictive checks for model validation. However, their use has largely been abandoned because they tend to be conservative: they often have poor sensitivity to detecting true departures from adequately fitting models (e.g., Robins et al. 2000).

A statistical model having a good fit of the data should be capable of generating distributions of values of the response variable and descriptive statistics similar to those actually observed in the data. Posterior predictive checks can then be used to evaluate the goodness of fit and the adequacy of Bayesian models. There is no such thing as "the" posterior predictive check, since many descriptive statistics could be used for the purpose of gauging a model's goodness of fit. Indeed, depending on one's inventiveness, it is always possible to find a particular feature of the data (say, the ratio of max/min per group, the rate of increase of a distribution of data at the 90th percentile) and use the posterior predictive checks to show the failings of a model to accurately reproduce it. But let's not miss the important point: posterior predictive checks are an ingenious and very powerful tool for assessing the adequacy of fit of Bayesian models. To avoid a lengthier detour from the WAIC, we defer the use of posterior predictive checks to later chapters of the book. We introduced posterior predictive checks here because of their close connection with the posterior predictive distribution of a model.

In closing, we should add that there is no direct equivalent of posterior predictive checks in the frequentist framework. Nevertheless, one could argue that using the parametric bootstrap (Chapter 13) to generate sampling distributions of data, and using some amount of R programming, one could calculate descriptive statistics to achieve evaluations of model goodness of fit similar to those of the posterior predictive checks.

7.6 Now back to the WAIC and LOO-CV

The posterior predictive distribution can be considered the best approximation to the true generative model that produce the observed data. Applying the same considerations used in Section 7.2 to define the Kullback–Leibler distance D_{KL} that quantified the distance between the true model and an approximation (the log posterior predictive distribution), we end up with something similar to Eq. (7.3):

$$D_{KL} \approx C - \int \Pr(Y_{NEW}) \times \log(\Pr(Y_{NEW} \mid Y))\, dY_{NEW}. \tag{7.16}$$

Equation (7.16) quantifies how close the posterior predictive distribution of Y_{NEW} would be to its true distribution $\Pr(Y_{NEW})$ for all values of Y_{NEW} (Lambert 2018). Its second term turns out to be the average log posterior predictive distribution. Equation (7.16) implies that as the average log posterior predictive distribution increases, D_{KL} becomes smaller because the posterior predictive distribution $\Pr(Y_{NEW} \mid Y)$ is a better approximation to $\Pr(Y_{NEW})$. By analogy with the sketches of the derivations of AIC and DIC above, Eq. (7.16) then suggests that we are getting closer to having one of the components of a fully Bayesian information criterion. Now, the problem is that we cannot compute Eq. (7.16) because we ignore what $\Pr(Y_{NEW})$ is. Let's not forget that we are after all trying to approximate it. However, as was the case for the AIC, there is a solution to the conundrum.

Gelman et al. (2014) suggested the solution by calculating the expected log posterior predictive density for n points (not just for one Y_{NEW} as in our sketch), taking one at a time to obtain the expected log pointwise predictive density (ELPPD) as

$$\begin{aligned} \text{ELPPD} &= \textstyle\sum_{i=1}^{n} \overline{\text{log posterior predictive density}\,(Y_{NEW}(i) \mid Y)} \\ &= \textstyle\sum_{i=1}^{n} \displaystyle\int \Pr(Y_{NEW}(i)) \times \log(\Pr(Y_{NEW}(i) \mid Y))\, dY_{NEW}(i). \end{aligned} \tag{7.17}$$

Equation (7.17) looks threatening. In practice, using Eq. (7.17) involves computing the (log) posterior probability density, Eq. (7.15), of each data point across the parameter values of the converged MCMC chains (Hoobs and Hooten 2015). The ELPPD can be seen as analogous to the frequentist deviance: the larger its value, the poorer the fit of the model to data.

Finally, Watanabe (2010) defined the WAIC as

$$\text{WAIC} = -2\text{ELPPD} + 2p_{WAIC}. \tag{7.18}$$

Much like the AIC and DIC, the WAIC has two terms: the first reflects the quality of a model fit (more precisely, a measure of the model's predictive accuracy, now using a truly Bayesian information criterion), and the second (p_{WAIC}) corresponds to the effective number of parameters, numerically estimated as $p_{WAIC} = \sum \text{Var}$ $(\text{log post predictive density}\,(Y_{NEW}(i)))$ (Gelman et al. 2014). Neither of the 2s in Eq. (7.18) are strictly needed, and they probably appear just to keep up the tradition initiated with the AIC. You should rest assured that there are dedicated R packages that calculate the WAIC after fitting a Bayesian model. As promised for the AIC above, we shall illustrate the use of the WAIC for model selection in later chapters.

Unlike the other metrics used in model comparison explained in this chapter, the WAIC is defined for each data point, and it is applicable for a very wide range of statistical models (including mixed and hierarchical models; Part III). Nevertheless, Eq. (7.16) assumes that all data points sequentially used in the computation of the WAIC are mutually independent (Hoobs and Hooten 2015). This assumption would hinder its correct estimation when there is spatial or temporal correlation in the data. We should not be too surprised to learn of the appearance in the near future of new and improved numerical methods to calculate the WAIC that take into account these realistic features of the data. Unlike the

DIC, the WAIC does not require the posterior distribution of a model to be multivariate normal, and it is not overly sensitive to the priors used when fitting the model (Vehtari et al. 2017). The WAIC is currently the preferred metric for Bayesian model comparison (Hooten and Hoobs 2015, Vehtari et al. 2017, Congdon 2020).

The WAIC is a fully Bayesian metric for model comparison that turns out to be asymptotically equivalent to LOO-CV (Gelman et al. 2014, 2017). The Bayesian LOO estimate for predicting each value given the data without using the ith point in the parameter estimation is

$$\text{ELPPD}_{\text{LOO}} = \sum_{i=1}^{n} \overline{\log\left(\text{post predictive density}\,(Y \mid Y_{-i})\right)}. \tag{7.19}$$

As for all cross-validation metrics, using Eq. (7.19) can be computationally demanding because it requires refitting the same model as many times as there are data points. We would need a computational shortcut to efficiently perform a Bayesian LOO-CV and to calculate its related WAIC. As you may guess by now, such a shortcut exists; it was implemented in the R package `loo` (Vehtari et al. 2017).

Let us at least give the gist of Vehtari et al.'s (2017) method. They evaluated Eq. (7.19) using a procedure known as importance sampling such that the posterior predictive distribution $(Y \mid Y_{-i})$ is replaced by

$$\text{post predictive density}\,(Y \mid Y_{-i}) = \frac{\sum r_i \times \text{post predictive density}\,(Y)}{\sum r_i}, \tag{7.20}$$

where r_i is the ratio of the posterior predictive distribution $(Y \mid Y_{-i})$ to the full posterior predictive distribution (Y) for each of the n data points. Intuitively, if this ratio or index of importance were very small, omitting the ith observation would be of relatively small importance for the predictive quality of the model. They also found that the importance ratio r_i of each point could be adequately approximated by

$$r_i = \frac{\text{post predictive density}\,(Y \mid Y_{-i})}{\text{post predictive density}\,(Y)} \propto \frac{1}{\text{Likelihood}\,(Y_i \mid \text{parameters})}. \tag{7.21}$$

While an involved series of approximations is employed to obtain Eqs. (7.20) and (7.21) (see Vehtari et al. 2017 for details), let us concentrate on their implications. The message of Eqs. (7.20) and (7.21) is that is we do not need to refit the model as many times as there are data points to obtain the Bayesian LOO-CV. This is because we can approximate the Bayesian LOO-CV as a weighted average of the model posterior predictive density over all data points, where the weights are the inverse of the model likelihood of each data point. And the posterior predictive density and the model likelihood are actually fast to compute.

But this is not all. Vehtari et al. (2017) also suggested that the r_i could be approximated using a generalized Pareto distribution to obtain reliable results. The Pareto probability distribution is defined for positive real values, has the shape of a negative exponential, and there are four forms of it depending on the number of parameters that determine the rate of decay of its upper tail (Johnson and Kotz 1994). These approximations moved Vehtari et al. (2017) to call their method for performing Bayesian cross validation PSIS (Pareto smoothed importance sampling).

What do we have at the end of this complex theory and the involved series of approximations that are carried out by the package `loo`? First, we have a numerically efficient method for Bayesian cross validation that estimates a model's WAIC and its standard error, both of which can be used in model selection. Second, by approximating the distribution of each r_i data point by a (type III, in case you're wondering) Pareto distribution, we have an estimate of a parameter called k. It turns out the the value of k for each data point can be interpreted much like the Cook distance in the frequentist framework (Chapters 4 to 6): data points with large values of k have a high importance or influence on the predictive ability of a Bayesian model. This procedure permits the flagging of data points with high k values that should be double checked and examined as they might possess undue influence on the model performance. Vehtari et al. (2017) gave ranges to interpret the estimated k value for each data point: $(-\infty, 0.5)$: OK; $(0.5, 0.7)$: good; $(0.7, 1.0)$: bad; and > 1: very bad. A model that fits well should have no or a very small proportion of data points with k values greater than 0.7. We use these ranges when fitting Bayesian models later in the book.

7.7 Prior predictive distributions: A relatively "new" kid on the block

While the initial idea of prior predictive distributions is not particularly new (Box 1980), their use to assess the suitability of priors is probably not older than 10–15 years. In a sense, dealing now with prior predictive distributions constitutes a break from the progression and the (hopefully) coherent flow of ideas that took us from frequentist model selection (AIC and the likelihood ratio test) to the more recent Bayesian approaches (DIC and WAIC) to accomplish the same task.

We made a detour to explain posterior predictive distributions because it was essential to understand the basis of the WAIC. Then, we quickly pointed out the connection between posterior predictive distributions and the posterior predictive checks that we use for the validation of Bayesian models later in the book. And, along the way, we highlighted the asymptotic equivalence between WAIC and Bayesian LOO-CV, and the fact that both are defined pointwise led to the recent development of PSIS, a Bayesian analogue of Cook's distance. In truth, we are discussing prior predictive distributions here just to take advantage of the explanation of posterior predictive distributions above.

Just as the posterior predictive distribution is the expected distribution of the response variable after observing the data and using them to obtain the posterior distribution, the prior predictive distribution is the expected distribution of the response variable before observing the data and fitting the model. Let's put the latter statement in equation form:

$$\Pr(Y) \approx \int \Pr(Y \mid \theta) \times \Pr(\theta)\, d\theta, \tag{7.22}$$

where θ denotes the set of model parameters. The first term on the right-hand side of Eq. (7.22) is the model likelihood, and the second is the prior distribution of model parameters. Equation (7.22) gives the probability distribution of the response variable resulting only from the information of parameters encoded in the prior distribution. Note that our data does not enter anywhere into Eq. (7.22). By integrating (i.e., summing) over the values of the set of parameters θ, we have the weighted contribution of how different values of θ produce the density plots of the response variable via the likelihood (cf. Figure 7.3).

Fine, but why should we care about the prior predictive distribution? The answer has to do with the eternal debate of how to define the priors of a Bayesian model. As explained in Chapters 4 to 6, there is no principled way of setting priors generally valid for common statistical models. Priors have been classified as "vague," "non-informative," "weakly informative," "reference," "informative," "strongly informative," "regularizing," and surely more. Because the degree of informativeness of a prior depends on (or at least cannot be separated from) the likelihood (Gelman et al. 2017), there might seem to be many (infinite?) shades of informativeness of priors. Gabry et al. (2019) and Gelman et al. (2017) argued that the prior predictive distribution can be used to visualize the generative character of the prior information. By generative they mean the extent to which the priors can generate plausible sets of values of the response variable and, by so doing, assess the effect of the prior before seeing the data to be analyzed.

The prior predictive distribution is a means of calibrating our previous knowledge on a problem encoded in the prior distributions. A very useful application of the prior predictive distribution is to show that the priors can produce impossible (such as negative values of a positive-definite variable) or inappropriate values (i.e., values far away from any plausible set) of the response variable. These infeasible or factually impossible predictions stemming from the prior would suggest that the latter needs be modified. It is critical to point out that judgment of the feasibility of the prior predictive distribution should be done without having looked at the data to be analyzed. Choosing or modifying the prior after seeing the data or after comparing its similarity with the prior predictive distribution is a form of cheating that leads to an incoherent form of inference (Gelman et al. 2017). We can rarely pretend that we are breaking into a topic devoid of previous knowledge applicable to our research question. It is the background knowledge on a given problem that should allow us to qualify data as implausible before actually examining the current data. In keeping with the promise of the future application of the previous material in this chapter, we will also employ the prior predictive distribution in the data analyses later in the book.

References

Akaike, H. (1973). Information theory and an extension of the maximum likelihood principle. In B. Petrov and F. Csáki (eds.) Proceedings 2nd International Symposium on Information Theory, pp. 267–281. Akadémiai Kiadó, Budapest.

Bayarri, P. and Berger, J. (2000). *p*-values for composite null models. *Journal of the American Statistical Association*, 95, 1127–1142.

Bolker, B. (2008). *Ecological Models and Data in R*. Princeton University Press, Princeton.

Box, G. (1976). Science and statistics. *Journal of the American Statistical Association*, 71, 791–799.

Box, G. (1980). Sampling and Bayes' inference in scientific inference and modeling. *Journal of the Royal Statistical Society A*, 143, 383–430.

Box, G. and Jenkins, G. (1970). *Time Series Analysis: Forecasting and Control*. Holden-Day, San Francisco.

Bürkner, P.-C. (2017). brms: Bayesian regression models using Stan. R package. *Journal of Statistical Software*, 80, 1–28.

Burnham, K. and Anderson, D. (2002). *Model Selection and Multimodel Inference: A Practical Information-Theoretic Approach*, 2nd edn. Springer, New York.

Burnham, K. and Anderson, D. (2004). Multimodel inference: Understanding AIC and BIC in model selection. *Sociological Methods and Research*, 33, 261–304.

Congdon, P. (2020). *Applied Bayesian Modelling*, 3rd edn. John Wiley and Sons, Chichester.

Fisher, R. (1922). On the mathematical foundations of theoretical statistics. *Proceedings of the Royal Society A*, 222, 309–368.

Gabry, J., Simpson, D., Vehtari, A. et al. (2019). Visualization in Bayesian workflow. *Journal of the Royal Statistical Society A*, 182, 389–402.

Gelman, A., Carlin, J. Stern, H. et al. (2004). *Bayesian Data Analysis*. Chapman and Hall / CRC Press, New York.

Gelman A., Hill, J., and Vehtari A. (2020). *Regression and Other Stories*. Cambridge University Press, Cambridge.

Gelman, A., Hwang, J., and Vehtari, A. (2014). Understanding predictive information criteria for Bayesian models. *Statistical Computing*, 24, 997–1016.

Gelman, A. and Loken, E. (2014). The statistical crisis in science. *American Scientist*, 102, 460–465.

Gelman, A., Simpson, D. and Betancourt M. (2017). The prior can often only be understood in the context of the likelihood. *Entropy*, 19, 555–568.

Hobbs, N. and Hooten, M. (2015). *Bayesian Models: A Statistical Primer for Ecologists*. Princeton University Press, Princeton.

Johnson, N. and Kotz, S. (1994). *Continuous Univariate Distributions*, 2nd edn. John Wiley and Sons, New York.

Kullback, S. and Leibler, R. (1951). On information and sufficiency. *Annals of Mathematical Statistics*, 22, 79–86.

Lambert, B. (2018). *A Student's Guide to Bayesian Statistics*. Sage Publishers, San Francisco.

Lazar, N. (2010). Ockham's razor. *WIREs Computational Statistics*, 2, 243–246.

McElreath, R. (2019). *Statistical Rethinking: A Bayesian Course with Examples in R and Stan*, 2nd edn. CRC Press, New York.

Nelder, J. and Wedderburn, R. (1972). Generalized linear models, *Journal of the Royal Statistical Society Series A*, 135, 370–384.

Robins, J., van der Vaart, A., and Ventura, V. (2000). Asymptotic distribution of *p*-values in composite models. *Journal of the American Statistical Association*, 95, 1143–1156.

Rubin, D. (1984). Bayesianly justifiable and relevant frequency calculations for the applied statistician. *The Annals of Statistics*, 12, 1151–1172.

Schielzeth, H. and Fortsmeier, W. (2011). Conclusions beyond support: Overconfident estimates in mixed models. *Behavioral Ecology*, 20, 416–420.

Schwarz, G. (1978) Estimating the dimension of a model. *Annals of Statistics*, 6, 461–464.

Shannon, C. (1948). A mathematical theory of communication. *Bell System Technical Journal*, 27, 379–423.

Simmons, J., Nelson, L., and Simonsohn, U. (2011). False-positive psychology: Undisclosed flexibility in data collection and analysis allows presenting anything as significant. *Psychological Science*, 22, 1359–1366.

Simonsohn, U., Nelson, L., and Simmons, J. (2014). *p*-curve and effect size: Correcting for publication bias using only significant results. *Perspectives on Psychological Science*, 9, 666–681.

Spiegelhalter, D., Best, N., Carlin B. et al. (2002). Bayesian measures of model complexity and fit. *Journal of the Royal Statistical Society B*, 64, 583–639.

Stigler, S. (1980). Stigler's law of epinomy. *Transactions of the New York Academy of Sciences*, 39, 147–158.

Stone, M. (1977). An asymptotic equivalence of choice of model by cross validation and Akaike's criterion. *Journal of the Royal Statistical Society B*, 39, 44–47.

Vehtari, A., Gelman, A., and Gabry, J. (2017). Practical Bayesian model evaluation using leave-one-out cross validation and WAIC. *Statistical Computing*, 27, 1413–1432.

Watanabe, S. (2010). Asymptotic equivalence of Bayes cross validation and widely applicable information criterion in singular learning theory. *Journal of Machine Learning Research*, 11, 3571–3594.

CHAPTER 8

The Generalized Linear Model

Packages required for this chapter:

`packages<-c("ggplot2","gridExtra","reshape2","brms")`
`lapply(packages.needed, FUN = require, character.only = T)` # loads these packages in the working session.

8.1 Introduction

The generalized linear model (GLM) by Nelder and Wedderburn (1972) was an important landmark in statistical modeling. Their paper synthesized, organized, and extended several disconnected modeling approaches. The GLM is an overarching framework that allows analysis of several types of response variables, including binary, counts, proportions (i.e., real numbers in [0, 1]), and other types of real-valued dependent variables from a common model (Table 2.1). This overarching feature lies behind the adjective "generalized" in its name. The approach of depicting the effects of categorical explanatory variables on the mean of the response variable as regression models (Chapter 5) also applies in the GLM. The "linear" part in GLM stems precisely from the additive composition of the effects of numerical and categorical explanatory variables.

This chapter starts by explaining the components of GLM. It then explains the methods used in the fitting and testing of GLMs, introducing new features such as the model deviance and the types of residuals employed in the assessment of their goodness of fit. GLMs can, of course, be fitted using both frequentist and Bayesian methods. The goal of this chapter is to provide a general and conceptual overview of GLMs, leaving their actual fitting to forthcoming chapters dedicated to different types of response variables. Accordingly, this chapter will include more theory than examples and applications.

8.2 What are GLMs made of?

All GLMs have three components: the random component, the linear predictor, and the link function. The random component refers to the probability distribution employed to model the response variable. The main attributes (general tendency, dispersion, shape) of these probability distributions are defined by their parameters, typically the mean and a second parameter that determines its dispersion and shape. The linear predictor defines the relation between categorical and explanatory variables with the mean of the response variable. The link function is the key component of a GLM that binds the random component and the linear predictor. Let's take each of these individually.

Statistical Modeling With R. Pablo Inchausti, Oxford University Press. © Pablo Inchausti (2023).
DOI: 10.1093/oso/9780192859013.003.0008

The *random component* of a GLM consists of a response variable Y with independent observations from a probability density or mass probability. Nelder and Wedderburn (1972) developed a general theory and estimation methods for GLMs for a specific set of probability distributions. In turn, these probability distributions can be derived from a "generating family" called the exponential family. By changing the components of the expression for the exponential family we can obtain the normal, binomial (for binary response variables), gamma and inverse Gaussian (both for positive, real response variables), and Poisson (for count response variables) probability distributions as special cases. These are probability distributions that are used to model different types of response variables (Table 2.1). We defer an explanation of these probability distributions to the chapters where we will discuss them in relation to different types of response variables.

Nelder and Wedderburn (1972) used the exponential family to obtain the likelihood equations, asymptotic distributions of estimators of model parameters, and an algorithm for frequentist fitting of models valid for all probability distributions that are special cases of the exponential family. Therefore, there is no need to "reinvent the wheel" many times to obtain specific maximum likelihood equations and develop algorithms for each special case.

The exponential family (also called the exponential dispersion family) is

$$f(Y_i; \theta, \phi) = \exp\left(\frac{Y_i\theta - b(\theta)}{a(\phi)} + c(Y_i, \phi)\right). \tag{8.1}$$

Equation (8.1) has three functions, $a(\phi)$, $b(\theta)$, and $c(Y_i, \phi)$, and the parameter θ whose meaning will be discussed shortly. The notion of the exponential family was independently formulated by Darmois (1935), Pitman and Wishart (1936), and Koopman (1936). It had limited practical implications for non-statisticians until it was used as the basis of Nelder and Wedderburn's (1972) formulation of GLMs.

Let's give an example. The expression for the normal probability density is $f(Y_i) = \frac{1}{\sqrt{2\pi\sigma_Y}}\exp\left(\frac{-(Y_i-\mu_Y)^2}{2\sigma_Y^2}\right)$, which, after a few algebraic manipulations, is written as

$$f(Y_i) = \exp\left[\frac{Y_i\mu_Y - \frac{1}{2}\mu_Y^2}{\sigma_Y^2} - \frac{1}{2}\log\left(2\pi\sigma_Y^2\right) - \frac{Y_i}{2\sigma_Y^2}\right]. \tag{8.2}$$

We can now use Eq. (8.2) to make a one-to-one identification with Eq. (8.1): $\theta = \mu_Y$, $a(\phi) = \sigma^2$, $b(\theta) = -\frac{1}{2}\mu_Y^2$, and $c(Y_i, \phi) = \frac{1}{2}\log\left(2\pi\sigma_Y^2\right) - \frac{Y_i}{2\sigma_Y^2}$. Therefore, by substituting these three specific $a(\phi)$, $b(\theta)$, and $c(Y_i, \phi)$ into Eq. (8.1), we can obtain the normal distribution as a special case of the exponential family. By using different functions $a(\phi)$, $b(\theta)$, and $c(Y_i, \phi)$ in Eq. (8.1) we can also obtain the other probability distributions mentioned above as special cases of the exponential family. θ is called the "natural parameter" and defines the third component of GLMs (the link function) that we will explain below. Table 8.1 summarizes the functions $a(\phi)$, $b(\theta)$, and $c(Y_i, \phi)$, the parameter θ, and the support range of response variables of probability distributions that are modeled as stemming from the exponential family (Eq. (8.1)).

Two comments about Table 8.1. First, we did not include the inverse Gaussian distribution since its main attributes, shape, and support range are so similar to the more familiar gamma distribution (Chapter 12) that is used far more in practice. Second, the binomial and Poisson are one-parameter distributions for which $a(\phi)$ is set equal to one

Table 8.1 Summary of probability distributions modeled as GLMs, showing the support ranges of the response variable, the parameter θ that defines the canonical or typical link function for each type of response variable, the functions $a(\phi)$, $b(\theta)$, and $c(y, \phi)$ in Eq. (8.1), and the variance function relating the mean and variance of the response variable. Each probability distribution is parameterized in terms of its mean μ and additional parameters that are explained in the respective chapters dealing with each data type. The meaning of the dispersion parameter ϕ differs according to the probability distribution. $\Gamma()$ and $\mathrm{B}()$ are the mathematical gamma and beta functions. The last two probability distributions do not formally stem from the exponential family of Eq. (8.1). However, for all intents and purposes they are considered as GLMs with extra parameter(s) to be estimated from the data.

Probability distribution	Support range	Link function (θ)	$a(\phi)$	$b(\theta)$	$c(y, \phi)$	Variance function (μ)	Canonical link	Chapter
Normal (μ, σ^2)	Real values, $(-\infty, +\infty)$	μ	$\frac{\sigma^2}{2}$	μ^2	$-\frac{1}{2}\log\left(2\pi\sigma_Y^2\right) - \frac{Y_i}{2\sigma_Y^2}$	σ^2	identity	4–6
Binomial (k, μ) ($\mu = \pi$ if $k = 1$)	Binary {0, 1} (integers)	$\log\left(\frac{\mu}{1-\mu}\right)$	1	$-k\log(1-\mu)$	$\log\begin{pmatrix} k \\ y \end{pmatrix}$	$\mu(1-\mu)$	logit	9
Poisson (μ)	Counts	$\log(\mu)$	1	$\exp(\mu)$	$-\log(y!)$	μ	log	10, 11
Gamma (μ, ϕ)	Real values, $(0, +\infty)$	$\frac{-1}{\mu}$	ϕ	$\log(\mu)$	$\frac{-\log\left(\frac{y}{\phi}\right)}{\phi} - \log(y) + \log\Gamma(\phi)$	μ^2	inverse	12
Negative binomial (μ, ϕ)	Counts	$\log\left(\frac{\mu}{\mu+\phi}\right)$	1	$-k\log(1-\mu)$	$\log\left(\frac{\Gamma(y+\phi)}{y!\Gamma(\phi)}\right)$	$\mu + \frac{\mu^2}{\phi}$	log	10, 11
Beta (μ, ϕ)	Real values, [0, 1]	$\log\left(\frac{\mu}{1-\mu}\right)$	$\frac{1}{\phi}$	$-\log(1-\mu)$	$\log\left(\frac{y}{(1-y)}\right) + \log\left(\frac{1}{\mathrm{B}(y,\phi)}\right)$	$\mu(1-\mu)$	logit	12

(see Chapters 9 and 10), while the normal and gamma are two-parameter distributions for which $a(\phi)$ is separately estimated from the data. Some authors reserve the term "exponential family" for cases where ϕ is not used, while using the term "exponential dispersion family" for cases where it is (Faraway 2016, p. 152). Table 8.1 includes the negative binomial (Chapter 10) and beta (Chapter 12) distributions that are two-parameter distributions which are not formally part of the exponential family. However, these two probability distributions can be reparameterized and rendered sufficiently close to the exponential family that we can use GLMs with minor modifications to account for the extra dispersion parameter. While it is not necessary to recall the details of Table 8.1 when fitting GLMs, it is helpful to have a general notion of the main theoretical basis of these models.

The functions $a(\phi)$, $b(\theta)$, and $c(Y_i, \phi)$ in Eq. (8.1) have two interesting properties (McCullaugh and Nelder 1989). First, $b'(\theta)$, the derivative of $b(\theta)$, determines the mean of the response variable: $\mu_Y = \frac{db(\theta)}{d\theta}$. And second, we can obtain the variance function $\sigma_Y^2 = \frac{d\mu_Y(\theta)}{d\theta} a(\phi) = b''(\theta)\, a(\phi)$, where $b''(\theta)$ is the second derivative of $b(\theta)$. The variance function indicates that the mean and conditional variance of the response variable for all probability distributions from Eq. (8.1) (except for the normal distribution, for which these two parameters are independent) have a precise relationship (Agresti 2015, Dunn and Smyth 2018, Faraway 2016) defined by the specific functions $a(\phi)$ and $b(\theta)$ (Table 8.1). Therefore, except for the normal distribution, GLMs modeling the effect of explanatory variables on the mean of the response variable will be accounting for their effect on its conditional variance.

The *linear predictor* describes how the mean of the response variable (μ_Y) is related to a linear combination of the categorical and explanatory variables, including their interactions. That is, $\text{E}(Y) = \mu_Y = \beta_0 + \beta_1 X_1 + \beta_2 X_2 + \cdots + \beta_{k-1} X_{k-1} + \beta_k X_k$, or, for brevity, $\text{E}(Y) \equiv \mu_Y = \text{X}\beta$. This is exactly the same as we did for all cases of the general linear model covered in Chapters 4 to 6.

The *link function* is the third component of a GLM that connects the random component with the linear predictor. This key feature of GLMs puzzles some readers on first meeting these statistical models. What GLMs do is model the relation between the mean of the response variable and the explanatory variables in the scale of the link function g. These link functions may differ among probability distributions, and among types of response variables (Table 2.1). In general, we will denote this relation as $g(\mu_Y) = \text{X}\beta$. It is very important to understand that GLMs *DO NOT* transform the response variable. Prior to the formulation of GLMs, scientists used common and arcane non-linear transformations (e.g., arcsin(sqrt), square root, logarithm, Box–Cox transformations) to render their response variables approximately normal in order to apply the general linear model. Nevertheless, these non-linear transformations are known to often result in highly biased and imprecise parameter estimates (e.g., O'Hara and Kotze 2010, Stroup 2014, Warton and Hui 2011).

To take one example, if the response variable Y were counts that followed a Poisson distribution with mean μ_Y (Chapter 10), we would use log as the link function (Table 8.1) and thus $\log(\mu_Y) = \text{X}\beta$. We would fit the Poisson GLM in the scale of the log link function where the effects of the explanatory variables are to be linear, and then invert the link function (the inverse is sometimes called the *response function*) to obtain model predictions in the scale of the data: $\mu_Y = g^{-1}(\text{X}\beta)$ or $\mu_Y = \exp(\text{X}\beta)$ in this case. The response variable Y remains untouched in the process of model fitting. The values of μ_Y predicted by the statistical model will be real and positive numbers, and thus comparable to the observed counts (which are, of course, positive integers). Therefore, GLMs have a data scale and a model scale where parameters are estimated in the scale of the link function. We can switch back to the data scale by inverting the link function. We will illustrate these general and now abstract notions when fitting GLMs for different types of data in the coming chapters.

It was possible to obtain "canonical link functions" with term-by-term identification of the expressions for different probability distributions and Eq. (8.1) (Table 8.1). "Canonical" here just means that these are the preferred, or more often used, link functions for each case. Other link functions are feasible, provided that they are invertible (for which they need to be monotonic, meaning that their first derivative does not change its sign) to allow back-transforming from the model scale, $g(\mu_Y) = X\beta$, to the data scale, $\mu_Y = g^{-1}(X\beta)$. In principle, any invertible, continuous mathematical function can be a valid link function, provided that it does not obtain nonsensical results (e.g., negative predicted counts, probabilities outside the unit interval [0, 1]) when back-transforming the predicted values of μ_Y to the data scale. In practice, however, nearly every scientist using GLMs sticks to the default or canonical link functions for each data type.

Response variables with a normal distribution have the identity function as the canonical link function (Table 8.1). The identity function $g(\mu_Y) = \mu_Y$ leaves its argument unchanged. This means that we could have fitted all the statistical models of Chapters 4 to 6 as GLMs using the command `glm(y ~ x1 + ... + xk, family = gaussian)` with results identical to `lm(y ~ x1 + ... + xk)`. The general linear model is a special case of the generalized linear model.

8.3 Fitting GLMs

The fitting of all GLMs in the frequentist framework is based on maximum likelihood (Chapter 3), using the iteratively (re-)weighted least squares algorithm (IWLS; Nelder and Wedderburn 1972). The IWLS algorithm was formulated using the exponential family of Eq. (8.1) and the canonical link functions for each probability distribution and type of response variable. Details of the IWLS algorithm can be found in Agresti (2015) and Faraway (2016). In a nutshell, the IWLS algorithm obtains the maximum likelihood estimates of the parameters β_i relating the explanatory variables X to μ_Y in the scale of the link function by making a quadratic approximation of the log-likelihood surface for all data (Wood 2017). The estimated β_i are then the parameter values that are most likely to have produced the observed data given the statistical model. Frequentist model fitting of GLMs also obtains the standard errors of the model parameters needed to evaluate their statistical significance. The confidence intervals of the model parameters are numerically obtained using profile likelihood (Chapter 3).

GLMs are also fitted in the Bayesian framework using MCMC-based algorithms to obtain posterior distributions of their parameters. Because the fitting of Bayesian GLMs does not rely on optimization algorithms based on the exponential family of Eq. (8.1), they can in principle be formulated for any probability distribution, including those of Table 8.1. Just like their frequentist analogues, Bayesian GLMs do have the linear predictor and employ a suitable link function. These are key distinctive features of GLMs regardless of how we fit them. The posterior distributions of model parameters are used to evaluate the relevance of explanatory variables and to obtain the credible intervals. Some additional care may be needed to define the priors in the link scale so that they reflect the intended content of information on the data (rather than on the model) scale. This last point will be discussed for each type of Bayesian GLM in the coming chapters.

8.4 Goodness of fit in GLMs

Deviance is a measure of the goodness of fit of a statistical model formulated by Nelder and Wedderburn (1972) in the context of the general theory of GLMs (Chapter 7). It compares the likelihood of our current model with the saturated model that achieves a perfect fit to the data. The model deviance can be used as an "objective function" to minimize when finding the parameter values of GLMs by maximum likelihood. Because the saturated model has, by definition, a residual deviance of zero, the likelihood ratio test statistic of our current model compared with the saturated model is the residual deviance $D(y, \hat{\mu})$ of our current model (Fig. 7.2). For a set of parameter estimates, the model deviance is the sum of the deviances for all i data points:

$$\text{Model deviance}\,(Y;\text{parameters}) = \sum_i \text{Deviance}\,(Y_i, \hat{\mu}_i\,(X, \beta)), \tag{8.3}$$

where $\hat{\mu}_i\,(X, \beta)$ are the predicted values of the mean of the response variable for the values of the explanatory variables X and the β parameter estimates. The actual expression for $\text{Deviance}\,(Y_i, \hat{\mu}_i\,(X, \beta))$ varies according to the probability distribution and type of response variable (Agresti 2015, Faraway 2016). The squared sum of the deviance residuals is, of course, the overall model deviance (Eq. (8.3)). The model deviance will be used to assess the correctness of assuming that the parameter ϕ in Eq. (8.1) is equal to one in binary and count GLMs (Chapters 10 and 11).

There is no direct equivalent of the R^2 statistic to depict the percentage of the variation of the response variable explained by a fitted GLM. Except for the normal distribution, the mean and the variance of all the probability distributions in Table 8.1 are not independent. Hence, when we model the effect of explanatory variables on the mean of the response variable, we are also modeling their effect on its variance. McFadden (1974) proposed a simple global metric to characterize the goodness of fit of a GLM:

$$\text{Dev.explained}_{\text{McF}} = 1 - \frac{\text{Residual deviance}}{\text{Null deviance}}, \tag{8.4}$$

where the residual deviance is the deviance left to explain after fitting the current GLM, and the null deviance is the deviance explained by the minimal GLM only fitting the grand mean. The better the fit of the GLM, the lower the residual deviance, and the higher the proportion of deviance explained. While there are other alternatives, such as the metrics of Cox and Snell (1989) and Nagelkerke (1991), Eq. (8.4) is simple and easy enough to calculate to assess the comparative goodness of fit of different GLMs to the same data.

The Pearson residuals $r_P = \frac{(Y_i - \hat{\mu}_i)}{\sqrt{(\text{Var}(\hat{\mu}_i))}}$, where $\text{Var}\,(\hat{\mu}_i)$ is the variance function for each probability distribution of the exponential family (Table 8.1), have been used to assess the goodness of fit of GLMs (McCullaugh and Nelder 1990, Faraway 2016, Dunn and Smyth 2018). The squared sum of the Pearson residuals corresponds to the familiar Pearson chi-square statistic that measures the overall goodness of fit of a GLM. The Pearson statistic and the overall model deviance of Eq. (8.3) asymptotically converge to a chi-square distribution with $n - k$ degrees of freedom, with n being the number of data points and k the number of model parameters (McCullaugh and Nelder 1990, Faraway 2016, Dunn and Smyth 2018). This asymptotic convergence hinders the use of both the overall Pearson and deviance statistics to assess the overall goodness of fit of a GLM for more moderate and realistic sample sizes. It is much more informative to carry out a full residual analysis to evaluate the different aspects of model fit, rather than these global goodness-of-fit statistics based on Pearson or deviance residuals.

Regardless of the probability distribution of the response variable, the deviance and Pearson residuals of a well-fitting model should asymptotically have a normal distribution in Q–Q plots (Pierce and Schaffer 1986, McCullagh and Nelder 1990). When the response variable has a normal distribution, the Pearson and deviance residuals should be exactly Gaussian for a well-fitting model. In addition, the residuals of a well-fitting model should also exhibit a random scatter with homogeneous variation when plotted against both the fitted values and the explanatory variables (cf. Chapters 4 to 6). Yet again, the normal distribution expected in Q–Q plots for deviance and Pearson residuals of well-fitting GLMs is valid only for "sample sizes" that may exceed those of typical data sets in life sciences. Extensive simulation studies have shown that neither of these residuals attain a normal distribution even for large sample sizes when the response variable has a binomial (Park et al. 2020), Poisson (Feng et al. 2020), beta (Pereira 2019), or gamma (Scudilio and Pereira 2020) distribution. Nevertheless, both deviance and Pearson residuals are still the recommended means for evaluating the goodness of fit of GLMs through residual analysis (e.g., Bolker 2008, Agresti 2015, Faraway 2016).

Dunn and Smyth (1996) proposed randomized quantile residuals as a better alternative to Pearson and deviance residuals. These are calculated using simulated data from a fitted model to obtain an empirical cumulative density function of the response variable. The residuals are generated as the values corresponding to the observed data along the cumulative density function. If the statistical model is correctly specified and fits the data well, then the observed data can be considered as a random draw from the fitted model. If this is the case, then the randomized quantile residuals thus calculated should have a uniform distribution (Dunn and Smyth 1996) for all statistical models based on the probability distributions that make up the extended set derived from the exponential family (Table 8.1). Yielding to nothing more substantial than tradition, we convert the uniform distribution of the residuals of a well-fitting model thus generated to a normal distribution. This is accomplished using the probability integral transform (Congdon 2014, Dunn and Smyth 2018).

For the continuous response variables in Table 8.1 (i.e., all but those with binomial, Poisson, and negative binomial distributions), the randomized quantile residuals for each data item i are defined as:

$$r_{Q,i} = \Phi^{-1}\left(F\left(y_i \mid \text{data, parameters}\right)\right), \tag{8.5a}$$

where F is the cumulative distribution function of the response variable of the fitted GLM for the ith data point given the data and the estimated parameters, and Φ^{-1} is inverse of the cumulative normal distribution function. Despite its seemingly threatening name, you should be familiar with the quantile function Φ^{-1} from previous statistic courses. It is the one corresponding to the lower tail for a given value z_0 (i.e., $\Pr(Z < z_0)$) of a normal distribution. Older readers may recall obtaining this function from the old-fashioned tables at the back of statistics books. We can now more reasonably use the function `qnorm` in R.

For discrete response variables and their associated binomial, Poisson, and negative binomial distributions (Table 8.1), an additional step is required to obtain a continuous distribution of residuals. The reason for the additional step is that by definition the cumulative distribution of a discrete response variable increases in jumps, thereby precluding obtaining a smooth, continuous distribution of residuals. Dunn and Smyth (1996, 2018) suggested jittering the cumulative distribution of discrete response variables by adding a value q_i randomly sampled from a uniform distribution in [0, 1]:

$$r_{Q,i} = \Phi^{-1}\left(F\left(y_i \mid \text{data, parameters}\right)\right) + \left(1 - q_i\right)\Phi^{-1}\left(F\left(y_i^- \mid \text{data, parameters}\right)\right), \quad (8.5b)$$

where y_i^- is the limit obtained by approaching the value of y_i from below.

A simpler way of writing Eq. (8.5b) (Park et al. 2020) to depict the random jittering is:

$$r_{Q,i} = \Phi^{-1}\left(\text{Uniform}\left(\lim_{y \to y_i} F\left(y_i \mid \text{data, parameters}\right), F\left(y_i \mid \text{data, parameters}\right)\right)\right), \quad (8.5c)$$

where Uniform(a, b) is a random value sampled from a uniform probability distribution in the interval [a, b]. An example of the calculation of randomized quantile residues using Eq. (8.5) is shown in Fig. 8.1.

The key idea behind randomized quantile residuals (Dunn and Smyth 1996, 2018) is to introduce random jittering in the discontinuity gap of the cumulative distribution function of the discrete response variable, and then invert the fitted distribution for each response value to obtain the equivalent standard normal quantile (Feng et al. 2020). As suggested by Dunn and Smyth (1996), it is necessary to make multiple realizations of the randomized quantile residuals to make sure that the resulting patterns do not result from the randomization procedure.

The adequacy of randomized quantile residuals to assess the goodness of fit of any GLM for even moderate sample sizes has been established by extensive simulations (e.g., Feng et al. 2020, Dunn and Smyth 2018, Park et al. 2020). These residuals are also suitable for detecting other specific causes of poor fit in GLMs, such as over-dispersion and an excess of zeros, that are common in count GLMs (Chapter 10). In addition, Feng et al. (2020) showed that randomized quantile residuals have desirable properties when evaluating the goodness of fit of GLMs such as high power and low Type I error (i.e., they are likely to detect true deviations and have low false positive error rates).

The approach we use in the coming chapters to obtain randomized quantile residuals for frequentist GLMs will be based on the package `DHARMa`. The approach will be somewhat similar to a parametric bootstrap (Chapter 13). We will: (a) simulate new data sets from a fitted GLM; (b) obtain the cumulative distribution function of each observed value of the response variable for the simulated data sets; and (c) calculate the randomized quantile residual of each observed value of the response variable using Eq. (8.5).

Bayesian GLMs also require that we assess their goodness of fit with residual analysis. The randomized quantile residuals can be calculated from the posterior predictive distribution (Chapter 7) from which we can obtain $\Pr(y_{\text{rep},i} \leq y_{\text{obs},i})$, which is the cumulative distribution function for a simulated value of the response variable $y_{\text{rep},i}$ being less than or equal to the observed value $y_{\text{obs},i}$. This cumulative distribution is also known as the posterior predictive p-value (Chapter 7), which is defined as the average of the tail probability of an observed value of the response variable (as a function of the model parameters) with respect to the posterior distribution of parameters (Gelman et al. 1996, Li et al. 2017). $\Pr(y_{\text{rep},i} \leq y_{\text{obs},i})$ can be used to obtain the randomized quantile residuals using Eq. (8.5)

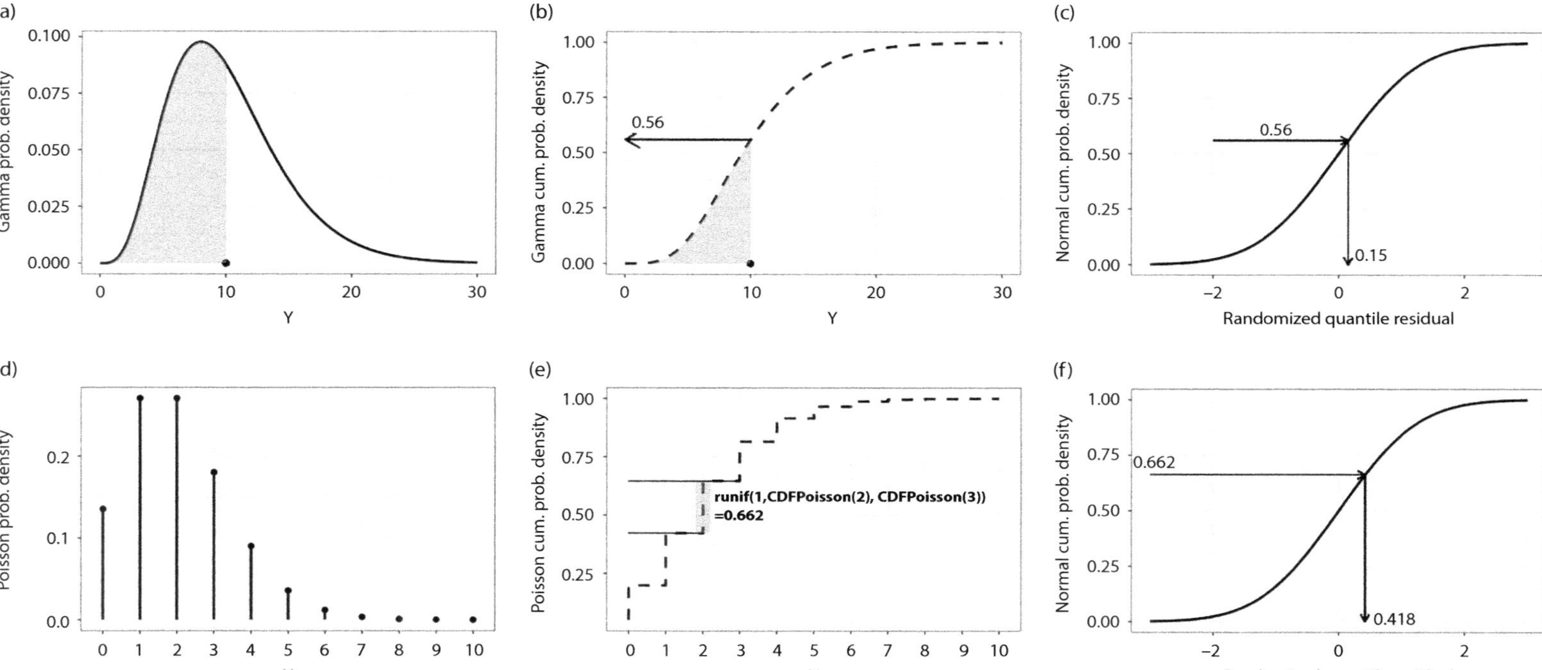

Fig 8.1 Calculation of the randomized quantile residuals for a response variable having a gamma (mean = 5, dispersion parameter = 2; see Chapter 12) or a Poisson (mean = 3; see Chapter 10) distribution (bottom row). From left to right, the three plots are the probability density/mass function of the response variable highlighting a single value ($Y = 10$ and $Y = 3$ for the gamma and Poisson distributions, respectively), the cumulative probability distribution functions, and the inverse of the cumulative normal distribution function that is used to obtain the randomized quantile residuals using Eq. (8.5). For the Poisson distribution, the continuous distribution of the residuals is obtained by jittering (i.e., drawing a uniform variate in the interval defined in Eq. (8.5b)) the value of the cumulative Poisson distribution (Eq. (8.5c)), and then inverting the cumulative normal distribution of the jittered value. Figure drawn after Dunn and Smyth (2018, pp. 301–303).

(Congdon 2014, Feng et al. 2020). In sum, we can obtain the randomized quantile residuals for all Bayesian GLMs and use them to assess goodness of fit in a residual analysis in the next chapters of Part II.

8.5 Statistical significance of GLM

There are three methods to assess the statistical significance of frequentist GLMs: the likelihood ratio test and its associated Wilks test, and the Wald test (Chapter 3).

We can fit to the same data two nested GLMs differing only in the term whose significance we wish to evaluate. The term by which the two models differ can be an explanatory variable or an interaction between variables. For example, let `m1 = glm(y ~ x1, family = ...)` and `m2 = glm(y ~ x1+x2, family = ...)` such that the more specific model `m1` is a special case of the more general model `m2`. We can now use the Wilks test based on the likelihood ratio of the two models to evaluate whether the two models differ in their explanatory capacity (Chapter 3). The Wilks test on the likelihood ratio of the two models differs depending on whether the dispersion parameter ϕ, Eq. (8.1), is either equal or set to one, or estimated from the data (McCullaugh and Nelder 1990, Fox 2015). In the first of these cases, corresponding to binary (Chapter 9) and count (Chapter 11) GLMs (Table 8.1), the Wilks test requires using the chi-square distribution with one degree of freedom to compare the two statistical models. In R, we just write `anova(m1, m2, test = "chisq")` to obtain a p-value that will allow us to select between the two models. In the second case, and applicable to all other cases in Table 8.1, the dispersion parameter ϕ is estimated from the data and we need to use the F-distribution for the comparison by writing `anova(m1, m2, test = "F")`. Should our statistical model contain many explanatory variables and interactions, we would need to fit a large number of separate models to assess the statistical significance of each term in the original GLM.

We can also use the Wald test (Chapter 3) to assess the statistical significance of each term of a GLM. The Wald statistics are always printed by default in frequentist GLMs; they can be obtained by fitting a single GLM, and are thus a great simplification compared to the Wilks test. The statistical significance of Wald statistics uses the normal distribution. However, tests for significance based on the Wald statistics and Wilks test rely on a quadratic approximation of the likelihood surface that is only valid for large sample sizes (Pawitan 2001, Millar 2011). For the more moderate sample sizes gathered by scientists, both tests may underestimate the true p-value and become "too liberal" by falsely detecting terms as statistically significant. The p-values from Wald tests tend to be less liberal than those of the Wilks tests (Dunn and Smyth 2018), and they should be preferred.

In the Bayesian framework, statistical significance is not an issue. Here, decisions about the relevance of model parameters are based on their posterior distributions and their associated credible intervals (see Chapter 3), neither of which are based on sampling distributions as in the frequentist framework.

References

Agresti, A. (2015). *Foundations of Linear and Generalized Linear Models*. Academic Press, New York.
Bolker, B. (2008). *Ecological Models and Data in R*. Princeton University Press, Princeton.
Congdon, P. (2014). *Applied Bayesian Modelling*, 2nd edn. John Wiley and Sons, Chichester.
Cox, D. and Snell, E. (1989). *Analysis of Binary Data*, 2nd edn. CRC Press, New York.
Darmois, G. (1935). Sur les lois de probabilités à estimation exhaustive. *Comptes Rendus de l'Academie des Sciences*, 200, 1265–1266.

Dunn, P. and Smyth, G. (1996). Randomized quantile residuals. *Journal of Computational and Graphical Statistics*, 5, 236–244.
Dunn, P. and Smyth, G. (2018). *Generalized Linear Models With Examples in R*. Springer, New York.
Faraway, J. (2016). *Extending the Linear Model with R: Generalized Linear, Mixed Effects and Nonparametric Regression Models*. CRC Press / Chapman and Hall, New York.
Feng, C., Li, L., and Sadeghpour, A. (2020). A comparison of residual diagnosis tools for diagnosing regression models for count data. *BMC Medical Research Methodology*, 20, 175.
Fox, J. (2015). *Applied Regression Analysis and Generalized Linear Models*, 3rd edn. Sage, New York.
Gelman, A., Meng, X., and Stern, H. (1996). Posterior predictive assessment of model fitness via realized discrepancies. *Statistica Sinica*, 6, 733–807.
Koopman, B. (1936). On distributions admitting a sufficient statistic. *Transactions of the American Mathematical Society*, 39, 399–409.
Li, L., Feng, C., and Qiu, S. (2017). Estimating cross-validatory predictive p-values with integrated importance sampling for disease mapping models. *Statistics in Medicine*, 36, 2220–2236.
McCullaugh, P. and Nelder, J. (1990). *Generalized Linear Models*. Chapman and Hall, London.
McFadden, D. (1974). Conditional logit analysis of qualitative choice behavior. In P. Zarembka (ed.), *Frontiers in Econometrics*, pp. 105–142. Academic Press, New York.
Millar, R. (2011). *Maximum Likelihood Estimation and Inference with Examples in R, SAS and ADMB*. John Wiley & Sons, New York.
Nagelkerke, N. (1991). A note on a general definition of the coefficient of determination. *Biometrika*, 78, 691–692.
Nelder, J., and Wedderburn, R. (1972). Generalized linear models, *Journal of the Royal Statistical Society Series A*, 135, 370–384.
O'Hara, R. and Kotze, D. (2011). Do not log-transfom your data. *Methods in Ecology and Evolu-tion*, 1, 118–122.
Park, K., Jung, D., and Kim, J. (2020). Control charts based on randomized quantile residuals. *Applied Stochastic Models in Business and Industry*, 36, 716–729.
Pawitan, J. (2001). *In All Likelihood: Statistical Modeling and Inference Using Likelihood*. Oxford University Press, Oxford.
Pereira, G. (2019). On quantile residuals in beta regression. *Communications in Statistics – Simulation and Computation*, 48, 302–316.
Pierce, D. and Schaffer, D. (1986). Residuals in generalized linear models. *Journal of the American Statistical Association*, 81, 977–986.
Pitman, E. and Wishart, J. (1936). Sufficient statistics and intrinsic accuracy. *Mathematical Proceedings of the Cambridge Philosophical Society*, 32, 567–579.
Scudilio, J. and Pereira, G. (2020). Adjusted quantile residual for generalized linear models. *Computational Statistics*, 35, 399–421.
Stroup, W. (2014). Rethinking the analysis of non-normal data in animal and plant science. *Agronomy Journal*, 116, 1–17.
Warton, D. and Hui, C. (2011). The arcsine is asinine: The analysis of proportions in ecology. *Ecology*, 92, 3–10.
Wood, S. (2017). *Generalized Additive Models*, 2nd edn. CRC Press / Chapman and Hall, New York.

CHAPTER 9

When the Response Variable is Binary

Packages needed in this chapter:

```
packages.needed<-c("ggplot2","arm","brms","broom","GGally",
"bayesplot","DHARMa","ggeffects","qqplotr","pROC")
lapply(packages.needed, FUN = require, character.only = T)
```
loads these packages in the working session.

9.1 Introduction

This chapter explains the fitting of GLMs when the response variable is binary. In this case, the binomial is the only possible probability distribution. The distribution was derived by Jacob Bernoulli in 1713 in a posthumous publication to calculate the probability of obtaining y "successes" (where success is defined as the desired outcome of a binary event) in k repetitions of a binary experiment involved the binomial expansion of the expression $(p + q)$, hence its name, where p is the probability of success and the probability of failure is $q = 1 - p$. This distribution is used in every introductory statistics course to model coin flips:

$$\Pr(Y = y) = \binom{k}{y} \pi^k (1 - \pi)^{k-y}, \tag{9.1}$$

where k is the number of observations from which we observed y "successes," and π is the probability of success. The first term in Eq. (9.1) is a combinatorial number denoting the number of ways in which y successes (say, heads in a coin toss, if that is our focus of interest) can occur in k independent observations of an event or experiment. A special case of the binomial distribution occurs when there is $k = 1$ observation (i.e., a single coin toss) and it becomes the Bernoulli distribution: $\Pr(Y = y) = \pi(1 - \pi)$.

The mean and variance of a response variable Y having a binomial (or a Bernoulli) distribution are $k\pi$ and $k\pi(1 - \pi)$, respectively. Of course, these parameters are not independent as there is a quadratic relation (a downward parabola) between them. The linear predictor of a GLM portrays the predicted linear relation between the categorical and/or numerical explanatory variables X and the mean or expected value of the response variable $\mu(Y) \equiv E(Y) = X\beta$ (Chapter 8).

At this point, it is important to distinguish between two types of binary GLMs, grouped and ungrouped analyses, whose difference is best explained by means of an example. Say

Statistical Modeling With R. Pablo Inchausti, Oxford University Press. © Pablo Inchausti (2023).
DOI: 10.1093/oso/9780192859013.003.0009

we are interested in establishing a dose–response relation between the dose of an insecticide and the numbers of pest insects killed. At the elementary level, the response variable is binary since each experimental unit (i.e., an individual insect) exposed to a dose X of the insecticide is either dead or alive after a given time of exposure. We could, however, set up the experiment in two different ways. We could give each individual insect a different insecticide dose and record its fate. In this case, each experimental unit ($k = 1$) would have a different value of the explanatory variable X, and our response variable would be the binary outcome. This is an ungrouped binary GLM because each experimental unit has a different value of the explanatory variable(s). It is enough that one explanatory variable differs among experimental units for the analysis to be ungrouped. In contrast, we could expose a known number k of insects to the same dose of insecticide, and we could record the numbers of dead insects after the exposure. At the elementary level, the response variable for each insect is still binary (dead/alive), but the set of k individuals would have the same value of the explanatory variable(s). The latter is called a grouped binary GLM. By default, R assumes that we are carrying out an ungrouped analysis, unless we indicate otherwise by entering the data in an alternative format (see Section 9.6). Both analyses use the binomial distribution for the response variable with the number of individuals exposed, k, being either assumed to be one in ungrouped analysis or entered as part of the data input.

9.2 Key concepts for binary GLMs: Odds, log odds, and additional link functions

The main goal of binary GLMs is to relate the explanatory variable(s) with the mean of the response variable. As we explained in Chapter 8, GLMs establish a relation between the mean of the response variable E(Y) and the explanatory variables X in the space of the link function g, such that $g(\mathrm{E}(Y)) = \mathrm{X}\beta$. Logit is the canonical link function for binary GLMs (see Chapter 8). However, to make sense of the logit link function, we first need to explain the odds ratio.

Say we are watching the championship final of your favorite sport, and that there is a probability $\pi = 0.75$ that your team wins (and, of course, $(1 - \pi) = 0.25$ that it loses). There cannot be ties in the final. There is a long tradition dating to at least the sixteenth-century Italian polymath and inveterate gambler Girolamo Cardano of measuring the relative likelihood of two events as the ratio of their probabilities: $\frac{\pi}{(1-\pi)} = \frac{0.75}{0.25} = 3{:}1$. This ratio is known as the odds ratio and it says that your team is three times more likely to win than to lose the championship final. Of course, from the odds ratio of 3:1 we can recover the probability: $\pi = \frac{3}{(3+1)} = 0.75$. The odds ratio of obtaining a six when rolling a die is $\frac{\pi}{(1-\pi)} = \frac{1/6}{5/6} = 1{:}5$, i.e., we are five times less likely to draw a six than to obtain any other number when rolling a fair die. Odds ratios are commonly used when expressing the relative chances of events when betting. In our context, the odds ratio maps the probability π onto the positive real line $(0, \infty)$.

Logit is simply the log of an odds ratio: $\mathrm{logit}(\pi) = \log\frac{\pi}{(1-\pi)}$. It maps the odds ratio onto the real line. Therefore, binary GLMs establish a linear relation between the mean of the response variable E(Y) and the explanatory variables as $\mathrm{logit}(\pi) = \mathrm{X}\beta$. Note that the response variable is *not* transformed, just that its mean is related to the explanatory variables in the space of the link function logit. Therefore, once the parameters β are estimated, we would obtain an equation such as $\mathrm{logit}(\pi) = 1.3 + 2.4\mathrm{X}$. The relation between π and X is linear only in the scale of the logit link function. To compare the observed data

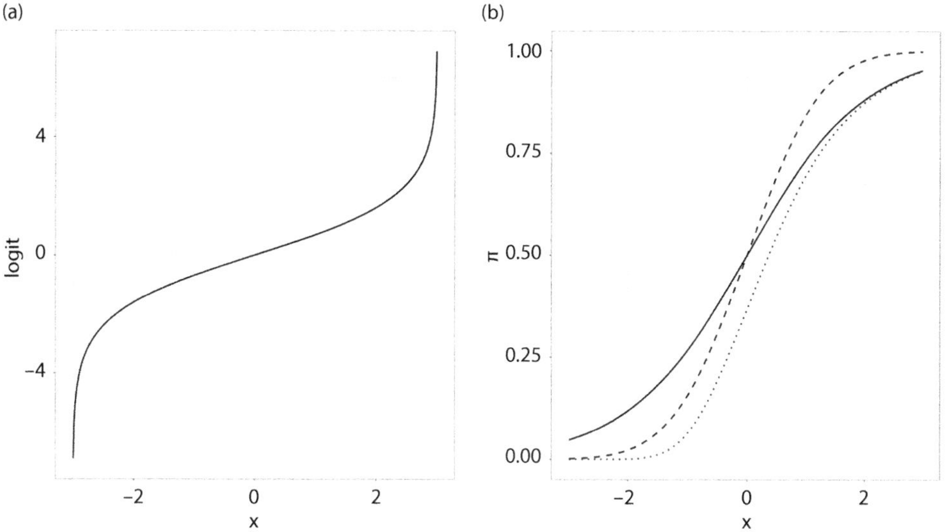

Fig. 9.1 Theoretical relations between (a) the logit link function (i.e., $\log(\pi / (1 - \pi))$) and (b) π (the probability of success) and a standardized explanatory variable for three common link functions used in binary GLMs: logit (continuous), complementary loglog (dashed), and probit.

(presences and absences) with model predictions, we invert the link function to obtain a predictive relation between π and the explanatory variable X. The inverse logit function, $\text{logit}^{-1}(\pi) = \frac{\exp(X\beta)}{1+\exp(X\beta)}$, takes the values of the explanatory variables X and their estimated parameters β and predicts the probability of observing a "success" for every value of X. If the logit function maps a probability onto the real line, its inverse function maps any value of X onto the unit interval [0, 1] to predict the parameter π of the binomial distribution (and, by extension, its mean and variance) as a function of the explanatory variables.

While logit is by far the most frequently used link function, any other invertible non-linear function mapping real values onto the unit interval could be used as a link function in binary GLMs. Other link functions sometimes used are probit and the complementary loglog (or Gumbel) link functions. Probit is the inverse of the cumulative normal distribution (`qnorm()` in R), and the complementary loglog function, which stems from the theory of extreme values (Coles 2001), is defined as $\log(-\log(\pi))$. The logit, probit, and complementary loglog link functions are all similarly S-shaped, and they only differ from each other in the rates of change either around their mean or towards the tails (Fig. 9.1). Since only for very special data sets would different link functions lead to qualitatively different results, we will use the logit link function when fitting binary GLMs in what follows.

9.3 Fitting binary GLMs

We will fit two binary GLMs, the first for ungrouped and the other for grouped data.

THE DATA IN CONTEXT: Hunter (2000), obtained from the package `DAAG`, wanted to determine which ecological and climatic variables could help predict the presence (`pres.abs`) of the corroboree frog in 212 sampling sites in the mountains of southeast Australia. The authors recorded the altitude of the sampling sites in meters, the distance to the nearest extant population in kilometers, the number of potential pools in each sampling site and of breeding sites in a range of 2 km from each sampling site, the average spring rainfall (`avrain`), and the average of the daily minimum (`meanmin`) and maximum (`meanmax`) spring temperature. This was an essentially exploratory or pattern-seeking analysis without clear hypotheses defined beforehand.

EXPLORATORY DATA ANALYSIS: We start by first importing and having a first look at the data.

```
> summary(DF1)
    pres.abs          altitude         distance          NoOfPools          NoOfSites
 Min.   :0.0000   Min.   :1280   Min.   :  250   Min.   :  1.00   Min.   : 0.000
 1st Qu.:0.0000   1st Qu.:1480   1st Qu.:  500   1st Qu.:  8.00   1st Qu.: 1.000
 Median :0.0000   Median :1580   Median : 1000   Median : 18.00   Median : 3.000
 Mean   :0.3726   Mean   :1547   Mean   : 1933   Mean   : 25.11   Mean   : 2.939
 3rd Qu.:1.0000   3rd Qu.:1625   3rd Qu.: 2000   3rd Qu.: 32.00   3rd Qu.: 4.000
 Max.   :1.0000   Max.   :1800   Max.   :18000   Max.   :232.00   Max.   :10.000
     avrain          meanmin          meanmax
 Min.   :124.7   Min.   :2.03   Min.   :11.60
 1st Qu.:141.7   1st Qu.:2.57   1st Qu.:12.97
 Median :148.8   Median :3.00   Median :13.38
 Mean   :148.1   Mean   :3.12   Mean   :13.67
 3rd Qu.:155.0   3rd Qu.:3.57   3rd Qu.:14.21
 Max.   :198.3   Max.   :4.33   Max.   :15.97
```

The response variable `pres.abs` was entered as values {0, 1} that R took as integers, and it is best summarized by a table of relative frequencies or proportions:

```
> table(DF1$pres.abs)/nrow(DF1)

    0     1
0.627 0.373
```

The response variable obviously has a binomial distribution.

We need to examine the covariation of the explanatory variables to avoid including numerical variables that are too strongly correlated (see Chapter 4). Figure 9.2 shows that because altitude and the three climatic variables are very strongly correlated, they would be largely redundant in explaining the response variable. Based on these correlations, we chose altitude, distance, avrain, NoOfPools, and NoOfSites as explanatory variables in the statistical models. These explanatory variables have very disparate scales. Thus, we standardized them to have a mean of zero and a unit standard deviation to compare their relative importance after fitting statistical models with:

```
DF1s=data.frame(cbind(scale(DF1[,2:6],center=T,scale=T),pres.abs= DF1$pres.abs))
```

The new data frame `DF1s` contains the standardized explanatory variables that will be used in the models and the original response variable (see Chapter 4).

When the response variable is binary, it is advisable to include either a smoothed or a fitted line to visualize its relation with the numerical explanatory variables. We see from Fig. 9.3 that the probability of presence declines with altitude and distance to potential breeding sites, increases with the number of pools and breeding sites, and is largely unrelated to the average spring rainfall.

THE STATISTICAL MODEL:

$$\begin{cases} \text{pres.abs} \sim \text{Binomial}(\pi), \\ \text{logit}(\pi) = X\beta = \beta_0 + \beta_1\text{altitude} + \beta_2\text{distance} + \beta_3\text{NoOfPools} + \beta_4\text{NoOfSites} + \beta_5\text{avrain}. \end{cases}$$

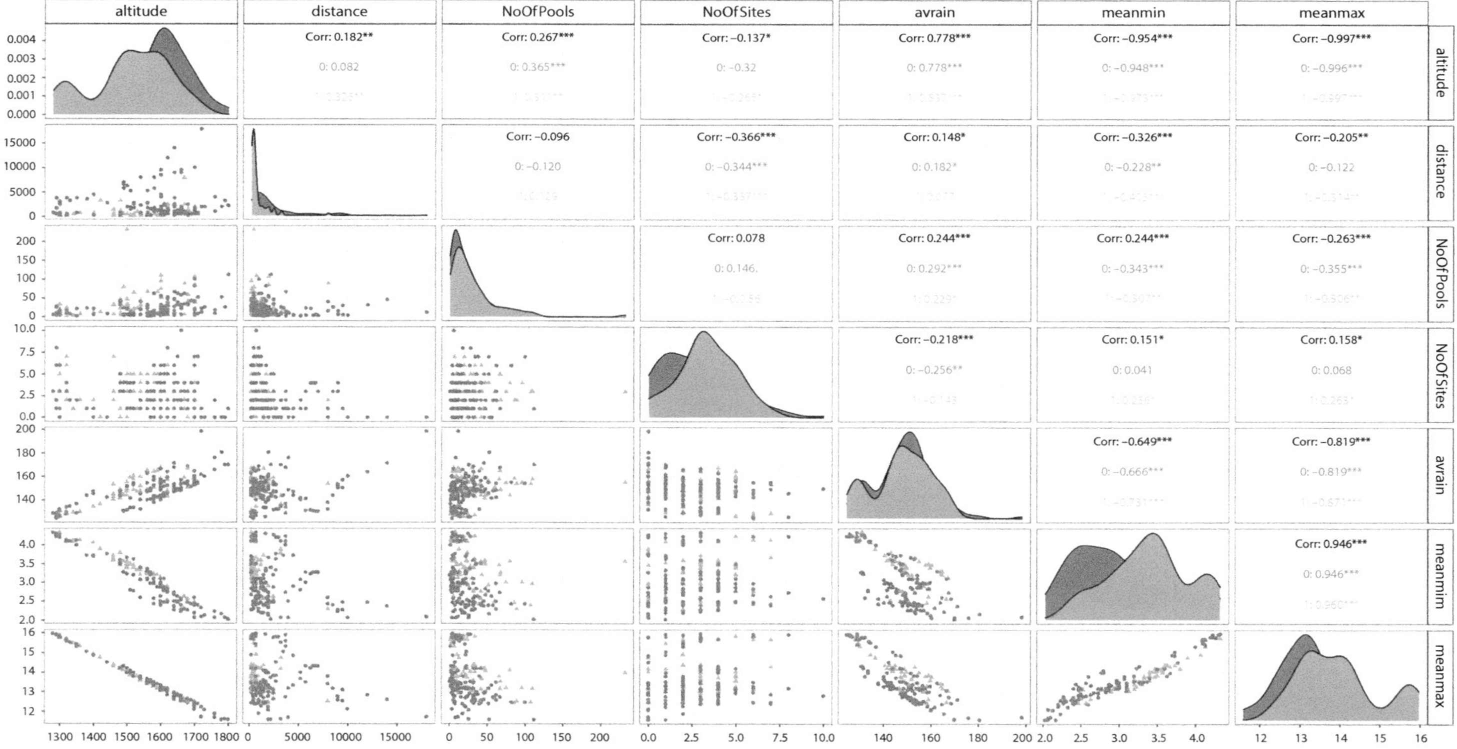

Fig. 9.2 Covariation between the explanatory variables in the binary GLM. The diagonal shows the density plots of each explanatory variable for the entire data set depending on the value of the response variable. Below the diagonal are pairwise *x*–*y* plots, with presences coloured in red and absences in green. Above the diagonal are the pairwise correlation of explanatory variables for the entire data sets and for each value of the response variable. The figure was produced with the package `GGally`.

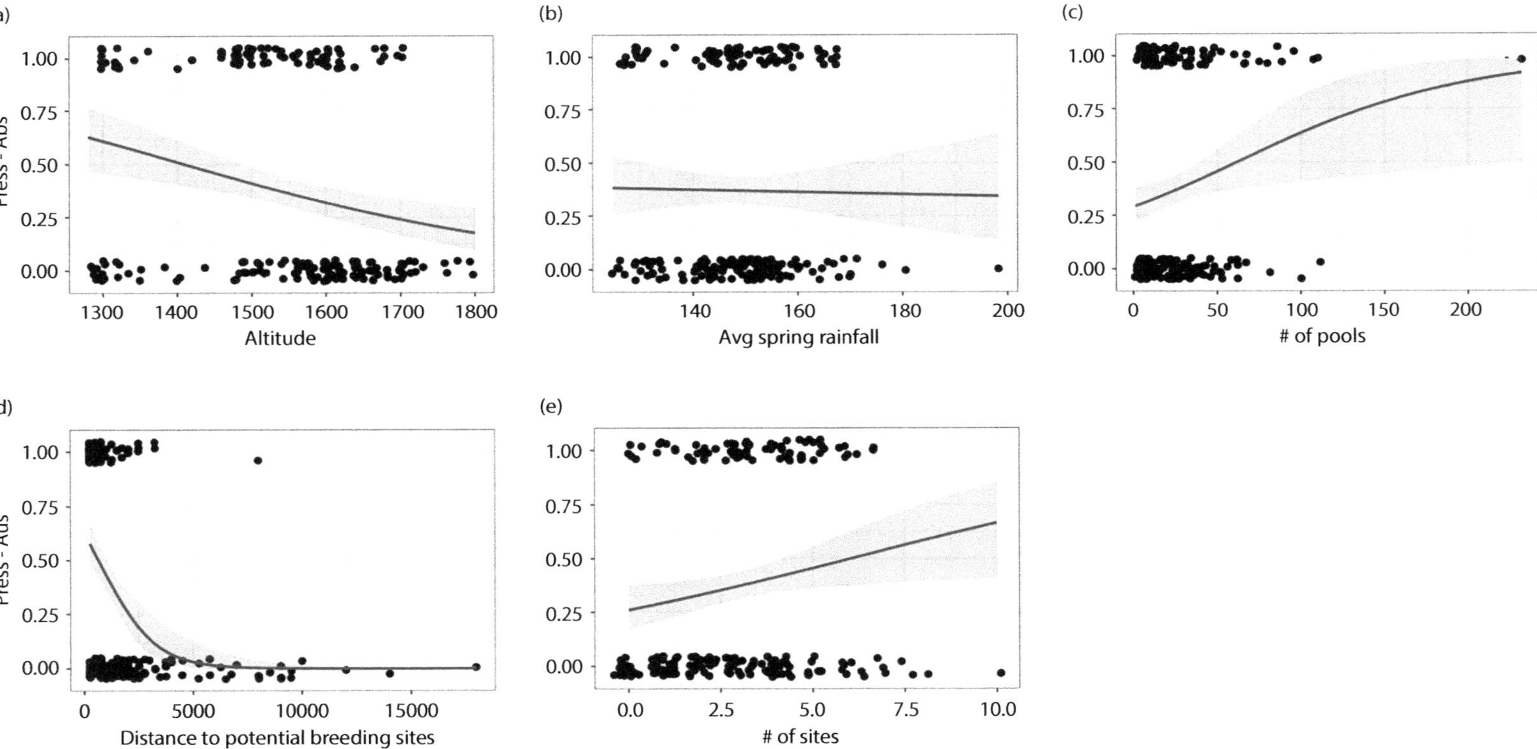

Fig. 9.3 Relations between the presence–absence response variable and the numerical explanatory variables for the frequentist binary GLM `m1`. All points are jittered by adding a random value to allow better visualization. The curves correspond to a fitted univariate logistic regression (blue) with a 95% confidence band (gray).

This is a multiple linear regression between logit(π) and the numerical explanatory variables. Everything we learned in Chapter 5 involving multiple regression will apply here. The only additional but fundamental feature is that we model the relation between the mean of the response variable and the explanatory variables in the scale of the link function.

9.4 Ungrouped binary GLM: Frequentist fitting

The command `glm` from the `stats` package (loaded by default when we start R) fits generalized linear models using the same syntax that we learned in Chapters 4 to 6 (Table 4.1):

```
m1=glm(pres.abs~altitude+ distance+ NoOfPools+ NoOfSites+ avrain, data=DF1s,
family=binomial)
```

The only difference is that we need to specify the probability distribution employed to model the response variable. Given `family = binomial`, the default link function is logit, but we could have chosen `family = binomial(link = "probit")` or `family = binomial(link = "loglog")`. The model summary is obtained as usual:

```
> summary(m1)

Call:
glm(formula = pres.abs ~ altitude + distance + NoOfPools + NoOfSites +
    avrain, family = binomial, data = DF1s)

Deviance Residuals:
   Min       1Q   Median       3Q      Max
-1.812   -0.829   -0.390    0.871    2.779

Coefficients:
            Estimate Std. Error z value Pr(>|z|)
(Intercept)  -0.9646     0.2304   -4.19  2.8e-05 ***
altitude     -1.5318     0.3239   -4.73  2.3e-06 ***
distance     -1.4986     0.5001   -3.00  0.00273 **
NoOfPools     0.6590     0.2219    2.97  0.00297 **
NoOfSites     0.0368     0.1975    0.19  0.85202
avrain        1.1720     0.3305    3.55  0.00039 ***
---
Signif. codes:  0 '***' 0.001 '**' 0.01 '*' 0.05 '.' 0.1 ' ' 1

(Dispersion parameter for binomial family taken to be 1)

    Null deviance: 279.99  on 211  degrees of freedom
Residual deviance: 210.56  on 206  degrees of freedom
AIC: 222.6
```

Just as for the general linear model (Chapter 4), we have the model intercept, the partial slopes for each variable, their standard errors, the Wald statistics (= estimate / SE(estimate); see Chapter 3), and their statistical significance indicating whether each coefficient differs from zero. At the bottom of the summary we have the null and residual deviances. The former is the deviance of a model having the model intercept, and the latter is the remaining deviance after introducing the five explanatory variables in the model (Chapter 8). McFadden's analogue of R^2 denoting the reduction from the null to the residual deviance is calculated as `1 - (m1$deviance / m1$null.deviance)` = 0.248, and it corresponds to the proportion of the deviance explained by model m1. We also see the model's AIC, which is of no interest until we compare it with that of other models fitted to the same data.

The model summary indicates that we have estimated the following predictive equation: logit(π) = –0.946 – 1.531altitude – 1.498distance + 0.659NoOfPools + 0.036NoOfSites + 1.173avrain. Because the explanatory variables were centered to have a mean of zero, the intercept of –0.964 is the value of logit(π) for a "typical" site having average values of all the explanatory variables. But how can we interpret logit(π) = –0.946? Recalling the definition logit (π) = $\log\frac{\pi}{(1-\pi)}$, we can transform the logit into an odds ratio by exp(–0.946) = 0.381. Because the odds ratio is smaller than one, the frog is clearly more likely to be absent than present in the typical site.

Besides centering the explanatory variables, we also scaled them to have the same standard deviation of one prior to the analysis. Therefore, we can now compare the magnitudes of their partial slopes in the logit scale, just as we did for the general linear model (Chapter 4). Just focusing now on the four variables having statistically significant effects, those that are negatively related to logit(π) have a roughly similar relative effect, whereas the relative effect of avrain was (1.173 / 0.659 =) 1.8 or 80% stronger than that of NoOfPools. We can also compare the absolute value of the partial slopes of different signs to gauge their relative importance on the logit of presence. Thus, the effect of altitude was (1.531 / 1.173 =) 1.31 or 31% stronger than that of the average spring rainfall. The partial slope in a binary GLM with standardized variables denotes the changes in logit(π) with a change in one standard deviation of an explanatory variable. To keep it concrete, let's calculate the standard deviation of the explanatory variables:

```
> sqrt(diag(var(DF1[,2:6])))
 altitude  distance NoOfPools NoOfSites    avrain
   125.06   2588.60     26.56      1.94     12.00
```

This complex notation is because the command `var()` when applied to the set of variables `DF1[,2:6]` yields a variance–covariance matrix, from which we extract the diagonal terms (the variances) to obtain the standard deviations using the square root function. Back to the problem. If we increase the distance between potential breeding sites by 2588.6 m, logit(π) will decrease by 1.498. But does this mean a decrease of logit(π) of 1.498 units? Again, exponentiating this partial slope will render an odds ratio that is simpler to interpret: exp(–1.498) = 0.224 is a multiplicative effect and hence the probability of presence is this much less likely (0.224) than absence with an increase of 2588.6 m in the distance between potential breeding sites. Likewise, an increase in the number of pools of 26.56 would make frog presence exp(0.659) = 1.93 times more likely than absence.

All binary GLMs predict a linear relation in the scale of the logit link function that becomes logistic (Fig. 9.1b) when the function is inverted to compare model predictions with the data. The non-linearity of the inverted link function in all (except Gaussian) GLMs implies that the partial slopes can no longer be interpreted as the change in the mean of the response variable per unit change of an explanatory variable. Consider Fig. 9.1b: the slope of the logistic curve is steeper in the central range of the explanatory variable, and flatter at the tails. This implies that a unit change of an explanatory variable (or "effect," if you wish) in the midpoint values of the explanatory variable entails a larger change in the probability of presence than when the change occurs towards the extreme values of the variable. Gelman et al. (2014) suggested the rough "divide by four" rule to interpret the values of partial slopes in binary GLMs in terms of changes in probabilities (not in logits or odds). This rule corresponds to the maximum difference in the probability per unit change of an explanatory variable. Using `summary(m1)` for `altitude`, –1.538 / 4 = –0.38 means that a change of one standard deviation (125.06 m) in `altitude` can at most produce a reduction of 38% in the probability of presence. This largest change in probability of presence would occur for altitude changes towards the midpoint values, as changes anywhere else in its range would entail smaller changes in the probability of presence.

The command `anova(m1)` shows the decrease in the model deviance starting with the null deviance by the progressive addition of explanatory variables in the order we defined them when we wrote the model equation:

```
> anova(m1, test="Chisq")
Analysis of Deviance Table

Model: binomial, link: logit

Response: pres.abs

Terms added sequentially (first to last)


          Df Deviance Resid. Df Resid. Dev Pr(>Chi)
NULL                        211        280
altitude   1     12.0       210        268  0.00052 ***
distance   1     34.1       209        234  5.2e-09 ***
NoOfPools  1      8.8       208        225  0.00306 **
NoOfSites  1      0.5       207        225  0.48077
avrain     1     14.0       206        211  0.00018 ***
```

The p-values just shown were obtained by Wilks' test. They assess whether each decrease in model deviance by progressively adding explanatory variables is statistically significant. Let's recall that the p-value of 5.2×10^{-9} evaluates whether the decrease of 34.1 in deviance from 268 to 234 by adding the variable distance to a model that already contained the variable altitude was statistically significant. This table is only useful to summarize the overall relevance of explanatory variables, but only by making the doubtful assumption of associating relevance with statistical significance. It is clearly better and more informative to assess the relevance of explanatory variables (and their interactions) by the summary of estimated parameters and their standard errors.

We can gather the parameter estimates and the likelihood confidence intervals into a table using the command `tidy` from the package `broom`:

```
> require(broom)
> tidy(m1, conf.int = T)
# A tibble: 6 x 7
  term        estimate std.error statistic    p.value conf.low conf.high
  <chr>          <dbl>     <dbl>     <dbl>      <dbl>    <dbl>     <dbl>
1 (Intercept)  -0.965      0.230    -4.19  0.0000282    -1.45     -0.544
2 altitude     -1.53       0.324    -4.73  0.00000226   -2.20     -0.919
3 distance     -1.50       0.500    -3.00  0.00273      -2.56     -0.648
4 NoOfPools     0.659      0.222     2.97  0.00297       0.238     1.11
5 NoOfSites     0.0368     0.198     0.187 0.852        -0.356     0.422
6 avrain        1.17       0.331     3.55  0.000391      0.545     1.85
```

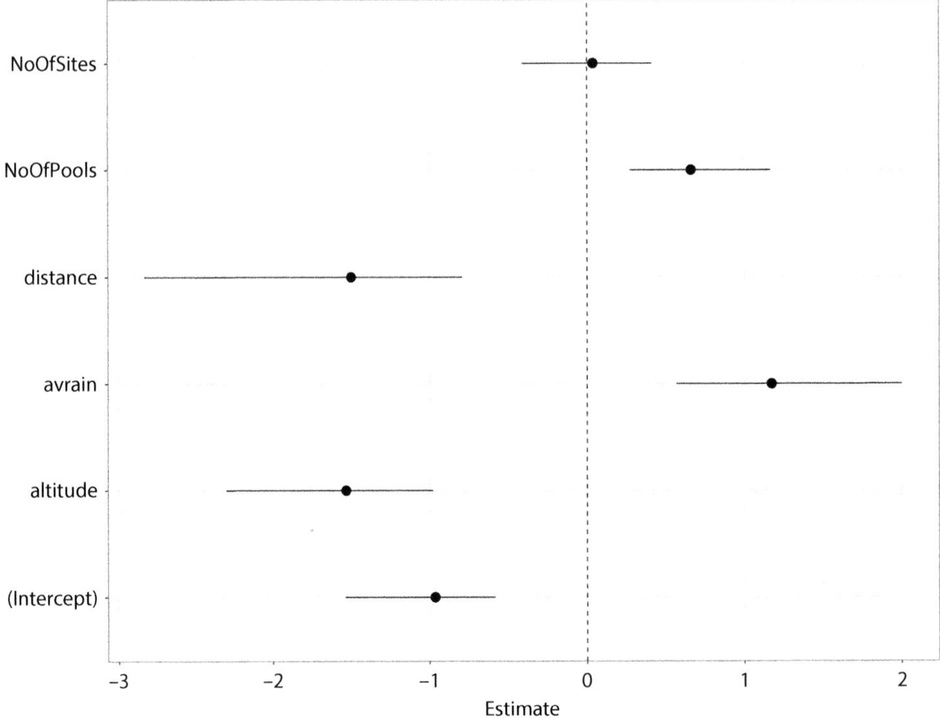

Fig. 9.4 Estimates and 95% confidence intervals of the intercept and partial slopes of the frequentist binary GLM `m1`.

The tibble `param.m1` can now be used to generate a figure that summarizes the above table (Fig. 9.4).

Other metrics (all of which were explained before) related to the goodness of fit of model `m1` can be obtained with:

```
> glance(m1)
# A tibble: 1 x 8
  null.deviance df.null logLik   AIC   BIC deviance df.residual  nobs
          <dbl>   <int>  <dbl> <dbl> <dbl>    <dbl>       <int> <int>
1          280.     211  -105.  223.  243.     211.         206   212
```

We can also obtain the predicted probabilities of presence for different values of the explanatory variables and combinations thereof using the command `ggpredict` from the package `ggeffects`. The `m1` model contained five explanatory variables; using it to predict the probability of presence for one of them requires assuming specific values for the others. The command `pred.m1.avrain = ggpredict(m1, terms = c("avrain [all]"))` uses `m1` to generate the predicted values of the probability of presence for a set of values of `avrain`, keeping by default all other explanatory variables at their means, which happen to be equal to zero given that we standardized the explanatory variables prior to the analysis:

```
> pred.m1.avrain
# Predicted probabilities of pres.abs
# x = avrain

    x | Predicted |         95% CI
-----------------------------------
-1.95 |      0.04 | [0.01, 0.13]
-1.48 |      0.06 | [0.02, 0.17]
-0.84 |      0.12 | [0.06, 0.23]
-0.26 |      0.22 | [0.15, 0.32]
 0.10 |      0.30 | [0.21, 0.40]
 0.46 |      0.40 | [0.28, 0.52]
 0.80 |      0.49 | [0.33, 0.65]
 4.19 |      0.98 | [0.77, 1.00]

Adjusted for:
*  altitude = -0.00
*  distance = -0.00
* NoOfPools =  0.00
* NoOfSites =  0.00
```

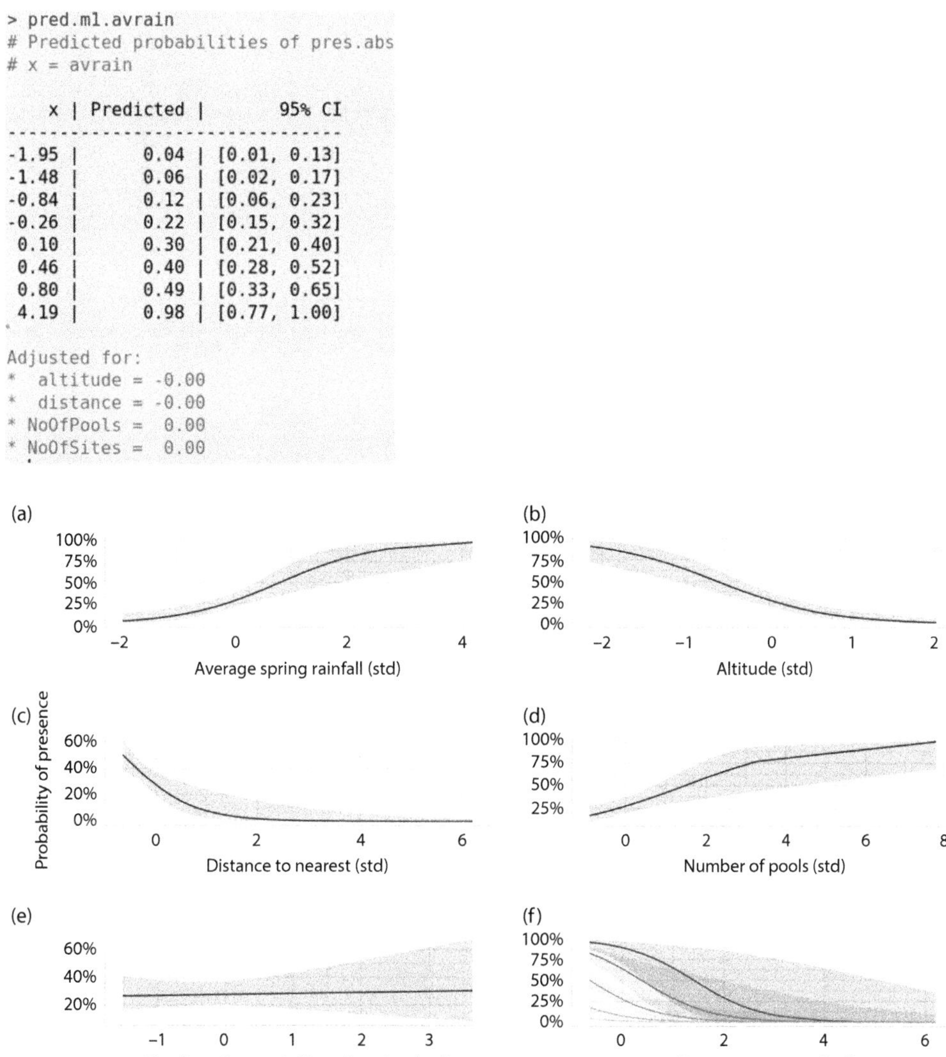

Fig. 9.5 Conditional plots of the frequentist binary GLM `m1` showing the predicted relations between the standardized explanatory variables (mean of zero and unit standard deviation) and the presence–absence response variable in the binary GLM model `m1`. The gray bands in all the plots correspond to the 95% confidence interval around each predicted relation. Plots (a) to (e) show the predicted relation for each variable, holding all other variables at zero (i.e., their mean values). Plot (f) shows the conditional plots for combinations of distance to the nearest population, and five values (mean, ±1, and ±2 standard deviations) of altitude.

We can do likewise for the other explanatory variables, and also consider combinations of them: `ggpredict(m1, terms = c("distance [all]", "altitude [−2:2]"));` these predicted relations (called conditional plots) are shown in Fig. 9.5. Note that all the predicted conditional curves are relations between the probability of presence and the

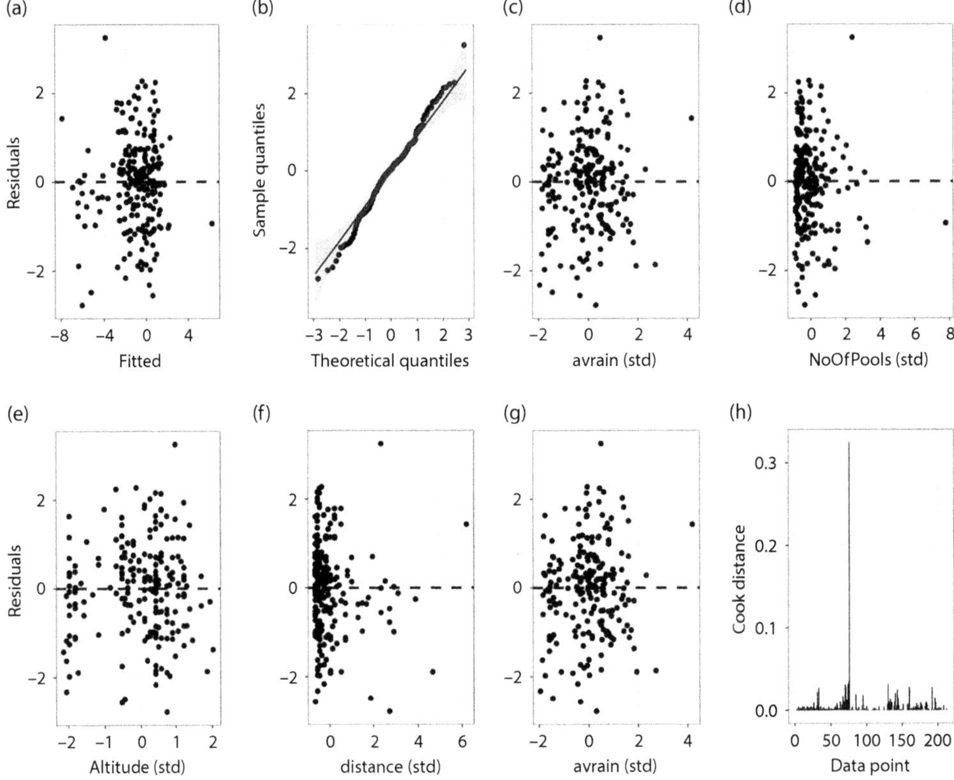

Fig. 9.6 Residual analysis of the frequentist binary GLM `m1` based on simulated randomized quantile residuals converted to normal residuals. Plots (a) to (f) show the relations between these residuals and the fitted values, and to the standardized numerical explanatory variables. Plot (g) shows the quantile–quantile plot (and the 95% confidence band) of the transformed, simulated residuals. Plot (h) shows the Cook distances.

explanatory variable(s), i.e., the command `ggpredict` used the inverse of the logit link function to draw these curves using the parameter estimates.

Residual analysis of a GLM is based on the randomized quantile residuals (Chapter 8) that can be obtained with the command `simulateResiduals` from the package `DHARMa`. We start by storing the basic information for the residual analysis collated by the command `augment` from the package broom in the data frame `resid.m1 = augment(m1)`. We then generate the randomized quantile residuals with `res.m1 = simulateResiduals(fittedModel = m1, n = 1e3, integerResponse = T, refit = F, plot = F)`, which renders a list from which we extract the simulated residuals that are converted to have a normal distribution when model fit is adequate. We next replace the default deviance residuals generated by R: `resid.m1$.std.resid = residuals(res.m1, quantileFunction = qnorm)`. We can now generate the plots of Fig. 9.6.

All plots of Fig. 9.6 show a random scatter of the residuals plotted against the fitted values and all explanatory variables. The Q–Q plot shows that the randomized quantile residuals are normally distributed. The plot of Cook distances shows one data point that may have a worryingly large influence on the parameter estimates of model `m1`. We can introduce the Cook distances in the original data frame `DF1` to identify this data point with `DF1$Cook = resid.m1$.cooksd`:

```
> DF1[DF1$Cook>0.2,]
   pres.abs altitude distance NoOfPools NoOfSites avrain meanmin meanmax  Cook
76        1     1670     8000        89         2    154     2.3    12.5 0.324
```

Comparing the information of the 76th row with the summary of the data frame shows that this row contains neither extreme nor obviously wrong values of the explanatory variables. We can fit another model without the 76th row and compare the parameter estimates with `m1`:

```
> m1.1=glm(pres.abs~ altitude+distance+NoOfPools+NoOfSites+avrain, data=DF1s[DF1$Cook<0.2,],
+          family=binomial)
> coef(m1)
(Intercept)    altitude    distance   NoOfPools   NoOfSites      avrain
    -0.9646     -1.5318     -1.4986      0.6590      0.0368      1.1720
> coef(m1.1)
(Intercept)    altitude    distance   NoOfPools   NoOfSites      avrain
     -1.239      -1.504      -2.266       0.564      -0.065       1.151
```

There are meaningful differences in the estimated parameters, including changes in sign for NoOfSites, which are now statistically significant and call into question the conclusion of model `m1`. If these were your data, you would probably know whether the information in the influential 76th row is reliable. However, in this case, all we can state is that we reach different conclusions depending on whether or not we use all the data when fitting the model. The only honorable conduct is to present both analyses as our results, and discuss their implications in the context of the original research.

9.5 Further issues about validating binary GLMs

After inverting the logit link function, an ungrouped binary GLM predicts the probability π as a function of the explanatory variables. Given that we did not observe or measure probabilities but binary outcomes, a plot of observed values against the probabilities predicted by a GLM can only look very bad (Fig. 9.1), even for a model perfectly fitting model. While we cannot generate real numbers out of a binary variable, we can discretize the predicted probabilities by applying a rule such as predicted probability $\leq$ some threshold $\Rightarrow$absent, otherwise present to obtain binary predictions. We can now compare the observed {0, 1} with the predicted {0, 1} in a 2 × 2 table (known as a "confusion matrix") to quantify the concordance between predictions and observations (Table 9.1).

The diagonal in the table contains the correct model predictions. We can calculate the proportion of true presences (called sensibility) as true presences / total presences, and the proportion of true absences (called specificity) as true absences / total absences. These are the proportions of correct predictions generated by our model and the discretization rule. All clinical procedures (for instance, the concentration of prostate antigen; Chapter 3) to evaluate an illness discretize a real variable, such that below or above a threshold

Table 9.1 Confusion matrix for predicted vs. observed presences and absences.

	Observed presence	Observed absence
Predicted presence	True presence	False absence
Predicted absence	False presence	True absence
	Total presences	Total absences

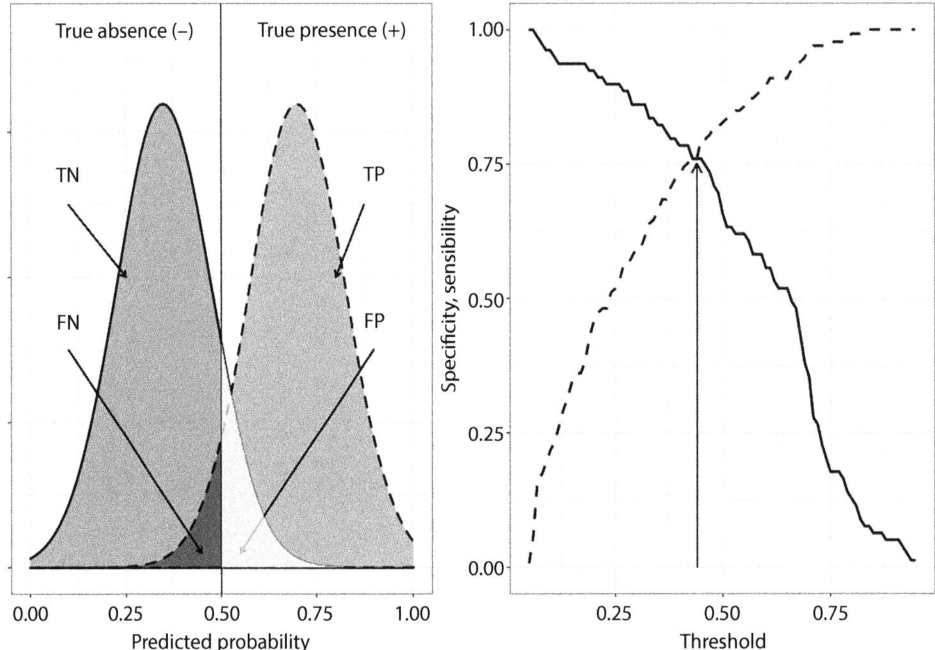

Fig. 9.7 Schematic representation of the receiver operating curves. On the left, a schematic representation of the probability functions of true presence (left) and true absence, showing the dependence of the true and false negative (TN, FN) and true and false positive (TP, FP) probabilities on the threshold (here set at 0.50) to discretize them. The Gaussian curves are used for convenience only. On the right, the dependence of the specificity (probability of true absence; dotted line) and sensitivity (probability of true presence; continuous line) with the threshold used to discretize the predicted probabilities.

we are declared ill or not ill. Any such determination entails false positive (the patient is declared ill when they are not) and false negative (the patient is declared fit when they are actually ill) rates. We ideally want clinical tests to possess high sensibility (and thus low false positive rates) and specificity (and hence low false negative rates) when ascertaining a patient's state of health. Nevertheless, neither the clinical tests nor our discretization of the predictions of binary GLMs can simultaneously achieve high sensibility and high specificity because they depend on a single threshold that determines both of these attributes (Fig. 9.7).

Minimizing the false positive rate unavoidably leads to increasing the false negative rate (Fig. 9.7); this is exactly the same trade-off involved in the Type I and Type II errors in hypothesis testing. Besides, there is no fundamental and defensible reason to choose 0.5 as the criterion for discretizing the predicted probabilities of binary GLMs.

The development of radar prior to World War II made it necessary to develop methods to enable operators to discriminate or classify between true and false detection of enemy planes. This is the origin of the receiver operating curve (ROC) that plots the sensitivity (i.e., true presence rate) vs. 1-specificity (i.e., false positive rate) for different values of the threshold used to discretize the GLM predictions. A ROC curve (see Fig. 9.8) allows us to make assessments of model fit independent of the arbitrariness of the discretization threshold. We use the command `roc` from the package `pROC`, `roc.m1 = roc(response`

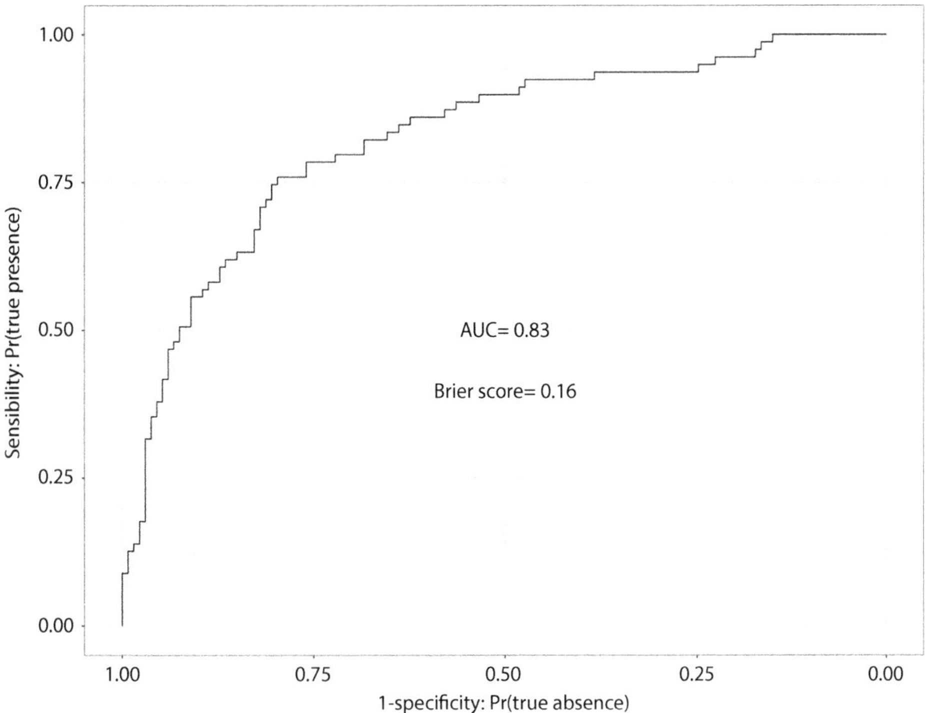

Fig. 9.8 Receiver operating curve of the frequentist binary GLM `m1`, showing the values of the area under the curve (AUC) and the Brier score, implying that the model has a reasonably high predictive ability.

`= DF1s$pres.abs, predictor = fitted(m1))`, to obtain the ROC curve that is plotted in Fig. 9.8.

The closer the ROC curve is to the upper left corner, the better the discriminating power and accuracy of the model. The area under the curve (AUC) is a statistic that summarizes the ROC curve. It ranges between 0.5 (the GLM discriminates as well as flipping a coin) and 1.0 (perfect discrimination of presences and absences). The AUC can be interpreted as the probability that a randomly chosen data point is correctly predicted by the binary GLM, and is designed to reflect the trade-off between sensitivity (correct predictions of presence) and specificity (correct predictions of absence), and to weight both of them equally. If equal weighting of both correct predictions might seem a reasonable default in the AUC, consider detecting the presence of a deadly type of cancer. Based on the extra statistical consequence of a wrong diagnosis, we would prefer to have a high sensitivity at the expense of a low specificity and hence we would not weight them equally as the AUC does. Leaving behind the AUC, meteorologists have developed a simple index (Brier 1953) that reflects the disparities between the binary observations (Y_{obs}) and probability predictions for a set of n observations:

$$\text{Brier score} = \frac{\sum\left(\text{Prob}_{\text{pred}} - Y_{\text{obs}}\right)^2}{n}. \tag{9.2}$$

The closer the Brier score is to zero, the better the predicted capacity of the binary GLM. Much like R^2 (i.e., the proportion of the variance of the response variable explained by a model; Chapters 4 to 6) and the McFadden R^2 (i.e., the proportion of the deviance explained by a model; Chapter 8), the AUC and the Brier score are only useful when compared to other models fitted to the same data. The ROC curve, AUC, and Brier score are independent of the theoretical framework employed to fit the binary GLM, and hence can also be calculated and interpreted in the Bayesian framework.

9.6 Ungrouped binary GLMs: Bayesian fitting

We use the package `brms` to fit this model. The same statistical model and link function with centered and standardized explanatory variables is used as for the frequentist fitting (see Section 9.4). We start by obtaining the list of model parameters whose priors we need to define:

```
> get_prior(formula=pres.abs~ altitude+distance+NoOfPools+NoOfSites+avrain,
+           family= bernoulli, data=DF1s)
                prior     class      coef group resp dpar nlpar lb ub       source
               (flat)         b                                            default
               (flat)         b  altitude                             (vectorized)
               (flat)         b    avrain                             (vectorized)
               (flat)         b  distance                             (vectorized)
               (flat)         b NoOfPools                             (vectorized)
               (flat)         b NoOfSites                             (vectorized)
 student_t(3, 0, 2.5) Intercept                                            default
```

These are weakly informative priors defined by default in `brms`. We should now define our own priors for the analysis. A key feature in non-Gaussian GLMs is that model parameters are estimated in a non-linear link function scale. Let's first recall that in our models of Chapters 4 to 6, the degree of informativeness of a prior was defined by its variance: the higher the variance, the wider the range of prior values that were judged to be compatible with previous knowledge. Nevertheless, the non-linear link functions in GLMs may produce counter-intuitive results (Hobbs and Hooten 2015, p. 96) whereby logit(π) ~ Normal(mean = 0, sd = 100) is less vague and much more informative than logit(π) ~ Normal(mean = 0, sd = 2). There is clearly an extra layer of complexity to defining priors for non-Gaussian GLMs that we need to bear in mind. It is therefore crucial to gauge the implications of priors not just in the link function scale where we define them, but more importantly in terms of their implications in the data scale.

How can we define the priors in the link function scale from the information gathered at the scale of the data? In a binary GLM, the intercept is logit(π) when the explanatory variables are zero (if we carry out the analysis with the original data), or at their mean values (if we carry out the analysis with standardized explanatory variables). We may possess information on the probability of frog presence in "average conditions" (whatever that may mean) from which we need to calculate its logit to define the features of the prior for model intercept. Say that someone had estimated the best baseline presence probability of the Australian frog to be 0.4 with a range from 0.2 to 0.6. We decide to use the normal distribution to define the intercept prior. Given that we are to set the prior on the logit link scale, we need to convert these values as `mean.interc = logit(0.4)` (and likewise for the range limits) using the `logit` command of package `arm` to obtain one of the hyperpriors. We may now use Wan et al.'s (2014) approach (see Chapter 5) to approximate the standard deviation of the intercept in the logit scale as `sd.interc = (logit(0.6) - logit(0.2)) / 4` as for the other hyperprior. We then end up with logit(intercept) ~ Normal(mean = –0.405, sd = 0.448), which looks both cryptic enough. To illustrate the implications of this prior distribution in terms of the background presence

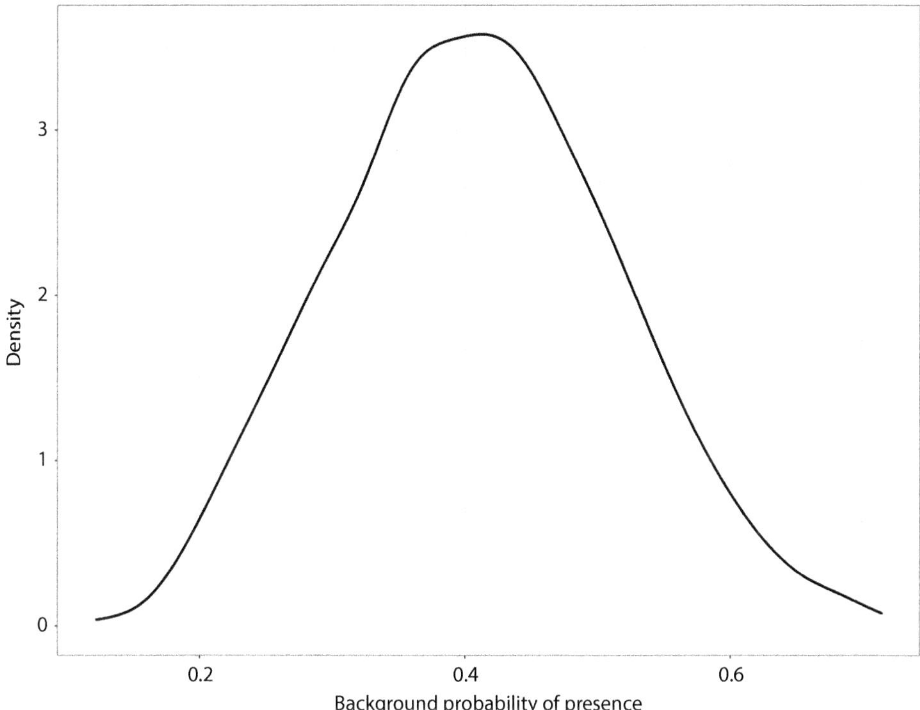

Fig. 9.9 Graphical implication of the prior for the intercept of the Bayesian binary GLM `m1.brms`. This is meant to act as an introduction to interpreting the prior predictive distribution shown in Fig. 9.10.

probability, we use the inverse logit command from package `arm`, `invlogit(rnorm(n = 1000, mean = mean.interc, sd = sd.interc))`, to simulate the 1,000 values of the intercept shown in Fig. 9.9. To keep things simple for now, we may (arbitrarily at this point) set Normal(mean = 0, sd = 0.5) priors for the partial slopes, implying that 95% of their values would be in [–0.98, 0.98] on the logit scale.

We will use the prior predictive distribution to assess the generative capacity of priors (Chapter 7). The goal here is a "negative" one: to assess that the previous knowledge embodied in the priors is unlikely to generate data that is clearly inconsistent with what we know about the problem (Gabry et al. 2019). We are *not* using the data in the assessment, but looking at the implications of the choice of priors. The distribution of the generated data should have some support around all plausible data sets, given our knowledge of the problem. The prior sensitivity analysis (Chapter 6) is complementary to the use of prior predictive distributions (Chapter 7).

To obtain the prior predictive distribution, we need to define the priors:

```
> prior.m1 = c(set_prior("normal(0,0.5)",class = "b"),
+              set_prior("normal(-0.484,0.448)", class = "Intercept"))
> prior.m1
                prior     class coef group resp dpar nlpar bound source
        normal(0,0.5)         b                                     user
 normal(-0.484,0.448) Intercept                                     user
```

Plotting the priors in non-Gaussian GLMs is not very helpful when considering their implications. This is because the predicted probabilities of frog presence would depend on specific values of each explanatory variable. This job is actually done by the prior predictive distribution. To obtain it:

```
m1.prior=brm(formula=pres.abs~ altitude+distance+NoOfPools+NoOfSites+avrain, data=DF1s,
            family=bernoulli, prior = prior.m1, warmup = 1000,sample_prior = "only",
            chains=3, iter=2000, future=T,control = list(adapt_delta = 0.9))
```

The only differences from the Bayesian models of Chapter 4 to 6 are that `family = bernoulli` (for ungrouped binary data) and that `sample_prior = "only"` (samples of parameters are solely obtained from the priors ignoring the likelihood, i.e., the data!). For the sake of comparison, we ran the same model with a second, less informative, set of priors:

```
prior2.m1 = c(set_prior("normal(0,5)",class = "b"),
              set_prior("normal(-0.484,0.448)", class = "Intercept"))
```

and stored the results in `m1.prior2`.

The command `posterior_predict` from the package `bayesplot` in this case obtains the prior predictive distribution: `m1.prior.pred = posterior_predict (m1.prior, nsamples=500)`. `m1.prior.pred` is obtained by applying Eq. (7.22) to get 500 samples for each of the 212 values of response variable:

```
> dim(m1.prior.pred)
[1] 500 212
```

The command `pp_check` performs posterior predictive checks for functions defined by the user and calculated over the prior posterior distribution just stored. This command also contains a wealth of plotting functions that can be used to compare the posterior distribution with the actual data. Here we apply the function `presences = function(x) mean(x == 1)`, which calculates the proportion of ones (i.e. presences) for the rows of the predicted distribution, `pp_check(m1.prior1, nsamples = 500, type = "stat", stat = "presences")`, to compare the prior predictive distributions for the two sets of priors (Fig. 9.10).

The prior predictive distributions having a smaller hyperprior variance for the partial slope (Fig. 9.10b) clearly predicted a higher probability[1] of presence than was observed (0.373), and thus the first set of priors is to be preferred. The goal of this exercise is not to define the priors so as to match the observed data. Rather, what we want is to avoid certain choices of priors that may generate data that are clearly inconsistent with the available knowledge prior to obtaining the data (Gabry et al. 2019).

```
> summary(apply(m1.prior.pred, 2, mean))
   Min. 1st Qu.  Median    Mean 3rd Qu.    Max.    5%   95%
 0.3340  0.3780  0.3940  0.3948  0.4100  0.4800 0.354 0.439
> summary(apply(m1.prior.pred2, 2, mean))
   Min. 1st Qu.  Median    Mean 3rd Qu.    Max.    5%   95%
  0.404   0.456   0.470   0.472   0.490   0.530 0.437 0.511
```

We can now fit the model using the first set of priors in the usual manner:

```
m1.brms=brm(formula=pres.abs~ altitude+distance+NoOfPools+NoOfSites+
avrain, data =DF1s, family=bernoulli, prior = prior.m1, warmup = 1000,
chains=3, iter=2000, future=T,control = list(adapt_delta = 0.9))
```

[1] These intervals were calculated with quantile(apply(m1.prior.pred, 2, mean), probs = c(0.05,0.95)).

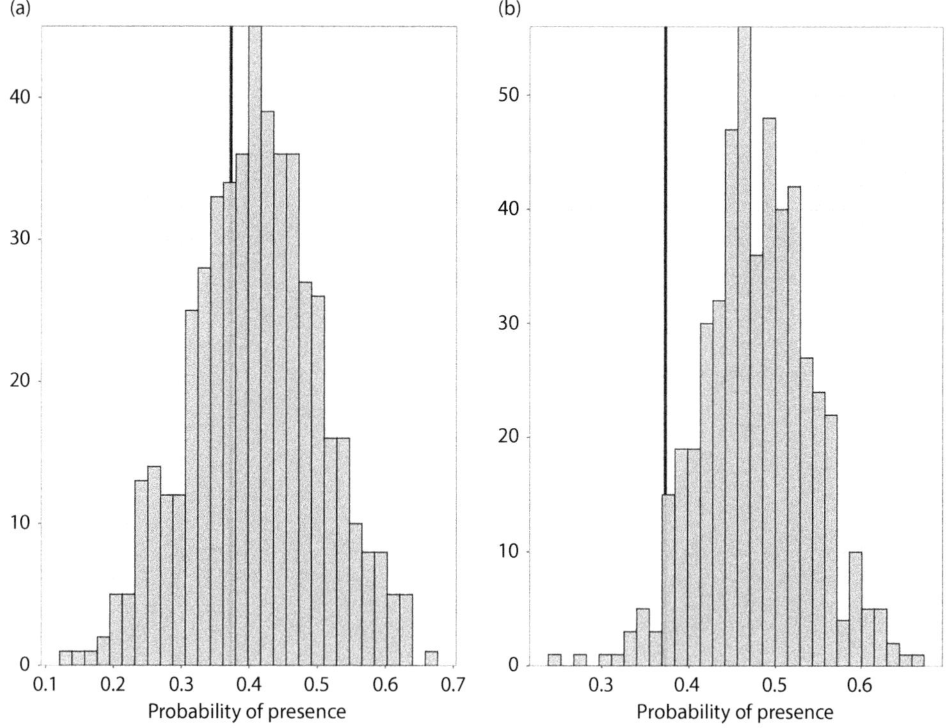

Fig. 9.10 Prior predictive distributions of the Bayesian binary GLM `m1.brms`, showing the distribution of predicted probability of presence for stronger (a; normal(0, 0.5)) and weaker (b; normal(0, 5)) informative priors for the model intercept. The information contained in the data is not used to obtain the prior predictive distributions. The stronger informative priors for the intercept predicted the observed proportion of presences (thick line) as a typical result, and hence should be preferred.

and obtain its summary as:

```
> summary(m1.brms)
 Family: bernoulli
  Links: mu = logit
Formula: pres.abs ~ altitude + distance + NoOfPools + NoOfSites + avrain
   Data: DF1s (Number of observations: 212)
Samples: 3 chains, each with iter = 2000; warmup = 1000; thin = 1;
         total post-warmup samples = 3000

Population-Level Effects:
           Estimate Est.Error l-95% CI u-95% CI Rhat Bulk_ESS Tail_ESS
Intercept     -0.73      0.17    -1.07    -0.41 1.00     2917     2069
altitude      -1.01      0.23    -1.48    -0.55 1.00     1833     2085
distance      -0.97      0.29    -1.55    -0.46 1.00     2654     1953
NoOfPools      0.50      0.19     0.15     0.86 1.00     2960     2162
NoOfSites      0.09      0.17    -0.24     0.42 1.00     2727     2126
avrain         0.67      0.24     0.21     1.14 1.00     1938     2087
```

The metrics of MCMC convergence (effective sample size and `Rhat`; Chapter 3) suggest that the MCMC algorithm converged to a stationary posterior distribution of parameter estimates. As usual, we have the `Estimate` (mean), `Est.Error` (standard error), and the bounds of the 95% credible intervals of each parameter (Fig. 9.11), all calculated from

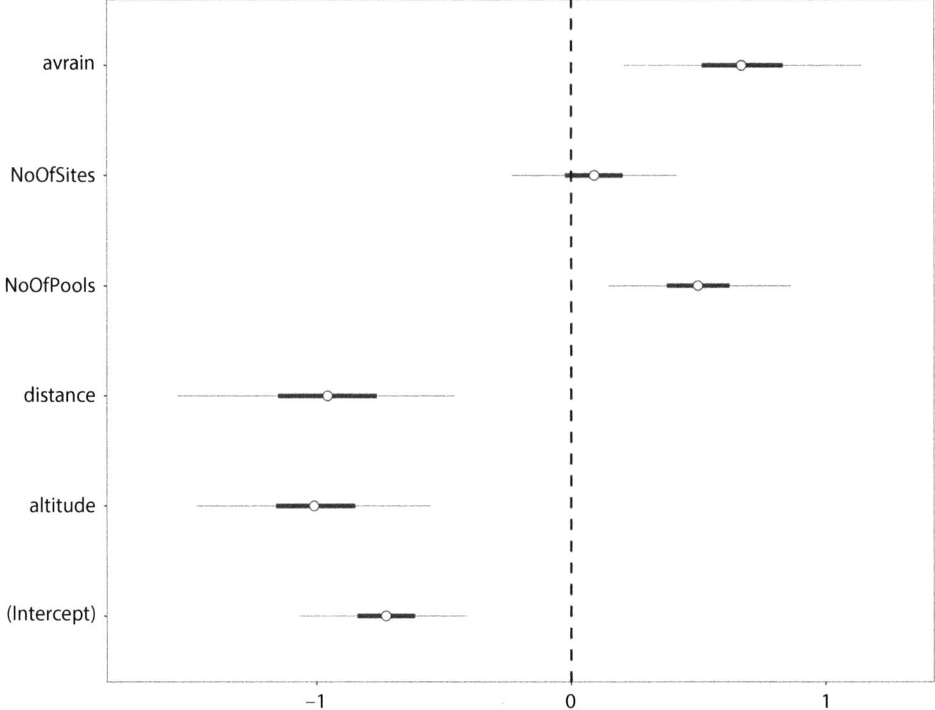

Fig. 9.11 The means and the 50% (thick lines) and 95% (thin lines) credible intervals of the parameters of the Bayesian binary GLM `m1.brms`.

their marginal posterior distribution. All these values are, of course, in the same logit link scale as their frequentist analogues. The percentage of variation explained by the model (Gelman et al. 2018) is:

```
> bayes_R2(m1.brms)
   Estimate Est.Error  Q2.5 Q97.5
R2     0.21    0.0356 0.138 0.276
```

The interpretation of the parameter estimates for standardized explanatory variables, and their comparison to assess their relative importance in their effects on the probability of presence, were described in Section 9.4, and the credible intervals in Chapter 3. Hence, none of that need be repeated here. While the signs of the coefficients are similar, the magnitudes of the best estimates, the standard errors, and the 95% credible/confidence intervals differ between the frequentist and Bayesian fittings of the same model to the same data. We had stored the frequentist parameter estimates and parametric bootstrap 95% confidence intervals in `param.m1`. We can obtain the analogous Bayesian features with `fixef(m1.brms)` and now show their percentage differences:[2]

[2] These were calculated by `100*(param.m1[,] - fixef(m1.brms)[,]) / fixef(m1.brms)[,]`, selecting the correct columns in both `param.m1` and `fixef(m1.brms)`.

```
           term estimate std.error conf.low conf.high
1 (Intercept)       32.9      41.7      49.2      46.6
2     altitude      52.9      41.1      60.4      70.2
3     distance      55.5      74.1      80.3      79.0
4    NoOfPools      33.6      19.5      86.9      43.3
5    NoOfSites     -57.3      22.3      71.9       5.2
6       avrain      74.2      40.6     191.9      80.0
```

We are not quibbling here about tiny differences. The Bayesian mean posterior values were always smaller, hence the differences were positive, with the exception of the partial slope for `NoOfPools`. The standard errors and 95% credible intervals are narrower and more precise than their frequentist analogues. There is no "true" framework in which to analyze these or any other data. We should not choose by trying both and selecting that which seems "best" to us. Each framework stems from different views of probability and employs different methods (Chapter 3). Only a divine miracle or undisclosed cheating would make them coincide to the third decimal digit. Before you are tempted to "choose Bayesian," let's see the same differences in parameter estimates with respect to a second Bayesian GLM fitted using `prior2.m1` (the fitted model is not shown):

```
        term estimate std.error conf.low conf.high
(Intercept)      6.31     20.52   18.720     6.870
   altitude     -1.38      2.11    5.526     0.616
   distance      1.97     15.59   17.647    12.160
  NoOfPools     -1.20      1.04   14.184     6.217
  NoOfSites    -27.79      4.46   30.651    -1.876
     avrain     -1.47      1.05    0.601     7.680
```

The parameter estimates are now more similar, again with the exception of the partial slope of `NoOfPools`. Should we continue to make the priors more and more diffuse (i.e., less informative), we would eventually converge to the frequentist estimates when we consider uniform priors (Chapter 3). This begs the question of the importance of having a principled approach to define the priors in Bayesian statistics. Inputting prior knowledge when analyzing the evidence is obviously reasonable and meaningful as the world of knowledge does not start from scratch again with each new analysis. Using more informative priors (such as `prior.m1`) did lead to more precise parameter estimates than the frequentist ones. The use of informative priors is strongly advocated by Gelman et al. (2014, 2017) when there is sufficient empirical information available. The real issue is how to translate previous knowledge into priors, the importance of which becomes greater for smaller data sets and when explanatory variables have weaker true effects. Until there is a principled way of obtaining and defining priors for general statistical models, every Bayesian data analysis should at least evaluate and discuss the extent to which different priors make a qualitative difference to the results.

We now turn to examine the convergence of the MCMC algorithm and the degree of autocorrelation of the sampled parameter values (see the details in Chapter 3). Figures 9.12 and 9.13 show that the MCMC algorithm converged (the three chains had nearly identical posterior distributions), and that the sampled parameters are essentially statistically independent, in agreement with the `Rhat` and `ESS` of the model summary. The trace plots (not shown) also indicated that the three chains mixed well (i.e., the chains did not get "stuck" in a narrow range of values) and converged to a common stationary distribution, in agreement with the metrics of model convergence.

We can also obtain the conditional plots with `m1.brms.cond.eff = conditional_effects(m1.brms)`. `m1.brms.cond.eff` is a list containing the information

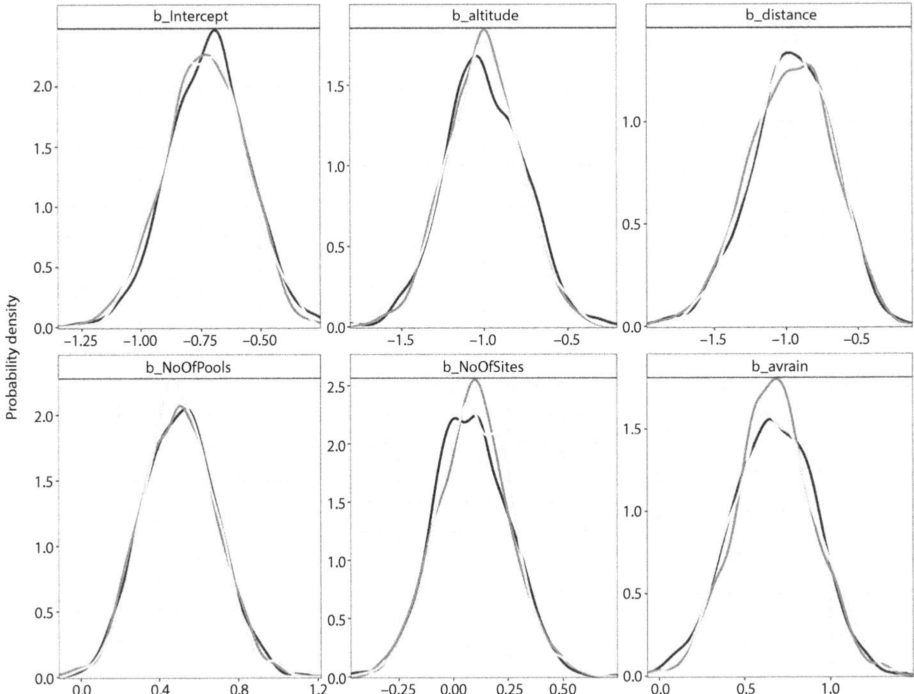

Fig. 9.12 Marginal posterior distributions of each model parameter for the three chains used to fit the Bayesian binary GLM `m1.brms`.

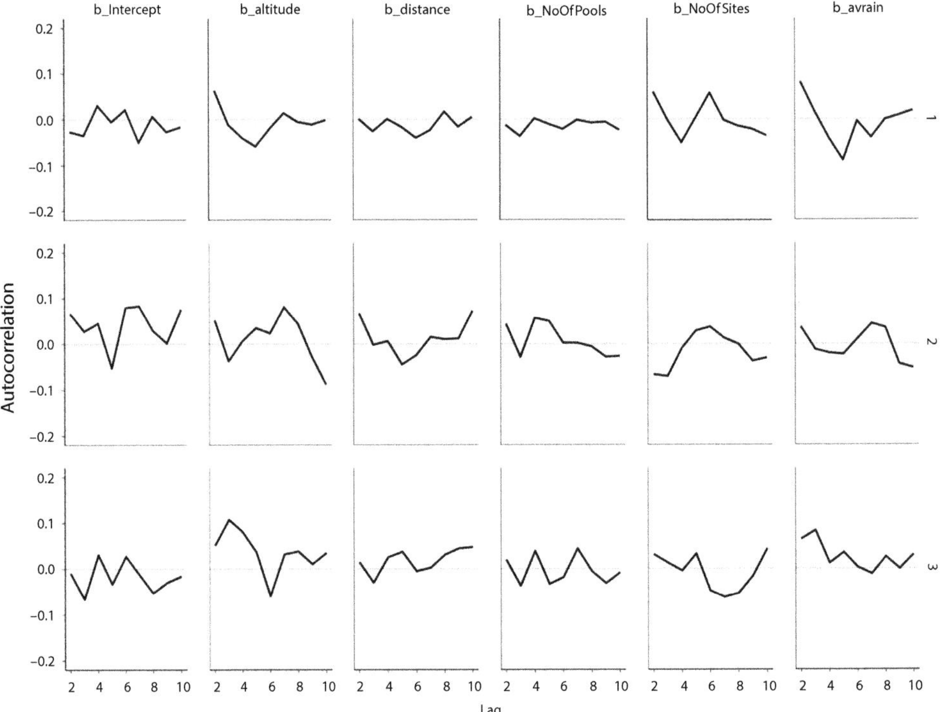

Fig. 9.13 Autocorrelation plots of each model parameter for the three chains used to fit the Bayesian binary GLM `m1.brms`.

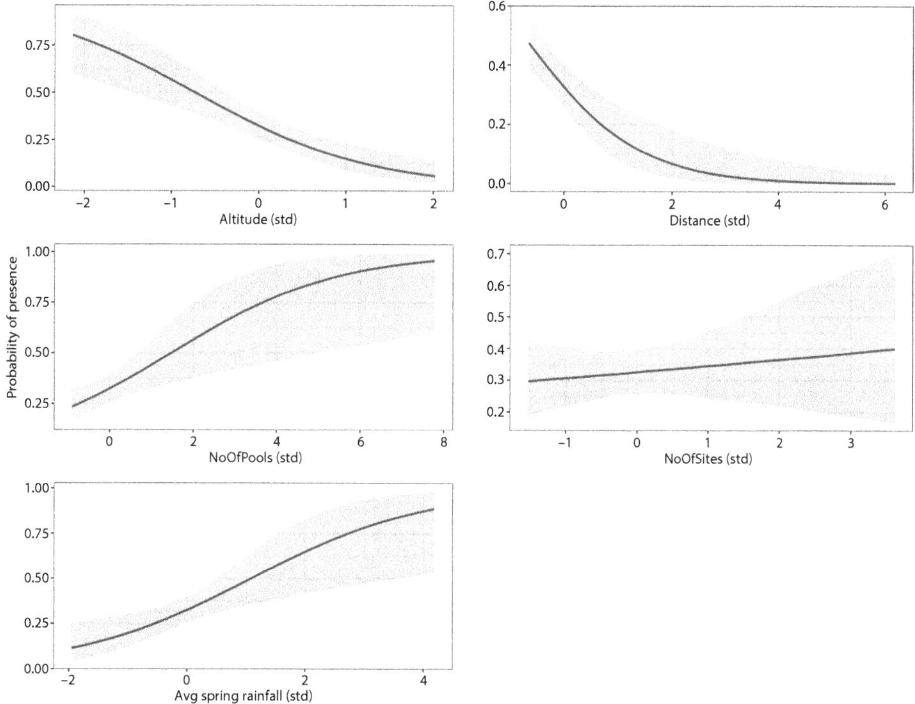

Fig. 9.14 Conditional plots of the Bayesian binary GLM `m1.brms` showing the predicted relations between the standardized explanatory variables (mean of zero and unit standard deviation) and the presence–absence response variable in binary GLM model `m1.brms`. The gray bands correspond to the 95% credible intervals around each predicted relation. All the plots show the predicted relation for each variable, holding all other variables at zero (i.e., their mean values in the original scale).

to make each of the five plots shown in Fig. 9.14. These plots show the predicted relation between each explanatory variable and the probability of presence while keeping all other explanatory variables at their means, in complete analogy to Fig. 9.5 obtained for the frequentist fitting.

Finally, we assess the goodness of fit of model `m1.brms` to the data using the randomized quantile residuals (Chapter 8). To obtain these residuals (Eq. (7.15)), we first obtain the posterior predictive distribution (Eq. (7.15)):

```
post.pred.m1.brms=predict(m1.brms, nsamples=1e3, summary=F)
```

These are 1,000 sets of 212 (as many as our original data set) presence/absence values predicted by Bayesian GLM. The posterior predicted distribution is used to generate uniformly distributed residuals with `createDHARMa`:

```
qres.m1.brms=createDHARMa(simulatedResponse = t(post.pred.m1.brms),
 observedResponse = DF1s$pres.abs,
 fittedPredictedResponse=apply(post.pred.m1.brms, 2,median),integerResponse = T)
```

The last command calculates the cumulative distribution function of the response variable, including random jittering (Eq. (8.5c)) for smoothing purposes (see Chapter 8). These simulated residuals are now converted into normally distributed residuals with:

```
res.m1.brms=data.frame(res=qnorm(residuals(qres.m1.brms)))
```

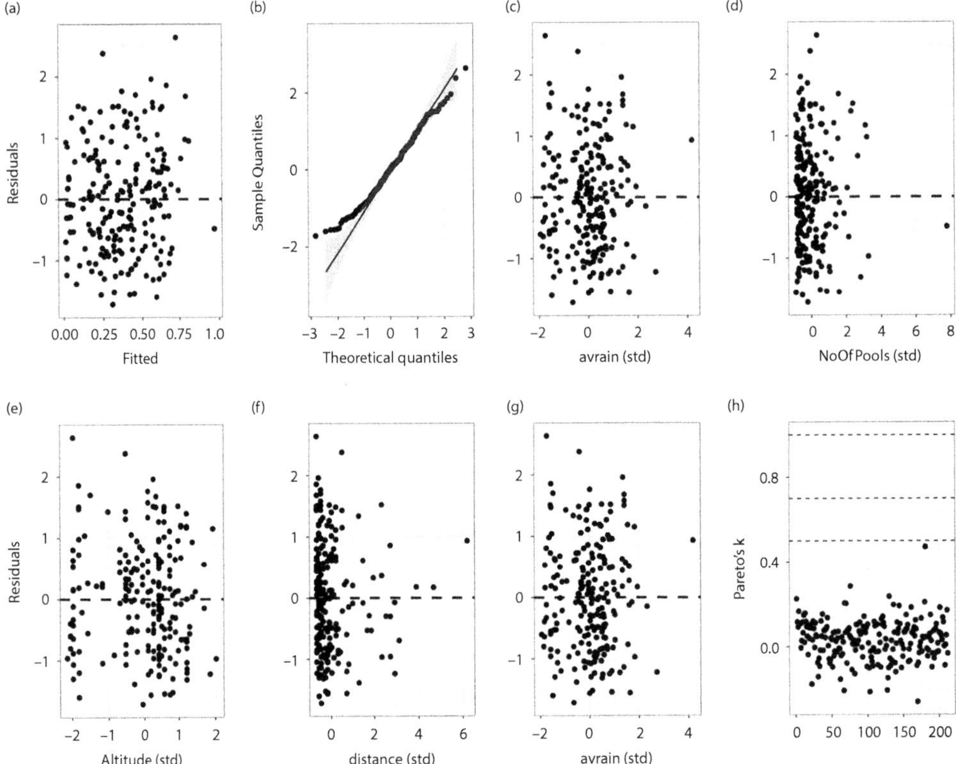

Fig. 9.15 Residual analysis of model `m1.brms` based on randomized quantile residuals converted to normal residuals, showing the mean of the Pearson residuals vs. the mean of the fitted value (a), the quantile–quantile plot of the randomized quantile residuals (b), the mean of the Pearson residuals vs. the explanatory variables in the model (c–g), and Pareto's *k* cross-validation statistic for each data point (h).

We then add the the explanatory variables, the average fitted values,[3] and the Pareto-*k* metric of leave-one-out cross validation (LOO-CV; see Chapter 7) for each of the 212 values of our data set (more on this in a second):

```
res.m1.brms=cbind(res.m1.brms,
   DF1s[,c("altitude", "distance", "NoOfPools", "NoOfSites", "avrain")],
   fitted=fitted(m1.brms, nsamples=1000)[,1], # average of predicted values
   pareto=loo(m1.brms, pointwise=T)$diagnostics$pareto_k) #LOO_CV Pareto k
```

We now have what we need to assess the goodness of fit of the model in a data frame that we can use to plot Fig. 9.15. We can see that residuals appear to be randomly distributed and without any discernible trends in the plots of randomized quantile residuals vs. fitted values and the (standardized) explanatory variables (Fig. 9.15a, c–g). The Q–Q plot shows some discrepancies (as did its frequentist counterpart; see Fig. 9.6) from the line and band of perfect fit (Fig. 9.15b). These plots suggest that the Bayesian GLM fitted the data reasonably well.

[3] Recall that Bayesian models do not produce a single predicted or fitted value for a set of values of the explanatory variables, but a true distribution of predicted values (see Chapter 4) that we summarize by its average.

Figure 9.15h requires an additional explanation. We explained in Chapter 7 how LOO-CV aims to evaluate the importance of each row of data on the estimated parameter values. This is analogous to the Cook distance used in the frequentist framework. We use the computationally efficient Pareto smoothed importance sampling in package `loo` (see Chapter 8; Vehtari et al. 2017, Gabry et al. 2019). This allows flagging points with potentially high effect on the estimated model parameters. The main command from the package `loo` is:

```
> loo(m1.brms, pointwise=T)

Computed from 3000 by 212 log-likelihood matrix

         Estimate   SE
elpd_loo   -112.6  6.0
p_loo         4.5  0.6
looic       225.2 12.0
------
Monte Carlo SE of elpd_loo is 0.0.

All Pareto k estimates are good (k < 0.5).
```

This generates `elpd` (expected log pointwise predictive density); Eq. (7.17)), `looic` (asymptotically equivalent to the WAIC), and `p_loo` (a complex index reflecting the number of effective parameters; see Vehtari et al. 2017), and their standard errors. Figure 9.15h shows that all the Pareto-k estimates are "good" because they are smaller than 0.5 (see Chapter 7). We should be slightly concerned when the Pareto-k pointwise estimates are in [0.5, 0.7], and more worried when they are greater than one. What should one do in these cases? First, refit the Bayesian GLM with more and longer chains, be more thorough in the evaluation of model convergence, and recalculate the `loo` statistics. We should also refit the original model without these "problematic" points to detect qualitative changes compared with the full data set. If you find important differences, remain honest, and discuss and report both outputs. The command `loo(m1.brms, pointwise = T)$diagnostics$pareto_k` allows us to obtain the pointwise estimates that were part of Fig. 9.15h.

The posterior predictive distribution (Eq. (7.15)) `post.pred.m1.brms` contains 1,000 predicted zeros and ones for each of the 212 observed frog presences and absences:

```
> dim(post.pred.m1.brms)
[1] 1000  212

> post.pred.m1.brms[1:6, 1:10]
     [,1] [,2] [,3] [,4] [,5] [,6] [,7] [,8] [,9] [,10]
[1,]    1    1    1    1    1    1    1    0    1     0
[2,]    1    0    1    1    1    1    1    1    0     0
[3,]    1    1    0    1    1    1    1    1    0     0
[4,]    1    1    0    1    0    1    0    0    1     1
[5,]    1    0    1    0    0    0    0    1    0     1
[6,]    1    1    1    1    1    1    0    1    1     0
```

If we calculate the proportion of zeros (or ones) for each of these 1,000 rows, we would have a distribution of predicted probabilities of presence for the 212 data points generated by our model, and see where the observed proportion of absences lies in such a distribution. Alternatively, we could use the command `pp_check` to make a calculation over the posterior predictive distribution and plot the result in just one step. Figure 9.16 shows that the observed probability of presence (the vertical line) is a typical value among many predicted realizations of our binary GLM model. Being able to reproduce the observed data

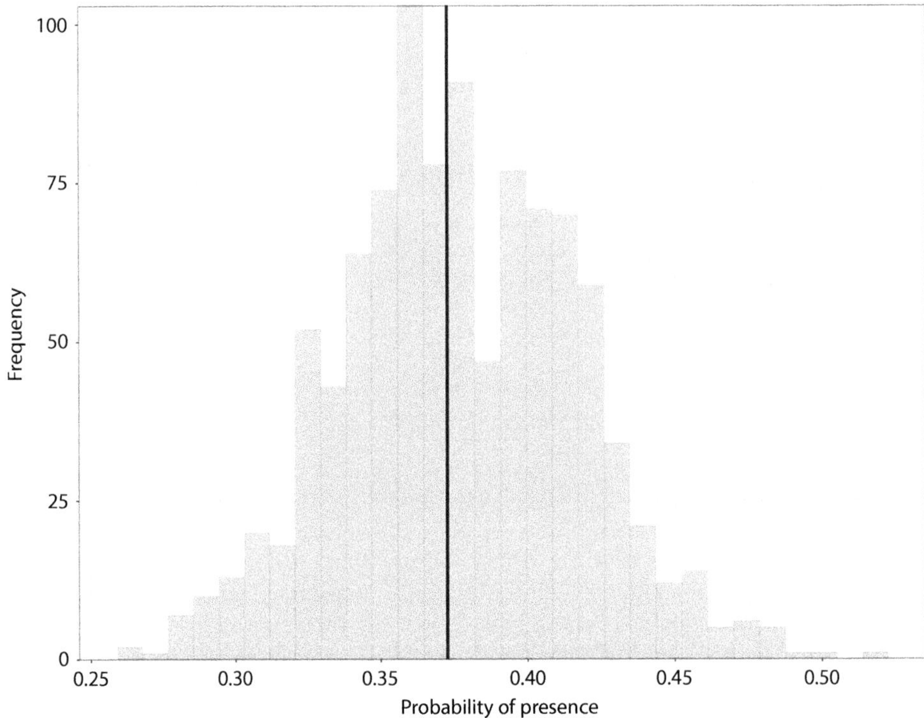

Fig. 9.16 Posterior predictive distribution of the Bayesian binary GLM `m1.brms`, showing the distribution of predicted probability of presence and the observed proportion of presences (thick line). The actual observations are produced as a "typical" prediction of the model.

(or a function calculated from it) is the sign of a well-fitting model. To avoid being fooled by a few lucky predictions that happen to match our observations, our posterior predictive distribution generated 1,000 predicted probabilities of presence. This is why posterior predictive distributions have been increasingly used to assess the quality of fit of Bayesian models (e.g., Gelman et al. 2000, 2014, Kruschke 2015).

We can use the posterior distribution `post.pred.m1.brms` to calculate the AUC and Brier score as follows:

```
nsamp=1e3 # number of sampled values
m1.brms.post=posterior_predict(m1.brms, nsamples=nsamp)
ROC.m1.brms=vector(length=nsamp) # creates an empty vector for results
Brier.m1.brms=vector(length=nsamp) # creates an empty vector for results
for (i in 1:nsamp){
  temp=auc(roc(predictor=m1.brms.post[i,],response=DF1s$pres.abs))
  ROC.m1.brms[i]=unlist(temp[grep("0", temp)])
  Brier.m1.brms[i]= mean((m1.brms.post[i,]-DF1s$pres.abs)^2)}
```

A few comments about this code. We use the command `roc` from the package `pROC` to obtain the ROC curve and its related AUC for each of the rows (i.e., samples) of the posterior distribution and store the result in the variable `temp`. It turns out that the output of the `auc` command contains the string "Area under curve" followed by the calculated AUC value. The command `grep` from the `base` package searches the string and allows us to save just the numeric value in the vector `ROC.m1.brms` (because the result of `grep` is a list, we need to `unlist` its result prior to storing it). The Brier score for each row of the posterior distribution `m1.brms.post` is calculated using the same formula employed in Section 9.5 for the frequentist ungrouped binary GLM. The difference is that we now have a distribution of values of the AUC and the Brier score rather than a single value as in the frequentist analogue model.

We can now characterize the distribution of estimated AUCs and Brier scores:

```
> summary(ROC.m1.brms)
   Min. 1st Qu.  Median    Mean 3rd Qu.    Max.
  0.505   0.590   0.615   0.614   0.637   0.727
> quantile(ROC.m1.brms, probs=c(0.025,0.975))
 2.5% 97.5%
0.545 0.683
```

```
> summary(Brier.m1.brms)
   Min. 1st Qu.  Median    Mean 3rd Qu.    Max.
  0.259   0.340   0.363   0.363   0.382   0.467
> quantile(Brier.m1.brms, probs=c(0.025,0.975))
 2.5% 97.5%
0.297 0.429
```

Note that the frequentist ROC was 0.83, and the frequentist Brier score 0.16. The average values of the Bayesian AUC and Brier scores suggest that model `m1.brms` has a modest capacity to make accurate predictions on plausible data. For the same observed data, the average values of the Bayesian AUC and Brier scores are clearly worse than the frequentist estimates, and their 95% credible intervals do not even contain the frequentist ones. This should not be a concern since the frequentist estimates are by no means true or gold standard values to be matched by the sets of Bayesian estimates. We are just making the comparison between frequentist and Bayesian metrics because of the relatively unusual circumstance of using both frameworks in the same book. We obtained estimates of these metrics of model performance from the posterior distribution of `m1.brms`. These are a set of 1,000 possible values of the response variable given the posterior distribution of model parameters and the observed data (Chapter 7). There is an interesting difference here. The frequentist AUC and Brier score point estimates were a retrospective assessment on the probability of correct assessment of randomly chosen points from the actual data. In contrast, the Bayesian estimates of both metrics are actually a prospective assessment of model behavior, as they reflect the distribution of plausible, unobserved values of the response variable contained in the posterior predictive distribution. To keep things clear and fair, measures of model behavior and goodness of fit are always relative, and they need to be compared with those of other models fitted to the same data within (rather than between) a statistical framework.

9.7 Grouped binary GLMs

The data in context: Bishop (1972) studied whether industrial melanic and non-melanic forms of the polymorphic peppered moth differed in abundance between smoke-polluted areas and unpolluted countryside around Liverpool. Melanism is controlled by a single dominant gene, with the non-melanic form being recessive. The peppered moth was studied by J. B. S. Haldane in the 1920s to demonstrate the changes in allele frequencies due to selective pressure. A series of experiments and observations allowed Bishop (1972) to capture (using pheromone traps), count, and release individuals of the melanic and non-melanic phenotypes at different distances in kilometers from Liverpool, whose factories were the source of industrial pollution. An individual moth of a given phenotype was captured, released at a given distance from Liverpool, and, if it survived, eventually recaptured. The explanatory variables (phenotype, distance from Liverpool) had the same values for the moths at a given sampling site.

We start by importing the data and examining the resulting data frame:

```
> DF2=read.csv("Ch 10 peppered moth.csv" , header=T)
> str(DF2)
'data.frame':    14 obs. of  4 variables:
 $ morph   : Factor w/ 2 levels "dark","light": 2 1 2 1 2 1 2 1 2 1 ...
 $ distance: num  0 0 7.2 7.2 24.1 24.1 30.2 30.2 36.4 36.4 ...
 $ placed  : int  56 56 80 80 52 52 60 60 60 60 ...
 $ removed : int  17 14 28 20 18 22 9 16 16 23 ...
```

We are going to calculate the number and proportion of surviving moths:

```
DF2$surv=DF2$placed-DF2$removed
DF2$prop_surv=(DF2$placed-DF2$removed)/DF2$placed
```

and plot the latter against the other explanatory variables (see Fig. 9.17).

The proportion of moths surviving and recaptured changed with the distance in different ways for each morph (Fig. 9.17), thus suggesting the need to evaluate the interaction `morph:distance` in the models. We obtain basic descriptive statistics with the package `doBy`:

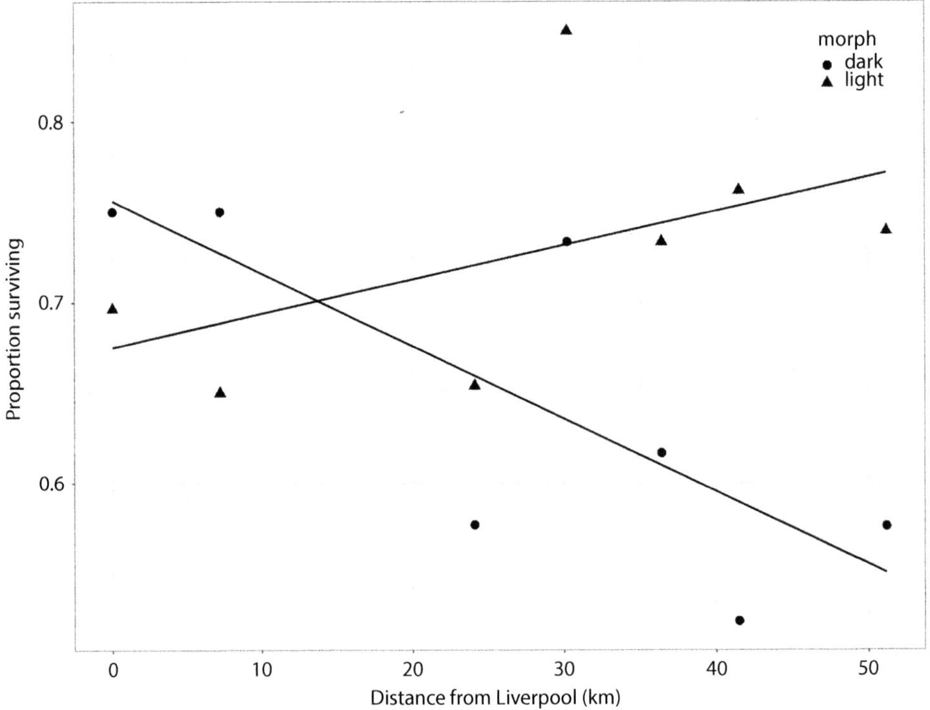

Fig. 9.17 Proportions of the dark and light *Biston betularia* morphs with the distance from Liverpool (data from Bishop 1972).

```
> summaryBy(prop_surv~morph, data=DF2, FUN=c(mean, median, sd, length))
   morph prop_surv.mean prop_surv.median prop_surv.sd prop_surv.length
1:  dark          0.647            0.617       0.0955                7
2: light          0.726            0.733       0.0692                7
```

They suggest that the light morph survived better than the dark one. We can also see that there are just seven data points per morph, one for each of the seven distances considered.

The response variable is the number of surviving moths $Y \sim \text{Binomial}(k, \pi)$, where k is the number of moths placed (i.e., released) and π is the probability of surviving and being recaptured. Using the logit link function, we can write, for each morph m, $\text{logit}(\pi) = \beta_0^m + \beta_1^m\text{distance}$. This equation is the GLM grouped binary equivalent of analysis of covariance (see Chapter 6).

We can fit the frequentist grouped binary GLM as:

```
m2=glm(cbind(surv,removed)~distance*morph, data=DF2, family=binomial)
```

With binary grouped data, there is a mandatory format for the entry of the response variable: the first column must be what we define as "success" or the event of interest, and the second one is the "failure" or complement of success. These two variables are put together by the command `cbind` into a matrix, and their sum is k, the total number of moths of each morph placed at each distance.

The model summary is:

```
> summary(m2)

Call:
glm(formula = cbind(surv, removed) ~ distance * morph, family = binomial,
    data = DF2)

Deviance Residuals:
    Min       1Q   Median       3Q      Max
-1.3124  -0.6829  -0.0115   0.3988   2.2118

Coefficients:
                   Estimate Std. Error z value Pr(>|z|)
(Intercept)         1.12899    0.19791    5.70  1.2e-08 ***
distance           -0.01850    0.00565   -3.28  0.00105 **
morphlight         -0.41126    0.27449   -1.50  0.13407
distance:morphlight  0.02779    0.00809    3.44  0.00059 ***
---
Signif. codes:  0 '***' 0.001 '**' 0.01 '*' 0.05 '.' 0.1 ' ' 1

(Dispersion parameter for binomial family taken to be 1)

    Null deviance: 35.385  on 13  degrees of freedom
Residual deviance: 13.230  on 10  degrees of freedom
AIC: 83.9
```

The reference group was (alphabetically) chosen to be `morphdark` and its fitted equation was $\text{logit}(\pi) = 1.18 - 0.016 \times \text{distance}$. The estimates `morphlight` and `distance:morphlight` are the differences in the intercept and slope for this morph with respect to the the reference group (cf. Chapter 6). The fitted equation for `morphlight` was $\text{logit}(\pi) = (1.18 - 0.41) + (-0.016 + 0.027) \times \text{distance}$. The interpretation of the intercept and the slope in the logit scale was given in Section 9.3. The results show that $\text{logit}(\pi)$ significantly declined with distance and that the slope of the light morph was significantly higher than that of the other morph. The model accounts for a large proportion, `1 - (m2$deviance / m2$null.deviance)` = 0.626, of the deviance. Notice that the small sample size (14) hinders obtaining precise parameter estimates and making reliable inferences from the analysis. Interpreting these results with a "grain of salt" would be overly generous in cases of very small sample sizes. It is then best to consider the analysis as no more than a description of broad trends in the data.

Nevertheless, we should evaluate the goodness of fit of model `m2` using randomized quantile residuals (Chapter 9):

```
resid.m2=augment(m2)
res.m2=simulateResiduals(fittedModel=m2,n=1e3, integerResponse=T, refit=F,plot=F)
resid.m2$.std.resid=residuals(res.m2, quantileFunction = qnorm)
```

The first line uses a command from the package `broom` to store in `resid.m2` the basic information needed for the residual analysis. The second line uses a command from the `DHARMa` package to simulate the randomized quantile residuals, which are converted into normally distributed residuals and stored in `resid.m2` to replace the original deviance residuals that are of dubious and limited use in GLMs (Chapter 8). Figure 9.18 shows plots of the residual analysis.

Although hardly any conclusions can be obtained with 14 data points, `m2` has a barely reasonable fit to data. The Cook distances suggest that there are four data points with large influence in the fitted parameters, which is not shocking when we look again at Fig. 9.18.

The conditional plots of `m2` show the predicted relations between the probability of surviving and the distance for each morph. We first obtain the predicted values with `pred.m2 = ggpredict(m2, terms = c("distance [all]", "morph"))` and then plot them in Fig. 9.19.

Grouped binary GLMs can also be fitted in the Bayesian framework with the package `brms`. As for its frequentist counterpart, the response variable needs be entered in a different manner from the usual statistical models. Let's first see the priors that need be defined:

```
> get_prior(formula=surv|trials(placed) ~distance*morph, data = DF2, family = binomial)
                prior     class                coef group resp dpar nlpar lb ub       source
               (flat)         b                                                      default
               (flat)         b            distance                             (vectorized)
               (flat)         b distance:morphlight                             (vectorized)
               (flat)         b          morphlight                             (vectorized)
 student_t(3, 0, 2.5) Intercept                                                      default
```

The expression `surv|trials(placed)` in the model formula indicates the response variable `(surv)` and, separated by a vertical bar, the variable `(placed)` denoting the number of sampling units (individual moths) exposed.

Without further ado, we are going to use the default non-informative `brms` default priors to run the model:

```
m2.brms=brm(surv|trials(placed) ~distance*morph, data = DF2, family = binomial,
  warmup = 1000,chains=3, iter=2000, future=T,control = list(adapt_delta = 0.9))
```

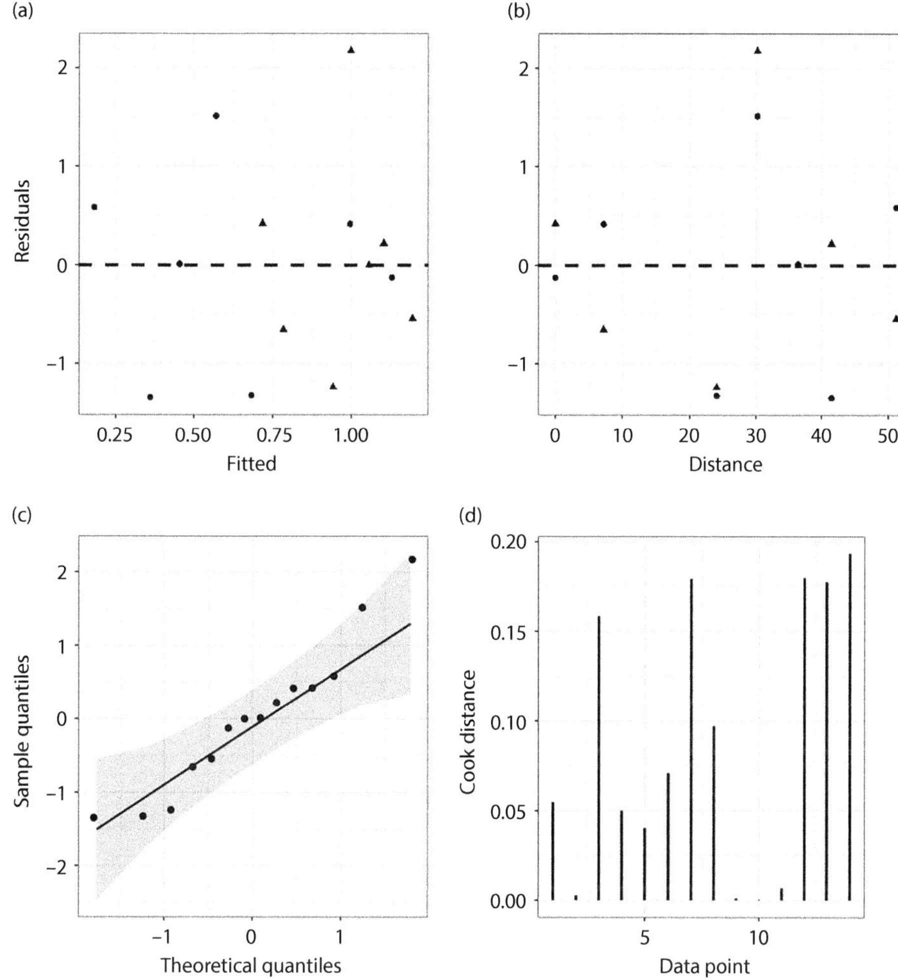

Fig. 9.18 Residual analysis of the frequentist binary GLM `m2` for grouped data based on randomized quantile residuals converted to normal residuals, showing the relation between these residuals and the fitted values (a), and to the standardized numerical explanatory variables (b). The dashed lines were added to help the visualization. Plot (c) shows the quantile–quantile plot and the 95% confidence band of the residuals. Plot (d) shows the Cook distances.

Note the use of `family = binomial` instead of `family = bernoulli` as in the ungrouped binary GLM (Section 9.6). The model summary is:

```
> summary(m2.brms)
 Family: binomial
  Links: mu = logit
Formula: surv | trials(placed) ~ distance * morph
   Data: DF2 (Number of observations: 14)
Samples: 3 chains, each with iter = 2000; warmup = 1000; thin = 1;
         total post-warmup samples = 3000

Population-Level Effects:
                   Estimate Est.Error l-95% CI u-95% CI Rhat Bulk_ESS Tail_ESS
Intercept              1.13      0.20     0.75     1.53 1.00     1389     1809
distance              -0.02      0.01    -0.03    -0.01 1.00     1399     1988
morphlight            -0.40      0.27    -0.93     0.16 1.00     1053     1434
distance:morphlight    0.03      0.01     0.01     0.04 1.00     1028     1253
```

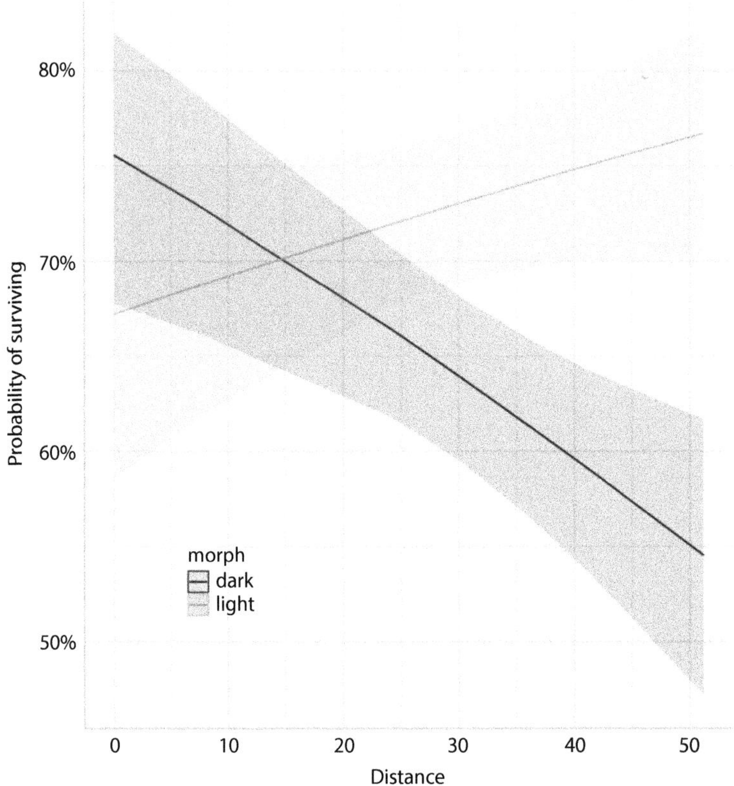

Fig. 9.19 Conditional plots of the frequentist binary GLM `m2` for grouped data showing the predicted relations between the distance from Liverpool and the probability of presence of each morph variable. The bands correspond to the 95% confidence intervals.

The `Rhat` and `ESS` suggest that the three MCMC chains have converged to a stationary posterior distribution, and that this contains a fairly large number of largely independent parameter estimates at our disposal. The interpretation of the model parameters is, of course, independent of the method used for their estimation (frequentist or Bayesian) and need not be repeated here.

The model explains a fairly large proportion of the variation of the response variable (Gelman et al. 2018):

```
> bayes_R2(m2.brms)
   Estimate Est.Error  Q2.5 Q97.5
R2    0.878    0.0214 0.819 0.902
```

We may roll over the usual plots to examine the convergence of the MCMC chains and the auto-correlation of sampled parameters that you have seen before in this chapter and in Chapters 4 to 6. Then, we would carry out residual analysis with randomized quantile residuals, obtain the conditional predictive plots, and also use the posterior distribution to assess the model goodness of fit as in Section 9.6. We have done all this for the more common ungrouped binary GLMs. But given that the current chapter has already become very long, we would rather drop our good intentions of rigor and stop here (should you be interested, you can find the code on the companion website).

9.8 Problems

Visit the companion website at https://www.oup.com/companion/InchaustiSMWR to obtain the data sets for these problems.

9.1 Bolger et al. (1997) wanted to know if the presence of native rodents (RODENTSP) in a fragmented landscape in coastal California was related to the distance (m) to the nearest source canyon (DISTX), the time (years) since the fragment was isolated by urbanization (AGE), and the percentage of the fragment area covered in shrubs (PERSHRUB). Data file: Pr9-1.csv

9.2 Bliss (1935) studied the dose–reponse relation of the flour beetle after a five-hour exposure at one of eight dosage concentrations to the toxic carbon sulphide. Sets of animals were exposed to eight different concentrations of the gas, and the number of survivors were counted at the end of the assay. Data file: Pr9-2.csv

References

Bishop, J. (1972). An experimental study of the cline of industrial melanism in *Biston betularia* (l.) (Lepidoptera) between urban Liverpool and rural North Wales. *Journal of Animal Ecology*, 20, 941–224.

Bliss, C. (1935). The calculation of the dosage–mortality curve. *Annals of Applied Biology*, 22, 134–167.

Bolger, D., Alberts, A., Sauvajot, R., et al. (1997). Response of rodents to habitat fragmentation in coastal southern California. *Ecological Applications*, 7, 552–563.

Brier, G. (1953). Verification of forecast expressed in terms of probability. *Monthly Weather Review*, 78, 1–3.

Coles, S. (2001). *An Introduction to Statistical Modeling of Extreme Values*. John Wiley and Sons, New York.

Gabry, J., Simpson, D., Vehtari, A., et al. (2019). Visualization in Bayesian workflow. *Journal of the Royal Statistical Society A*, 182, 389–402.

Gelman, A., Carlin, J., Stern, H. et al. (2014). *Bayesian Data Analysis*, 3rd edn. Chapman and Hall / CRC Press, New York.

Gelman, A., Goegebeur, Y., Tuerlinckx, F., et al. (2000). Diagnostic checks for discrete data regression models using posterior predictive simulations. *Applied Statistics*, 49, 247–268.

Gelman, A., Goodrich, B., Gabry, J. et al. (2018). R-squared for Bayesian regression models. *American Statistician*, 73, 307–309.

Gelman, A., Simpson, D., and Betancourt, M. (2017). The prior can often only be understood in the context of the likelihood. *Entropy*, 19, 555–568.

Hobbs, N. and Hooten, M. (2015). *Bayesian Models: A Statistical Primer for Ecologists*. Princeton University Press, Princeton.

Hunter, D. (2000). The conservation and demography of the southern corroboree frog (*Pseudophryne corroboree*). M.Sc. thesis, University of Canberra.

Kruschke, J. (2015). *Doing Bayesian Data Analysis: A Tutorial with R, JAGS and Stan*, 2nd edn. Academic Press, New York.

Vehtari, A., Gelman, A., and Gabry, J. (2017). Practical Bayesian model evaluation using leave-one-out cross validation and WAIC. *Statistical Computing*, 27, 1413–1432.

Wan, X., Wang, W., Liu, J. et al. (2014). Estimating the sample mean and standard deviation from the sample size, median, range and/or interquartile range. *BMC Medical Research Methodology*, 14, 135–147.

CHAPTER 10

When the Response Variable is a Count, Often with Many Zeros

Packages needed in this chapter:

```
packages.needed<-c("ggplot2","fitdistrplus","brms","broom","GGally",
"doBy","arm","gamlss.dist","broom.mixed","ggeffects","DHARMa",
"qqplotr","bayesplot","glmmTMB")
lapply(packages.needed, FUN = require, character.only = T)# loads these packages in the working session.
```

10.1 Introduction

This chapter explains the fitting of GLMs when the response variable is a count, i.e., a natural number, including zero. When the response variable is a count, the two main probability distributions that can be used are the Poisson and negative binomial distributions. These are not the only probability distributions for discrete random variables but they are by far the most often used in applied statistics.

The Poisson distribution was introduced by the French mathematician Simon Denis Poisson. It has a single parameter, its mean μ_Y, which is sometimes called λ. A discrete random variable has a Poisson distribution if three assumptions hold. First, events occur at a constant rate μ_Y per unit interval of time t (or of space s) such that the expected number of counts to be observed is $\mu_Y t$ (or $\mu_Y s$); since the rate μ_Y of a Poisson process is constant over time or space, there is no temporal or spatial heterogeneity in the event occurrence rate. Second, events occurring in a time or space interval have no effect on events occurring in a second interval; this is called the independence assumption. Third, the probability of observing two or more events in a time or space interval is very small, tending to zero. The first two assumptions are shared with the binomial distribution, for which the Poisson distribution arises as the limit when the third assumption holds (Poisson 1837). The Poisson distribution is

$$\Pr(Y = y) = \frac{\exp(-\mu_Y)\,\mu_Y^y}{y!}, \tag{10.1}$$

and the mean or expected value of a random variable with a Poisson distribution is equal to its variance: $\mu_Y \equiv \mathrm{E}(Y) = \sigma_Y^2$. The mean and the variance of a Poisson random variable are not only not independent, they are equal. Therefore, the single parameter μ_Y of the

Statistical Modeling With R. Pablo Inchausti, Oxford University Press. © Pablo Inchausti (2023).
DOI: 10.1093/oso/9780192859013.003.0010

Poisson distribution performs a double duty: it describes both the central tendency of the response variable Y and its variability about its mean.

The negative binomial distribution was probably derived by the French mathematician Pierre de Montmort in 1713 in the context of its feature as the number of failures y before the kth success in a series of binary trials (Hilbe 2011). It seems safer and clearer to associate the negative binomial distribution with Student (1907). There are several equivalent ways of deriving and parameterizing the negative binomial distribution as a function of its two parameters (Hilbe 2011). We are going to use one of them symbolized as NB2 in which the negative binomial distribution is derived from a Poisson–gamma mixture distribution (Hilbe 2011):

$$\Pr(Y = y) = \frac{\Gamma(y + \phi)}{y!\Gamma(\phi)} \left(1 - \frac{\phi}{\phi + \mu_Y}\right)^{\phi} \left(\frac{\mu_Y}{\phi + \mu_Y}\right)^{y}. \tag{10.2}$$

Equation (10.2) is expressed in terms of two real, positive parameters, μ_Y and ϕ. $\Gamma()$ is the mathematical gamma function that is the analogue of the factorial function for real numbers. Equation (10.2) now becomes superficially similar to Eq. (9.1) of the binomial distribution. The other parameter, ϕ, governs the relation between the mean and the variance of a random variable Y with a negative binomial distribution, such that $\mu_Y \equiv \mathrm{E}(Y)$ and $\mathrm{Var}(Y) = \mu_Y + \mu_Y^2/\phi$ (Table 8.1). Therefore, the mean and the variance of of a random variable with a negative binomial distribution have a positive, quadratic relation modulated by the parameter ϕ, which is sometimes called the "inverse index of aggregation parameter" or "aggregation parameter" (Krebs 1999). The relation between the mean and the variance of the negative binomial distribution implies that $\phi = \mu_Y^2/(\mathrm{Var}(Y) + \mu_Y)$. Let's consider two extreme cases. First, when ϕ goes to infinity, and thus the aggregation of events goes to zero over a time or space interval, the mean and variance become equal and the negative binomial reduces to the Poisson distribution. Second, when ϕ goes to zero, it denotes more aggregation or clumping of events over a time or space interval than would be expected in a Poisson process. Figure 10.1 show the shapes of the Poisson and negative binomial distributions for different combinations of their parameters.

The Poisson distribution stems from the exponential dispersion family that lies at the heart of GLMs (Chapter 8). The negative binomial distribution can be forced into the exponential dispersion family if the aggregation parameter ϕ is treated as a constant to be separately estimated from data (Hilbe 2011, 2014). If so, the IWLS algorithm (Chapter 8) can also be used for the negative binomial distribution (Fox 2015). Count GLMs can, of course, be fitted in the Bayesian framework using MCMC-based methods that do not rely on either using or tweaking an optimization algorithm linked to the exponential dispersion family (Chapter 8).

The canonical link function to be used for count GLMs is log (Table 8.1). Therefore, the count GLM statistical model can be expressed as $Y \sim \mathrm{Poisson}(\mu_Y)$ or $Y \sim \mathrm{NegativeBinomial}(\mu_Y, \phi)$ and $\log(\mu_Y) = \mathrm{X}\beta$, where X contains the numerical and/or categorical explanatory variables, and β holds the coefficients to be estimated that reflect the effects of the explanatory variables on $\log(\mu_Y)$. While other link functions might be used (e.g., square root, identity), the inverse of the log link function (i.e., the exponential) guarantees that the predicted values of count GLMs remain positive, real values that can be compared with the observed counts.

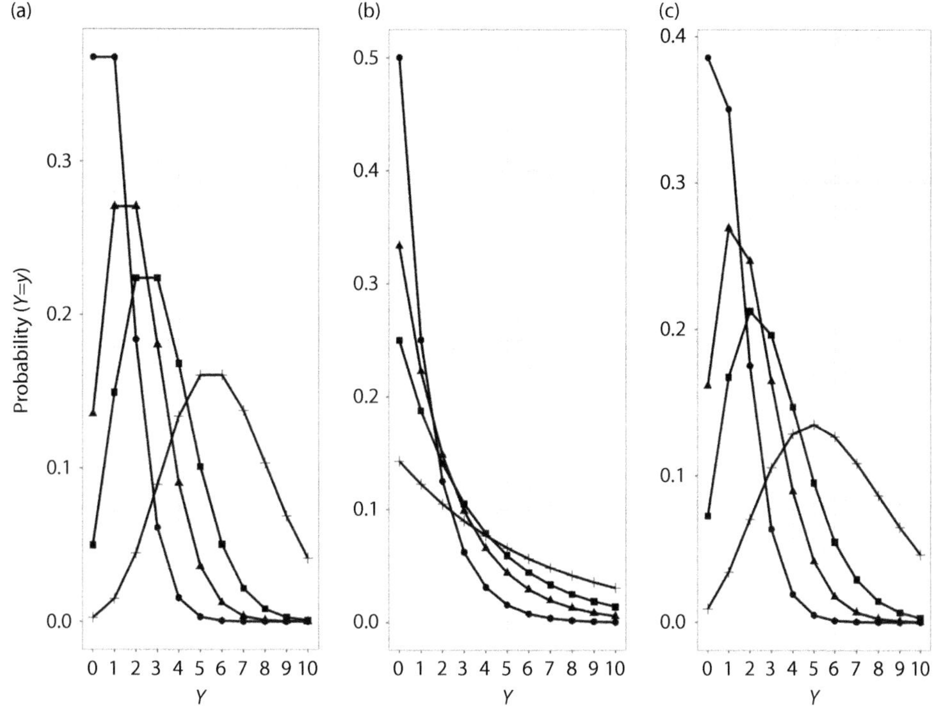

Fig. 10.1 Mass probability functions for Poisson (a) and negative binomial distributions (b and c) for means of 1 (circles), 2 (triangles), 3 (squares), and 6 (crosses). Plots (b) and (c) show the negative binomial distributions for a dispersion parameter of 1 and 10, respectively.

10.2 Over-dispersion: A common problem with many causes and some solutions

With the exception of the normal distribution, all members of the exponential dispersion family have a specific relation between their mean and their variance, known as the variance function (Table 8.1). Therefore, when a non-Gaussian GLM models the effect of the explanatory variables on its mean in the scale of the link function, it is also establishing a relation between these variables and its variance, subject to the functional constraint imposed by a specific variance function (Chapter 8). The simplest count GLM presented in every statistics textbook is based on the Poisson distribution, probably for no other reason than most people are less familiar with the negative binomial. The Poisson distribution has the unique attribute that its mean and variance are equal, a feature known as equidispersion.

Over-dispersion is by far the most frequent problem encountered when fitting count GLMs. It just means that the variance of the response variable is appreciably higher than the given mean of the probability distribution chosen in the count GLM. The problem is that most real count data do not satisfy the equidispersion feature of the Poisson distribution. Of course, the problem is not that the data are faulty, but that we are using inadequate models to analyze them. Table 10.1 lists the main, non-exclusive causes of over-dispersion and their likely remedies, most of which are explained in this chapter.

Table 10.1 Main causes of over-dispersion in count GLMs and the common solutions. The "excess" of zeros and of overall variation are relative to those predicted by the probability distribution (Poisson or negative binomial) fitted to the data. The common solutions to over-dispersion are listed in increasing order of complexity. The causes of over-dispersion are not mutually exclusive, and thus several common solutions may be incorporated in the same GLM.

Causes of over-dispersion	Common solutions to over-dispersion
Inadequate probability distribution in the GLM specification	Use the negative binomial or quasi-Poisson rather than Poisson distribution
Insufficient or incomplete set of explanatory variables	Add more explanatory variables or interactions to the model, if available and justifiable
There might be outliers	Carry out a residual analysis to detect potential outliers using Cook distances or Pareto-*k* LOO-CV; refit the model without the potential outliers and compare
Excess of zeros	Use a mixture model (zero-inflated or zero-augmented GLM)
Excess of variation in the response variable	Change the model structure (GLMM; Chapter 13), often adding a random effect at the level of individual observations

It is very important to bear in mind that the detection of over-dispersion never points to the true underlying cause(s) of the phenomenon. Over-dispersion is actually the main reason for using the negative binomial rather than Poisson distribution in count GLMs. The negative binomial distribution has an extra parameter (ϕ) that gives additional, but not unlimited, flexibility to account for the variation frequently encountered in count data.

Given the prevalence of over-dispersion, it becomes useful to have a metric to evaluate it. The two main metrics of over-dispersion are derived from the deviance and Pearson residuals that have been used to assess the overall goodness of fit of count GLMs (Chapter 8). These two residuals are used to calculate different measures of the overall goodness of fit of a GLM that reflect the residual variability in the data after fitting a GLM. It is from the readily available deviance and Pearson overall measure of goodness of fit that we can estimate the metrics of over-dispersion.

The residual deviance D (Eq. (8.3)) is a measure of the variability in the data that remains to be explained. Thus, any good statistical model should have a small value of this measure of "badness of fit." The expression of the deviance residual changes with the probability distribution used to model the variation of the response variable in a GLM (Faraway 2016). For GLMs having a dispersion parameter ϕ to be estimated, we need to compute a scaled residual deviance $D^* = D/\phi$. The ratio of the scaled residual deviance D^* and the number of degrees of freedom (= number of data points minus the number of parameters estimated) should be close to one for a well-fitting GLM (Hilbe 2011, Dunn and Smyth 2016). The sum of the squared Pearson residuals (Chapter 8) equals the familiar Pearson χ^2 statistic (Hilbe 2011). The ratio of the Pearson statistic and the model degrees of freedom should also be close to one for a well-fitting GLM (Hilbe 2011, Dunn and Smyth 2016). For both metrics of over-dispersion, "close to one" is meant to tolerate random variation that will never allow even Poisson simulated data to be perfectly equidispersed. For moderate-sized (100 points?) data sets, common practice based on nothing more than tradition and folklore tolerates over-dispersion metrics up to 1.30. For even larger data sets (1,000 points?) the tolerance is lower and values of 1.05 might trigger remedial action (Table 10.1). Which of the two metrics of over-dispersion should be used? An informally appointed jury seems divided on this issue. On the one hand, Dunn and Smyth (2016, p. 277) recommend the

Pearson-based estimator for being less biased (provided that most counts are not very small) but more variable than the deviance estimator. On the other, Hilbe (2014, p. 78) has the opposite view based on the same grounds. While both the deviance and Pearson statistics are asymptotically χ^2 distributed, the usual large-sample asymptotics employed in overall goodness-of-fit tests do not hold for small sample sizes (Dunn and Smyth 2016, p. 275).

Under-dispersion, having a smaller variance than predicted by the variance function, occasionally occurs in count GLMs for some data sets. Again, the problem is not the data but the statistical model employed in their analysis. In these cases, the Conway–Maxwell–Poisson model (Sellers and Shumeli 2010, Lynch et al. 2014, Huang and Kim 2019) can account for under- and over-dispersion in count data.

The main consequences of over-dispersion in count GLMs are obtaining slightly biased and, more importantly, overly precise parameter estimates (e.g., Agresti 2015). In the frequentist framework, under-estimation of the standard errors leads to much smaller *p*-values than are warranted (i.e., "liberal tests") in the Wald test. In the Bayesian framework, large over-dispersion similarly leads to biased and unduly precise standard errors and credible intervals. When fitting a count GLM, we should adjust for under- or over-dispersion. One "quick and dirty" approach is to use the quasi-Poisson distribution[1] that has an extra parameter: the over-dispersion. In this approach, the standard errors of parameter estimates are divided by the square root of the over-dispersion, thus leading to higher and approximately correct *p*-values (Richards 2008). The quasi-Poisson distribution does not model over-dispersion; it merely estimates an extra parameter with respect to the Poisson distribution. Using the negative binomial distribution in a count GLM appears to be a more elegant way of proceeding (e.g., ver Hoef and Boveng 2007, Linden and Mantyniemi 2011). Beyond statistical significance, the presence of over-dispersion is a clear sign of a model with an inadequate fit to the data that requires our attention in terms of remedial action (Table 10.1).

10.3 Plant species richness and geographical variables

THE DATA IN CONTEXT: Johnson and Raven (1973) collected data about the total number of plant species and several geographical variables (area in km^2, distance to the nearest and to Santa Cruz island, elevation in meters) in the Galapagos archipelago. The goal of their analysis was to determine which of these explanatory variables is related to the total species richness, in the light of the classic theory of island biogeography (MacArthur and Wilson 1967). The response variable Y is a count to be modeled using either the Poisson or the negative binomial distribution. The statistical model is:

$$Y \sim \text{Poisson}(\mu_Y) \text{ or } Y \sim \text{NegativeBinomial}(\mu_Y, \phi),$$
$$\log(\mu_Y) = \mathrm{X}\beta,$$

with X being the numerical explanatory variables and β the coefficients (intercept and partial slopes) expressing the relation between the explanatory variables and the mean of the response variable in the scale of the logarithmic link function (the default choice for count GLMs).

[1] This is not a proper probability distribution because it does not sum to one.

EXPLORATORY DATA ANALYSIS: We start by importing and summarizing the data:

```
> DF=read.csv("chap11 Galapagos plants.csv", header=T)
> summary(DF)
         island        spp            endemics       area_km2           elev_m        dist_nearest    dist_StaCruz
 Baltra      : 1   Min.   :  2.00   Min.   : 1   Min.   :   0.01   Min.   :   5.0   Min.   : 0.2   Min.   :  0.00
 Bartolome   : 1   1st Qu.: 12.00   1st Qu.: 8   1st Qu.:   0.34   1st Qu.:  90.0   1st Qu.: 0.7   1st Qu.: 10.70
 Caldwell    : 1   Median : 44.00   Median :19   Median :   2.85   Median : 186.0   Median : 2.8   Median : 47.40
 Champion    : 1   Mean   : 87.34   Mean   :27   Mean   : 270.73   Mean   : 360.8   Mean   :10.2   Mean   : 58.53
 Coamano     : 1   3rd Qu.: 97.00   3rd Qu.:33   3rd Qu.:  59.56   3rd Qu.: 458.0   3rd Qu.:10.7   3rd Qu.: 85.90
 Daphne Major: 1   Max.   :444.00   Max.   :95   Max.   :4669.32   Max.   :1707.0   Max.   :47.4   Max.   :290.20
 (Other)     :23
```

We can see that the area of the islands and the elevation span six and four orders of magnitude. It is difficult to envisage that the relation between $\log(\mu_Y)$ and these two explanatory variables may be linear over such wide ranges. This is why we are going to log-transform these two explanatory variables before the analysis: `DF$log.area = log(DF$area_km2); DF$log.elev = log(DF$elev_m)`.

The next step is to assess the best probability distribution to model the response variable using the command `fitdist` from the package `fitdistrplus` to compare the Poisson and negative binomial distributions:

```
pois=fitdist(DF$spp,"pois")
negbin=fitdist(DF$spp,"nbinom")
```

with which we produce Fig. 10.2.

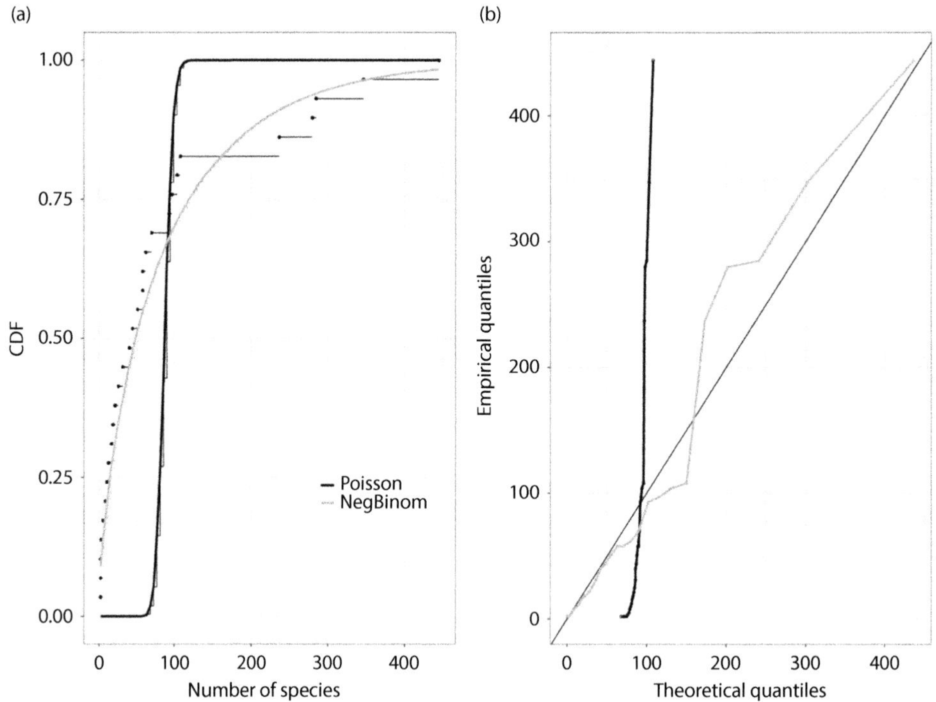

Fig. 10.2 Goodness-of-fit assessment of the response variable number of plant species in Galapagos showing the cumulative distribution function (a) and the quantile–quantile plot comparing the Poisson and negative binomial distributions obtained using the package `fitdistrplus` (b).

There cannot be any argument that the negative binomial distribution is the best candidate to model the variation of the response variable. This should not be terribly surprising: it is hard to imagine that a one-parameter distribution may adequately fit data spanning 2–444 species and having a hugely skewed distribution.

Given that we have four numerical explanatory variables (`log.area`, `log.elev`, `dist_nearest`, and `dist_StaCruz`), we should examine their relations with the response variable and their degree of covariation using the command `ggpairs` from the package package `GGally`. The goal is to avoid including explanatory variables that are so strongly correlated that they are redundant in their explanatory capacity and cause numerical problems in the fitting (Dormann et al. 2013; Chapter 4).

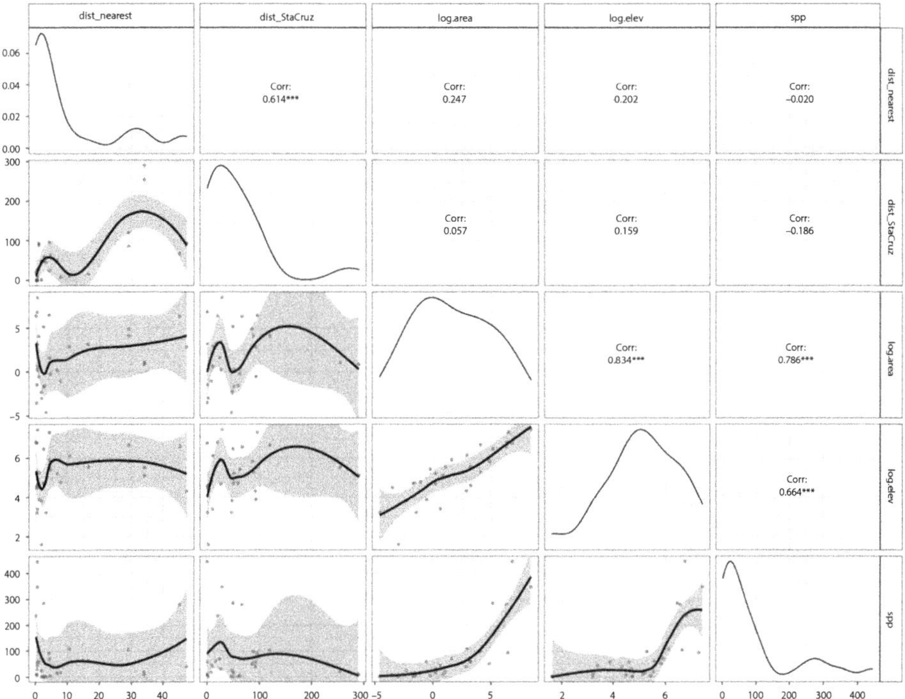

Fig. 10.3 Covariation between the explanatory variables in the frequentist count GLM `m1`. The diagonal shows the density plots of each explanatory variable for the entire data set and depending on the value of the response variable. Below the diagonal are shown all pairwise *x*–*y* plots, with presences colored in red and absences in green. Above the diagonal are the pairwise correlation of explanatory variables for the entire data set and for each value of the response variable. The figure was produced with the package `GGally`.

Figure 10.3 shows that the response variable seems to be related to `log.area` and `log.elevation`, and much less so to the other explanatory variables (bottom row of plots). We can also see that `log.area` is strongly correlated with `log.elevation` and thus, following Dormann et al.'s (2013) heuristic criterion, we will only include one of them on the statistical model (Chapter 4). On purely statistical grounds we could choose either explanatory variable, but we will choose `log.area` using the non-statistical criterion of its relevance for the theory of island biogeography in ecology. It is always the scientific context that should take primacy in selecting the explanatory variables to be entered in statistical models.

Because we have three numerical explanatory variables, it is convenient to standardize them to have a mean of zero and a unit standard deviation so as to compare the relative magnitudes of their partial slopes (Chapter 4) in the scale of the link function:

```
DFs=data.frame(scale(DF[c("dist_nearest","dist_StaCruz","log.area")],scale=T,
      center=T), spp=DF$spp)
```

The resulting data frame `DFs` will be used to assess the importance of the explanatory variables on the average number of plant species in the Galapagos archipelago.

10.3.1 *Frequentist fitting of the count GLM*

We fitted the model `m1 = glm.nb(spp ~ log.area + dist_nearest + dist_StaCruz,data = DFs)` with the package `MASS`; its summary is:

```
> m1=glm.nb(spp~log.area+dist_nearest+dist_StaCruz, data=DFs)
> summary(m1)

Call:
glm.nb(formula = spp ~ log.area + dist_nearest + dist_StaCruz,
    data = DFs, init.theta = 2.983093189, link = log)

Deviance Residuals:
   Min      1Q  Median      3Q     Max
-2.016  -0.859  -0.234   0.456   2.331

Coefficients:
             Estimate Std. Error z value Pr(>|z|)
(Intercept)     3.754      0.114   32.85   <2e-16 ***
log.area        1.305      0.122   10.71   <2e-16 ***
dist_nearest   -0.112      0.147   -0.76     0.45
dist_StaCruz   -0.141      0.148   -0.95     0.34
---
Signif. codes:  0 '***' 0.001 '**' 0.01 '*' 0.05 '.' 0.1 ' ' 1

(Dispersion parameter for Negative Binomial(2.98) family taken to be 1)

    Null deviance: 148.082  on 28  degrees of freedom
Residual deviance:  31.483  on 25  degrees of freedom
AIC: 274.9

Number of Fisher Scoring iterations: 1


              Theta:  2.983
          Std. Err.:  0.891
```

Using McFadden's (1974) metric, `m1` accounts for a large fraction of the deviance in the data, `1 – (m1$deviance / m1$null.deviance)` = 0.787, and it has an acceptable level of over-dispersion, `m1$deviance / m1$df.residual` = 31.438 / 25 = 1.26. The predicted equation is log(E(spp)) = 3.724 + 1.305log.area – 0.112dist_nearest – 0.141distStaCruz. By exponentiating this equation we obtain the effects of the explanatory variables in a multiplicative fashion: E(spp) = exp(3.754) × exp(1.305log.area) × exp(–0.112dist_nearest) × exp(–0.141distStaCruz). Using Wald's test (Chapter 3), only `log.area` has an important (if by that we mean statistically significant) effect on log(E(spp)). Given that all the numerical explanatory variables were standardized to zero mean and unit standard deviation, we can compare their partial slopes on the log link scale and see that the relative effect of `log.area` is between 9 and 11 times higher than those of the other explanatory variables. The intercept exp(3.754) = 42.7 is the expected number of plant species in a typical or "average" Galapagos island that has an average `log.area` (1.69, or exp(1.69) = 5.42 km^2) and average distances to the nearest island (10.2 km) and to Santa

Cruz island (58.5 km). The partial slope for `log.area` (1.305) indicates that changing this variable by one standard deviation (`sd(DF$log.area)` = 3.48, or exp(3.48) = 32.3 km^2) leads to an increase of 1.305 in log(E(spp)). That is, an increase in island area in the Galapagos of 32.3 km^2 (keeping all other explanatory variables at their average values) is predicted to lead to exp(1.305) = 2.82 more plant species on average. Another useful way to interpret the slope of a count GLM is that a unit (or a standard deviation in this case) change in the explanatory variable leads to a 100 × (exp(slope) – 1) change in the mean of the response variable. This is valid for any GLM with a log link function (Chapters 11 and 12), regardless of whether it is fitted in the frequentist or Bayesian framework.

We can also summarize the overall effects of the explanatory variables in the progressive reduction of deviance as terms are sequentially entered in the model (Chapter 8):

```
> anova(m2, test="Chisq")
Analysis of Deviance Table

Model: Negative Binomial(2.98), link: log

Response: spp

Terms added sequentially (first to last)


             Df Deviance Resid. Df Resid. Dev Pr(>Chi)
NULL                           28    148.082
log.area      1  113.335       27     34.747   <2e-16 ***
dist_nearest  1    2.413       26     32.334   0.1203
dist_StaCruz  1    0.851       25     31.483   0.3562
```

While model `m1` with standardized explanatory variables is useful in assessing their relative effect on log(E(spp)), it cannot be used to predict the expected number of plant species on a Galapagos island not already included in the data set employed to fit it. The reason is that we would need to include the new value to recalculate the values of all the standardized explanatory variables and then refit the model with the new data set. If we just wish to have a predictive equation, we should have a statistical model fitted with the raw data. Predicting and understanding the effect of explanatory variables need not be accomplished with a single statistical model. The new model is:

```
> m2=glm.nb(spp~log.area+dist_nearest+dist_StaCruz, data=DF)
> summary(m2)

Call:
glm.nb(formula = spp ~ log.area + dist_nearest + dist_StaCruz,
    data = DF, init.theta = 2.983093189, link = log)

Deviance Residuals:
   Min      1Q  Median      3Q     Max
-2.016  -0.859  -0.234   0.456   2.331

Coefficients:
              Estimate Std. Error z value Pr(>|z|)
(Intercept)    3.31686    0.16562   20.03   <2e-16 ***
log.area       0.37555    0.03507   10.71   <2e-16 ***
dist_nearest -0.00773    0.01015   -0.76     0.45
dist_StaCruz -0.00206    0.00216   -0.95     0.34
---
Signif. codes:  0 '***' 0.001 '**' 0.01 '*' 0.05 '.' 0.1 ' ' 1

(Dispersion parameter for Negative Binomial(2.98) family taken

    Null deviance: 148.082  on 28  degrees of freedom
Residual deviance:  31.483  on 25  degrees of freedom
AIC: 274.9
```

Note that the AIC, the dispersion parameter of the negative binomial distribution, the index of over-dispersion, and McFadden's (1974) proportion of the deviance explained are all identical to `m1`. It is the interpretation of the parameters that clearly differs between models. In `m2`, the intercept is the value of log(E(spp)) = 3.316 when all the explanatory variables are zero, which, of course, has no factual meaning in the context of this problem. The partial slope of `log.area` denotes that an increase of one unit in this variable leads to an increase of 0.375 in log(E(spp)) or, putting it in better terms, that an increase in `log.area` of 1 unit (or exp(1) = 2.72 km^2) leads to increase in E(spp) of exp(0.375) = 1.45 plant species. Before trusting either `m1` or `m2` to perform any reliable inference, we would need to assess their goodness of fit through the customary residual analysis.

If the goal of the research was to assess the importance of the explanatory variables on plant species richness, either `m1` or `m2` is the final answer to the question. However, if our goal had been to generate the simplest or most parsimonious statistical model that can be fitted to the data, we should try to simplify `m2` (or `m1`, whichever you prefer) through model selection. Now, frequentist model selection can be either based on the differences in AIC, or in the sequential application of Wilks' likelihood ratio test (Chapter 7). These are two different theoretical frameworks to accomplish the same goal of model selection, and you are free to choose whichever you prefer. We are going to use Wilks' likelihood ratio test for no better reason than that the much-maligned significance tests at least provide a clear-cut criterion to discern between alternative models that is largely missing in the use of the AIC (Chapter 7).

Using `m2` as the starting point, we are going to generate models `m3` to `m5`, each of which will be a special case of `m2` but lacking just one term (here, one explanatory variable), and then compare each model with `m2` using Wilks' test. These models can be generated by `m3 = update(m2, ~. dist_nearest)`; `update` deletes (this is what the "`-`"means—we could also make other changes; see `?update`) the variable `dist_nearest` from model `m2` and generates model `m3`. We proceed likewise for the other explanatory variables in `m2` to generate models `m4` and `m5`. The three model comparisons are:

```
> anova(m3,m2)
Likelihood ratio tests of Negative Binomial Models

Response: spp
                                     Model theta Resid. df    2 x log-lik.   Test    df LR stat. Pr(Chi)
1              log.area + dist_StaCruz  2.93         26            -266
2 log.area + dist_nearest + dist_StaCruz  2.98         25            -265 1 vs 2    1       0.6   0.439
> anova(m4,m2)
Likelihood ratio tests of Negative Binomial Models

Response: spp
                                     Model theta Resid. df    2 x log-lik.   Test    df LR stat. Pr(Chi)
1             log.area + dist_nearest  2.87         26            -266
2 log.area + dist_nearest + dist_StaCruz  2.98         25            -265 1 vs 2    1     0.833   0.361
> anova(m5,m2)
Likelihood ratio tests of Negative Binomial Models

Response: spp
                                     Model theta Resid. df    2 x log-lik.   Test    df LR stat.  Pr(Chi)
1          dist_nearest + dist_StaCruz 0.697         26            -311
2 log.area + dist_nearest + dist_StaCruz 2.983         25            -265 1 vs 2    1      45.8 1.33e-11
```

Because only the deletion of `log.area` from `m2` generated a significant difference, we conclude that `dist_nearest` and `dist_StaCruz` have limited explanatory power and may be omitted from the final model `m6`:

```
> m6=glm.nb(spp~log.area, data=DF)
> summary(m6)

Call:
glm.nb(formula = spp ~ log.area, data = DF, init.theta = 2.634
    link = log)

Deviance Residuals:
   Min      1Q  Median      3Q     Max
-1.902  -0.822  -0.182   0.330   2.119

Coefficients:
            Estimate Std. Error z value Pr(>|z|)
(Intercept)    3.150      0.140    22.4   <2e-16 ***
log.area       0.367      0.036    10.2   <2e-16 ***
---
Signif. codes:  0 '***' 0.001 '**' 0.01 '*' 0.05 '.' 0.1 ' ' 1

(Dispersion parameter for Negative Binomial(2.63) family taken

    Null deviance: 132.140  on 28  degrees of freedom
Residual deviance:  31.295  on 27  degrees of freedom
AIC: 274
```

`m6` has an acceptable level of over-dispersion, `m1$deviance / m1$df.residual` = 1.12, and it accounts for a sizable fraction of the deviance: `1 – (m1$deviance / m1$null.deviance)` = 0.763.

The predicted equation is log(E(spp)) = 3.15 + 0.367log.area, which, after exponentiating and applying the properties of logarithms and exponentials, becomes the well-known power law E(spp) = exp(3.15)area$^{0.367}$ from island biogeography (MacArthur and Wilson 1967).

We can now assess the goodness of fit of the final model `m6`. We start by gathering the main input for the residual analysis with the command `augment` from package `broom`: `resid.m6 = augment(m6)`. Next, we simulate the randomized quantile residuals (Chapter 8) with the package `DHARMa` as `res.m6 = simulateResiduals(fittedModel = m6, n = 1e3, integerResponse = T, refit = F, plot = F)`, and then replace the original residuals with a normal transformation of the randomized quantile residuals (Chapter 8) as `resid.m6$.std.resid = residuals(res.m6, quantileFunction = qnorm)`. Figure 10.4 shows that `m6` has an excellent fit to the data: largely random scatters of residuals without any discernible trends, most residuals close to the expected normal distribution, and no points with large Cook distances.

Finally, we can display the estimated parameters and their 95% confidence intervals in the scale of the log link function using the command `ggcoef` from the package `GGally`. We should also display the predicted relation between the response and explanatory variables using the command `ggpredict` from the package `ggeffects`, `pred.m6.log.area = ggpredict(m6, terms = c("log.area"))`, to generate the values plotted in Fig. 10.5. Inverting the linear relation fitted in the scale of the log link function in `m6` generates the exponential predicted curve in the original data scale.

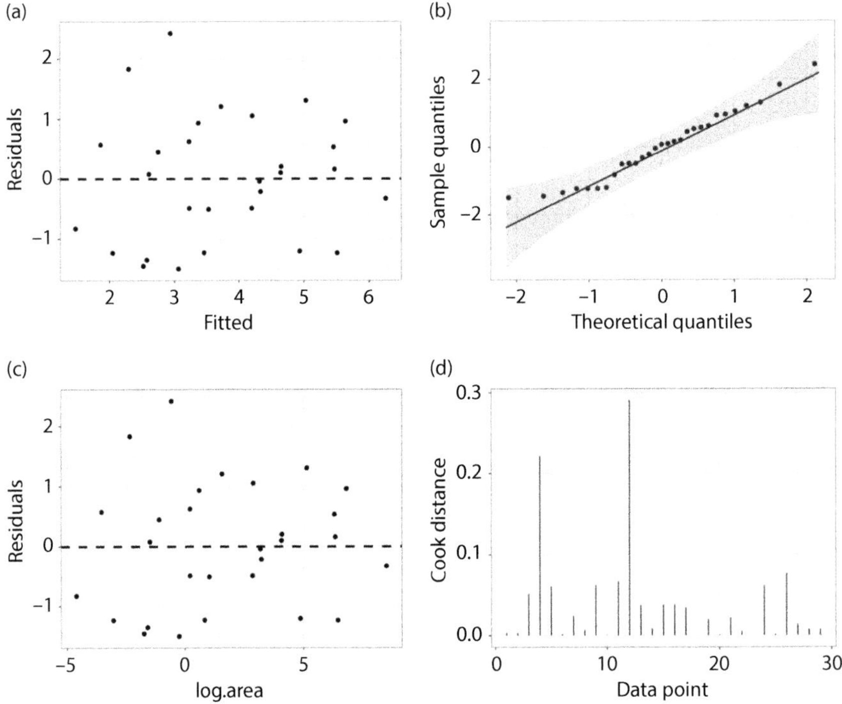

Fig. 10.4 Residual analysis of the frequentist count GLM `m1` based on simulated randomized quantile residuals converted to normal residuals, showing the relation between these residuals and the fitted values (a), the quantile–quantile plot (and the 95% confidence band) of residuals (b), the residuals vs. the standardized numerical explanatory variable (c), and the Cook distances (d).

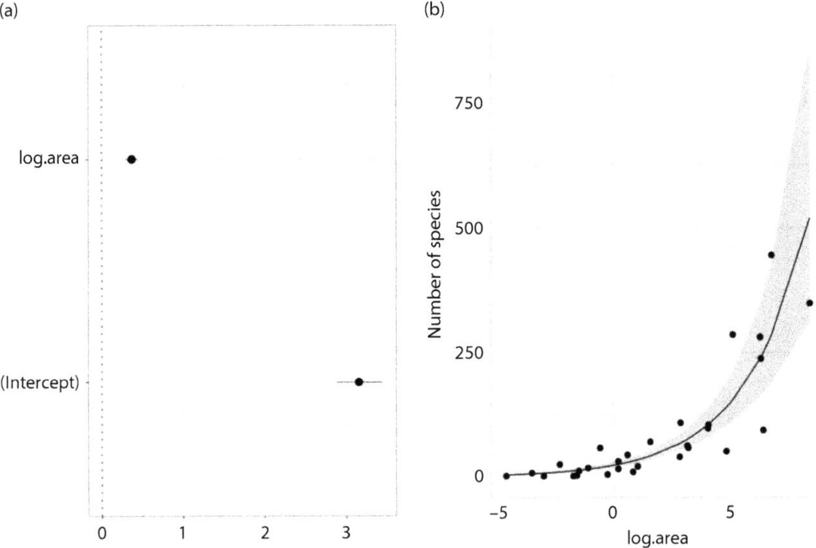

Fig. 10.5 (a) Estimates and 95% confidence intervals of the intercept and slopes of the frequentist count GLM `m1`. (b) Conditional relation between the response variable and the numerical explanatory variable, with a 95% confidence band.

10.3.2 *Bayesian fitting of count GLMs*

We use the package brms to fit a Bayesian count GLM equivalent of model m2. We first obtain the list of priors that we need to define:

```
> get_prior(formula=spp~log.area+dist_nearest+dist_StaCruz, family= negbinomial, data=DF)
                prior     class         coef group resp dpar nlpar bound       source
               (flat)         b                                               default
               (flat)         b dist_nearest                             (vectorized)
               (flat)         b dist_StaCruz                             (vectorized)
               (flat)         b     log.area                             (vectorized)
 student_t(3, 3.8, 2.5) Intercept                                              default
      gamma(0.01, 0.01)     shape                                              default
```

(the last line is ϕ, the dispersal parameter of the negative binomial in Eq. (10.2), whose name unfortunately changes in different books and software).

Let's first recall that priors need be defined in the log link scale. What do we know about the model parameters before seeing the data? We are tentatively going to define the following priors (Fig 10.6):

```
prior.m1 = c(set_prior("normal(0,2.5)", class="b"),
             set_prior("normal(log(50),5", class = "Intercept"),
             set_prior("gamma(0.01,0.01)", class = "shape"))
```

Ecologists would be familiar with the species–area relationship spp = $k$area^z from island biogeography (MacArthur and Wilson 1967), with the partial slope being a small positive number to generate a decelerating power-law curve. We may expect similar relations for the partial slopes of the other explanatory variables. The prior for the partial slopes would be compatible with any slope in [–5; +5] on the log link scale. The intercept (log(k) in the above power-law relationship) has to be a positive value that might have been reported in previous fits of the species–area relationships. Let's imagine that there might be, on average, 50 species on a "typical" island, but we have a large uncertainty about this parameter (hence sd = 5). The final parameter (shape) in the above set is the inverse aggregation parameter of the negative binomial distribution (Eq. (10.2)). We have already reached the preliminary conclusion that the response variable likely had a negative binomial distribution (Fig. 10.2). If so, the parameter ϕ (Eq. (10.2)) should have a small, positive value. Herein lies a possible contradiction. Strict, hardcore Bayesians are not supposed to look at the data while setting up the priors. Yet we must postulate a probability distribution (Poisson or negative binomial) for the likelihood component of the fundamental Bayesian equation (Eq. (3.14)). There are two possibilities: either we examine the response variable to choose the best probability distribution (Fig. 10.2) for the likelihood component, or we make an arbitrary choice, fit the model, evaluate its goodness of fit, and if it is inadequate we return to square one to fit a second statistical model with the other option (assuming there are only two) having already seen the results of our first inadequate choice. We feel that the first option expedites the work involved, even though it may not be "pure enough" for the theoretical canons.

We attempted to calibrate the generative character of our priors using the prior predictive distribution (see Chapter 7) using the code for m1.brms shown below but inserting sample_prior = "only". The resulting distribution was poor (and hence is not shown) due to the very extreme values that the parameter ϕ might assume that were generated in the absence of empirical information from data.

We fitted the Bayesian count GLM model with the original values of the standardized values of the explanatory variables:

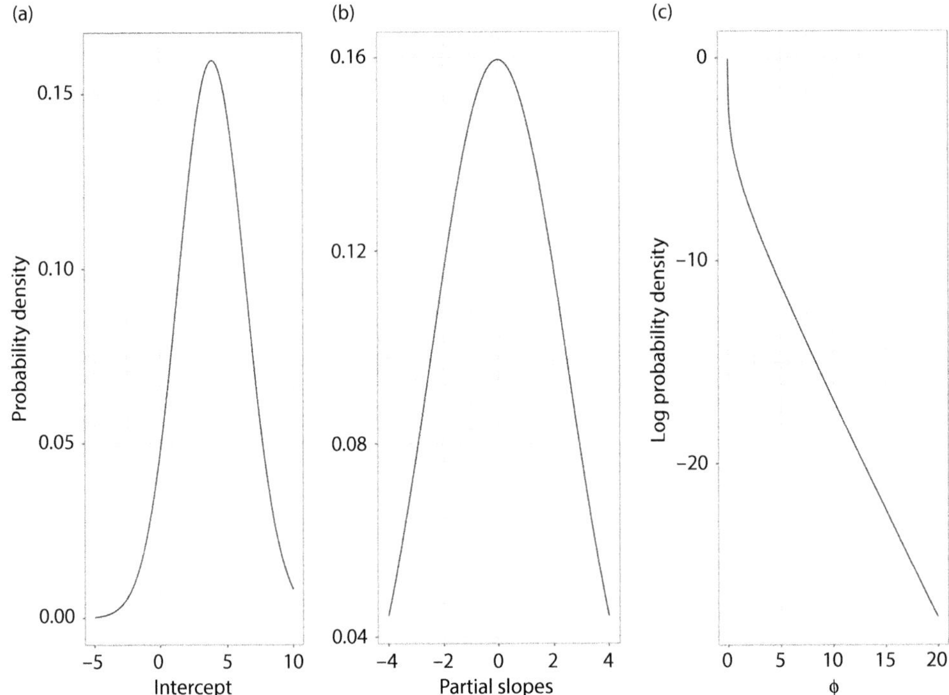

Fig. 10.6 Prior distributions of model parameters for the Bayesian count GLM `m1.brms`. Note that plot (c) shows the log of the probability density.

```
m1.brms=brm(formula=spp~log.area+dist_nearest+dist_StaCruz, data=DFs, family=
    negbinomial, prior = prior.m1, warmup = 1000,chains=3, iter=2000,
future=T, control = list(adapt_delta =0.99))
```

whose (abridged) summary is:

```
> summary(m1.brms)
 Family: negbinomial
  Links: mu = log; shape = identity
Formula: spp ~ log.area + dist_nearest + dist_StaCruz
   Data: DFs (Number of observations: 29)
Samples: 3 chains, each with iter = 2000; warmup = 1000; thin = 1;
         total post-warmup samples = 3000

Population-Level Effects:
             Estimate Est.Error l-95% CI u-95% CI Rhat Bulk_ESS Tail_ESS
Intercept        3.78      0.12     3.55     4.03 1.00     2415     1930
log.area         1.31      0.15     1.02     1.60 1.00     1938     1950
dist_nearest    -0.10      0.17    -0.42     0.24 1.00     1888     1972
dist_StaCruz    -0.14      0.17    -0.46     0.21 1.00     1977     2116

Family Specific Parameters:
      Estimate Est.Error l-95% CI u-95% CI Rhat Bulk_ESS Tail_ESS
shape     2.59      0.80     1.32     4.42 1.00     2269     2148
```

We can readily observe that all the `Rhat` values are close to one, and the effective sample size (ESS) metrics are very large, suggesting the convergence of the three chains to a single posterior stationary distribution. The estimates and the "Est. error" are the means and standard deviations of the marginal posterior distributions of each model parameter in the log link scale. The 95% credible intervals denote the probability that each model parameter lies between its lower and upper limits. We note that only `log.area` had a substantial effect on the expected number of plant species. Because they were standardized prior to the analysis, we can also compare the relative importance of the explanatory variables. We note that `log.area` had an effect one order of magnitude higher than the other variables. The exponential of the intercept (exp(3.78) = 43.87) is the expected number of plant species of the "typical" Galapagos island with average `log.area`, `dist_StaCruz`, and `dist_nearest`. Of course, there is no such thing as an "average" Galapagos island. Exponentiating the partial slopes gives the change in the expected number of plant species per change of a standard deviation in each explanatory variable. Compared with the explanation given in Chapter 4, the only addition is that we need to remember to invert the log link function to interpret the model parameters in the scale of the data, not in the scale of the link function where they were estimated. We can again interpret the partial slopes as a percentage change in the mean of the response variable given by 100(exp(slope) – 1). For `m1.brms`, they would imply 270.6%, –9.51%, and –13.1% changes in the mean `spp` when `log.area`, `dist_nearest`, and `dist_StaCruz` each change by one standard deviation. The value of the shape parameter ϕ was also estimated in the log link scale but its interpretation is of limited interest and will not be pursued here. The summary table for model `m1.brms` is displayed in Fig. 10.7.

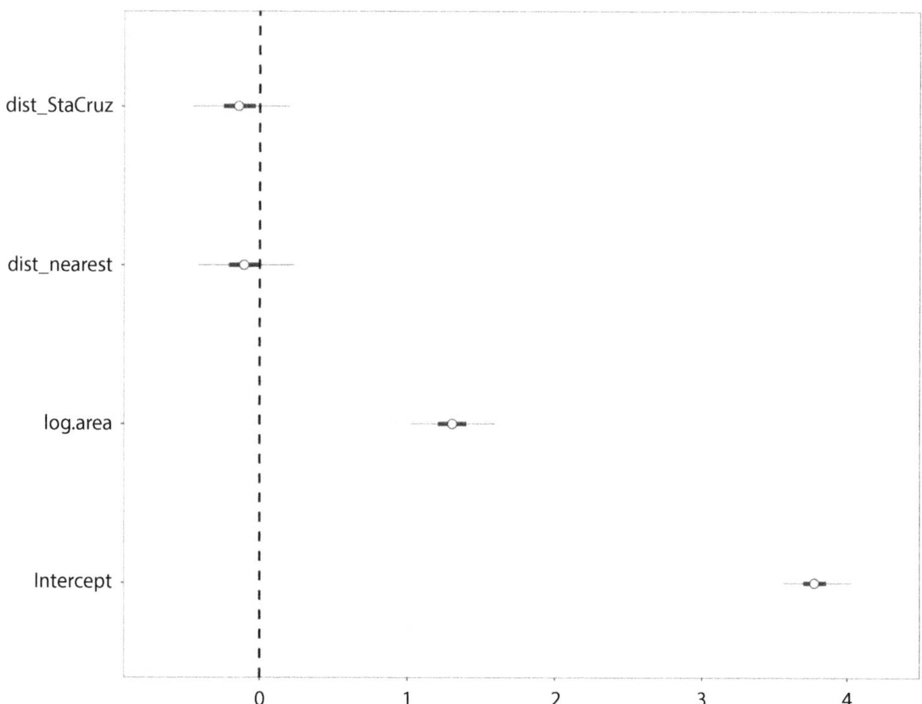

Fig. 10.7 Means and 50% (thick lines) and 95% (thin lines) credible intervals of the parameters of the Bayesian count GLM `m1.brms`.

We now turn to visually assessing the convergence of the chains using the package `bayesplot`, as shown in Fig. 10.8. The sampled parameter values of each chain can be largely considered statistically independent values since their auto-correlations are all smaller than |0.2| (Fig. 10.9). Thus far we can state that Hamiltonian Monte Carlo algorithm has converged to a posterior stationary distribution, and their sampled parameter values can be used to make inferences.

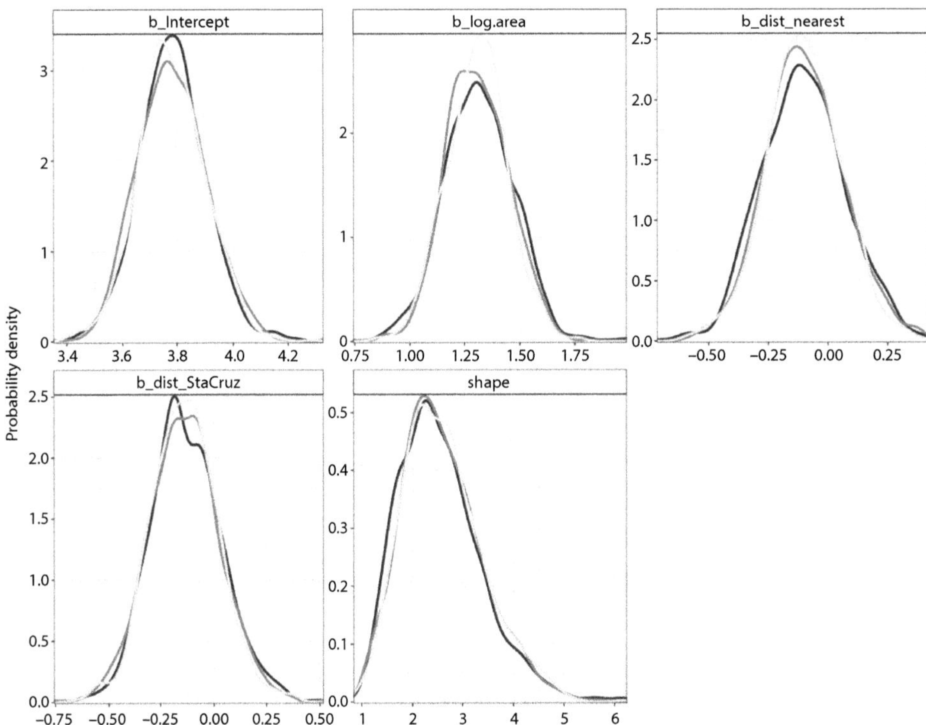

Fig. 10.8 Marginal posterior distributions of each model parameter for the three chains used to fit the Bayesian binary GLM `m1.brms`. The great degree of coincidence of the chains suggests their convergence to a posterior stationary distribution.

Gelman et al.'s (2018) metric shows that the model accounts for a large proportion of the variation of the response variable:

```
> bayes_R2(m1.brms)
   Estimate Est.Error  Q2.5 Q97.5
R2    0.747    0.0562 0.614 0.824
```

The assessment of the goodness of fit of `m1.brms` also requires computing the randomized quantile residuals (Chapter 8). The first step is to obtain its posterior predictive distribution, `post.pred.m1.brms = predict(m1.brms, nsamples = 1e3, summary = F)`, and then use the `DHARMa` package to obtain the residuals:

```
qres.m1.brms=createDHARMa(simulatedResponse  = t(post.pred.m1.brms),
                  observedResponse  = DFs$spp,
             fittedPredictedResponse = apply(post.pred.m1.brms, 2, median),
                  integerResponse  = T)
```

These commands were explained in our first encounter with these residuals in Chapter 9. Finally, we convert the quantile randomized residuals to have a normal distribution with `res.m1.brms = data.frame(res = qnorm(residuals(qres.m1.brms)))`. The final

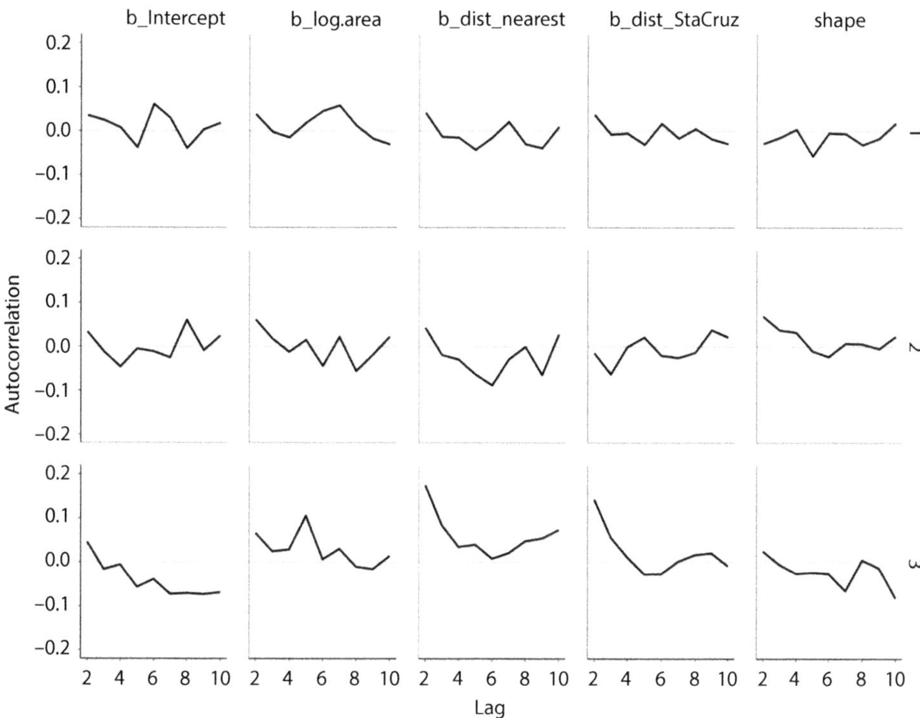

Fig. 10.9 Autocorrelation plots of each model parameter for the three chains used to fit the Bayesian binary GLM `m1.brms`. The very low autocorrelation for all parameters and chains suggest that the sampled values of the posterior distributions can be considered largely statistically independent.

elements for the residual analysis are putting into `res.m1.brms` the mean of the fitted values, the values of the explanatory variables, and Pareto's *k* values for the LOO-CV (Chapter 7), which may be thought of as analogues of the frequentist Cook distances:

```
res.m1.brms=cbind(res.m1.brms,
                  DFs[,c("log.area", "dist_nearest", "dist_StaCruz")],
                  fitted=fitted(m1.brms, nsamples=1000)[,1],
                  pareto=loo(m1.brms, pointwise=T)$diagnostics$pareto_k)
```

Turning to the plots (Fig. 10.10), we see that the `m1.brms` model has a reasonable but not excellent goodness of fit to the data. The points in Fig. 10.10a, d, and e have a sort of triangular shape rather than the random scatter observed in Fig. 10.10b. The most plausible explanation lies in the rather skewed distribution of the non-transformed explanatory variables and of the response variable, rather than a problem with model fitting. You might check this by comparing the residual plots with another model having log-transformed `dist_nearest` and `dist_StaCruz`. The randomized quantile residuals have a normal distribution, and all but two data points have Pareto's *k* values in the "good" range of Vehtari et al. (2017).

We can also calculate the over-dispersion parameter to assess whether there is an excess variation of the response variable not accounted for by `m1.brms`. Over-dispersion is an issue for all GLMs whose probability distributions possess a fixed relation between the mean and variance of the response variable, regardless of the statistical framework

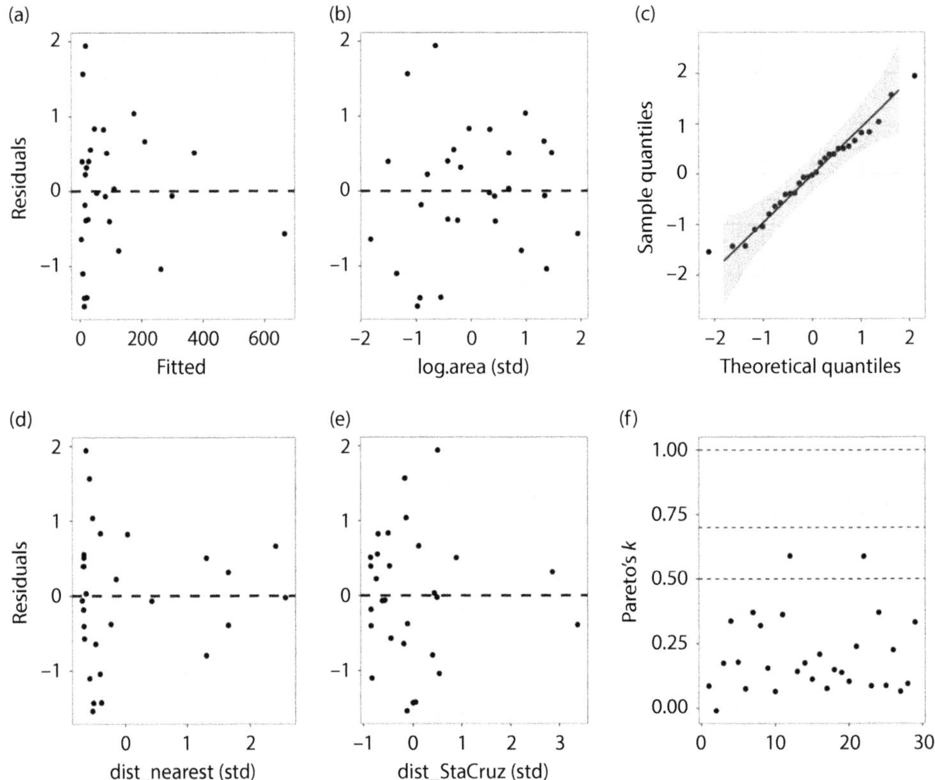

Fig. 10.10 Residual analysis of model `m1.brms` based on randomized quantile residuals converted to normal residuals, showing the mean of the Pearson residuals vs. the mean of the fitted value (a), the quantile–quantile plot of the randomized quantile residuals (c), the mean of the Pearson residuals vs. the explanatory variables in the model (b, d, e), and the Pareto *k* cross-validation statistic for each data point (f).

employed to fit them. We are going to calculate the index of over-dispersion (Chapter 8) as the sum of the squared Pearson residuals of each point divided by a rough equivalent of the degrees of freedom (that do not exist in the Bayesian framework) as follows:

```
Res.Pears.m1.brms=residuals(m1.brms, type='pearson',summary=F)
index.overdispersion=apply(Res.Pears.m1.brms^2,1,sum)/
                        (nrow(DFs)-loo(m1.brms)$estimates [2,1])
```

We first obtain the distribution of Pearson residuals for each of the 29 data points:

```
> dim(Res.Pears.m1.brms)
```

```
[1] 3000   29
```

The 3,000 is because we set 3 chains × ((`iter` = 2000) – (`warmup` = 1000)).

Next, we sum the squared residuals by row (this is what the "`1`" in `apply` means), and then divide the sum by the difference between the number of data points and the effective number of model parameters. Besides the Pareto-*k* statistics used in Fig. 10.10, the output of the leave-one-out cross validation also contains the means and standard errors of the expected log pointwise predictive density (`elpd_loo`), the effective number of parameters (`p_loo`), and the LOO information criterion (`looic`) that is involved in the calculation of the WAIC (Chapter 7):

```
> loo(m1.brms)$estimates
          Estimate    SE
elpd_loo   -137.72  7.80
p_loo         4.35  1.29
looic       275.45 15.61
```

`loo(m1.brms)$estimates[2,1]` simply extracts the 4.35 from this table and subtracts it from the number of data points (`nrow(DFs)`) to finally obtain a distribution of values of the index of over-dispersion. This distribution can be summarized by descriptive statistics as:

```
> summary(index.overdispersion)
   Min. 1st Qu.  Median    Mean 3rd Qu.    Max.
   0.83    0.89    0.93    0.99    1.02    3.72
```

Given that the average and third quantile are close to one, we can conclude that the model `m1.brms` does not have any substantial unaccounted variation of the response variable.

We can perform model selection across the same set of GLM counts fitted in the frequentist framework. Using `m1.brms` as the starting point, we first generate a set of simpler models `m2.brms`, `m3.brms`, and `m4.brms` by eliminating only one explanatory variable:

```
m2.brms=update(m1.brms, ~. -dist_nearest)
m3.brms=update(m1.brms, ~. -dist_StaCruz)
m4.brms=update(m1.brms, ~. -log.area)
```

And then, to each newly fitted model and to `m1.brms` we should add `m4.brms = add_criterion(m4.brms, criterion = "loo")` so that we can now compare them:

```
> print(loo_compare(m1.brms,m2.brms,m3.brms,m4.brms), digits=4)
        elpd_diff se_diff
m2.brms   0.0000   0.0000
m3.brms  -0.0032   1.6497
m1.brms  -0.6974   0.7676
m4.brms -22.2780   4.6749
```

This table gives the estimate and standard error of the pairwise differences of expected log pointwise predictive density between each model and the one with the lowest `elpd_loo` (`m2.brms`). What the table shows is that three models are very similar and that `m4.brms` is the odd one out with a much lower expected log pointwise predictive density. The key difference between `m4.brms` and the other three models is that it does not have `log.area`, the only explanatory variable with important effects. That is, a model without `log.area` had a much lower predictive ability for the average number of plant species in the Galapagos. The most parsimonious model for these data should then only contain `log.area` as a predictor.

We can use the posterior predictive distribution to gauge a model's ability to generate the observed data, or any specific statistic that may be calculated from it, into what are called the posterior predictive checks (Chapter 7). We are going to be progressively explore some of the many options available for posterior predictive checks. A well-fitting model should generate the observed values of the metric used for the posterior predictive checks as a "typical" or "average" realization. Two examples are illustrated in Figure 10.11, which shows that `m1.brms` can generate the overall density curve of observed plant species, and also that all the 95% credible intervals (and 19 out of 29 of the 50% credible intervals) enclosed the observed values of the response variable. It is not that a few lucky simulations from the model happen to match certain features of the observed data, but that the average performance of the model could have generated (and hence explain or account for) the available evidence. Such agreement, together with the goodness-of-fit assessment, should enhance our trust in the reliability of the fitted model.

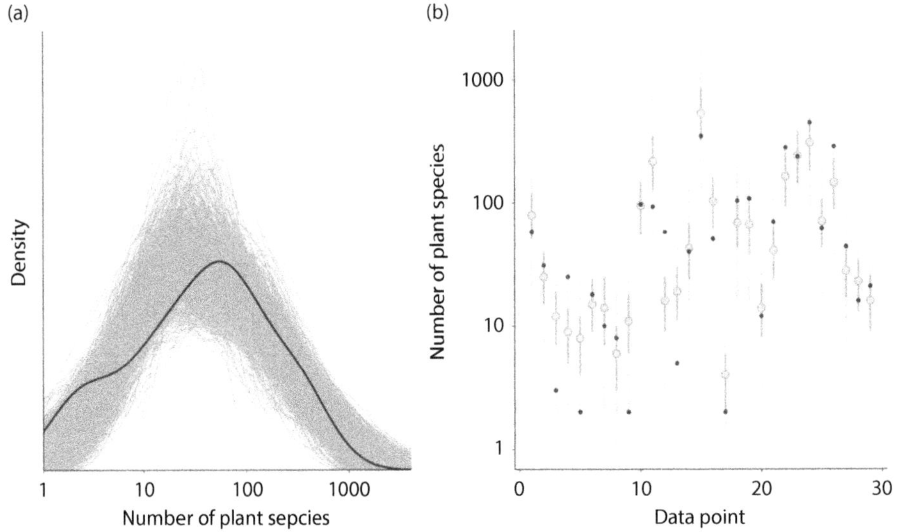

Fig 10.11 Posterior predictive distribution of the Bayesian count GLM `m1.brms`, showing (a) the distribution of predicted (thin lines) and observed (thick line) density curves, and (b) the mean and 95% credible intervals for each value, and the actual observations (dots).

10.4 Modeling of counts with an excess of zeros: Zero-inflated and hurdle models

An "excess" of zeros is a frequent cause of over-dispersion in count GLMs (Table 10.1) using either Poisson or negative binomial distributions. "Excess" here refers to having a larger number of zeros compared to those predicted by a probability distribution for the estimated parameters. For a given mean, the negative binomial predicts more zeros than the Poisson distribution (Fig. 10.1). Hence, using the former distribution may account for or resolve a moderate amount of over-dispersion due to an excess of zeros. However, we often require a more complex approach to modeling the data. Vuong (1989) developed a statistical test to detect an excess of zeros, but Wilson (2015) has cautioned against its use because of the incorrect formulation of the test.

Mixture models are two-part count GLMs involving use of the binomial distribution and either the Poisson or negative binomial distribution to model the response variable. Mixing two probability distributions obviously entails having more parameters overall, and hence a greater flexibility to model the vagaries of the data compared to a simpler count GLM. There are two main classes of mixture models for count GLMs: zero inflated (Lambert 1992) and zero augmented or hurdle models (Mullahy 1986), depending on the presence of "true" and "false" zeros in the counts. False zeros are due to observer error or deficiencies in the experimental or sampling design, and they need to be removed prior to the analysis (Martin et al. 2005). The true zeros are in turn divided into structural and random zeros, with the former being related to structural constraints requiring certain experimental units to always have a zero count, and the latter resulting from sampling variability (Blasco-Moreno et al. 2019).

Zero-inflated (ZI) models (Lambert 1992) postulate that the observed counts result from a mixture of two latent (i.e., non-observable) classes of observations: some are structural zeros for which the response variable will always be zero, with the rest of the counts sometimes giving zero just by chance. The underlying theory of zero-inflated models then divides zero counts between structural and random zeros, and considers that only the former contribute to over-dispersion. In practice, however, it is often difficult to make a clear distinction between the two classes of zero counts. In zero-inflated models the structural zeros are modeled with a binomial distribution, and the random zeros and all non-zero counts with either a Poisson or negative binomial distribution, giving rise to the ZIP and ZINB models (Table 10.2).

Table 10.2 Mixture distributions for count GLMs arising from combinations of zero-inflated/zero-augmented and Poisson / negative binomial distributions. The resulting distributions are shown in terms of their parameters, where μ_Y is the mean of the response variable, π is the probability of absence (i.e., of obtaining a zero), and ϕ is the dispersion parameter of the negative binomial distribution.

	Zero-inflated Poisson (μ_Y, π)	Zero-inflated negative binomial (μ_Y, π, ϕ)
$\Pr(Y = 0)$	$\pi + (1 - \pi)\exp(-\mu_Y)$	$\pi + (1 - \pi)\left(1 - \frac{\phi}{\phi+\mu_Y}\right)^{\phi}$
$\Pr(Y = y \mid Y > 0)$	$\frac{(1-\pi)\exp(-\mu_Y)\mu_Y^y}{y!}$	$(1 - \pi)\frac{\Gamma(y+\phi)}{y!\Gamma(\phi)}\left(1 - \frac{\phi}{\phi+\mu_Y}\right)^{\phi}\left(\frac{\mu_Y}{\phi+\mu_Y}\right)^{y}$
	Zero-augmented Poisson (μ_Y, π)	Zero-augmented negative binomial (μ_Y, π, ϕ)
$\Pr(Y = 0)$	π	π
$\Pr(Y = y \mid Y > 0)$	$\frac{(1-\pi)}{1-\exp(-\mu_Y)}\exp(-\mu_Y)\frac{\mu_Y^{yi}}{y_i!}$	$\frac{\Gamma(y+\phi)}{(y+1)!\Gamma(\phi)}\left(\frac{\phi}{\phi+\mu_y}\right)^{\phi}\left(1 - \frac{\phi}{\phi+\mu_Y}\right)^{\phi}\left(\frac{\mu_Y}{\phi+\mu_Y}\right)^{y}$

Zero-augmented (ZA) models (Mullahy 1986) consider that the zero and non-zero counts are fully observed (rather than latent), and there are separate processes determining the absences and the non-zero counts. Zero-augmented models assume that all zero counts are structural. The zero counts are modeled here only by a binomial distribution, and the non-zero counts by a truncated Poisson or negative binomial distribution, giving rise to the ZAP and ZANB models. ZA model parameters have a cleaner interpretation than those of ZI models. Because both distributions can always yield a zero count, "truncated" here means that the probability distribution used for the non-zero counts has to be redefined so that it still sums to one (Table 10.2). Mixture models are by no means restricted to count GLMs, and most use the same approach as ZA models (see Chapter 12).

Should you use ZI or ZA models in your analysis? It is hard to give a definitive answer in most cases, because of the difficulty of neatly distinguishing between structural and random zeros without having detailed knowledge of the experimental or sampling setup and its execution, and about the fundamental nature of the processes giving rise to the observations. Making a rough assessment, it would seem that ZI models are more popular in biological sciences, and ZA models predominate in social sciences and economics (Cameron and Trevedi 1998). Of course, apparent popularity provides no good reason for making a defensible choice between ZI and ZA models.

The mixture models applied to count GLMs use a logit link function for the zero counts (more exactly, the structural zeros in ZI, and all zeros in ZA), and a log link function for all other values of the response variable. We can incorporate numerical and/or categorical explanatory variables in either or both two-part components of the model, and the fitted parameters are to be interpreted in the scale of the corresponding link function.

10.4.1 *Frequentist fitting of a zero-inflated model*

THE DATA IN CONTEXT: Sofaer et al. (2014) investigated the effect of competition and nest predation on the number of fledglings per nest of the orange-crested warbler, a small passerine nesting in the deciduous shrubs of Santa Catalina island, California, between 2009 and 2013. The data is at the scale of each nest, and it mostly comes from ringed birds.

EXPLORATORY DATA ANALYSIS: We start by importing and summarizing the data:

```
> DF3= read.csv(file="Chapter 11 warblers.csv", header=T)
> summary(DF3)
      Year        NumFledged   BreedingDensity      Precip
 Min.   :2003   Min.   :0.0   Min.   :3.43    Min.   : 9.1
 1st Qu.:2004   1st Qu.:0.0   1st Qu.:4.93    1st Qu.:18.5
 Median :2005   Median :0.0   Median :4.95    Median :20.6
 Mean   :2006   Mean   :1.4   Mean   :5.23    Mean   :30.4
 3rd Qu.:2007   3rd Qu.:3.0   3rd Qu.:5.59    3rd Qu.:39.2
 Max.   :2009   Max.   :6.0   Max.   :6.32    Max.   :62.0
```

Having 52.5% of entries as zeros suggests (but does not assure) that there might be an excess of zeros in the response variable:

```
> table(DF3$NumFledged==0)/nrow(DF3)

FALSE  TRUE
0.475 0.525
```

We next assess the best probability distribution to model the response variable using the command `fitdist` from the package `fitdistrplus` to compare the relative fits to Poisson, negative binomial, ZIP, and ZINB distributions; we must also use the package `gamless.dist` to compare the latter two distributions:

```
poiss=fitdist(DF3$NumFledged,"pois")
negbin=fitdist(DF3$NumFledged,"nbinom")
ZI.Poisson=fitdist(DF3$NumFledged, "ZIP",discrete=T,
             start = list(mu=mean(DF3$NumFledged),
             nu=sum(DF3$NumFledged == 0)/length(DF3$NumFledged)))
ZI.NB=fitdist(DF3$NumFledged, "ZINBI", discrete=T,
              start = list(mu=mean(DF3$NumFledged),
        sigma=mean(DF3$NumFledged)^2 /(var(DF3$NumFledged)-mean(DF3$NumFledged)),
        nu=sum(DF3$NumFledged == 0)/length(DF3$NumFledged)))
```

(note that we must provide starting values for the two or three parameters (μ, ϕ, and π; see below) involved in the two ZI distributions). From this we can produce Fig 10.12.

There seems to be a marginal difference for the ZI Poisson, though it is not very clear (at least to the author's eyes) which is the best-fitting distribution. We can also use the AIC (Chapter 8):

```
> gofstat(list(poiss,negbin,ZI.Poisson,ZI.NB))$aic
  1-mle-pois 2-mle-nbinom    3-mle-ZIP  4-mle-ZINBI
         658          592          524          577
```

Even for AIC's skeptics, the ZI Poisson is the very clear winner here.

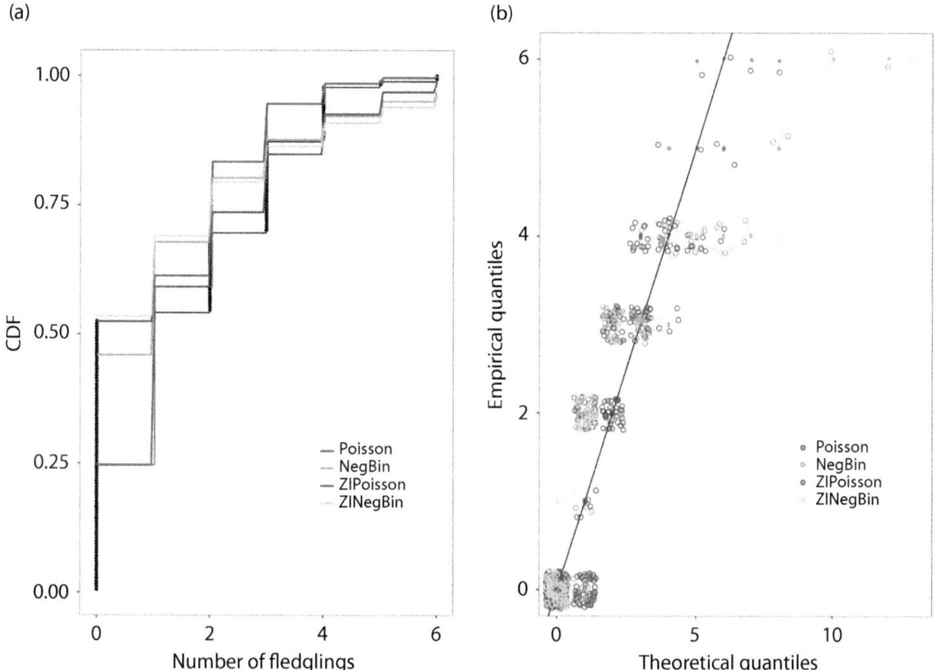

Fig 10.12 Goodness-of-fit assessment of the response variable number of fledglings showing the cumulative distribution function (a) and the quantile–quantile plot comparing the Poisson, negative binomial, zero-inflated Poisson, and zero-inflated negative binomial distributions obtained using the package `fitdistrplus` (b).

In the interests of space, we are not going to show the rather unclear `ggpairs` graph showing the relations between the response and the explanatory variables (the associated code is still given on the website). Suffice it to say that the explanatory variables have a very low correlation (–0.062) and thus can both be included in the analysis. We first standardize them, `DF3s = data.frame(scale(DF3[,c("BreedingDensity", "Precip")], center = T, scale = T), NumFledged = DF3$NumFledged)`, so as to be able to compare their partial slopes.

THE STATISTICAL MODEL:

$$Y \sim \text{ZIP}(\mu_Y, \pi) \text{ or } Y \sim \text{ZINB}(\mu_Y, \pi, \phi),$$
$$\log(\mu_Y) = \text{X}\beta \text{ and } \text{logit}(\pi) = \text{X}\beta,$$

with X being the numerical explanatory variables and β the coefficients (intercept and partial slopes) denoting the relation between them and the mean of the response variable in the scale of the link function used for each part of the mixture model. The magnitudes and signs of the partial slopes for each explanatory variable may, of course, differ between the two parts of the model.

We use the package `glmmTMB` to fit our model:

```
m6=glmmTMB(NumFledged ~ BreedingDensity+ Precip, family=poisson,
            ziformula=~Precip+BreedingDensity,data=DF3s)
```

Note while that we are including the same explanatory variables for the ZI part as for the count part; we could have just estimated the ZI coefficient by setting `zi ~ 1`. The summary is:

```
> summary(m6)
 Family: poisson  ( log )
Formula:          NumFledged ~ BreedingDensity + Precip
Zero inflation:             ~Precip + BreedingDensity
Data: DF3s

     AIC      BIC   logLik deviance df.resid
     462      482     -225      450      175


Conditional model:
                Estimate Std. Error z value Pr(>|z|)
(Intercept)       0.9051     0.0958    9.45   <2e-16 ***
BreedingDensity  -0.1318     0.0670   -1.97    0.049 *
Precip            0.1032     0.0808    1.28    0.201
---
Signif. codes:  0 '***' 0.001 '**' 0.01 '*' 0.05 '.' 0.1 '

Zero-inflation model:
                Estimate Std. Error z value Pr(>|z|)
(Intercept)      -0.0064     0.2104   -0.03     0.98
Precip           -1.0546     0.2185   -4.83  1.4e-06 ***
BreedingDensity   1.1722     0.2953    3.97  7.2e-05 ***
```

We have a two-part summary, one for the counts and the other for the excess of zeros, each with a different link function (log and logit, respectively). `BreedingDensity` (related to the strength of intraspecific competition) has relevant negative effects for both the number of fledglings per nest and for breeding failures. `Precip` was only strongly related to breeding failures. These explanatory variables had effects of similar magnitudes in each of the two parts of model `m6`. This model had a sizable over-dispersion (450 / 175 = 2.57), suggesting that there is a fair amount of variation of the response data that was not adequately modeled. We tried a variant of `m6` by changing `poisson` to `nbinom2` (i.e., a zero-inflated negative binomial) in the model specification, and obtained a rather similar degree of over-dispersion (results not shown). Therefore, `m6` had an over-dispersion over and above the excess of zeros. The model would require adding other explanatory variables, if we had them, or, if adequate according to the sampling design, including a random effect (see Chapter 13) as suggested by Harrison (2014). Can the current results be trusted? Tentatively yes, but with a fair amount of skepticism.

We can extract the estimated parameters and 95% confidence intervals of `m6` using the package `broom.mixed` (required for `glmm.TMB`) as `m6.res = tidy(m6, conf.int = 0.95)`, and the resulting data frame is:

```
> m6.res
# A tibble: 6 x 9
  effect component term             estimate std.error statistic  p.value conf.low conf.high
  <chr>  <chr>     <chr>               <dbl>     <dbl>     <dbl>    <dbl>    <dbl>     <dbl>
1 fixed  cond      (Intercept)         0.905    0.0958     9.45  3.30e-21   0.717   1.09
2 fixed  cond      BreedingDensity    -0.132    0.0670    -1.97  4.92e- 2  -0.263  -0.000467
3 fixed  cond      Precip              0.103    0.0808     1.28  2.01e- 1  -0.0551  0.262
4 fixed  zi        (Intercept)        -0.00640  0.210     -0.0304 9.76e- 1 -0.419   0.406
5 fixed  zi        Precip             -1.05     0.219     -4.83  1.39e- 6  -1.48   -0.626
6 fixed  zi        BreedingDensity     1.17     0.295      3.97  7.19e- 5   0.594   1.75
```

which serves as the input for Fig. 10.13.

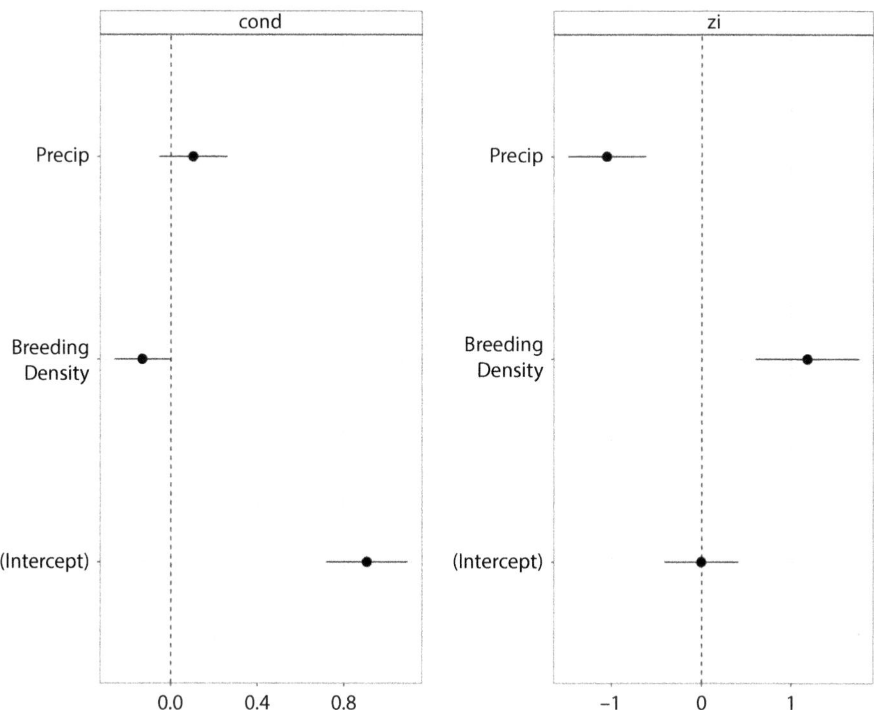

Fig 10.13 Estimates and 95% confidence intervals of the parameters of the frequentist zero-inflated model `m6` for the conditional and zero-inflated parts of the mixture model.

We now use the package `ggeffects` to first produce the marginal plots based on the fitted relations for each explanatory variable:

```
pred1.m6=ggpredict(m6, terms = c("BreedingDensity [all]"))
pred2.m6=ggpredict(m6, terms = c("Precip [all]"))
```

and then to produce Fig. 10.14.

We can see that the 95% confidence bands of the marginal plots are very wide, suggesting that `m6` makes very imprecise predictions. Accordingly, McFadden's (1974) percentage of deviance explained is also low:

```
m6.null=update(m6,~1, ziformula=~1) # fits a model with no explanatory variables
1-(as.numeric(-2*logLik(m6))/as.numeric(-2*logLik(m6.null)))
```

```
[1] 0.134.
```

This is admittedly a bit convoluted, but it does render the desired metric.

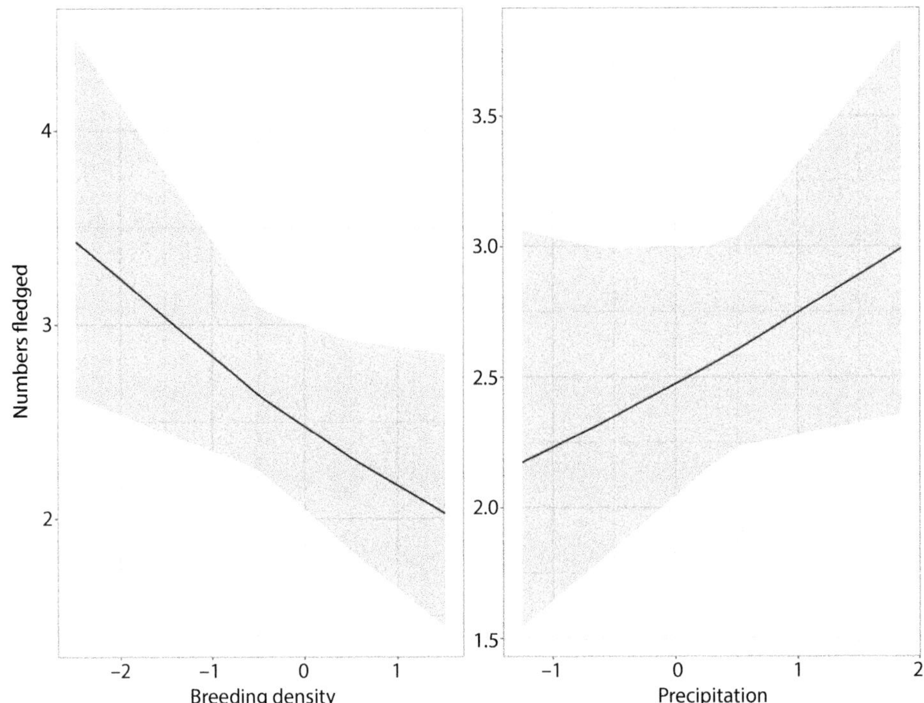

Fig 10.14 Marginal plots of the frequentist zero-inflated model `m6` showing the predicted relations between the standardized explanatory variables (mean of zero and unit standard deviation) and the number of fledglings. The gray bands in all plots correspond to the 95% credible intervals around each predicted relation. Both plots show the predicted relation for each variable, holding all other variables at zero (i.e., their mean values in the original scale).

Finally, we use the package `DHARMa` to obtain the randomized quantile residuals (Chapter 7) needed to assess the goodness of fit of model `m6`: `res.m6 = simulateResiduals(fittedModel = m6, n = 500, refit = T, plot = F)`. Next, we gather all the information required for the residual analysis with `resid.m6 = augment(m6)`, and then convert these residuals to have a normal distribution and store them in `resid.m6`: `resid.m6$.std.resid = residuals(res.m6, quantileFunction = qnorm)`. Figure 10.15 shows the plots used in the residual analysis. We can see that `m6` fitted the data well: the residuals were randomly distributed and without any discernible trends, and they have a normal distribution, all signs of a well-fitting model. The only problem with `m6` was its above-mentioned over-dispersion, unrelated to the excess of zeros, that requires substantial changes in the model.

The ZIP model could be fitted in the Bayesian framework with the `brms` package:

```
m6=brm(bf(formula= NumFledged ~ BreedingDensity + Precip), data=DF3s,
      family='zero_inflated_poisson',    warmup=1000,chains=3,    iter=2000,
      future=T,control = list(adapt_delta =0.99))
```

after setting appropriate priors. Visualizing the model output, assessing the convergence of the Hamiltonian Monte Carlo algorithm, obtaining the marginal plots for each explanatory variable, and evaluating the model goodness of fit can all be accomplish with minor adaptations (i.e., change the names of variables

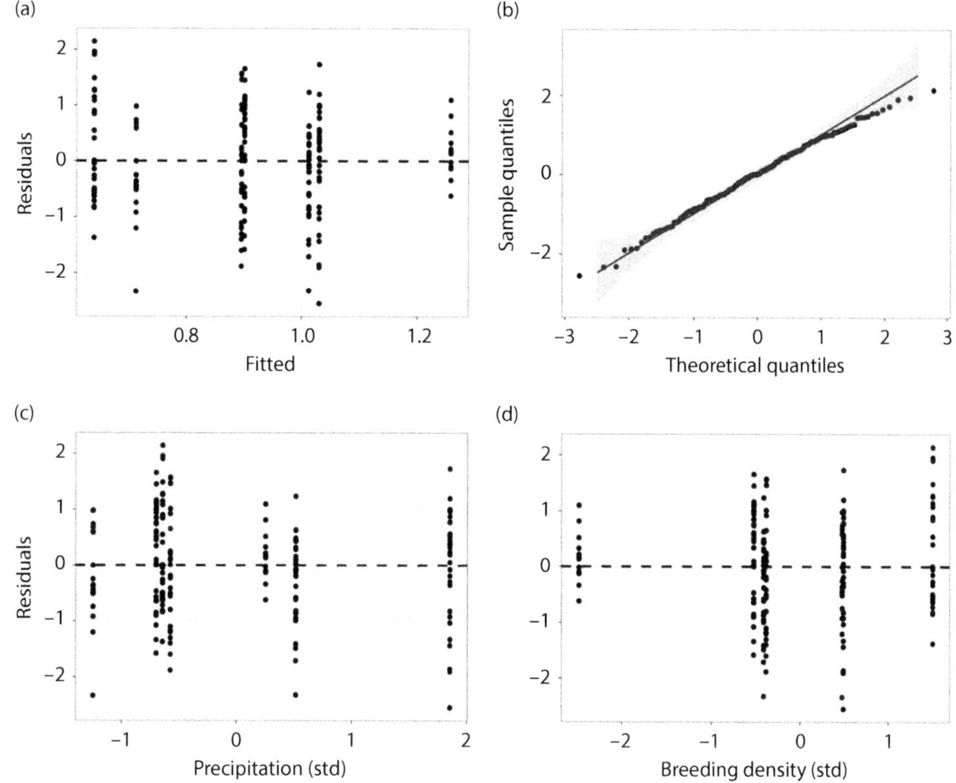

Fig. 10.15 Residual analysis of the frequentist zero-inflated model m6 based on randomized quantile residuals converted to normal residuals, showing the mean of the residuals vs. the mean of the fitted value (a), the quantile–quantile plot of the residuals (b), and the mean of the residuals vs. the explanatory variables in the model (c and d).

and a few other glitches) of the code employed for the Bayesian count GLM (Section 10.3.2).

10.4.2 *Bayesian fitting of a zero-augmented model*

THE DATA IN CONTEXT: Yuen and Dudgeon (2015) investigated proximate factors determining the local abundance of spiders of the genus *Draconarius* (there are at least 260 species in East Asia) in pristine riparian systems in Hong Kong. Spiders were sampled by hoovering them up in 70 × 70 cm^2 quadrats. Accordingly, all zeros may be considered as structural. It is this specific detail of the data collection methodology that underlies the potential choice of a ZA model in this case. This is a rarity since we more often lack a principled argument to choose between ZI and ZA models. The explanatory variables measured at the scale of each quadrat were the terrestrial arthropod abundance and dry weight, the total dry weight of Blattodea (the taxonomic order of cockroaches and mites) and Orthoptera, the percentage of soil moisture, the leaf litter dry weight in a log scale (LogLeaf_dw), and the two seasons (dry, wet) that the authors wished to compare.

THE STATISTICAL MODEL:

$$Y \sim \text{ZAP}(\mu_Y, \pi) \text{ or } Y \sim \text{ZANB}(\mu_Y, \pi, \phi)$$
$$\log(\mu_Y) = \text{X}\beta \text{ and } \text{logit}(\pi) = \text{X}\beta$$

(the meaning of X, β, and the two link functions are explained in Section 10.4.1).

EXPLORATORY DATA ANALYSIS: As usual, we start by importing and summarizing the data:

```
> DF4= read.csv(file="Chapter 11 Spiders HongKong.csv", header=T)
> summary(DF4)
 Season      Site      Draconarius     Terr.Arth.abund  Blattodea_dw     Orthoptera_dw       Terr.Arth.dw      SoilMoisture       LogLeaf_dw
 dry:136   A:135   Min.   : 0.00   Min.   :  0    Min.   :0.000   Min.   :0.0000   Min.   :0.000   Min.   :0.000   Min.   :0.307
 wet:138   B:139   1st Qu.: 0.00   1st Qu.:  5    1st Qu.:0.000   1st Qu.:0.0000   1st Qu.:0.012   1st Qu.:0.090   1st Qu.:0.972
                   Median : 0.00   Median : 10    Median :0.000   Median :0.0006   Median :0.032   Median :0.170   Median :1.566
                   Mean   : 0.88   Mean   : 19    Mean   :0.017   Mean   :0.0046   Mean   :0.059   Mean   :0.207   Mean   :1.491
                   3rd Qu.: 1.00   3rd Qu.: 21    3rd Qu.:0.006   3rd Qu.:0.0044   3rd Qu.:0.069   3rd Qu.:0.290   3rd Qu.:1.950
                   Max.   :12.00   Max.   :368    Max.   :0.380   Max.   :0.0954   Max.   :0.619   Max.   :0.840   Max.   :2.875
```

The median of the response variable (`Draconarius`) being zero suggests that there are at least 50% of zeros (it is actually 69%), which would suggest a ZA model.

We next evaluate four candidate distributions (Poisson, negative binomial, and their ZA variants) to model the response variable, again using the package `fitdistrplus` aided by the package `gamless.dist` for the two ZA distributions:

```
poiss2=fitdist(DF4$Draconarius,"pois")
negbin2=fitdist(DF4$Draconarius,"nbinom")
ZA.Poisson=fitdist(DF4$Draconarius, "ZAP",discrete=T,
                   start = list(mu=mean(DF4$Draconarius),
                   nu=sum(DF4$Draconarius == 0)/length(DF4$Draconarius)))
ZA.NB=fitdist(DF4$Draconarius, "ZANBI", discrete=T,
               start = list(mu=mean(DF4$Draconarius),
    sigma=mean(DF4$Draconarius)^2/(var(DF4$Draconarius)-mean(DF4$Draconarius)),
    nu=sum(DF4$Draconarius == 0)/length(DF4$Draconarius)))
```

Figure 10.16 would suggest that the ZA negative binomial may be the best option to model the response variable.

Examining the relations between the response and explanatory variables in this data set with so many zeros is best done by selecting the 31% of the data frame where `Draconarius` was present to avoid the relations being overwhelmed by the absences. The bottom row of Fig. 10.17 shows that `Draconarius` and the explanatory variables would be weakly related, and that the effect of `Terr.Arth.dw` would differ between seasons, which may suggest a possible interaction between these variables. We can also see that the numerical explanatory variables are weakly correlated, both overall and for each season, thus suggesting that they should all be incorporated into our model.

As always, we standardized the numerical explanatory variables prior to the analysis to compare the relative magnitudes of their coefficients in the two link scales of the model:

```
DF4s=data.frame(scale(DF4[,c("Orthoptera_dw","Terr.Arth.dw","SoilMoisture",
"LogLeaf_dw")],center=T,scale=T),Draconarius=DF4$Draconarius, Season=DF4=Season)
```

Let's look at the priors that need to be specified:

```
> get_prior(formula=Draconarius~Orthoptera_dw+Terr.Arth.dw*Season+SoilMoisture+LogLeaf_dw,
+   data=DF4s,family='hurdle_negbinomial')
                   prior     class                    coef group resp dpar nlpar bound       source
                  (flat)         b                                                          default
                  (flat)         b              LogLeaf_dw                             (vectorized)
                  (flat)         b           Orthoptera_dw                             (vectorized)
                  (flat)         b               Seasonwet                             (vectorized)
                  (flat)         b            SoilMoisture                             (vectorized)
                  (flat)         b            Terr.Arth.dw                             (vectorized)
                  (flat)         b  Terr.Arth.dw:Seasonwet                             (vectorized)
              beta(1, 1)        hu                                                          default
 student_t(3, -2.3, 2.5) Intercept                                                          default
       gamma(0.01, 0.01)     shape                                                          default
```

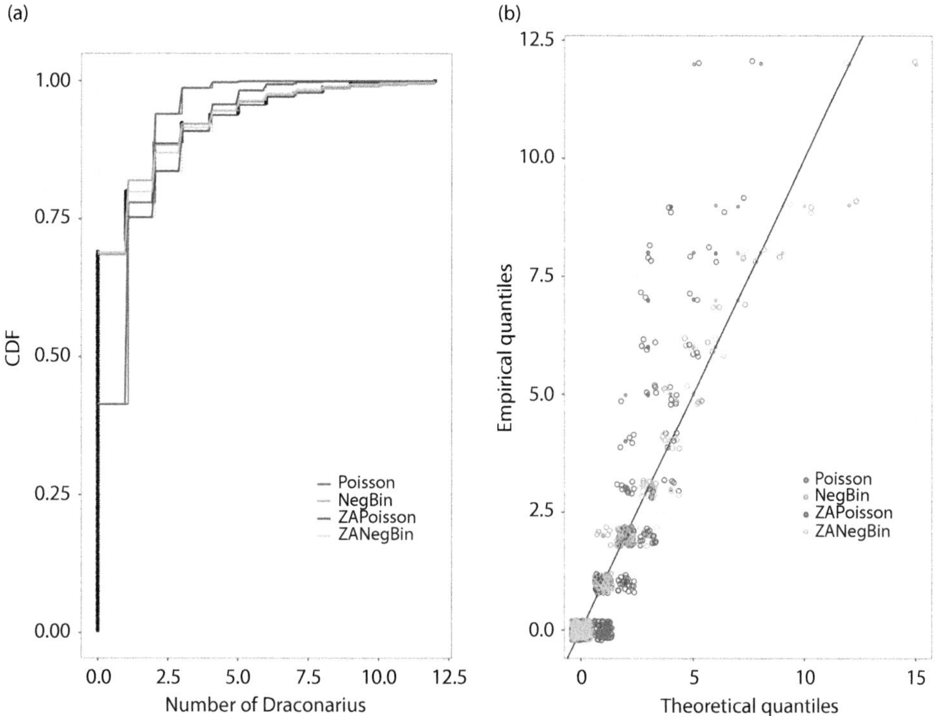

Fig. 10.16 Goodness-of-fit assessment of the response variable number of *Draconarius spp* spiders showing the cumulative distribution function (a) and the quantile–quantile plot comparing the Poisson, negative binomial, zero-augmented Poisson, and zero-augmented negative binomial distributions obtained using the package `fitdistrplus` (b).

The parameters of class `hu` and `shape` are the probability of zero counts and the parameter ϕ of the negative binomial distribution (Eq. (10.2)), respectively. The other parameters are the intercept (here for `season` = wet, taken as the "group of reference") and the partial slopes for the explanatory variables not involved in interactions, while `Seasonwet` and `Terr.Arth.dw:Seasonwet` are the differences in intercept and slopes for the wet season and these parameters and those of the group of reference. This is exactly the same approach as used for the analysis of covariance (Chapter 6).

We then ran the model:

```
m7=brm(bf(formula=Draconarius~Orthoptera_dw+Terr.Arth.dw*Site+SoilMoisture+
     LogLeaf_dw),family='hurdle_negbinomial', data=DF4s, warmup=1000,chains=3,
  iter=2000, future=T, control = list(adapt_delta =0.99))
```

using the default uninformative `brms` priors, and obtained (results not shown) warning messages of poor convergence and very low (< 400) effective sample size for the `shape` and `hu` parameters. We then changed the default priors of these two parameters to make them somewhat more informative than the default priors (Fig. 10.18):

```
prior.m7 = c(set_prior("beta(2,2)", class = "hu"),
             set_prior("gamma(1,1)", class = "shape"))
```

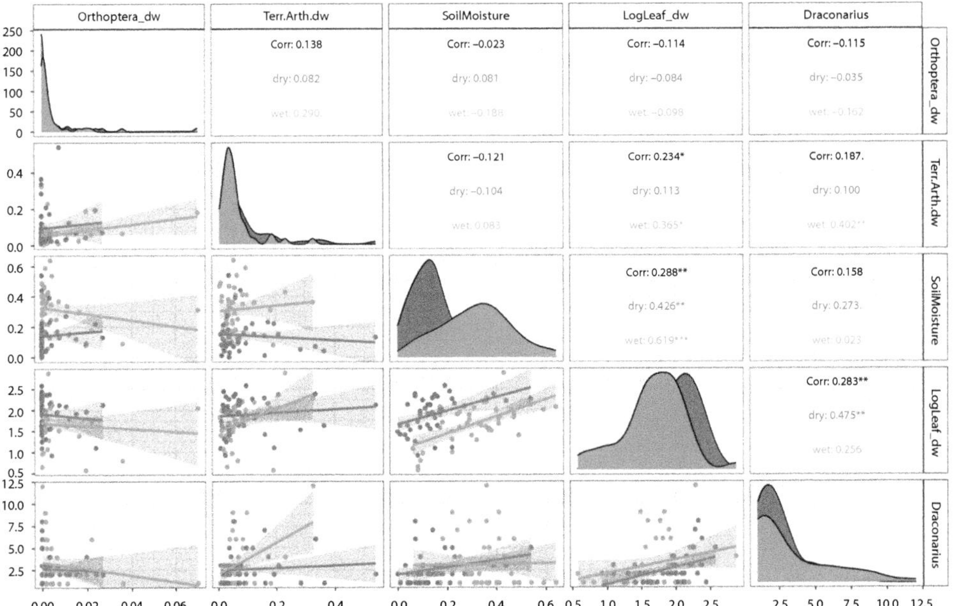

Fig. 10.17 Covariation between the explanatory variables in the Bayesian zero-augmented GLM `m7`. The diagonal shows the density plots of each explanatory variable for the entire data set and depending on the value of the response variable. Below the diagonal are shown all pairwise *x*–*y* plots, with presences colored in red and absences in green. Above the diagonal are the pairwise correlation of explanatory variables for the entire data set and for each value of the response variable. The figure was produced with the package `GGally`.

and reran the same model:

```
m7=brm(bf(formula=Draconarius~Orthoptera_dw+Terr.Arth.dw*Season+SoilMoisture+
LogLeaf_dw), prior = prior.m7, data=DF4s, family='hurdle_negbinomial',
chains=3, warmup=1000, iter=2000, control = list(adapt_delta =0.99,
max_treedepth = 15))
```

whose summary is:

```
> summary(m7)
 Family: hurdle_negbinomial
  Links: mu = log; shape = identity; hu = identity
Formula: Draconarius ~ Orthoptera_dw + Terr.Arth.dw * Season + SoilMoisture + LogLeaf_dw
   Data: DF4s (Number of observations: 274)
Samples: 3 chains, each with iter = 2000; warmup = 1000; thin = 1;
         total post-warmup samples = 3000

Population-Level Effects:
                      Estimate Est.Error l-95% CI u-95% CI Rhat Bulk_ESS Tail_ESS
Intercept                -0.47      0.63    -1.96     0.44 1.00      964      706
Orthoptera_dw            -0.20      0.22    -0.63     0.25 1.00     1957     1664
Terr.Arth.dw              0.02      0.20    -0.35     0.44 1.00     2082     1705
Seasonwet                 0.69      0.47    -0.21     1.66 1.00     1409     1249
SoilMoisture             -0.08      0.25    -0.55     0.42 1.00     1420     1641
LogLeaf_dw                0.63      0.29     0.09     1.25 1.00     1451     1637
Terr.Arth.dw:Seasonwet    0.41      0.42    -0.29     1.35 1.00     1663     1261

Family Specific Parameters:
      Estimate Est.Error l-95% CI u-95% CI Rhat Bulk_ESS Tail_ESS
shape     0.79      0.54     0.08     2.05 1.00      959      849
hu        0.69      0.03     0.63     0.74 1.00     1959     1638
```

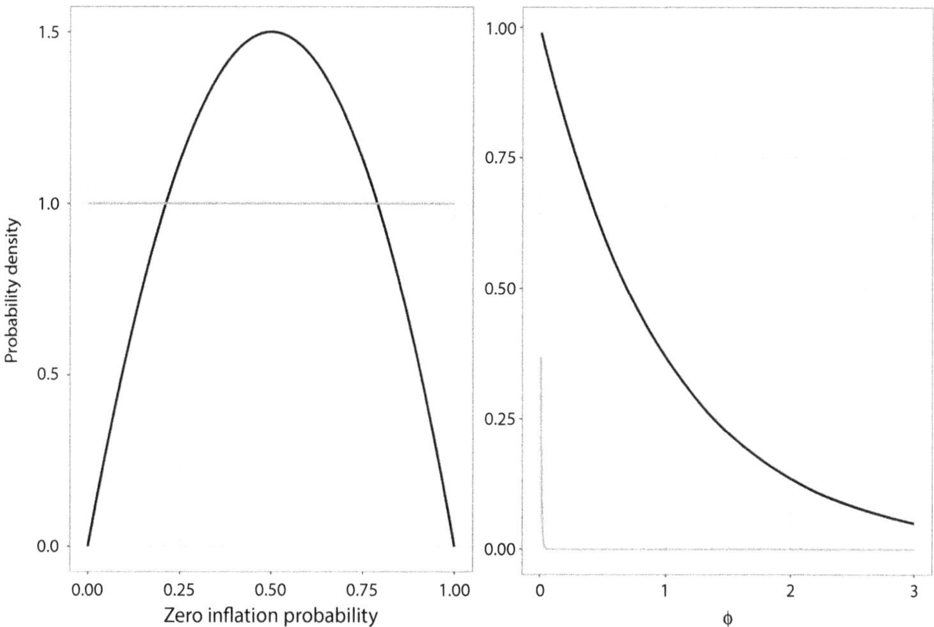

Fig. 10.18 Probability distributions of weakly (gray) and more informative (black) priors for the zero-inflation probability (left) and the dispersion parameter of the Bayesian zero-augmented GLM `m7`.

All values of Rhat were equal to one, suggesting the convergence of the three chains to a stationary posterior distribution, and the ESS values were very large, perhaps with the exception of the intercept and the shape parameter. In the interests of space, we are not going to evaluate these features here, but rest assured that they range from acceptably good to very good; see the code on the companion website.

The intercept is the value of log(E(`Draconarious`)) for the average values of the explanatory variables. The means of the posterior distributions of `Seasonwet` and `Terr.Arth.dw:Seasonwet` were both relatively large (compared to those of the other parameters), but their 95% credible intervals were so wide and consistent with both negative and positive differences that we should conclude that the interaction between these explanatory variables was not very relevant to explaining the data, and that the differences in log(E(`Draconarious`)) between seasons were rather modest (Fig. 10.18). Focusing now on the other numerical explanatory variables, we can see that the effect of `LogLeaf_dw` on log(E(`Draconarious`)) was (0.63 / 0.02) = 31.5, (0.63 / 0.08) = 7.88, and (0.63 / 0.20) = 3.15 times more important than those of `Terr.Arth_dw`, `SoilMoisture`, and `Orthoptera_dw`, respectively. However, the 95% credible interval for the partial slope of `LogLeaf_dw` was very wide, and consistent with both small and large positive effects on the response variable. The average probability of absence can be estimated as `invlogit(0.69)` = 0.66.

The percentage of variation explained by the model (Gelman et al. 2018) was unsurprisingly rather modest:

```
> bayes_R2(m7)
   Estimate Est.Error   Q2.5 Q97.5
R2    0.161     0.154 0.0153 0.528
```

Because model `m7` had rather limited explanatory power, there is little interest in showing the marginal plots depicting the fitted effects of each explanatory variable as we did for the other models.

We end by assessing of the goodness of fit of `m7` using randomized quantile residuals. The procedure is the same as that used in Section 10.3.2, just changing the name of the model (`m1.brms` to `m7`) and the names of the explanatory variables. We first obtain the posterior predictive distribution, `post.pred.m7 = predict(m7, nsamples = 1e3, summary = F)`, and then use the package `DHARMa` to obtain the residuals:

```
qres.m7=createDHARMa(simulatedResponse = t(post.pred.m7),
            observedResponse = DF4s$Draconarius,
            fittedPredictedResponse = apply(post.pred.m7, 2, median),
            integerResponse = T)
```

We then convert the quantile randomized residuals to have a normal distribution with `res.m7 = data.frame(res = qnorm(residuals(qres.m7)))`. The final elements for the residual analysis involve putting into `res.m7` the mean of the fitted values, the values of the explanatory variables, and Pareto's k values for the LOO-CV:

```
res.m7=cbind(res.m7,
        DF4s[,c("Orthoptera_dw", "Terr.Arth.dw", "SoilMoisture", "LogLeaf_dw")],
        fitted=fitted(m7, nsamples=1000)[,1],
        pareto=loo(m7, pointwise=T)$diagnostics$pareto_k)
```

We see in Figs. 10.19 and 10.20 that the `m7` model has a reasonable but not excellent goodness of fit. We warn you not to over-interpret the right-hand side of Figs. 10.19A, B and D that have very few data points; the left half of each plot does show a reasonably random scatter of points. The randomized quantile residuals have a normal distribution, and all but three data points have Pareto k values in the "good" range of Vehtari et al. (2017).

Model `m7` does, however, have over-dispersion, which we calculate using the Pearson statistic as in Section 10.3.2, just changing the name of the model. The summary statistics of the over-dispersion index are:

```
> summary(index.overdispersion.m7)
   Min. 1st Qu.  Median    Mean 3rd Qu.    Max.
   1.33    1.40    1.42    2.16    1.47 1121.45
```

It is then clear that the there is much larger variation in the data than accounted for by `m7`, since 98.1% of the distribution of the over-dispersion is in [1.3, 2.0]. Further, the over-dispersion is unrelated to the excess of zeros because the latter are explicitly modeled. The only simple options we have to improve the quality of fit would be to include other explanatory variables (`Site` and `Blattodea_dw`, for instance). Or we could just have used a negative binomial distribution instead of ZANB (left as an exercise for the reader; then compare the two models with `loo_compare` as in Section 10.3.2). The point here is not just to show you "success stories" of data fitting, but to illustrate a reasoned process of building, validating, and interpreting statistical models.

In closing, we could have fitted the same zero-augmented negative binomial model in the frequentist framework using the package `glmmTMB`:

```
glmmTMB(Draconarius~ Orthoptera_dw + Terr.Arth.dw*Site +SoilMoisture+LogLeaf_dw,
        zi= 1,family="truncated_nbinom2", data=DF4s)
```

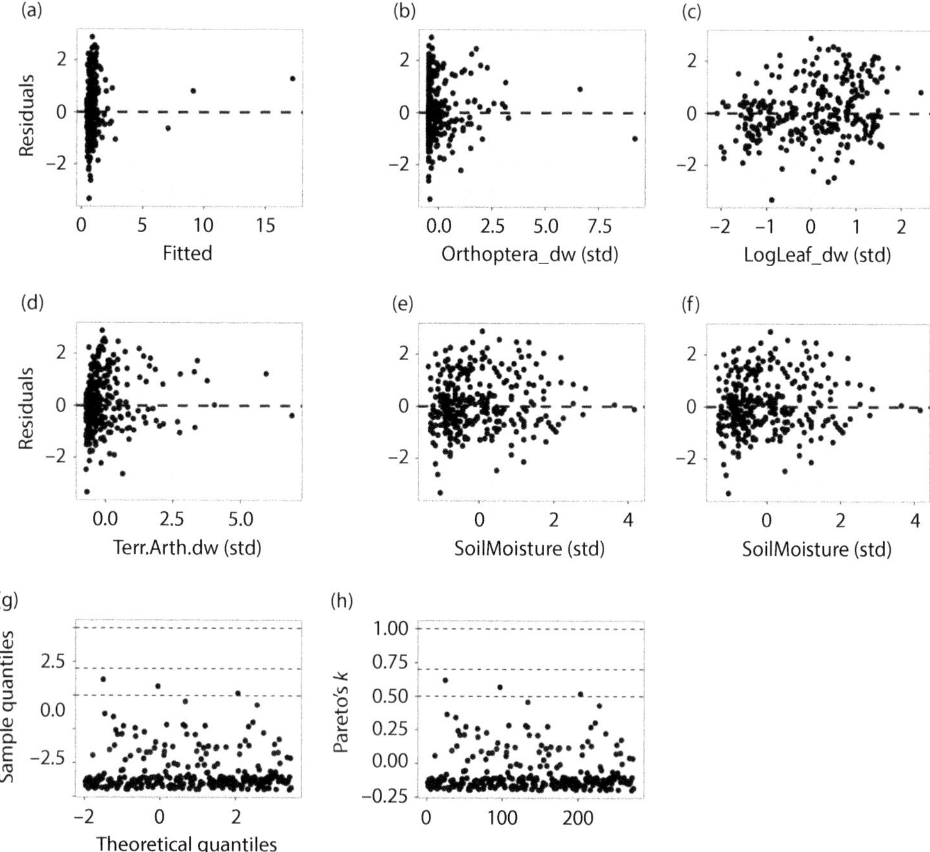

Fig. 10.19 Residual analysis of the Bayesian zero-augmented GLM `m7` based on randomized quantile residuals converted to normal residuals, showing the relation between these residuals and the fitted values (a), and to the standardized numerical explanatory variables (b–f), the quantile–quantile plot (and the 95% confidence band) of the residuals showing that most of these residuals follow a normal distribution (g), and the Pareto *k* cross-validation statistic for each data point (h). The dashed lines were added to help the visualization. All plots exhibit the random scatter expected from a well-fitting model.

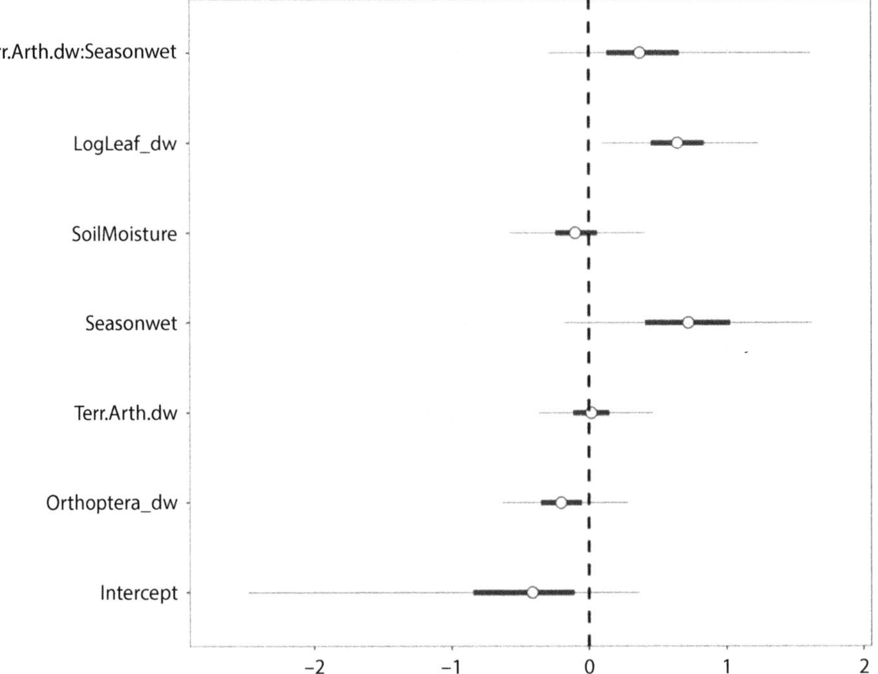

Fig. 10.20 Means and 50% (thick lines) and 95% (thin lines) credible intervals of the parameters of the Bayesian zero-augmented GLM `m7`.

10.5 Problems

Visit the companion website at https://www.oup.com/companion/InchaustiSMWR to obtain the data sets for these problems.

10.1 Flaherty et al. (2012) studied the habitat use of the red squirrel in forests of northwest England in relation to the structure of forest stands. Their response variable was the number of cones removed (SqCones_removed) over a fixed time period and sampling area, and their explanatory variables were the number of trees in the sampling plot (Ntrees), their average diameter at breast height (DBH; cm), TreeHeight (m), and the canopy coverage (CanopyCover; %). File: Pr10-1.csv

10.2 Espinoza et al. (2014) studied the patterns of abundance of the tiger shark using baited remote underwater video stations along the entire Great Barrier Reef Marine Park over a ten-year period in northeastern Australia. They used the hard coral cover at each sampling site (HardCoral; %), the distance to the nearest reef feature (DistReef; km) and a zoning categorical variable (Zoning; referring to pre-/post-2004 re-zoning of areas in the park opened and closed to fishing). The authors standardized the sampling effort by summing the total hours of video (soak_time) for each site. File: Pr10-2.csv.

References

Agresti, A. (2015). *Foundations of Linear and Generalized Linear Models*. Academic Press, New York.

Blasco-Moreno, A. Pérez-Casany, M., Puig, P. et al. (2019). What does a zero mean? Understanding false, random and structural zeros in ecology. *Methods in Ecology and Evolution*, 10, 949–959.

Cameron, A. and Trevedi, P. (1998). *Regression Analysis of Count Data*. Cambridge University Press, Cambridge.

Dormann, C., Elith, J., Bacher, S., et al. (2013). Collinearity: A review of methods to deal with it and a simulation study evaluating their performance. *Ecography*, 36, 27–46.

Dunn, P. and Smyth, G. (2016). *Generalized Linear Models With Examples in R*. Springer, New York.

Espinoza, M., Cappo, M., Heupel., M. et al. (2014). Quantifying shark distribution patterns and species–habitat associations: Implications of marine park zoning. *PLOS One*, 9, e106885.

Faraway, J. (2016). *Extending the Linear Model with R: Generalized Linear, Mixed Effects and Nonparametric Regression Models*. CRC Press / Chapman and Hall, New York.

Flaherty, S., Patenaude, G., Close, A. et al. (2012). The impact of forest stand structure on red squirrel habitat use. *Forestry*, 85, 437–444.

Fox, J. (2015). *Applied Regression Analysis and Generalized Linear Models*, 3rd edn. Sage, New York.

Gelman, A., Goodrich, B., Gabry, J. et al. (2018). R-squared for Bayesian regression models. *American Statistician*, 73, 307–309.

Harrison, X. (2014). Using observation-level random effects to model overdispersion in count data in ecology and evolution. *PeerJ*, 3, peerj.616.

Hilbe, J. (2011). *Negative Binomial Regression*, 2nd edn. Cambridge University Press, Cambridge.

Hilbe, J. (2014). *Modeling Count Data*. Cambridge University Press, Cambridge.

Huang, A. and Kim, A. (2019). Bayesian Conway–Maxwell–Poisson regression models for overdispersed and underdispersed counts. *Communications in Statistics Theory and Methods*, 48, 1–12.

Johnson, K. and Raven, P. (1973). Species number and endemism: The Galapagos archipelago revisited. *Science*, 179, 893–895.

Krebs, C. (1999). *Ecological Methodology*, 2nd edn. Addison-Wesley, New York.

Lambert, D. (1992). Zero-inflated Poisson regression, with applications to defects in manufacturing. *Technometrics*, 34, 1–14.

Linden, A. and Mantyniemi, S. (2011). Using the negative binomial distribution to model overdispersion in ecological count data. *Ecology*, 92, 1414–1421.

Lynch, H., Thorson, J., and Shelton, A. (2014). Dealing with under- and over-dispersed count data in life history, spatial, and community ecology. *Ecology*, 95, 3173–3180.

MacArthur, R. and Wilson, E. (1967). *The Theory of Island Biogeography*. Princeton University Press, Princeton.

Martin, T., Wintle, B., Rhodes, J. et al. (2005). Zero tolerance in ecology: Improving ecological inference by modelling the source of zero observations. *Ecology Letters*, 8, 1235–1246.

McFadden, D. (1974). Conditional logit analysis of qualitative choice behavior. In P. Zarembka (ed.) *Frontiers in Econometrics*, pp. 105–142. Academic Press, New York.

Mullahy, J. (1986). Specification and testing of some modified count models. *Journal of Econometrics*, 33, 341–365.

Poisson, S. (1837). *Recherches sur la probabilité des jugements en matière criminelle et en matière civile*. Bachelier, Paris.

Richards, S. (2008). Dealing with overdispersed count data in applied ecology. *Journal of Applied Ecology*, 45, 218–227.

Sellers, K. and Shumeli, G. (2010). A flexible regression model for count data. *Annals of Applied Statistics*, 4, 943–961.

Sofaer, H., Sillett, T., Langin, G., et al. (2014). Partitioning the sources of demographic variation reveals density-dependent nest predation in an island bird population. *Ecology and Evolution*, 4, 2738–2748.

Student (W. S. Gosset) (1907). On error of counting with an haemocytometer. *Biometrika*, 5, 351–360.

Vehtari, A., Gelman, A., and Gabry, J. (2017). Practical Bayesian model evaluation using leave-one-out cross validation and WAIC. *Statistical Computing*, 27, 1413–1432.

ver Hoef, J. and Boveng, P. (2007). Quasi-Poisson vs negative binomial regression: How should we model overdispersed count data? *Ecology*, 88, 2766–2772.

Vuong, Q. (1989). Likelihood ratio tests for model selection and non-nested hypotheses. *Econometrica*, 57, 307–333.

Wilson, P. (2015). The misuse of the Vuong test for non-nested models to test for zero-inflation. *Economics Letters*, 127, 51–53.

Yuen, E. and Dudgeon, D. (2015). Spatio-temporal variability in the distribution of ground-dwelling riparian spiders and their potential role in water-to-land energy transfer along Hong Kong forest streams. *PeerJ*, 3, e1134.

CHAPTER 11

Further Issues Involved in the Modeling of Counts

Packages needed in this chapter:

```
packages.needed<-c("ggplot2","fitdistrplus","brms","broom",
"GGally","qqplotr","gamlss.dist","broom.mixed","ggeffects","DHARMa",
"bayesplot","glmmTMB","ggmosaic","multcomp","MASS")
lapply(packages.needed, FUN = require, character.only = T)
```
loads these packages in the working session.

11.1 "The more you search, the more you find"

When the response variable is either binary or a count, the total number of observed presences often depends on the sampling effort. For a fixed rate of occurrence, the larger the area of search and the longer the time we cover, the higher the number of occurrences that will be observed. When counts are observed over time or across space, the lengths of time or the extent of the area searched may importantly vary among observations. Therefore, the heterogeneity in the sampling effort may lead us astray when we model the mean of counts across samples in terms of the effects of explanatory variables. When the heterogeneity of the sampling effort is important, it may make more sense to work with the rate of occurrence (i.e., the number of counts per unit area or per unit of time; more generally, *Y* / sampling effort), rather than to model the raw count *Y* as the response variable.

A standard approach to modeling the rate of occurrence (*Y* / sampling effort) while still treating the model as a count GLM is to include the offset in the model. This, of course, requires that we record the observed count along with the sampling effort for each observation(s). When we model rates of occurrence, the basic statistical model is

$$\log(\mathrm{E}(rate)) = \log\left(\frac{\mathrm{E}(Y)}{effort}\right) = X\beta \quad \Rightarrow \quad \log(\mathrm{E}(Y)) = \log(effort) + X\beta.$$

The offset (*effort*) is an explanatory variable for which the coefficient is not estimated and is constrained to be equal to one.

Including an offset in a frequentist count GLM is very simple: `glm(Y ~ X1 + X2 + offset(effort), family = poisson, data = DF)`. Because we use the log link function for count GLMs, the effort should be also log-transformed. And for Bayesian GLMs, the notation is just the same: `brm(y ~ x + offset(effort), family = negbinomial)`. No prior distribution need be established for the offset variable.

Statistical Modeling With R. Pablo Inchausti, Oxford University Press. © Pablo Inchausti (2023).
DOI: 10.1093/oso/9780192859013.003.0011

The offset aims to correct the estimated parameters whenever there is known or suspected heterogeneity in the sampling effort among observations. The use of the offset in count GLMs is a trick that allows us to use Poisson or negative binomial distributions to model the rate of occurrence as if they were counts. The resulting model of rates will be interpreted as just a count GLM. When the interpretation of the response variable as a rate of occurrence is not important, but controlling for the lack of equivalence in the sampling effort is, it is perfectly legitimate to include the sampling effort as an explanatory variable. In the latter cases, a partial slope will be estimated for the sampling effort in the scale of the link function.

We may also include an offset when the response variable is Gaussian but we do not need to transform it because we use the identity link function: `glm(Y ~ X1 + X2 + offset(effort), family = gaussian, data = DF)`. Grouped binary GLMs (Section 9.7) are actually a kind of count GLM with the response variable being the number of "successes" and the offset being k, the number of observations for each value of the response variable. For many ungrouped binary GLMs it is not possible to include an offset because it cannot be written in the scale of the logit link function. The key point is to understand that the offset must always be in the same scale as the link function used in the GLM.

The analysis of grouped binary data (Chapter 9) can be equivalently achieved in two other ways. First, we can consider the response variable to be the number of "successes" (defined as the count of events that interests us) to be modeled using either a Poisson or a negative binomial distribution, and the total ("successes" + "failures") number of events as the offset. Second, we could calculate the proportion of "successes" for each value of the explanatory variables, and use these proportions as the response variable to be modeled using a beta distribution (Chapter 12).

11.2 Log-linear models as count GLMs

Log-linear analysis were developed in the early 1970s before the appearance of GLMs as a discrete analogue of analysis of variance for frequency data (Von Eye and Mun 2012). They were mostly used to detect association patterns between categorical variables involved in the cross-classification of counts. The counts in a cross-classified or double-entry table can be transformed into relative cell frequencies (called joint probabilities) dividing the cell frequencies by the overall frequency total. The marginal probabilities are calculated as the ratio of the summed cell counts across rows or columns by the overall frequency total. Let's make up an example with a 2 × 2 table where the cell frequencies are the counts for the combinations of the levels of categorical explanatory variables A and B (see Table 11.1).

Conditional on the observed total frequency ($Freq_{\text{tot}} = 66$), each cell frequency $Freq_{ij}$ has a Poisson distribution with mean $\mu_{A_iB_j}$ that can be obtained under a statistical model. For instance, the expected frequencies under the hypothesis of independence of effects of the categorical variables A and B can be expressed as

$$\mathrm{E}\left(Freq_{ij}\right) = Freq_{\text{tot}}\pi_j\pi_i = Freq_{\text{tot}}\left(\frac{Freq_{ij}}{Marg.tot_i}\right)\left(\frac{Freq_{ij}}{Marg.tot_j}\right).$$

Table 11.1 Example 2×2 table where the cell frequencies are the counts for the combinations of the levels of categorical explanatory variables *A* and *B*.

	B_1	B_2	Marginal total		B_1	B_2	Marginal probabilities
A_1	15	27	42	A_1	0.23	0.41	0.64
A_2	18	6	24	A_2	0.27	0.09	0.36
Marginal total	33	33	66	Marginal probabilities	0.50	0.50	1.00

Taking the log of both sides transforms the multiplicative model into an additive one:

$$\log\left(\mathrm{E}\left(Freq_{ij}\right)\right) = \log\left(Freq_{\mathrm{tot}}\right) + \log\left(\frac{Freq_{ij}}{Marg.tot_i}\right) + \log\left(\frac{Freq_{ij}}{Marg.tot_j}\right).$$

This can be expressed in terms of model parameters as a Poisson GLM,

$$\log\left(\mathrm{E}\left(Freq_{ij}\right)\right) = \mu_Y + \lambda_A + \lambda_B,$$

where λ_A and λ_B are the main effects of the categorical variables *A* and *B*. While there is more on log-linear models than sketched here (see Van Eye and and Mun 2012), the point was just to show that they can be see as equivalent to analysis of variance (Chapters 5 and 6) for counts.

THE DATA IN CONTEXT: Schoener (1970) investigated whether the main determinants of microhabitat use are different between two species of small lizards (*Anolis grahamii* and *Anolis opalinus*) coexisting on Caribbean islands. The categorical explanatory variables registered were `sun` (with two levels: Sun and Shade), `height` of the resting area (with two levels: High and Low), `time` of the day (with three levels: Afternoon, Mid.day, and Morning), the width of the perching area (`perch`, with two levels: Broad and Narrow), and, of course, the two `species`. This sort of low-tech endeavor is typical of many observational, natural history–based field studies by ecologists. The response variable (`n`) is the total number of lizards observed for the combinations of levels of the five categorical explanatory variables.

We start by importing the data and examining the resulting data frame:

```
> liz=read.csv ("Chapter 11 lizards.csv", header=T)
> str(liz)
'data.frame':   48 obs. of  6 variables:
 $ n      : int  20 13 8 6 34 31 17 12 8 8 ...
 $ sun    : Factor w/ 2 levels "Shade","Sun": 1 1 1 1 2 2 2 2 1 1 ...
 $ height : Factor w/ 2 levels "High","Low": 1 2 1 2 1 2 1 2 1 2 ...
 $ perch  : Factor w/ 2 levels "Broad","Narrow": 1 1 2 2 1 1 2 2 1 1 ...
 $ time   : Factor w/ 3 levels "Afternoon","Mid.day",..: 3 3 3 3 3 3 3 3 2 2 ...
 $ species: Factor w/ 2 levels "grahamii","opalinus": 2 2 2 2 2 2 2 2 2 2 ...
```

EXPLORATORY DATA ANALYSIS: We next assess which probability distribution (Poisson or negative binomial) can be used to model the response variable using, as usual, the package `fitdistrplus`. Figure 11.1 clearly shows that the negative binomial is the best option.

The exploratory graphics for this data set are not trivial. To start with, there are five categorical explanatory variables, some of which have pairwise interactions and perhaps third-order (or even fourth) interactions on the effect on the response variable. There

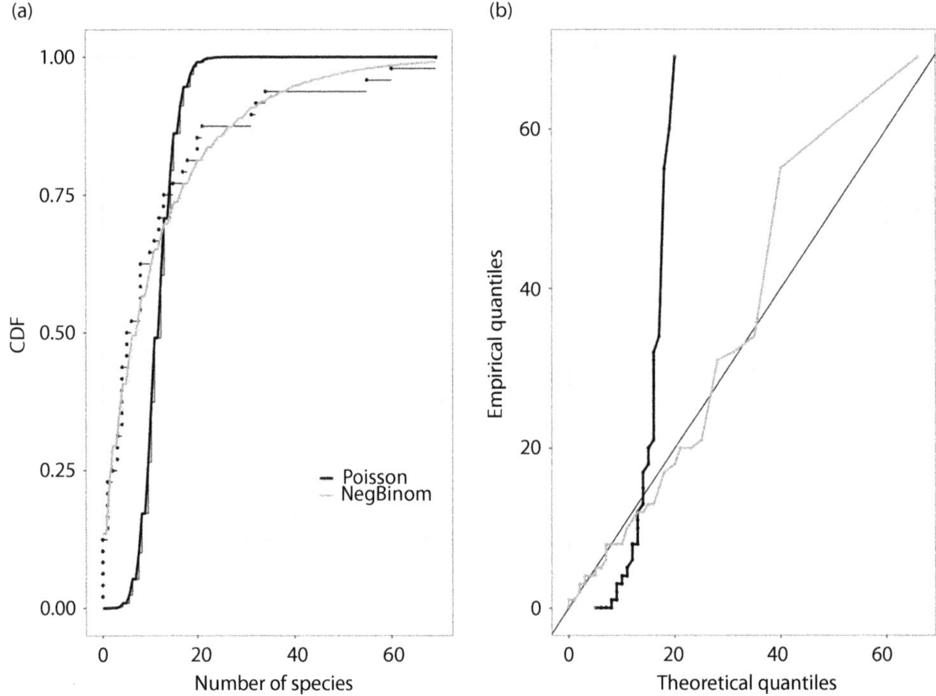

Fig. 11.1 Goodness-of-fit assessment of the response variable number of lizard individuals in the Bahamas showing the cumulative distribution function (a) and the quantile–quantile plot (b) comparing the Poisson and negative binomial distributions obtained using the package `fitdistrplus`.

could be at most ten third-order and ten pairwise interactions with five categorical explanatory variables. We also started the analysis without clear hypotheses (i.e., in a completely exploratory manner) that could guide the inclusion of certain effects in the model. Besides, this is a very unbalanced data set in that there are certain combinations of explanatory variables with very small (or even zero) counts, and others with much larger frequencies. We use the package `ggmosaic` to make some appropriate graphs for these data.

Figure 11.2 has a lot of information. Each plot displays the three-way interactions among explanatory variables, and the area for each combination of levels is proportional to the value of the response variable. Thus, by comparing the areas we can visualize both the magnitudes of the differences between levels of an explanatory variable and the putative existence of interactions among them. For instance, in Fig. 11.2a the four rectangles for `height` = `Low` have similar sizes to those for `height` = `High`, thereby suggesting that the pairwise interaction `species:sun` does not vary according to `height`. In other words, that there would not be a `species:sun:height` interaction. In the same plot, `species` = `grahamii` seems more abundant than `species` = `opalinus` in the `Sun`, but similarly abundant in the `Shade`, thereby suggesting the need to include a pairwise interaction `species:sun` in the model. While we could proceed to write similar assessments for the remaining three plots of Fig. 11.2, we trust that this introduction gives you a hint for interpreting these plots.

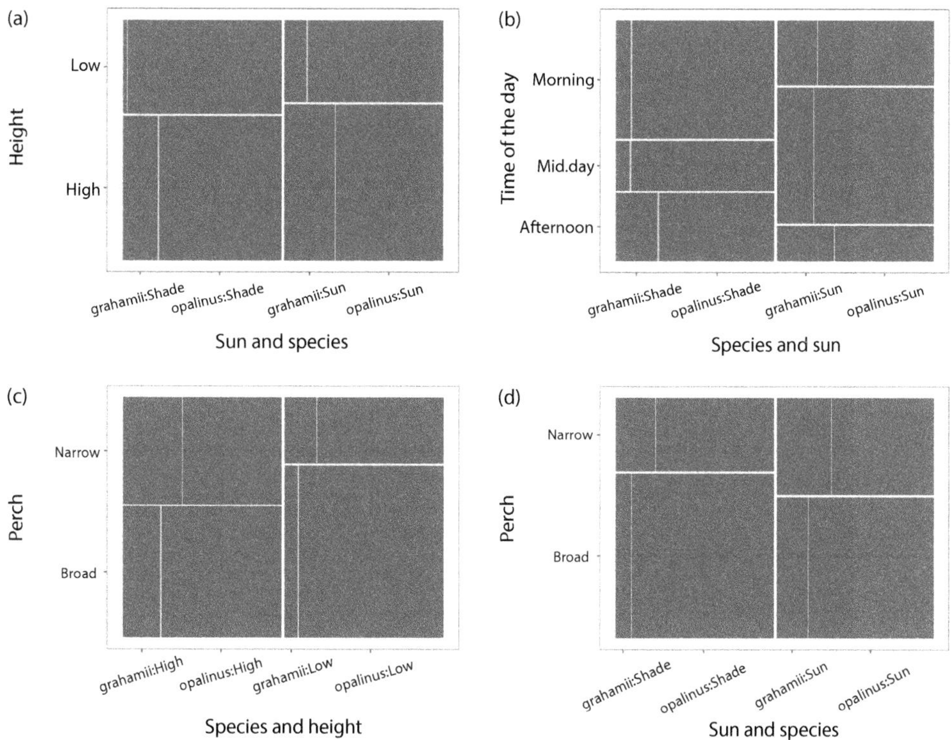

Fig. 11.2 Covariation between the categorical explanatory variables in the frequentist count GLM `liza1`. The area for each combination of explanatory variables is proportional to the value of the response variable. The figure was produced with the package `ggmosaic`.

To cut a potentially long story short, our model will consider all single-level effects, and all pairwise and third-order interactions among the explanatory variables. This approach is, of course, only reasonable in the context of an exploratory (not a confirmatory) analysis.

THE STATISTICAL MODEL:

$Y \sim \text{NegativeBinomial}(\mu_Y, \phi)$, with

$\log(\mu_Y) =$ `species` + `sun` + `height` + `perch` + `time` + all pairwise interactions + all third-order interactions

Fortunately, there is a shortcut to specify such a complex set of effects that remains valid in all R models: `(species + sun + height + perch + time)^3`.

11.3 Frequentist fitting of a log-linear model

We fit the model `liza1 = glm.nb(n ~ (species + sun + height + perch + time)^3, data = liz)` using the package `MASS`; its long summary is:

```
> summary(liza1)
Call:
glm.nb(formula = n ~ (species + sun + height + perch + time)^3,
    data = liz, init.theta = 143015.2672, link = log)
Deviance Residuals:
    Min       1Q   Median       3Q      Max
-1.3366  -0.2793  -0.0147   0.2364   1.1157
Coefficients:
                                           Estimate Std. Error z value Pr(>|z|)
(Intercept)                                 1.16914    0.50075    2.33    0.020 *
speciesopalinus                             0.57361    0.59812    0.96    0.338
sunSun                                      1.06103    0.56016    1.89    0.058 .
heightLow                                  -1.89075    1.11195   -1.70    0.089 .
perchNarrow                                 0.20156    0.61216    0.33    0.742
timeMid.day                                -1.10639    0.89088   -1.24    0.214
timeMorning                                -0.40977    0.70226   -0.58    0.560
speciesopalinus:sunSun                      0.07353    0.66359    0.11    0.912
speciesopalinus:heightLow                   2.50713    1.14361    2.19    0.028 *
speciesopalinus:perchNarrow                -0.80338    0.69289   -1.16    0.246
speciesopalinus:timeMid.day                 1.53283    0.94621    1.62    0.105
speciesopalinus:timeMorning                 1.56374    0.75848    2.06    0.039 *
sunSun:heightLow                            1.04523    1.14948    0.91    0.363
sunSun:perchNarrow                         -0.28681    0.65631   -0.44    0.662
sunSun:timeMid.day                          1.84017    0.91652    2.01    0.045 *
sunSun:timeMorning                          0.73710    0.73405    1.00    0.315
heightLow:perchNarrow                      -0.51073    0.76486   -0.67    0.504
heightLow:timeMid.day                      -1.52553    0.93387   -1.63    0.102
heightLow:timeMorning                      -1.71314    0.91713   -1.87    0.062 .
perchNarrow:timeMid.day                    -0.38252    0.82317   -0.46    0.642
perchNarrow:timeMorning                     0.05158    0.71718    0.07    0.943
speciesopalinus:sunSun:heightLow           -1.97869    1.20048   -1.65    0.099 .
speciesopalinus:sunSun:perchNarrow          0.13848    0.68861    0.20    0.841
speciesopalinus:sunSun:timeMid.day         -0.91258    0.97210   -0.94    0.348
speciesopalinus:sunSun:timeMorning         -1.24659    0.78940   -1.58    0.114
speciesopalinus:heightLow:perchNarrow      -0.30053    0.53230   -0.56    0.572
speciesopalinus:heightLow:timeMid.day       0.70694    0.66911    1.06    0.291
speciesopalinus:heightLow:timeMorning       0.88721    0.72977    1.22    0.224
speciesopalinus:perchNarrow:timeMid.day     0.00771    0.56818    0.01    0.989
speciesopalinus:perchNarrow:timeMorning    -0.04432    0.60776   -0.07    0.942
sunSun:heightLow:perchNarrow                0.36315    0.58333    0.62    0.534
sunSun:heightLow:timeMid.day                0.91692    0.72209    1.27    0.204
sunSun:heightLow:timeMorning                1.06643    0.63843    1.67    0.095 .
sunSun:perchNarrow:timeMid.day              0.99036    0.75022    1.32    0.187
sunSun:perchNarrow:timeMorning              0.05693    0.62453    0.09    0.927
heightLow:perchNarrow:timeMid.day          -0.39701    0.57579   -0.69    0.491
heightLow:perchNarrow:timeMorning           0.15358    0.59275    0.26    0.796
Signif. codes:  0 '***' 0.001 '**' 0.01 '*' 0.05 '.' 0.1 ' ' 1
(Dispersion parameter for Negative Binomial(143015) family taken to be 1)
    Null deviance: 737.476  on 47  degrees of freedom
Residual deviance:  13.231  on 11  degrees of freedom
```

The model summary is atrociously long. The dispersion parameter of the negative binomial distribution was very poorly estimated: the algorithm "went astray" towards positive infinity for which the distribution becomes Poisson. We should be skeptical of the parameter estimates. The three curly brackets at the left help to organize (from top to bottom) the summary into single effects, pairwise interactions, and third-order interactions. The intercept is the log(E(Y)) for the reference levels for each categorical variable: `grahamii` for `species`, `Shade` for `sun`, `Broad` for `perch`, and `Afternoon` for `time` (i.e., those not appearing in the single effects). The other parameters denote the differences in log(E(Y)) between either levels or combinations of levels of the categorical explanatory variables. We will interpret these coefficients later after simplifying this complex initial model.

The table of analysis of deviance provides a slightly better depiction of the results:

```
> anova(liza1, test="Chisq")
Analysis of Deviance Table

Model: Negative Binomial(143015), link: log

Response: n

Terms added sequentially (first to last)

                      Df Deviance Resid. Df Resid. Dev Pr(>Chi)
NULL                                     47        737
species                1    165.7        46        572  < 2e-16
sun                    1    242.7        45        329  < 2e-16
height                 1     49.6        44        279  1.9e-12
perch                  1     28.4        43        251  9.9e-08
time                   2     98.5        41        153  < 2e-16
species:sun            1      6.5        40        146   0.0110
species:height         1     26.9        39        119  2.2e-07
species:perch          1     18.5        38        101  1.7e-05
species:time           2      6.4        36         94   0.0414
sun:height             1      0.2        35         94   0.6366
sun:perch              1      2.2        34         92   0.1406
sun:time               2     50.3        32         42  1.2e-11
height:perch           1     10.6        31         31   0.0011
height:time            2      3.3        29         28   0.1900
perch:time             2      2.7        27         25   0.2617
species:sun:height     1      2.2        26         23   0.1375
species:sun:perch      1      0.1        25         23   0.7023
species:sun:time       2      1.3        23         21   0.5278
species:height:perch   1      0.4        22         21   0.5108
species:height:time    2      0.9        20         20   0.6294
species:perch:time     2      0.1        18         20   0.9501
sun:height:perch       1      0.0        17         20   0.8374
sun:height:time        2      2.9        15         17   0.2320
sun:perch:time         2      2.3        13         15   0.3189
height:perch:time      2      1.5        11         13   0.4769
```

Again, the curly brackets help to differentiate the single effects and the pairwise and third-order interactions. Using the *p*-value as a rough criterion of "relevance," we see that none of the third-order interactions appear to be important to explaining the data, thus suggesting that `liza1` could be (urgently, please!) simplified by model selection.

We are going to use `liza1` as the baseline to generate a set of simpler models differing from it by just one of the third-order interactions, and then compare each model with `liza1` using Wilks' likelihood ratio test. Each of these 10 models would then be a special case of `liza1`. We will show here just one of these "descendant" models, and its contrast with `liza1`:

```
liza1.1=update (liza1, ~. -species:sun:height)
anova(liza1,liza1.1, test="Chisq")
   theta Resid. df    2 x log-lik.   Test    df LR stat. Pr(Chi)
1 170343        12            -180
2 143015        11            -176 1 vs 2     1     3.71   0.054
```

Since the difference in deviance between the two models was not statistically significant, we may exclude the interaction `species:sun:height`. The same process is repeated for the remaining nine models, with identical conclusions. We can then formulate a baseline model including up to pairwise interactions between categorical variables: `liza2 = glm.nb(n ~ (species+sun+height+perch+time)^2, data = liz)`.

```
> anova(liza2, test="Chisq")
Analysis of Deviance Table

Model: Negative Binomial(140674), link: log

Response: n

Terms added sequentially (first to last)


               Df Deviance Resid. Df Resid. Dev Pr(>Chi)
NULL                              47        737
species         1    165.7        46        572  < 2e-16
sun             1    242.7        45        329  < 2e-16
height          1     49.6        44        279  1.9e-12
perch           1     28.4        43        251  9.9e-08
time            2     98.5        41        153  < 2e-16
species:sun     1      6.5        40        146   0.0110
species:height  1     26.9        39        119  2.2e-07
species:perch   1     18.5        38        101  1.7e-05
species:time    2      6.4        36         94   0.0414
sun:height      1      0.2        35         94   0.6366
sun:perch       1      2.2        34         92   0.1406
sun:time        2     50.3        32         42  1.2e-11
height:perch    1     10.6        31         31   0.0011
height:time     2      3.3        29         28   0.1900
perch:time      2      2.7        27         25   0.2617
```

Using the same criterion again, we may simplify `liza2` by making a set of 10 "descendant" models, each without one of the 10 pairwise interactions, and compare them with `liza2`. The procedure is just the same (you can find the code on the companion website), and we will just show the results in the final model `liza3`:

```
liza3=glm.nb(n~(species+sun+height+perch+time)^2
           -perch:time-sun:perch-height:time-perch:time-sun:height, data=liz)
```

whose summary is:

```
> summary(liza3)

Call:
glm.nb(formula = n ~ (species + sun + height + perch + time)^2 -
    perch:time - sun:perch - height:time - perch:time - sun:height
    data = liz, maxit = 1e+05, epsilon = 1e-10, init.theta = 13318
    link = log)

Deviance Residuals:
    Min       1Q   Median       3Q      Max
-2.0299  -0.7384  -0.0711   0.4247   1.6708

Coefficients:
                            Estimate Std. Error z value Pr(>|z|)
(Intercept)                   0.8044     0.3313    2.43  0.01519 *
speciesopalinus               1.3311     0.3446    3.86  0.00011 *
sunSun                        1.4299     0.3215    4.45  8.7e-06 *
heightLow                    -1.2512     0.2468   -5.07  4.0e-07 *
perchNarrow                   0.3060     0.1787    1.71  0.08682 .
timeMid.day                  -0.9683     0.3609   -2.68  0.00730 *
timeMorning                  -0.0263     0.3317   -0.08  0.93668
speciesopalinus:sunSun       -0.8697     0.3113   -2.79  0.00522 *
speciesopalinus:heightLow     1.0845     0.2527    4.29  1.8e-05 *
speciesopalinus:perchNarrow  -0.7336     0.2065   -3.55  0.00038 *
speciesopalinus:timeMid.day   0.7804     0.2667    2.93  0.00344 *
speciesopalinus:timeMorning   0.6881     0.2866    2.40  0.01636 *
sunSun:timeMid.day            1.7903     0.3177    5.64  1.7e-08 *
sunSun:timeMorning            0.1717     0.2801    0.61  0.53992
heightLow:perchNarrow        -0.6305     0.1949   -3.23  0.00122 *
---
Signif. codes:  0 '***' 0.001 '**' 0.01 '*' 0.05 '.' 0.1 ' ' 1

(Dispersion parameter for Negative Binomial(1.33e+10) family taken

    Null deviance: 737.555  on 47  degrees of freedom
Residual deviance:  33.303  on 33  degrees of freedom
AIC: 228.6
```

Model `liza3` is not over-dispersed (33.303 / 33 ≈ 1) and the dispersion parameter of the negative binomial distribution is still poorly estimated. The model accounts for a large fraction of the deviance: `1 - (liza3$deviance / liza3$null.deviance)` = 0.954. `liza3` contains all single effects of the explanatory variables, and six pairwise interactions: `species:sun`, `species:height`, `species:perch`, `species:time`, `sun:time`, and `height:perch`. The lack of third-order interactions implies that we can separately interpret each of the six pairwise interactions. The single effects of the explanatory variables cannot be interpreted because all variables are involved in at least one pairwise interaction, and hence we cannot separately interpret their effects on log(E(Y)) (see Chapter 6).

Two of the pairwise interactions (`sun:time` and `height:perch`) in `liza3` did not involve the variable `species`. Hence, these effects would be equally valid for both (or better, there is no evidence to believe that they substantially differ between) lizard species. Let's consider the interaction `perch:height` at the top of Fig. 11.3. In the sub-table at the left, the intercept is the value of log(E(Y)) for the reference levels of these variables (High and Broad). For `height` = `High`, 0.30 is the increase in log(E(Y)) due to changing from `perch` = `Broad` to `perch` = `Narrow`. For `perch` = `Broad`, –1.25 is the decrease in log(E(Y)) due to changing from `height` = `High` to `height` = `Low`. That is, exp(–1.25) is the multiplier that decreases (i.e., how many times smaller is) the lizard frequency from `height` = `High` to `height` = `Low`. Obtaining the value of log(E(Y)) for the bottom right cell of this 2 × 2 table requires adding the single effects of these explanatory variables (–1.25 and 0.30) and their interaction (–0.63) to the model intercept. The central sub-table gives the estimated values of log(E(Y)), and the right sub-table the expected number of lizards predicted by model `liza3` in each combination of the 2 × 2 table. What can we learn from these expected values? First, lizards are predicted to be more abundant in high resting places. Second, the ratio (High / Low) of predicted lizard frequencies (2.23 / 0.64 = 1.56) in Broad perching areas differs from that of Narrow ones (3.00 / 0.46 = 6.52). That is, the preference of lizards for high resting places differs between broad and narrow perching areas. This is what the significant interaction between these factors actually implies (cf. Chapter 6), and we can quantify that there is a four-fold (6.52 / 1.56 = 4.18) difference in preference for high resting places between narrow and broad areas. The rest of Fig. 11.3 for the remaining pairwise interactions was obtained similarly. You may consider that showing so many small tables in Fig. 11.3 is overkill (please accept our apologies), but this way you hopefully understand better the connection between the model summary and what can be extracted from just a few calculations.

The best way to summarize the effects predicted by `liza3` is with conditional plots. The six plots of Fig 11.4 for each pairwise interaction were produced using the package `ggeffects`. Each plot renders the predicted marginal means and 95% confidence interval for the levels of a pair of categorical variables, setting all other categorical explanatory variables to the reference values (Fig. 11.4a). Each of the conditional predictions is obtained with `ggpredict(liza3, terms = c("sun", "perch"))`; the names of the explanatory variables are then changed as needed to generate the other plots.

The final step is to assess the goodness of fit of model `liza3` with randomized quantile residuals (Chapter 8) obtained with the package `DHARMa`: `res.liza3 = simulateResiduals(fittedModel = liza3, n = 500, refit = T, plot = F)`. Next, we gather all the information required for the residual analysis with the command `augment` of package `broom`, `resid.liza = augment(liza3)`, and then convert the simulated residuals to have a normal distribution and store them as `resid.liza3$.std.resid = residuals(res.liza3, quantileFunction = qnorm)`. Figure 11.5 shows the plots used in the residual analysis.

log(E(Y))		perch	
		broad	narrow
height	high	0.80	0.80+0.30
	low	0.80-1.25	0.80-1.25+0.30-0.63

log(E(Y))		perch	
		broad	narrow
height	high	0.80	1.10
	low	-0.45	-0.78

E(Y)		perch	
		broad	narrow
height	high	2.23	3.00
	low	0.64	0.46

log(E(Y))		perch	
		high	low
time	afternoon	0.80	0.80+1.42
	mid.day	0.80-0.96	0.80+1.42-0.96+1.79
	morning	0.80-0.03	0.80+1.42-0.03+0.17

log(E(Y))		perch	
		high	low
time	afternoon	0.80	2.22
	mid.day	-0.16	3.05
	morning	0.77	2.36

E(Y)		perch	
		high	low
time	afternoon	2.23	9.21
	mid.day	0.85	21.12
	morning	2.16	10.59

log(E(Y))		perch	
		broad	narrow
species	grahamanii	0.80	0.80+0.30
	opalinus	0.80+1.33	0.80+1.33+0.30-0.73

log(E(Y))		perch	
		broad	narrow
species	grahamanii	0.80	1.10
	opalinus	2.13	1.70

E(Y)		perch	
		broad	narrow
species	grahamanii	2.23	3.00
	opalinus	8.41	5.47

log(E(Y))		height	
		high	low
species	grahamanii	0.80	0.80+0.30
	opalinus	0.80+1.33	0.80+1.33-1.25+1.08

log(E(Y))		height	
		high	low
species	grahamanii	0.80	1.10
	opalinus	2.13	1.96

E(Y)		height	
		high	low
species	grahamanii	2.23	3.00
	opalinus	8.41	7.10

log(E(Y))		sun	
		shade	sun
species	grahamanii	0.80	0.80+1.42
	opalinus	0.80+1.33	0.80+1.42+1.33-0.87

log(E(Y))		sun	
		shade	sun
species	grahamanii	0.80	2.22
	opalinus	2.13	2.68

E(Y)		sun	
		shade	sun
species	grahamanii	2.23	9.21
	opalinus	8.41	14.59

log(E(Y))		species	
		grahamanii	opalinus
time	afternoon	0.80	0.80+1.33
	mid.day	0.80-0.96	0.80+1.33-0.96+0.78
	morning	0.80-0.03	0.80+1.33-0.03+0.68

log(E(Y))		species	
		grahamanii	opalinus
time	afternoon	0.80	2.13
	mid.day	-0.16	3.05
	morning	0.77	2.36

E(Y)		species	
		grahamanii	opalinus
time	afternoon	2.23	8.41
	mid.day	0.85	21.12
	morning	2.16	10.59

Fig. 11.3 Step-by-step explanation of the summary of the frequentist count GLM `liza3` for each pair of interacting explanatory variables: (`perch`, `height`), (`time`, `perch`), (`species`, `perch`), (`species`, `height`), (`species`, `sun`), and (`species`, `time`) as rows. From left to right, the columns denote the coefficients of `summary(liza3)` estimated in the log link scale, the totals of the coefficients referring to log(E(*Y*)), and the inverted log link (i.e., exp(log(E(*Y*))) that give the predicted mean counts (E(*Y*)) in the data scale. Because the model did not contain any third-order interaction, each of the six pairwise interactions of the effects of explanatory variables can be separately interpreted.

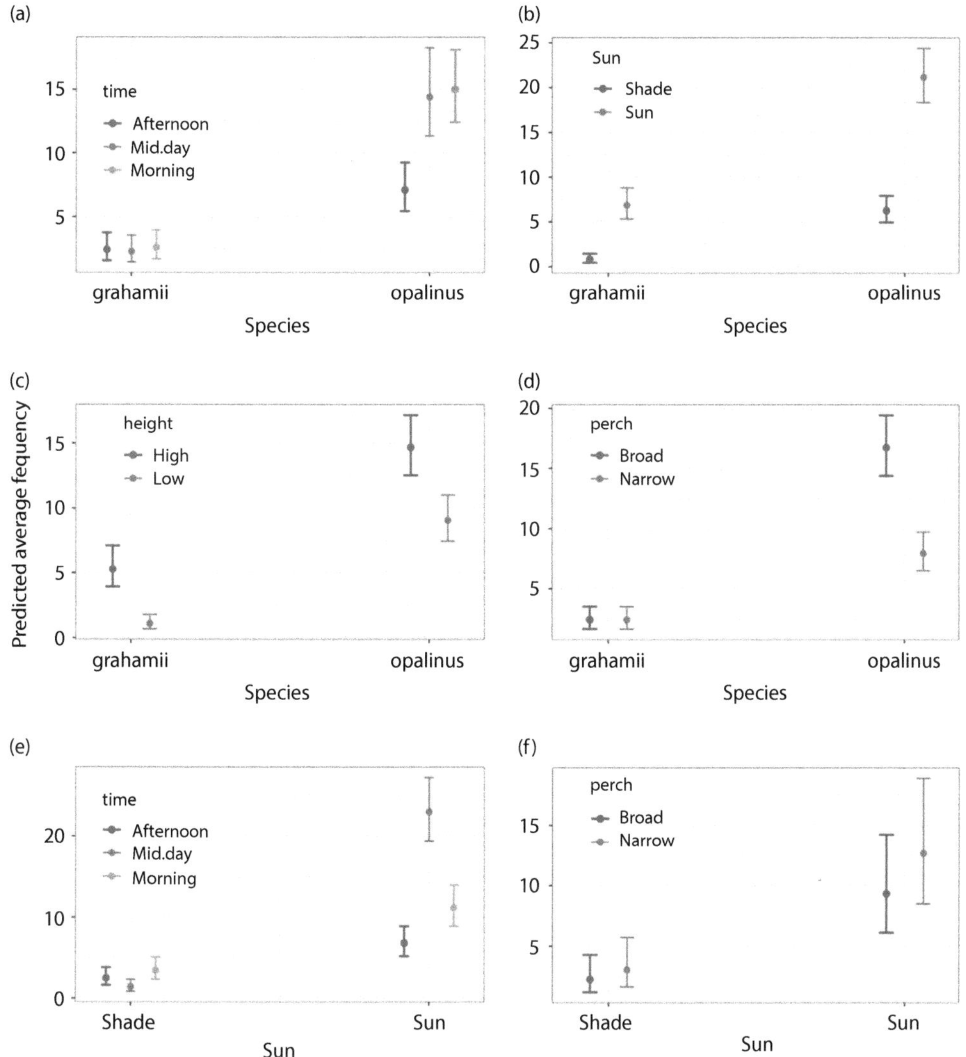

Fig. 11.4 Conditional relations for the number of individuals (response variable) and the interacting categorical variables in the frequentist count GLM `liza3` obtained after model selection, showing the predicted values of the means and their 95% confidence intervals.

Model `liza3` seems to fit that data reasonably well: the points in Fig. 11.5a–d show a random scatter without any clear trend or differences in variation among combinations of levels of explanatory variables, and the simulated residuals have a normal distribution. There are a couple of points in every plot (2 / 46 ≈ 4%) that have larger residuals, and there is one point (#15) that had a large Cook distance.

```
> resid.liza3[resid.liza3$.cooksd>0.4,]
# A tibble: 1 x 12
      n species  sun   height perch  time     .fitted .resid .std.resid  .hat .sigma .cooksd
  <int> <fct>    <fct> <fct>  <fct>  <fct>      <dbl>  <dbl>      <dbl> <dbl>  <dbl>   <dbl>
1    60 opalinus Sun   High   Narrow Mid.day     3.87   1.67       1.75 0.579  0.913   0.656
```

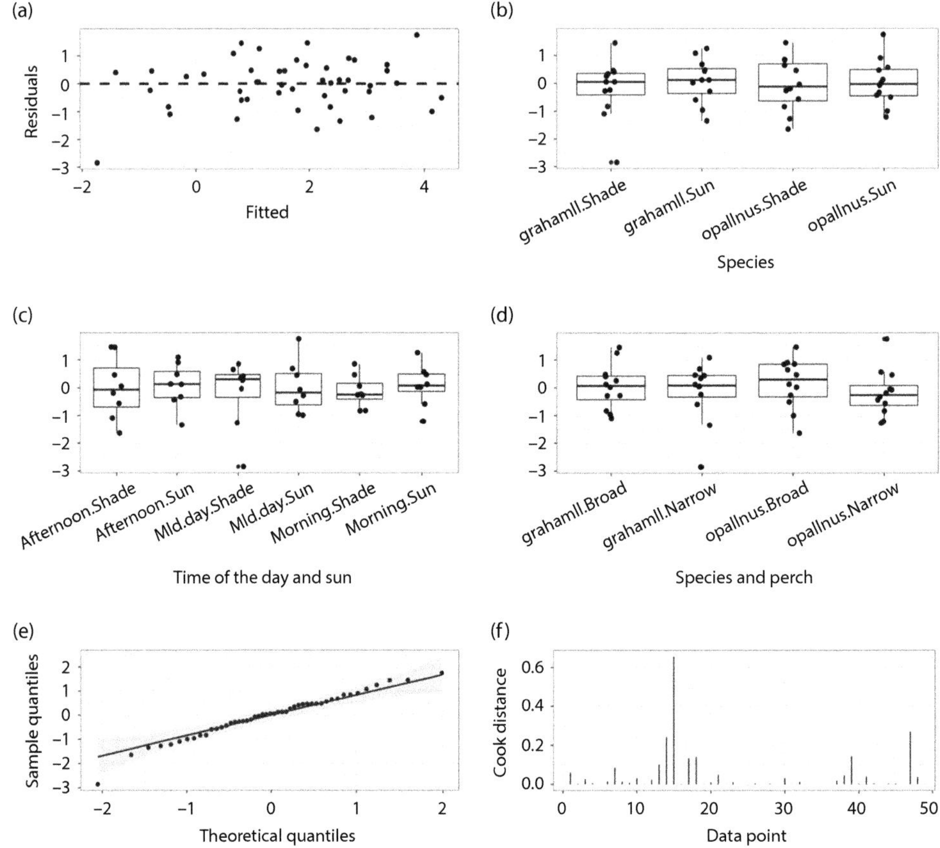

Fig. 11.5 Residual analysis of the frequentist count GLM `liza3` based on simulated randomized quantile residuals converted to normal residuals. Plots (a) to (d) show the relations between these residuals and the fitted values, and to the standardized numerical explanatory variables. The dashed lines were added to help the visualization. Plot (e) shows the quantile–quantile plot (and the 95% confidence band) of the residuals. Plot (f) shows the Cook distances.

Although we cannot know if this datum is correct (it is the second largest count), we could rerun model `liza3` without it to assess its importance on parameter estimates:

```
liza31=glm.nb(n~(species+sun+height+perch+time)^2
           -perch:time-sun:perch-height:time-perch:time-sun:height,
           data=liz[resid.liza3$.cooksd<0.4,])
```

We will not go through the model summary and analysis of deviance again. Let's just see the percentage difference between the parameter estimates of both models:

```
> 100*(coef(liza31)-coef(liza3))/coef(liza3)
                 (Intercept)              speciesopalinus                       sunSun
                       0.930                        4.036                        1.508
                   heightLow                  perchNarrow                  timeMid.day
                       9.363                      -14.268                      -11.453
                 timeMorning        speciesopalinus:sunSun    speciesopalinus:heightLow
                      12.330                        3.322                       12.336
speciesopalinus:perchNarrow speciesopalinus:timeMid.day speciesopalinus:timeMorning
                      33.061                      -16.369                        0.198
         sunSun:timeMid.day            sunSun:timeMorning          heightLow:perchNarrow
                      -6.645                        1.878                      -39.795
```

There are 10 out of 15 parameters with a difference of ±10% between these two models differing in one datum, and the interactions `species:time` and `height:perch` are no longer statistically significant in `liza31`. This is not to say that the point excluded in `liza31` is "wrong" or an outlier. However, our conclusions about some effects of the explanatory variables may not be as firm as we would wish. What should we do? We should probably report and interpret the results of `liza3`, and add a cautionary note in our discussion mentioning the consequences of its residual analysis.

Model `liza3` contained 10 pairwise interactions that were statistically significant. Thus, we cannot separately interpret the effects of the categorical explanatory variables involved. Take the `species:sun` interaction: we cannot assess the differences in abundance between species without taking into account whether the counts were made in the sun or in the shade. Nevertheless, there might be some occasions when we wish to know which of the observed counts in the 2 × 2 combinations of `species:sun` actually differ from one another. This is, of course, an a posteriori test (Chapters 5 and 6). The reasons to correct the significance level due to multiple testing are the same in count GLMs and in the general linear model. As a first approximation and for illustration purposes we fit a simpler count GLM using a single explanatory variable combining `species` and `sun`: `liz$sp.sun = factor(paste(substr(liz$species, 1, 3), liz$sun, sep = "."))`; `substr` simply shortens the variable names by taking just the first three letters of their names. We can then fit the model with `liz.sp.sun = glm.nb(n ~ sp.sun, data = liz)`, whose shortened summary is:

```
Coefficients:
               Estimate Std. Error z value Pr(>|z|)
(Intercept)       0.223      0.331    0.67      0.5
sp.sungra.Sun     2.063      0.402    5.13  2.8e-07
sp.sunopa.Shade   1.781      0.405    4.39  1.1e-05
sp.sunopa.Sun     3.127      0.395    7.92  2.4e-15
```

We now use the package `multcomp` to make all the pairwise comparisons: `mult.comp.liz = glht(liz.sp.sun, linfct = mcp(sp.sun = "Tukey"), adjust.method = "fdr")`.

Figure 11.6 displays the results, showing that only one of the five comparisons failed to show a significant difference between the predicted mean frequencies (in the log link scale, of course). A similar procedure could be used for the other pairwise comparisons, eventually using contrasts of specific interest (Chapter 6).

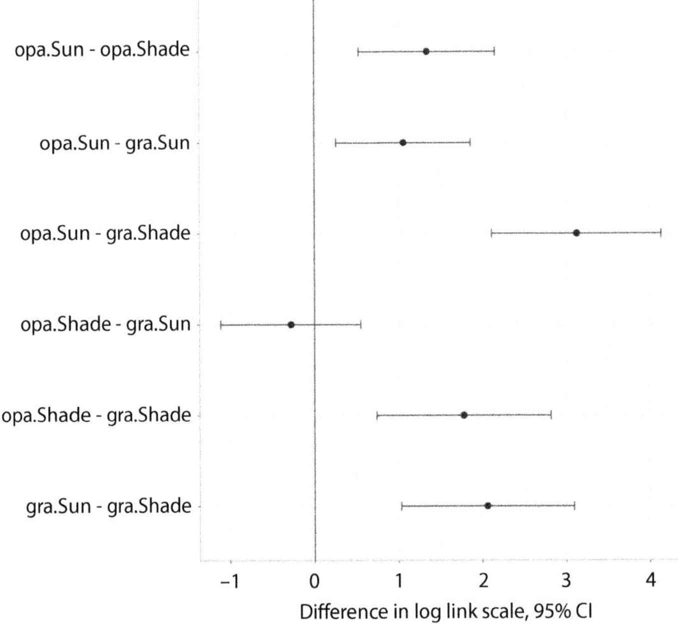

Fig 11.6 A posteriori tests comparing all pairwise mean numbers of individuals for the interaction `species*sun` in the frequentist count GLM `liza3`. The comparisons and the figure were produced with the package `multcomp`.

11.4 Bayesian fitting of a log-linear model

We use the package `brms` to fit the same Bayesian log-linear model. We will fit a model containing all single effects and their pairwise interactions for Schoener's lizards; the statistical model was outlined in Section 11.3. As always, we start by examining the priors that need to be defined for the model we intend to fit:

```
> get_prior(formula=n~(species+sun+height+perch+time)^2,
+          family= negbinomial, data=liz)
               prior     class                         coef
              (flat)         b
              (flat)         b                    heightLow
              (flat)         b         heightLow:perchNarrow
              (flat)         b         heightLow:timeMid.day
              (flat)         b         heightLow:timeMorning
              (flat)         b                  perchNarrow
              (flat)         b       perchNarrow:timeMid.day
              (flat)         b       perchNarrow:timeMorning
              (flat)         b               speciesopalinus
              (flat)         b     speciesopalinus:heightLow
              (flat)         b speciesopalinus:perchNarrow
              (flat)         b        speciesopalinus:sunSun
              (flat)         b speciesopalinus:timeMid.day
              (flat)         b speciesopalinus:timeMorning
              (flat)         b                       sunSun
              (flat)         b             sunSun:heightLow
              (flat)         b           sunSun:perchNarrow
              (flat)         b           sunSun:timeMid.day
              (flat)         b           sunSun:timeMorning
              (flat)         b                  timeMid.day
              (flat)         b                  timeMorning
 student_t(3, 1.7, 2.5) Intercept
     gamma(0.01, 0.01)     shape
```

We need to define the priors for the 22 parameters determining the mean of the response variable through the effects of the categorical explanatory variables, and the shape or dispersion parameter of the negative binomial distribution. It is a tall order to gather empirical information to parametrize the priors for the six pairwise interactions and the single effects of these explanatory variables. Even a true specialist in lizard or animal behavior would probably need to make a large and thorough literature review. We are going (arbitrarily) to use the following weakly informative priors:

```
prior.liz4=c(set_prior("normal(5,2)", class = "Intercept"),
             set_prior("normal(0,2)", class = "b"),
             set_prior("gamma(0.1,0.1)", class = "shape"))
```

They are shown in Fig. 11.7.

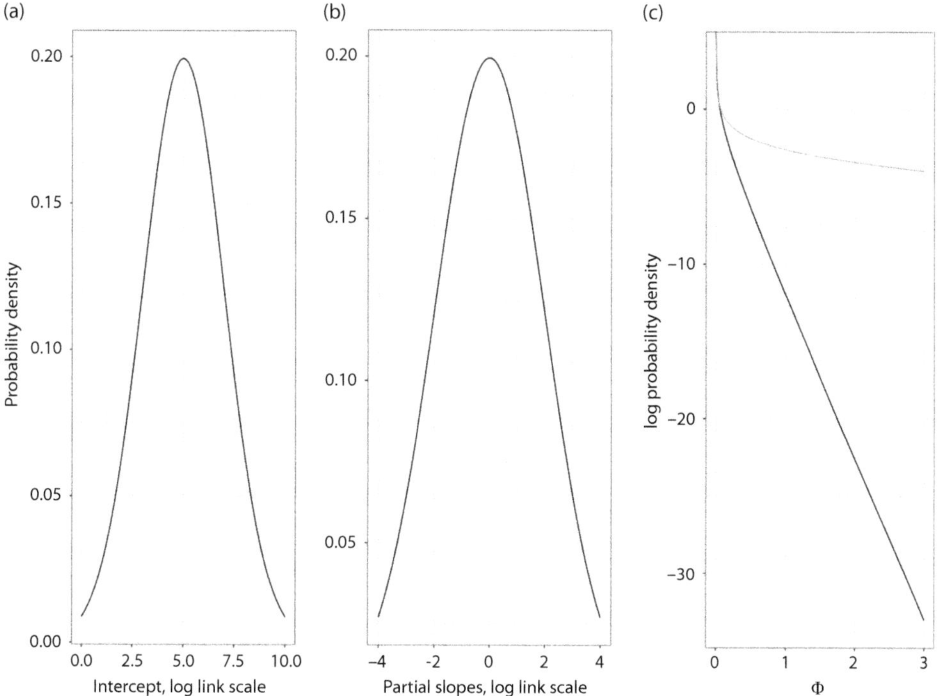

Fig 11.7 Prior distributions of model parameters for the Bayesian count GLM `liza4.brms`. Plot (c) shows a weakly (gray) and more informative (black) prior for the dispersion parameter of the negative binomial distribution.

The prior for the intercept suggests that we would expect with probability 0.95 that the frequencies are in [1, 9] in the log link scale, or [exp(1), exp(9)] in the data scale. The prior for the single effects and interactions would be in [−4, 4] in the log link scale. The prior for the inverse aggregation parameter of the negative binomial distribution (Eq. (10.2)) is the same as used in Section 11.3. Let's not forget that we have 46 data points in the data frame `liz` and are hoping to estimate 22 parameters, i.e., very high ambition from comparatively little data. Most statisticians would rightly call our attempt an over-fit of the data. When fitting complex models with limited data, the prior distributions may override the likelihood (i.e., the information contained in the data) in determining the posterior distribution. The problem then becomes to adequately justify the chosen priors based on sound background knowledge on the research question (which we do not have for the lizards). In cases of complex models with limited data, we should always make a prior sensitivity analysis to assess whether our somewhat arbitrary choice of priors has an undue influence on the results (cf. Chapters 6 and 7).

After all these caveats, we finally run the model:

```
liza4.brms = brm(formula = n ~ (species + sun + height + perch + time)^2,
family = negbinomial,data = liz, prior = prior.liz4, warmup = 1000,
chains = 3, iter = 2000, future = T, control = list(adapt_delta = 0.9))
```

The summary is:

```
> summary(liza4.brms)
 Family: negbinomial
  Links: mu = log; shape = identity
Formula: n ~ (species + sun + height + perch + time)^2
   Data: liz (Number of observations: 48)
Samples: 3 chains, each with iter = 2000; warmup = 1000; thin = 1;
         total post-warmup samples = 3000

Population-Level Effects:
                            Estimate Est.Error l-95% CI u-95% CI Rhat Bulk_ESS Tail_ESS
Intercept                       0.96      0.45     0.04     1.84 1.00     2212     2070
speciesopalinus                 1.18      0.48     0.22     2.14 1.00     2462     2038
sunSun                          1.32      0.47     0.40     2.23 1.00     2572     1843
heightLow                      -1.16      0.52    -2.18    -0.15 1.00     2311     2221
perchNarrow                    -0.01      0.48    -0.97     0.90 1.00     2499     2338
timeMid.day                    -0.96      0.56    -2.08     0.11 1.00     1796     1743
timeMorning                    -0.07      0.52    -1.13     0.93 1.00     2028     2348
speciesopalinus:sunSun         -0.82      0.44    -1.69     0.02 1.00     2613     1638
speciesopalinus:heightLow       1.18      0.40     0.39     1.95 1.00     3619     2342
speciesopalinus:perchNarrow    -0.81      0.39    -1.59    -0.04 1.00     3434     2226
speciesopalinus:timeMid.day     0.94      0.47     0.03     1.90 1.00     2663     2333
speciesopalinus:timeMorning     0.86      0.46    -0.02     1.79 1.00     2749     2481
sunSun:heightLow                0.13      0.41    -0.69     0.95 1.00     3074     2172
sunSun:perchNarrow              0.27      0.39    -0.49     1.04 1.00     3305     2049
sunSun:timeMid.day              1.78      0.48     0.85     2.72 1.00     2509     2017
sunSun:timeMorning              0.16      0.45    -0.69     1.01 1.00     2581     2205
heightLow:perchNarrow          -0.61      0.37    -1.35     0.11 1.00     4368     2137
heightLow:timeMid.day          -0.46      0.46    -1.33     0.41 1.00     3595     2337
heightLow:timeMorning          -0.19      0.44    -1.05     0.68 1.00     3320     2239
perchNarrow:timeMid.day         0.13      0.45    -0.75     1.02 1.00     3076     1868
perchNarrow:timeMorning         0.03      0.44    -0.85     0.86 1.00     3087     2366

Family Specific Parameters:
      Estimate Est.Error l-95% CI u-95% CI Rhat Bulk_ESS Tail_ESS
shape     5.41      1.87     2.40     9.79 1.00     2143     1725
```

A first thing to notice is that for all model parameters `Rhat` is equal to one, and that the effective sample sizes are all very large. These both suggest the convergence of the chains to a stationary posterior distribution containing many statistically independent sampled values, and the convergence is confirmed by Fig. 11.8.

Just like its frequentist analogue, the model `liza4.brms` contains the single effects and all pairwise interactions of the explanatory variables. In other words, it is almost equivalent to carrying out 10 separate models, each of which with just one pair of categorical explanatory variables and their interactions. It would be extremely tedious to provide a lengthy verbal explanation of all the parameters. The command `mcmc_plot(liza4.brms, type = "intervals", pars = "^b", ...)` would plot the means and the 50% and 95% credible intervals for all the parameters of `liza4.brms` shown in the summary, but this would not be a useful display in a model with so many parameters. The interpretation of the `perch:height` interaction is exactly the same as in Section 11.3 for the frequentist model `liza3`, and summarized in Table 11.1, except

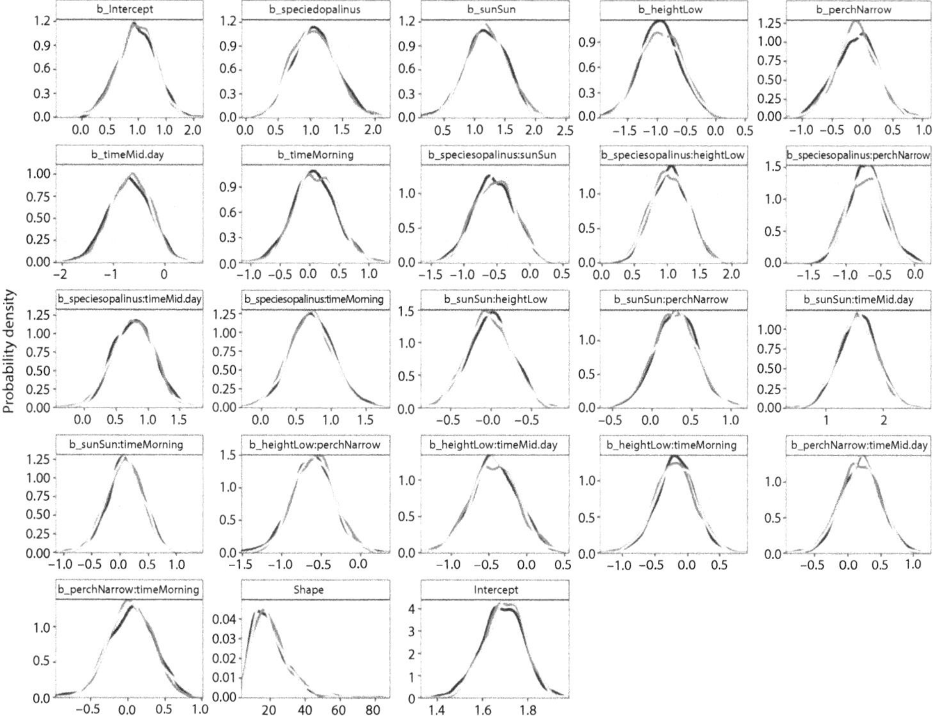

Fig. 11.8 Marginal posterior distributions of each model parameter for the three chains used to fit the Bayesian count GLM `liza4.brms`. The great degree of coincidence of the chains suggests their convergence to a posterior stationary distribution.

that we would now use the means of the posterior distributions rather than the maximum likelihood estimates. The command `bayes_R2` calculates Gelman et al.'s (2018) metric showing that the model accounts for a large proportion (0.914, with 95% credible interval [0.811, 0.966]) of the variation of the lizard frequencies.

We now turn to assessing the goodness of fit of the model `liza4.brms` using randomized quantile residuals (Chapter 8). As already done in Chapters 9 and 10, we start by obtaining the posterior predictive distribution, `post.pred.liza4.brms = predict(liza4.brms, nsamples = 1e3, summary = F)`, that is a required input to simulate the randomized quantile residuals:

```
qres.liza4.brms = createDHARMa(simulatedResponse = t(post.pred.liza4.
brms), observedResponse = liz$n, fittedPredictedResponse = apply(post.
pred.liza4.brms, 2, median), integerResponse = T)
```

which are then converted to have a normal distribution: `res.liza4.brms = data.frame(.std.resid = qnorm(residuals(qres.liza4.brms)))`. We finally calculate the LOO-CV Pareto's *k* index (Chapter 7), and put all the components needed to generate the plots of Fig. 11.9 into the data frame `res.liza4.brms`:

```
res.liza4.brms=cbind(res.liza4.brms,
      liz[,c("species", "height", "perch", "time", "sun")],
      fitted=fitted(liza4.brms, nsamples=1000)[,1],
      pareto=loo(liza4.brms, pointwise=T,moment_match = T)$diagnostics$pareto_k)
```

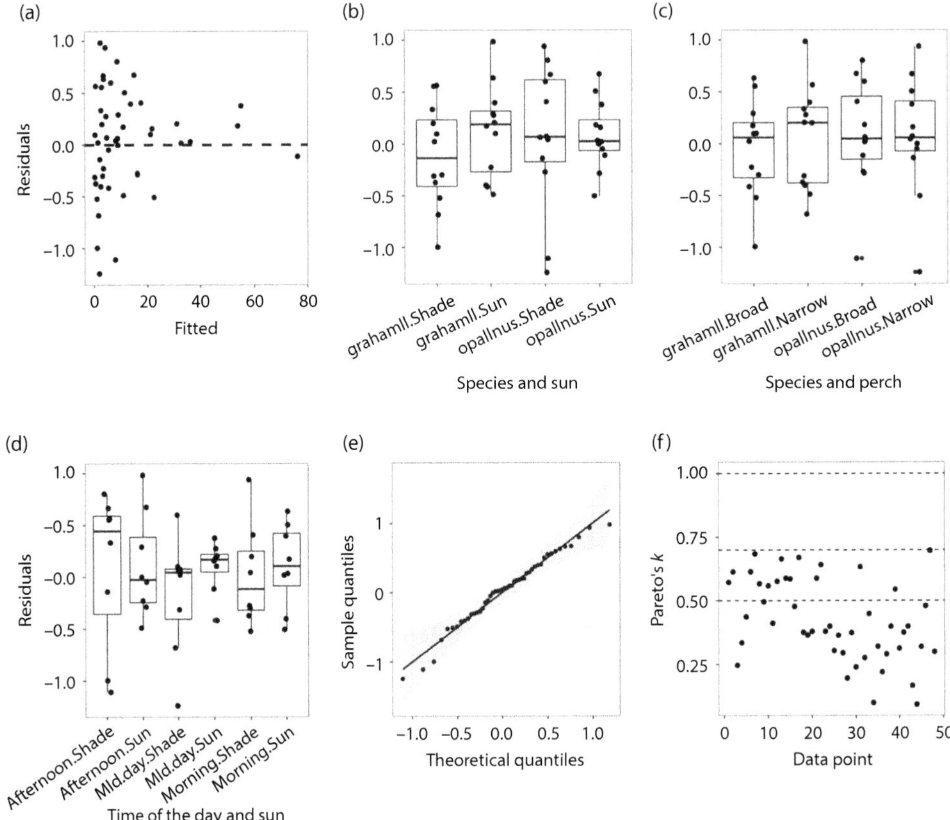

Fig. 11.9 Residual analysis of the Bayesian count GLM `liza4.brms` based on randomized quantile residuals converted to normal residuals, showing the mean of the Pearson residuals vs. the mean of the fitted value (a), the mean of the Pearson residuals vs. the combination of explanatory variables in the model (b to d), the quantile–quantile plot of the randomized quantile residuals (e), and the Pareto *k* cross-validation statistic for each data point (f).

Model `liza4.brms` has a barely decent fit to the data. The points in plots Fig. 11.9a–d show a random scatter without a clear trend or differences in variation among combinations of levels of explanatory variables, and the simulated residuals have a normal distribution (Fig.11.9e). A few warnings: we should not over-interpret the right-hand side of Fig.11.9a because there are very few points; the boxplots showing some (but not all!) combinations of categorical explanatory variables are based on very few points and thus assessments of residuals having homogeneous variability are tenuous. The overall goodness of fit of `liza4.brms` is perhaps not too surprising, given that we are fitting an overly complex model using limited data.

We can also calculate a distribution of values for the same index of over-dispersion for Bayesian count GLMs based on the Pearson residuals (Chapter 10) to compute the Pearson statistic (Chapter 8):

```
Res.Pears.liza4.brms=residuals(liza4.brms, type='pearson',summary=FALSE)
index.overdispersion.liza4.brms =apply(Res.Pears.liza4.brms^2,1,sum)/
            (nrow(liz)-loo(liza4.brms)$estimates [2,1])
```

whose summary is:

```
> summary(index.overdispersion.liza4.brms)
   Min. 1st Qu.  Median    Mean 3rd Qu.    Max.
  0.341   0.563   0.658   0.708   0.798   3.046
```

Taking the median (because it has a skewed distribution) value of the index of over-dispersion, we see that the model `liza4.brms` appears to be under-dispersed. What are the possible non-mutually-exclusive causes of under-dispersion? We may list an inadequate choice of the negative binomial distribution, an incorrectly specified model missing important terms or effects, and the unreasonable influence of poorly specified and unjustifiable priors.

Let's change the priors for the shape parameter of the negative binomial distribution to make it less informative (i.e., more spread out, Fig 11.7) than in our first model:

```
prior.liz4.1=c(set_prior("normal(5,2)", class = "Intercept"),
               set_prior("normal(0,2)", class = "b"),
               set_prior("gamma(0.1,5)", class = "shape"))
```

We rerun the same model, and show the percentage differences between the means of the posterior distributions of fitted parameters and of the shape parameters of the negative binomial distribution between the two models:

```
liza4.1.brms=brm(formula=n~(species+sun+height+perch+time)^2,
          family= negbinomial, data=liz, prior = prior.liz4.1, warmup = 1000,
          chains=3, iter=2000, future=T,control = list(adapt_delta =0.95),
          save_pars = save_pars(all = TRUE))
```

```
> 100*(fixef(liza4.1.brms)[,1]-fixef(liza4.brms)[,1])/fixef(liza4.brms)[,1]
                    Intercept                 speciesopalinus                          sunSun
                         8.46                           -7.72                           -6.82
                    heightLow                     perchNarrow                     timeMid.day
                        -3.76                           -9.09                          -35.00
                  timeMorning          speciesopalinus:sunSun       speciesopalinus:heightLow
                       -11.71                          -32.27                          -10.33
  speciesopalinus:perchNarrow speciesopalinus:timeMid.day speciesopalinus:timeMorning
                       -16.30                          -18.08                            4.63
             sunSun:heightLow              sunSun:perchNarrow              sunSun:timeMid.day
                      -592.54                           18.74                          -10.94
           sunSun:timeMorning           heightLow:perchNarrow           heightLow:timeMid.day
                        10.95                            4.86                            4.95
        heightLow:timeMorning         perchNarrow:timeMid.day         perchNarrow:timeMorning
                        20.20                         -108.81                         -152.25
> summary(liza4.brms)$ spec_pars[,1]
[1] 20.5
> summary(liza4.1.brms)$ spec_pars[,1]
[1] 1.95
```

We can readily see that there are huge differences between the two models, particularly in the shape parameter of the negative binomial distribution that governs the relation between the mean and the variance of the response variable (Chapter 10). Calculating the index of over-dispersion of the latest model using the same procedure renders:

```
> summary(index.overdispersion.liza4.1.brms)
   Min. 1st Qu.  Median    Mean 3rd Qu.    Max.
   0.11    0.25    0.34    0.44    0.51    5.22
```

The under-dispersion of `liza.4.1.brms` is clearly worse than that of `liza.4.brms`. This is not at all pleasing. These results suggest that we should be extremely wary of the parameter estimates of `liza.4.1.brms`, and of any eventual interpretation arising from them.

This is not to say that the results are "wrong," but they have limited reliability. We are using a small and very unbalanced data set to estimated a very complex model. We should really go back to the "drawing board" and think very carefully about which very few effects we might be able to assess with the limited information at hand. You may find it unsettling that changing priors may have such an impact on posterior distributions. If so, you would not be alone, for this has been the fundamental point of contention when using Bayesian methods in statistics. An unduly strong influence of priors is more likely to happen for small data sets conveying limited empirical information to fit statistical models. Large data sets often, but not always, contain enough information to swamp most priors. In defense of Bayesian methods, the choice of priors is open, nearly always explicit, and it often requires some justification based on actual (or the absence) of prior knowledge. That is, we can always perform a prior sensitivity analysis to evaluate their putative influence on posterior distributions. The entire procedure in Bayesian statistics, ranging from the selection of priors, the choice of the probability distribution for the likelihood, and all details of the MCMC algorithm, is open and available for critical re-evaluation. The same cannot always be said of frequentist methods that often involve a large amount of personal decision-making that is often not transparent (Gelman and Hennig 2017, DePaoli and van de Schoot 2017). Should you read the frequentist analysis of the same count GLM you would find a series of warnings about the inferential reliability of modeling results. Indeed, there is no "silver statistical bullet" that can provide an aura of reliability to a limited data set when used to fit complex models.

Despite all of these warnings, we could always display the conditional plots for the different pairwise interactions included in `liza4.brms` with `liza4.brms.cond.eff = conditional_effects(liza4.brms)`. The resulting object `liza4.brms.cond.eff` is a list whose contents can be examined:

```
> names(liza4.brms.cond.eff)
 [1] "species"        "sun"            "height"         "perch"          "time"
 [6] "species:sun"    "species:height" "species:perch"  "species:time"   "sun:height"
[11] "sun:perch"      "sun:time"       "height:perch"   "height:time"    "perch:time"
```

These are the single effects and the pairwise interactions between explanatory variables; `conditional_effects(liza4.brms)[[6]]` will render the marginal plot for the `sun:species` interaction, and so on.

Figure 11.10 shows nine conditional effects (there was not enough room to display the tenth!) predicted by the model `liza4.brms`. These conditional effects for pairwise interactions were obtained assuming that all other categorical explanatory variables were at their reference levels (see Section 11.3). Each plot in Fig. 11.10 displays the degree to which the interaction between each pair of explanatory variables affects the expected lizard counts. Needless to say, given the previous discussion of the goodness of fit and the high impact of unjustified priors, we should take these conditional plots with more than a grain of salt when using them to interpret the results.

In closing, we do not feel at all disappointed for having shown you modeling results that were not entirely trustworthy. While most statistics books typically show "success stories" of well-fitting models, we all know that the reality of dealing with limited and noisy data often leads to murkier results such as the ones just discussed.

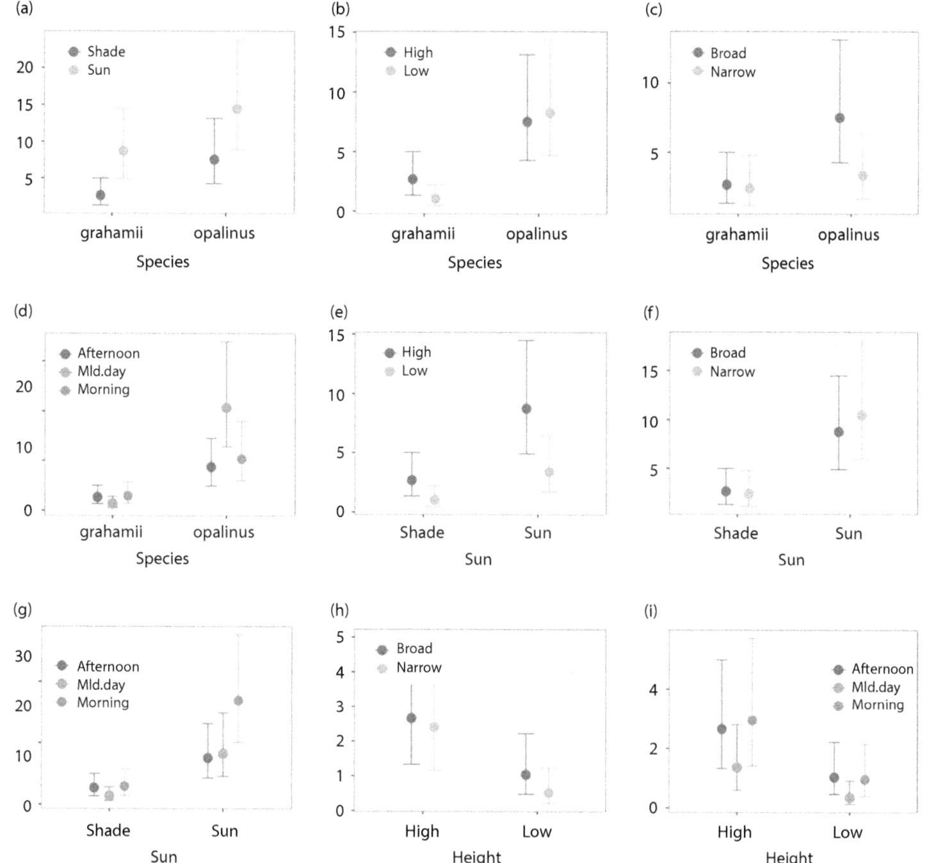

Fig. 11.10 Conditional plots of the relations between the response and interacting explanatory variables for the Bayesian count GLM `liza4.brms`, showing the means and the 95% credible intervals.

11.5 Problems

Visit the companion website at http://www.oup.com/companion/InchaustiSMWR to obtain the data sets for these problems.

11.1 Sinclair and Arcese (1995) studied the association between sex, health, and cause of death in wildebeest in eastern Africa. They cross-classified wildebeest carcasses from the Serengeti according to sex, cause of death (predation, non-predation), and bone marrow type (solid white fatty, opaque gelatinous, translucent gelatinous). The color and consistency of bone marrow is apparently a reasonably good indication of the health status of an animal, even after death, with the first type indicating a healthy animal that was not undernourished. File: Pr11-1.csv

11.2 Ozgul et al. (2009) investigated whether the number of shells from freshly dead gopher tortoises (shells) varied in relation to the mycoplasmal prevalence (prev) and the population density (dens) at different sites and over time. Note that population density does not change over time for each site. You should consider both the area of search and the population density at each site in the analysis. File: Pr11-2.csv

References

DePaoli, S. and van de Schoot, R. (2017). Improving transparency and replication in Bayesian statistics: The WAMBS-Checklist. *Psychological Methods*, 22, 240–261.

Gelman, A., Goodrich, B., Gabry, J. et al. (2018). R-squared for Bayesian regression models. *American Statistician*, 73, 307–309.

Gelman, A. and Hennig, C. (2017). Beyond subjective and objective in statistics. *Journal of the Royal Statistics Society A*, 180, 1–31.

Ozgul, A., Oli, M., Bolker, B. et al. (2009) Upper respiratory tract disease, force of infection, and effects on survival of gopher tortoises. *Ecological Applications*, 19, 786–798.

Schoener, T. (1970). Nonsychronous spatial overlap of lizards in patchy habitats. *Ecology*, 51, 408–418.

Sinclair, A. and Arcese, P. (1995) Population consequences of predation-sensitive foraging: The Serengeti wildebeest. *Ecology*, 76, 882–891.

von Eye, A. and Mun, E. (2012). Log-linear Modeling: Concepts, Applications and Interpretations. John Wiley & Sons, New York.

CHAPTER 12

Models for Positive, Real-Valued Response Variables

Proportions and others

Packages needed in this chapter:

```
packages.needed<-c("ggplot2","fitdistrplus","brms","broom","GGally",
"DHARMa","gamlss.dist","broom.mixed","ggeffects","qqplotr",
"bayesplot", "glmmTMB)
lapply(packages.needed, FUN = require, character.only = T)# loads these packages in the working session.
```

12.1 Introduction

This chapter explains the modeling of real-valued and bounded response variables using probability distributions that are less often taught, and hence are under-used by scientists. We first discuss the modeling of proportions (i.e., real values in the unit interval), and then consider other strictly positive real-valued response variables. All models are fitted in the frequentist and Bayesian frameworks. We finish by discussing (but not analyzing data for) other still less common cases involving positive, real-valued response variables including zeros, and complex relations between mean and variances that might arise for some data sets.

12.2 Modeling proportions

Proportional data can arise either from counts, or from continuous numbers (Douma and Weldon 2019, Smithson and Merkle 2014). When they arise from counts, there can be two cases. First, if we have a binary outcome we can compute the proportion of "successes" (defined as whatever we are interested in) for a set of values of the explanatory variables, and perform either a grouped binary GLM (Chapter 9) or a count GLM with an offset (Chapter 11). Second, the response variable can, of course, be more than two mutually exclusive categories (out of which we can always compute the proportions of each category). In this case, the corresponding analysis would be a either multinomial or an ordinal regression depending on whether an ordering of the categories is meaningful. The latter cases may be thought of as a generalization of binary GLMs (Chapter 9) that use different link functions. Multinomial and ordinal models are not considered in this book,

Statistical Modeling With R. Pablo Inchausti, Oxford University Press. © Pablo Inchausti (2023).
DOI: 10.1093/oso/9780192859013.003.0012

and we refer you to Agresti (2010) and Long (2007) for details. In general, proportional data arising from counts are just a re-expression of the counts in another scale. Therefore, they should be modeled as the underlying counts using the appropriate type of GLM.

When the response variables are proportions resulting from continuous numbers, they can be analyzed as a GLM. The key difference with the previous case is that we now truly register (rather than calculate) proportions as the response variables. When there are two mutually exclusive categories, we must use the beta distribution. When there are more than two categories, we should use the Dirichlet distribution (Douma and Weldon 2019). An example of the latter are compositional data (i.e., a set of more than two proportions adding up to one) that arise in chemistry, dietary studies, economics, behavioral ecology, psychology, and other fields of research. These compositional data should be analyzed using the Dirichlet distribution, which will not be considered in this book. Because proportions often have skewed, heteroskedastic (i.e., non-constant variance) distributions, different non-linear transformations (such as the arcsine of the square root of the proportion) were used to tame them in order to use the general linear model. In general, these transformations of proportions do not achieve their goal: they lead to biased and unreliable parameter estimates, and thus should be avoided (Warton and Hui 2011, Douma and Weldon 2019).

The more modern and correct approach is to model the response variable using the beta distribution (Chapter 8). The origins of the beta distribution are rather murky, and it is likely to have been independently invented several times before obtaining its current name in the early 1940s. Thomas Bayes used it in his historic, posthumous 1763 paper; Karl Pearson obtained it in 1895 as a solution to a differential equation that became a Pearson type I distribution, and the Italian statistician Conrado Gini deduced it in 1911 in his work with actuarial data. Originally defined in terms of two shape parameters (a, b) pulling the probability density towards either of the two alternative categories, the beta distribution was reformulated in terms of its mean μ_Y and a dispersion parameter ϕ to make it more like other GLMs (Ferrari and Cribari-Neto 2004):

$$\Pr(Y = y) = \frac{\Gamma(\phi)}{\Gamma(\mu_Y \phi)\,\Gamma((1-\mu_Y)\,\phi)} y^{\mu_Y \phi - 1} (1-y)^{(1-\mu_Y)} \phi - 1, \tag{12.1}$$

where Γ() is the gamma function, an analogue of factorials for real numbers. The variance of a beta distribution is related to its mean and to the dispersion parameter ϕ, $\mathrm{Var}(Y) = \frac{\mu_Y(1-\mu_Y)}{(1+\phi)}$, such that we can think of the latter as an inverse index of precision: the lower ϕ is, the higher the variance. The beta distribution has great plasticity, assuming a wide variety of shapes, which makes it suitable for modeling proportions (Fig. 12.1).

Strictly speaking, the beta distribution is not part of the exponential family that lies at the heart of McCullagh and Nelder's (1990) GLM formulation because, unlike other distributions of the family, the parameters μ_Y and ϕ of the beta distribution are not strictly orthogonal (Smithson and Verkuilen 2006). An alternative is to consider the exponential dispersion models (or family) of Smyth (1989) and Jorgensen (1997) that extend and generalize the exponential family to include other probability distributions. In the framework of the exponential dispersion family, Ferrari and Cribari-Neto (2004) developed a quasi-likelihood method that permits the formulation of a GLM for proportions. Quasi-likelihood models allow separate specification of the mean and the variance structure of a GLM, and estimation of the relations between μ_Y and ϕ and the explanatory variables as part of different sub-models in the scale of different link functions (Smyth 1989). It is to be noted that μ_Y and ϕ place no restriction on each other, so they may be modeled

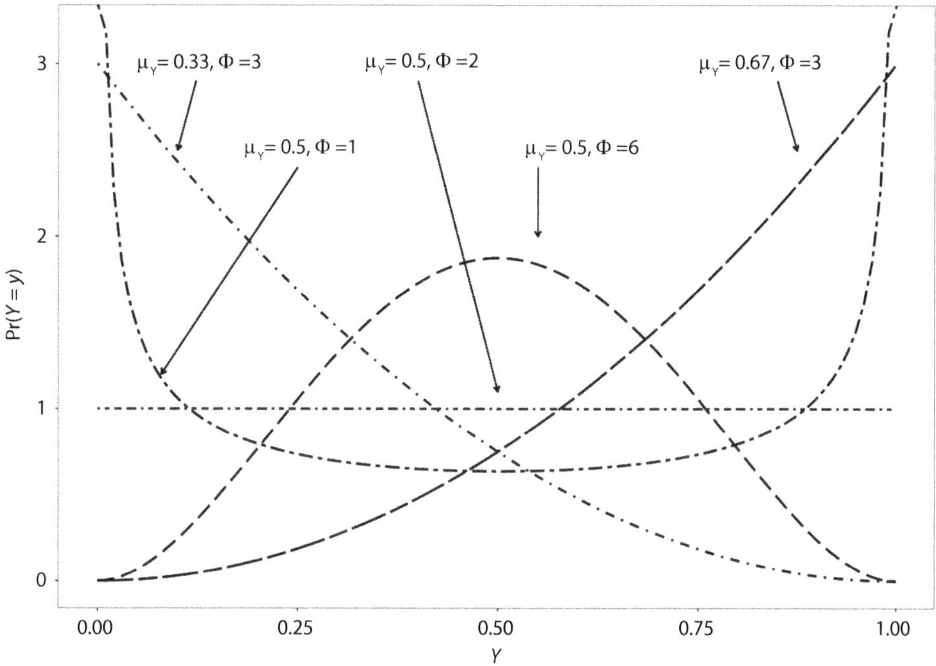

Fig. 12.1 Probability density for different combinations of parameters of the beta distribution.

separately. We are going to use a logit link function for μ_Y and a log link function for the dispersion parameter ϕ, such that each inverted link function renders predictions in the correct range for each parameter: [0, 1] for μ_Y and strictly positive values for ϕ. Following Smyth's (1989) formulation of two sub-models, Smithson and Verkuilen (2006) suggested that ϕ should not be treated as an "overall nuisance parameter" to be estimated and dismissed. This is because modeling ϕ in terms of explanatory variables may be of great help to model the variation observed in many proportion data sets (Douma and Weldon 2019).

Using the logit link function for μ_Y prevents having values of the response variable that are exactly zero or one, for which the link $\log(Y \ / \ (1 - Y))$ function is undefined. The beta distribution is defined for both zero and one, but the logit link function is not. One solution in such cases is to rescale the response variable by $Y^* = \frac{Y(n-1)+1/2}{n}$ such that Y* is in [0.005, 0.995] when $n = 100$ and in [0.0005, 0.9995] when $n = 1000$ (Smithson and Verkuilen 2006). Finally, whenever a response variable has known theoretical minimum and maximum values, we may transform it into the unit interval with $Y^* = \frac{(Y-\min)}{(\max-\min)}$ and then model it with a beta distribution. This extends the use of the beta distribution to modeling other bounded, real-valued response variables.

12.3 Plant cover, grazing, and productivity

The data in context: Oñatibia et al. (2018) studied the effect of sheep grazing pressure in three rangeland communities located along a regional productivity gradient in Southern Argentina. They wanted to know how total plant cover changed with grazing pressure, and whether these effects changed with primary productivity. The data came from a total of 53 transects laid in three areas of different primary productivity, where the authors also measured the number of patches and the average distance between vegetation patches. The total plant cover expressed as a proportion is particularly useful in non-destructive samplings, and when it is impossible to differentiate plant individuals as in perennial grassland communities. The response variable is a proportion that will be modeled using the beta distribution, $Y \sim \text{Beta}(\mu_Y, \phi)$ with $\text{logit}(\mu_Y) = X\beta$, $\log(\phi) = X\beta$, where X are the numerical and categorical explanatory variables, and β the parameters to be estimated. We will fit models with and without explicit modeling of the effect of the explanatory variables on the dispersion parameter ϕ, and compare them afterwards.

EXPLORATORY DATA ANALYSIS: We start by importing and summarizing the data:

```
> cover=read.csv ("Plant cover Chapter 12.csv", header=T)
> str(cover)
'data.frame':   53 obs. of  6 variables:
 $ Prod             : Factor w/ 3 levels "High","Low","Medium": 2 2 2 2 2 2 2 2 2 2 ...
 $ Gr.pressure      : num  0.551 0.391 0.161 0.201 0.237 ...
 $ patch.size       : num  17.5 20.4 15.8 20.1 20.3 16.5 18.7 27.6 21.4 30.7 ...
 $ inter.patch.dist: num  31.5 24.4 25.6 21.8 22.8 32.3 26.7 23.5 24.8 22.2 ...
 $ Nb.patches       : int  204 222 240 238 230 204 222 194 216 188 ...
 $ Totalcover       : num  0.356 0.452 0.379 0.477 0.47 0.337 0.414 0.536 0.461 0.576 ...
```

We see that the response variable (`Totalcover`) has neither zero nor one values that need to be dealt with. Just for the sake of consistency with other chapters, we explore the best probability distribution for the response variable using the package `fitdistrplus`. Despite the plasticity of the beta distribution (Fig. 12.2) that makes it unlikely that it may not fit the response variable well, we are also going to consider the gamma and normal distributions:

```
beta.cover=fitdist(cover$Totalcover,"beta")
normal.cover=fitdist(cover$Totalcover,"norm")
gamma.cover=fitdist(cover$Totalcover,"gamma")
```

Figure 12.2 shows that the three distributions are indistinguishable. The beta distribution has the smallest AIC, but the difference from the other three distributions is not decisive by any means:

```
> gofstat(list(beta.cover, normal.cover, gamma.cover))$aic
 1-mle-beta  2-mle-norm 3-mle-gamma
      -89.1       -88.5       -87.1
```

Given the advantages of using the logit link function for proportions, which guarantees that model predictions will be in the unit interval, we reach the foretold conclusion of using the beta distribution to model the variation of the response variable.

We next explore the relations between the response and explanatory variables using the package `GGally` as shown in Fig. 12.3. The bottom row of Fig. 12.3 shows that `Totalcover` is negatively related to `patch.dist`, `Gr.pressure`, and `Nb.patches`, and positively related to `patch.size`, and that the response variable also differed with `Prod` (top right). We can also observe (bottom row of Fig. 12.3) that the trend lines suggest that the relations between `Totalcover` and `Gr.pressure` and `patch.size` might differ depending on `Prod` (i.e., a possible interaction between these explanatory variables), but not so clearly for `Nb.patches` and `inter.patch.dist`. Because the latter two variables were strongly correlated (–0.840), we are going to follow Dormann et al.'s (2013) heuristic suggestion and exclude one of them (`Nb.patches`) from the analysis. We standardize the numerical explanatory variables prior to the analysis to be able to compare their relative effects, and include `Totalcover` and `Prod` in a new data frame: `cover.s = data.frame(scale(cover[c("Gr.pressure", "patch.size", "inter.patch.dist")], scale = T, center = T), cover`

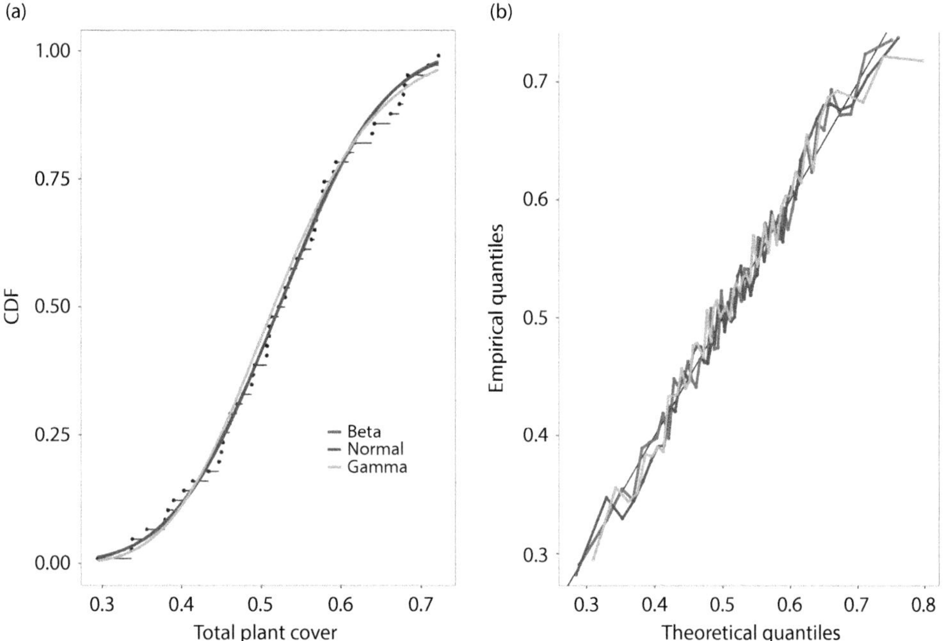

Fig. 12.2 Goodness-of-fit assessment of the response variable total plant cover showing the cumulative distribution function (a) and a quantile–quantile plot (b) comparing the beta, normal, and gamma distributions using the package `fitdistrplus`.

`= cover$Totalcover, Prod = cover$Prod)`. The fitted models will contain the intercept (corresponding to logit(μ_Y) for the reference group of the categorical explanatory variable `Prod`), the partial slopes for the reference group of `Prod` and the numerical explanatory variables (`Gr.pressure` and `patch.size`) involved in interactions with it and for `inter.patch.dist`, the changes in the intercept and in partial slopes between the reference level of `Prod` and the other two levels of this explanatory variable. This model is equivalent to multiple (because it contains several explanatory variables) analysis of covariance (because it has both categorical and numerical explanatory variables) fitted in the logit link scale (cf. Chapters 6 and 9).

12.4 Frequentist fitting of a GLM on proportions

We use the package `glmmTMB` to fit the models:

```
m1=glmmTMB(Totalcover~Gr.pressure*Prod+patch.size*Prod + inter.patch.dist,
      dispformula = ~1,    data=cover.s, family=beta_family)
```

and modeling the dispersal parameter:

```
m1.1=glmmTMB(Totalcover~Gr.pressure*Prod+patch.size*Prod + inter.patch.dist,
             dispformula= ~Gr.pressure*Prod+patch.size*Prod +inter.patch.
             dist, data=cover.s, family=beta_family)
```

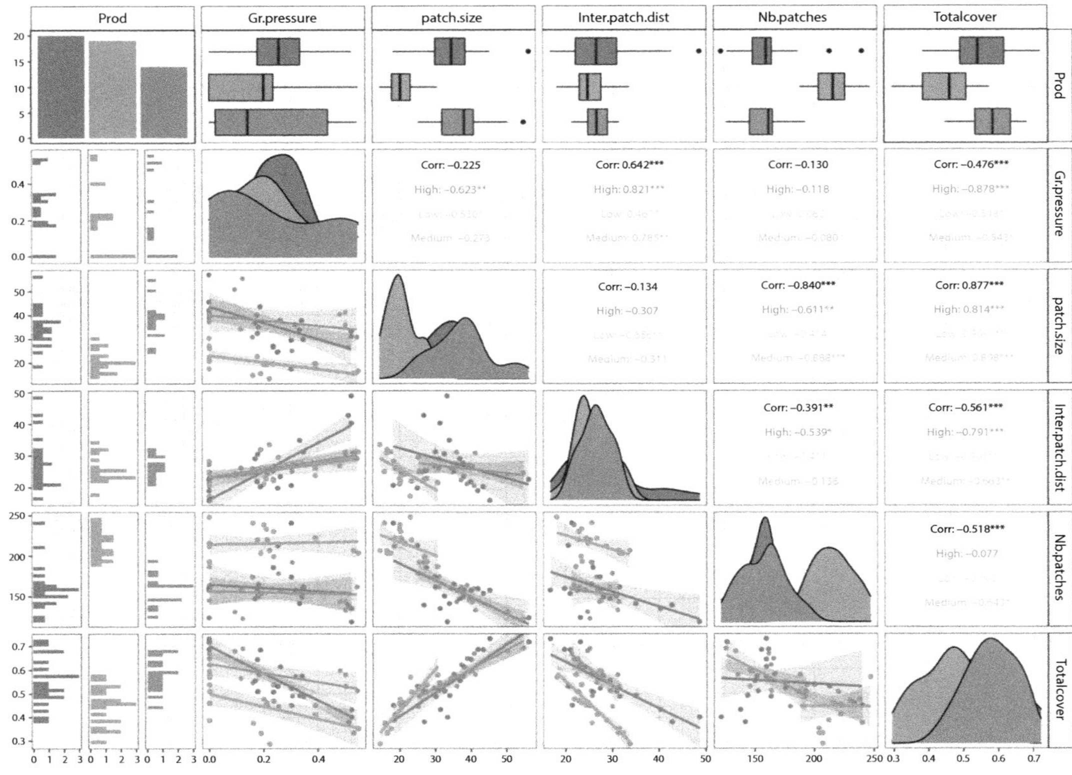

Fig. 12.3 Covariation between the explanatory variables in the proportion GLM. The diagonal shows the density plots of each explanatory variable for the entire data set depending on the value of the response variable. Below the diagonal are shown all pairwise x–y plots, with presences colored in red and absences in green. Above the diagonal are the pairwise correlations of explanatory variables for the entire data set and for each value of the response variable. The figure was produced with the `GGally` package.

We can use either the AIC or the likelihood ratio test (Chapter 7) to compare these models:

```
> anova(m1,m1.1)
     Df  AIC  BIC logLik deviance Chisq Chi Df Pr(>Chisq)
m1   11 -307 -285    164     -329
m1.1 20 -319 -280    180     -359  30.3      9    0.00038 ***
```

It is clear that despite being more complex, `m1.1` has a much smaller AIC, and that the difference between their deviances is statistically significant, and thus `m1.1` should be preferred. The summary of the preferred model is:

```
> summary(m1.1)
 Family: beta  ( logit )
Formula:          Totalcover ~ Gr.pressure * Prod + patch.size * Prod + inter.patch.dist
Dispersion:                  ~Gr.pressure * Prod + patch.size * Prod + inter.patch.dist
Data: cover.s

     AIC      BIC   logLik deviance df.resid
    -319     -280      180     -359       33

Conditional model:
                      Estimate Std. Error z value Pr(>|z|)
(Intercept)            0.10274    0.01090    9.43  < 2e-16 ***
Gr.pressure            0.06298    0.01857    3.39  0.00069 ***
ProdLow                0.26770    0.03709    7.22  5.3e-13 ***
ProdMedium             0.01714    0.03894    0.44  0.65978
patch.size             0.41799    0.01802   23.20  < 2e-16 ***
inter.patch.dist      -0.20912    0.00726  -28.79  < 2e-16 ***
Gr.pressure:ProdLow   -0.06512    0.01811   -3.60  0.00032 ***
Gr.pressure:ProdMedium -0.12044   0.02669   -4.51  6.4e-06 ***
ProdLow:patch.size     0.20613    0.03807    5.42  6.1e-08 ***
ProdMedium:patch.size -0.16779    0.03852   -4.36  1.3e-05 ***
---
Signif. codes:  0 '***' 0.001 '**' 0.01 '*' 0.05 '.' 0.1 ' ' 1

Dispersion model:
                      Estimate Std. Error z value Pr(>|z|)
(Intercept)             9.1811     0.6477   14.18  < 2e-16 ***
Gr.pressure            -1.1750     1.0806   -1.09    0.277
ProdLow                -5.1767     1.2222   -4.24  2.3e-05 ***
ProdMedium             -1.4609     1.2169   -1.20    0.230
patch.size             -2.3153     0.9330   -2.48    0.013 *
inter.patch.dist        0.0337     0.4974    0.07    0.946
Gr.pressure:ProdLow     1.1533     1.1030    1.05    0.296
Gr.pressure:ProdMedium -0.3366     1.0616   -0.32    0.751
ProdLow:patch.size     -2.6422     1.7873   -1.48    0.139
ProdMedium:patch.size   2.3177     1.6981    1.36    0.172
---
Signif. codes:  0 '***' 0.001 '**' 0.01 '*' 0.05 '.' 0.1 ' ' 1
```

The first part of the summary ("Conditional model") show the relation between μ_Y and the explanatory variables and their interactions in the logit link scale. The second part ("Dispersion model") does likewise with the dispersion parameter ϕ in the log link scale. The effects of the model terms in the two parts of the summary can be separately interpreted. However, the predictions and goodness of fit of both models critically depend on the adequacy of the modeling of the conditional effects and the dispersion.

Considering the conditional part that uses the logit link function, the reference level for the categorical variable `Prod` is High, and hence the parameter estimates `ProdLow` and `ProdMedium` correspond to differences in the intercept. Therefore, logit(μ_Y) of `Prod` = Low when all other variables have their mean values (let's recall that they were standardized prior to the analysis) is 0.102 + 0.268 = 0.370, and for `Prod` = Medium is 0.102 + 0.017 = 0.119. The partial slope for `inter.patch.dist` is the same for the three levels of `Prod`, as the model did not contain an interaction between these explanatory variables. For the two explanatory variables involved in interaction with `Prod`, the model shows the differences in partial slopes between each level of the latter variable and the reference level `Prod` = High. We can easily calculate the partial slopes for `Gr.pressure` for each level of `Prod`: 0.063 (High), 0.063 – 0.065 = –0.002 (Low), and 0.063 – 0.120

= −0.057 (Medium). The corresponding partial slopes for `patch.size` are 0.418 (High), 0.624 (Low), and 0.250 (Medium). If we were aiming to obtain the most parsimonious model that can be fitted to the data, we could have attempted to simplify terms in both the conditional and dispersion parts of `m1.1` using either the Wilks test or the AIC (see an example in Chapter 11).

While every term in the conditional model was statistically significant, not every term had similarly sized importance on the mean of the response variable in the logit link scale. In Table 12.1 we can gauge the relative effects of changes in the numerical explanatory variables on the response variable by comparing their partial slopes because they were standardized prior to the analysis. The magnitudes of the partial slopes convey the effects of same relative changes (one standard deviation) on logit(μ_Y) for each level of `Prod`: `patch.size` had the strongest effect, and `Gr.pressure` the weakest. We could compute ratios of partial slopes to discern the relative effects of explanatory variables within and across the levels of `Prod`. For instance, the effect of changes in the standard deviation of the patch size on logit(μ_Y) was 1.5 (= 0.624 / 0.418) times stronger and 0.6 (= 0.250 / 0.418) weaker when `Prod` = Low or High compared with `Prod` = Medium. Being able to make such statements is far more interesting and revealing than merely declaring that "all variables have statistically significant effects." It is also of great help in interpreting the results of statistical models in a research context.

Table 12.1 Parameter estimates for frequentist beta GLM of sheep grazing data.

	Grazing	Inter-patch distance	Patch size
Low	−0.002	−0.201	0.624
Medium	−0.057	−0.201	0.418
High	0.063	−0.201	0.250

The relative changes of the explanatory variables that we have been talking about are changes of a unit standard deviation of their values:

```
> sqrt(diag(var(cover[,-c(1,6)])))
    Gr.pressure        patch.size inter.patch.dist         Nb.patches
          0.178            10.175            5.881             34.491
```

Gelman et al.'s (2014) "rule of four" (Chapter 9) for models with a logit link function can help interpret these partial slopes. Let's just take those of `patch.size`: 0.418 / 4 = 0.104 (High), 0.624 / 4 = 0.156 (Low), and 0.250 / 4 = 0.062 (Medium). Therefore, a change of one standard deviation of `patch.size` (= 10.175) implies a change of 0.104, 0.156, and 0.062 in the average `Totalcover` when `Prod` is High, Low, or Medium, respectively.

To calculate McFadden's (1974) percentage of deviance explained by `m1.1`, we must first fit a null model (i.e., without explanatory variables) to the same data. This is because the package `glmmTMB` does not provide the null model deviance, in contrast to the `glm` command used in Chapters 10 and 11. The null model is fitted with `m1.2.null = update(m1.2, ~1, dispformula = ~1)`, and then the deviances are extracted to compute:

```
> 1-(as.numeric(-2*logLik(m1.2.null))/as.numeric(-2*logLik(m1.2)))
[1] 0.741
```

(note that we inverted the ratio in the expression because both deviances were negative). The model `m1.1` accounts for 74.1% of the deviance in the data.

A simple way of visualizing the partial slopes for each level of `Prod` would be to fit the following variant of model `m1.1`:

```
m1.2=glmmTMB(Totalcover~Gr.pressure*Prod+ patch.size*Prod+inter.patch.dist-
                    patch.size-Gr.pressure-1,
           dispformula=~Gr.pressure*Prod+patch.size*Prod+inter.patch.dist-
                    patch.size-Gr.pressure-1, data=cover.s, family=beta_family)
```

Compared with `m1.1`, this model incorporates two extra columns in the design matrix and deletes the global intercept. These changes allow us to obtain the intercepts and partial slopes (rather than the differences of intercepts and partial slopes from the reference level of `Prod`, as in `m1.1`) of each explanatory variable for the three levels of `Prod`. Although formally equivalent to `m1.1`, `m1.2` is mostly useful as the input for the plots of parameter estimates that most users seem to prefer (Fig. 12.4).

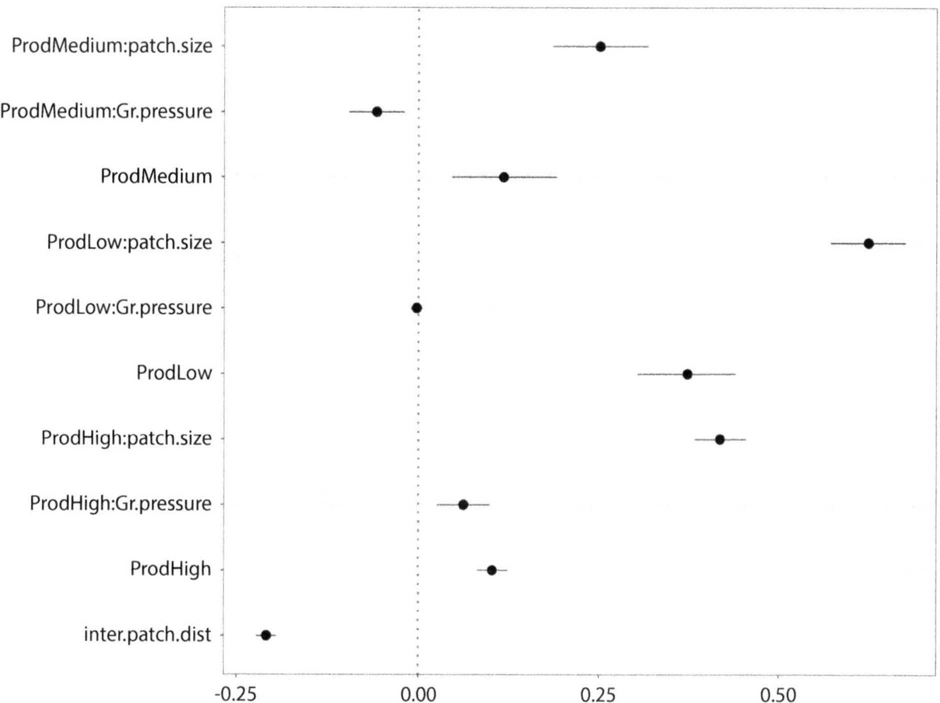

Fig. 12.4 Estimates and 95% confidence intervals of the intercept and partial slopes of the frequentist proportion GLM `m1`.

We can obtain plots of the marginal predictions of model `m1.1` using the package `ggeffects`. For that, we first need to obtain a set of predicted values for the each explanatory variable (in interaction with `Prod`, as needed) with:

```
pred.m1.1a=ggpredict(m1.1, terms = c("Gr.pressure [all]", "Prod"))
pred.m1.1b=ggpredict(m1.1, terms = c("patch.size [all]","Prod"))
pred.m1.1c=ggpredict(m1.1, terms = c("inter.patch.dist [all]"))
pred.m1.1d=ggpredict(m1.1, terms = c("Prod [all]"))
```

Model `m1.1` requires values of the five explanatory variables to generate a prediction. Each plot in Fig. 12.5 renders the predicted conditional means and 95% confidence interval, setting the values of all other variables not included to zero (i.e., to their means).

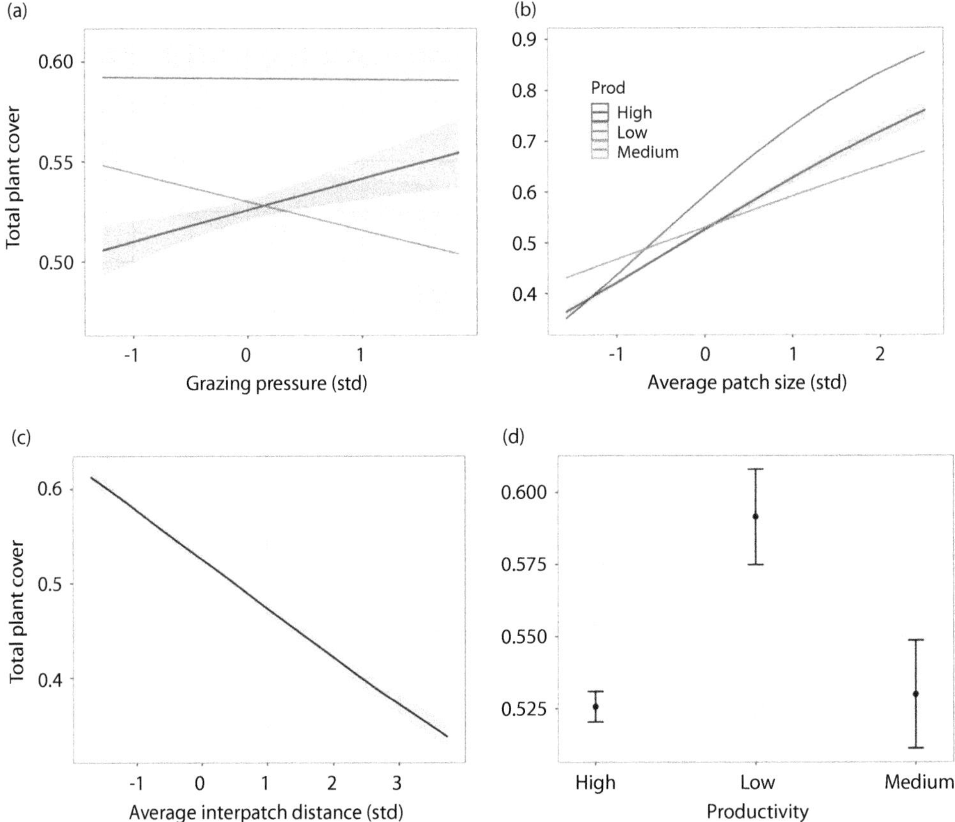

Fig. 12.5 Conditional plots of the frequentist proportion GLM `m1` showing the predicted relations between the standardized explanatory variables (mean of zero and unit standard deviation) and the response variable. The bands or intervals in all the plots correspond to the 95% confidence interval around each predicted relation. The plots show the predicted relation for each variable, holding all other variables at zero (i.e., their mean values).

The final step is to assess the goodness of fit of model `m1.2` with randomized quantile residuals (Chapter 8) obtained with the package `DHARMa`. This requires us to first simulate these residuals, `res.m1.1 = simulateResiduals(fittedModel = m1.1, n = 500, refit = F, plot = F)`, then to gather the data needed for the plots using the package `broom.mixed`, `resid.m1.1 = augment(m1.1)`, and finally to convert the simulated residuals to have a normal distribution with `resid.m1.1$.std.resid = residuals(res.m1.1, quantileFunction = qnorm)`.

Figure 12.6 shows that the goodness of fit of model `m1.1` to the data is acceptably good: the residuals have a seemingly random scatter and homogeneous variation when plotted against the fitted values and explanatory variables, and they are not too far from having a normal distribution. We did not obtain the Cook distances because the package `glmmTMB` does not produce a projection matrix (or "hat matrix") that would speed up their computation (Fox 2015). Hence, the Cook distances would need to be computed by the brute-force refitting of the model after deleting each data point at a time. If you are interested in computing the Cook distances for models fitted with the package `glmmTMB`, you will find a suitable R function on the companion website.

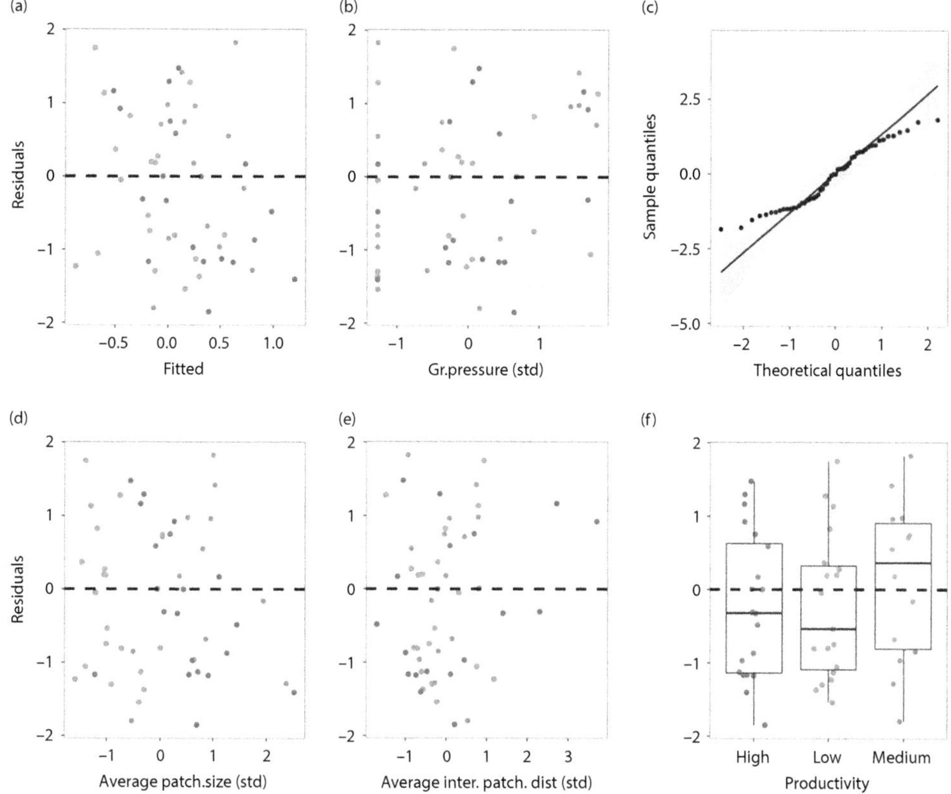

Fig. 12.6 Residual analysis of the frequentist proportion GLM `m1` based on simulated randomized quantile residuals converted to normal residuals. Plots (a), (b), (d–f) show the relations between these residuals and the fitted values, and to the explanatory variables. The dashed lines were added to help the visualization. Plot (c) shows the quantile–quantile plot (and the 95% confidence band) of the residuals.

12.5 Bayesian fitting of a GLM on proportions

We use the package `brms` to fit two models that either model or not the effect of explanatory variables for the dispersion parameter ϕ of the beta distribution. To that end, we need to write the model as:

```
formula.m1.1brms=bf(Totalcover~Gr.pressure*Prod+patch.size*Prod+inter.patch.dist)+
                    lf(phi~Gr.pressure*Prod + patch.size*Prod+inter.patch.dist)
```

and the other model as:

```
formula.m1.brms=bf(Totalcover~Gr.pressure*Prod+patch.size*Prod+inter.patch.dist)+
                   lf(phi~1)
```

A few comments: `bf` is shorthand for `brms-formula`, which allows writing separate formulae for μ_Y and for the ϕ of the beta distribution as strings of text separated by `+ lf`. Thus, `formula.m1brms` makes it explicit that ϕ is not modeled in terms of explanatory variables. The second model, `formula.m1.1brms`, does incorporate explanatory variables in the modeling of the variation of the dispersion parameter of the beta distribution. As in their frequentist analogues, the effect of explanatory variables on μ_Y and ϕ are to be respectively modeled with logit and log link functions to guarantee that the respective inverse functions generate predictions in the correct scale of each parameter: $0 < \mu_Y < 1$ and $\phi > 0$.

Let's examine the priors to be defined for the more complex model, on the understanding that the simpler `formula.m1brms` will not have the priors for the explanatory variables for ϕ:

```
> get_prior(formula.m1.1brms, data=cover.s,family='Beta')
                prior     class                   coef group resp dpar nlpar bound       source
               (flat)         b                                                          default
               (flat)         b            Gr.pressure                              (vectorized)
               (flat)         b    Gr.pressure:ProdLow                              (vectorized)
               (flat)         b Gr.pressure:ProdMedium                              (vectorized)
               (flat)         b        inter.patch.dist                             (vectorized)
               (flat)         b             patch.size                              (vectorized)
               (flat)         b                ProdLow                              (vectorized)
               (flat)         b     ProdLow:patch.size                              (vectorized)
               (flat)         b             ProdMedium                              (vectorized)
               (flat)         b  ProdMedium:patch.size                              (vectorized)
 student_t(3, 0, 2.5) Intercept                                                          default
               (flat)         b                               phi                   (vectorized)
               (flat)         b            Gr.pressure        phi                   (vectorized)
               (flat)         b    Gr.pressure:ProdLow        phi                   (vectorized)
               (flat)         b Gr.pressure:ProdMedium        phi                   (vectorized)
               (flat)         b        inter.patch.dist       phi                   (vectorized)
               (flat)         b             patch.size        phi                   (vectorized)
               (flat)         b                ProdLow        phi                   (vectorized)
               (flat)         b     ProdLow:patch.size        phi                   (vectorized)
               (flat)         b             ProdMedium        phi                   (vectorized)
               (flat)         b  ProdMedium:patch.size        phi                   (vectorized)
 student_t(3, 0, 2.5) Intercept                               phi                        default
```

The priors for model parameters are always on the scale where the model is linear, i.e., on the logit scale for μ_Y and on the log scale for ϕ in this case. As you might expect, the priors for the model not including explanatory variables for ϕ would not have the parameters enclosed by the lower bracket and labeled as `phi`. We have already set priors on the logit scale for binary GLMs (Chapter 9), and in the log scale for count GLMs (Chapters 10 and 11). One important thing to recall is the counter-intuitive effect of hyperpriors in the logit scale such that a larger standard deviation implies a more, rather than a less, informative prior in the data scale (Chapter 9).

Here we are going to assume the following priors:

```
prior.m1.1brms = c(set_prior("normal(0,0.5)",class = "b"),
   set_prior("normal(logit(0.5),((logit(0.8)-logit(0.2))/4",class="Intercept"))
```

The priors for the partial slopes and for the differences between slopes and intercepts imply that 95% of the plausible values would be in [–0.98, 0.98] on the logit scale. It should be noted that standardizing the explanatory variables prior to the analysis greatly simplifies the setting of priors when we have limited knowledge concerning the specific case study. The prior for the intercept deserves further comment: `logit(0.5)` (which equals zero) means that before seeing the data, we expect the average plant cover to be 0.5, which is converted to the logit scale. The hyperprior for the standard deviation of the intercept, `((logit(0.8) – logit(0.2)) / 4` = 0.693, says that we expect plant cover to range between 0.2 and 0.8, and we use the Wan et al. (2014) approach to obtain a plausible value for this parameter in the logit scale (see Chapter 9). Note also that by not setting priors for the dispersion parameter ϕ, we are accepting the default weakly informative priors defined by `brms`.

We are going to fit two models, one modeling the parameter ϕ in terms of the explanatory variables:

```
m1.1brms=brm(formula=formula.m1.1brms, family=Beta, warmup = 1000,data=cover.s,
           prior=prior.m1.1brms,chains=3, iter=3000, future=T,
           control = list(adapt_delta =0.9999, max_treedepth=15))
```

and the other simply estimating its value:

```
m1.brms=brm(formula=fTotalcover~Gr.pressure*Prod + patch.size*Prod+
inter.patch.dist, family=Beta, warmup = 1000,data=cover.s,
            family=Beta, warmup = 1000,data=cover.s, prior=prior.m1.1brms,
            chains=3, iter=3000, future=T, control = list(adapt_delta =0.9999,
            max_treedepth=15))
```

After fitting the models we add the `loo` criterion to be able to compare them:

```
m1.1brms=add_criterion(m1.1brms, "loo")
m1.brms=add_criterion(m1.brms, "loo")
```

```
> loo_compare(m1.brms,m1.1brms)
         elpd_diff se_diff
m1.1brms  0.0       0.0
m1.brms  -2.1       5.2
```

The difference in the expected log pointwise predictive density (`elpd`) between the two models is less than half of its standard error. Hence, we could say that the two models are equivalent, and choose the simpler (`m1.brms`) to make inferences. We note that the comparison between the frequentist analogue models reached the opposite conclusion. This is by no means incorrect as model selection in the frequentist and Bayesian frameworks (Chapter 7) has different theoretical underpinnings, and thus does not necessarily need to coincide, although it is reassuring when they do.

The summary of the selected `m1.brms` model is:

```
> summary(m1.brms)
 Family: beta
  Links: mu = logit; phi = log
Formula: Totalcover ~ Gr.pressure * Prod + patch.size * Prod + inter.patch.dist
         phi ~ 1
   Data: cover.s (Number of observations: 53)
Samples: 3 chains, each with iter = 3000; warmup = 1000; thin = 1;
         total post-warmup samples = 6000

Population-Level Effects:
                      Estimate Est.Error l-95% CI u-95% CI Rhat Bulk_ESS Tail_ESS
Intercept                 0.10      0.02     0.07     0.13 1.00     2383     3827
phi_Intercept             7.36      0.22     6.88     7.77 1.00     3578     3683
Gr.pressure               0.06      0.03     0.01     0.11 1.00     1464     2502
ProdLow                   0.09      0.04     0.02     0.17 1.00     3265     3519
ProdMedium                0.03      0.02    -0.02     0.08 1.00     2934     3623
patch.size                0.35      0.02     0.31     0.39 1.00     1831     3096
inter.patch.dist         -0.21      0.01    -0.23    -0.18 1.00     2161     3403
Gr.pressure:ProdLow      -0.07      0.03    -0.12    -0.01 1.00     1664     2886
Gr.pressure:ProdMedium   -0.07      0.03    -0.12    -0.02 1.00     1650     2743
ProdLow:patch.size        0.13      0.04     0.04     0.21 1.00     2086     3174
ProdMedium:patch.size    -0.07      0.03    -0.12    -0.01 1.00     2101     3039
```

As always, the first thing to examine are `Rhat` and the ESS metrics: they suggest that the Hamiltonian Monte Carlo algorithm has converged to a stationary posterior distribution, and that the chains contain many statistically independent parameter estimates. The line for `phi_Intercept` contains the mean, standard error, and 95% credible interval based on the marginal posterior distribution of the parameter ϕ in the log scale. Likewise, the other lines denote the effects of explanatory variables from their marginal posterior distributions on logit(μ_Y). The interpretation of the latter parameter was discussed in Section 12.4 for the frequentist model and need not be repeated here. We will just summarize the estimated effects by writing the fitted equations for each level of the variable `Prod`:

$$\text{logit}(\mu_Y)^{\text{HIGH}} = 0.10 + 0.06\ \text{Gr.pressure} + 0.35\ \text{patch.size} - 0.21\ \text{inter.patch.distance}$$
$$\text{logit}(\mu_Y)^{\text{MEDIUM}} = 0.13 - 0.01\ \text{Gr.pressure} + 0.28\ \text{patch.size} - 0.21\ \text{inter.patch.distance}$$
$$\text{logit}(\mu_Y)^{\text{LOW}} = 0.19 - 0.01\ \text{Gr.pressure} + 0.48\ \text{patch.size} - 0.21\ \text{inter.patch.distance}$$

Please make sure that you understand how these equations were obtained. Just as we did for the frequentist analogue (Section 12.4), we could compare the relative magnitudes of the partial slopes of the explanatory variables both within and among the different levels of `Prod` to aid our interpretation of model findings. Let's recall that these partial slopes denote the effect of changes in one standard deviation of each explanatory variable on logit(μ_Y).

Let's now visually check the convergence of the chains to a stationary posterior distribution, and the autocorrelation of the sampled parameter estimates (Figs. 12.7 and 12.8). For all the model parameters, these figures show that the three chains have converged to a common marginal stationary posterior distribution, and that the sampled values in each chain have a very low (< 0.3) autocorrelation and hence can be considered statistically independent estimates.

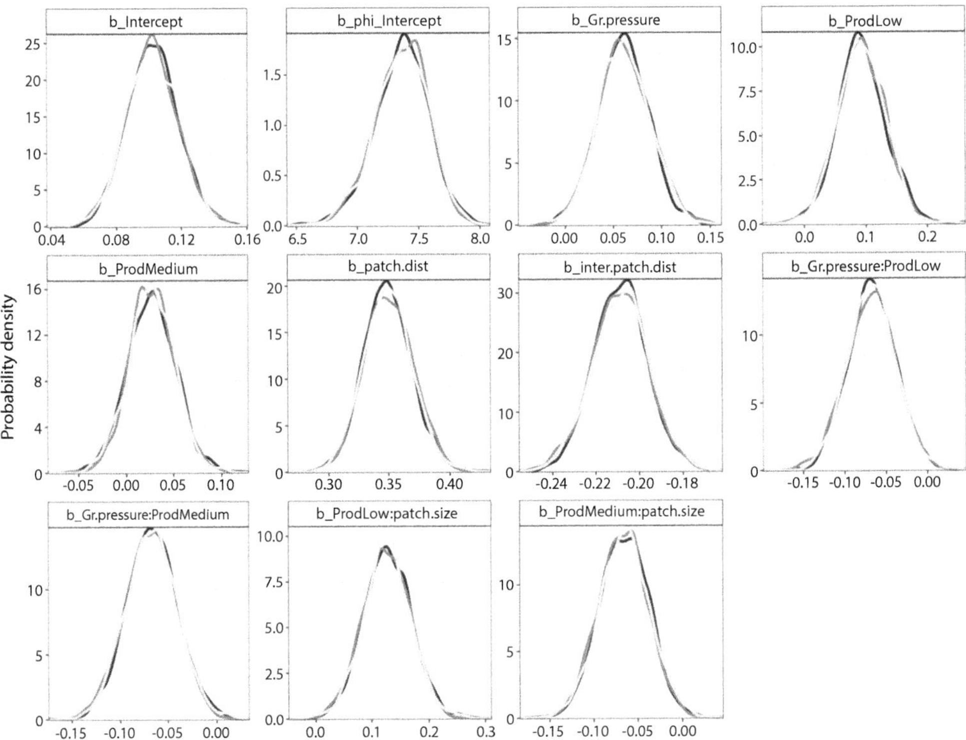

Fig. 12.7 Marginal posterior distributions of each model parameter for the three chains used to fit the Bayesian proportion GLM `m1.brms`. The great degree of coincidence of the chains suggests their convergence to a posterior stationary distribution.

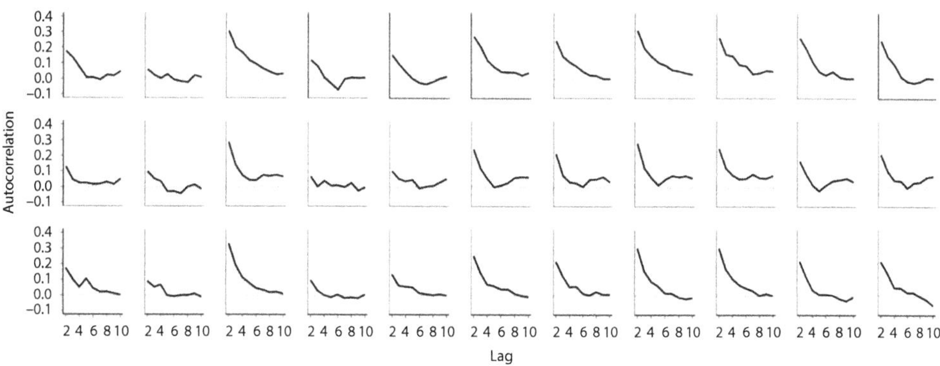

Fig 12.8 Autocorrelation plots of each model parameter for the three chains (in each row) used to fit the Bayesian proportion GLM `m1.brms`. Parameters (in columns) are given in the same order as in Fig 12.7. The very low autocorrelation for all the parameters and chains suggests that the sampled values of the posterior distributions can be considered largely statistically independent.

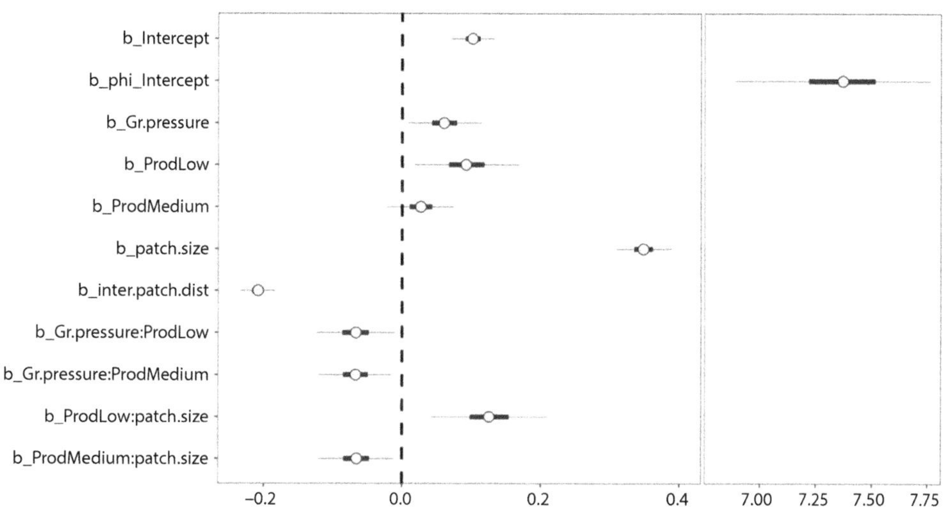

Fig. 12.9 Means and 50% (thick lines) and 95% (thin lines) credible intervals of the parameters of the Bayesian proportion GLM `m1.brms`.

The summary of model `m1.brms` can depicted as in Fig. 12.9, which is gentler than a table of numbers. The model accounts for a large proportion of the variance of the response variable (Gelman et al. 2018):

```
> bayes_R2(m1.brms)
   Estimate Est.Error  Q2.5 Q97.5
R2    0.986   0.00146 0.983 0.988
```

The means and 95% credible intervals for the differences in partial slope of `Gr.Pressure` between `Prod` = High and the other levels of this categorical explanatory variable were −0.07 and [−0.12, −0.01] for `Prod` = High, and −0.07 and [−0.12, −0.02] for `Prod` = Low (Fig. 12.9). We can obtain the sampled estimates of all model parameters as `post.samples.m1brms = posterior_samples(m1.brms, pars = "^b")`. The columns of this data frame contain the 6,000 estimates of each parameter determining logit(μ_Y). We can now calculate the marginal posterior distribution of the partial slope of Gr.Pressure between Prod = Medium and Prod = Low as:

```
diff.GrPressureMed.Low=post.samples.m1brms$`b_Gr.pressure:ProdMedium`-
                        post.samples.m1brms$`b_Gr.pressure:ProdLow`)
```

and obtain descriptive statistics from the newly obtained marginal posterior distribution:

```
> summary(diff.GrPressureMed.Low)
   Min. 1st Qu.  Median    Mean 3rd Qu.    Max.
-0.0812 -0.0129 -0.0001 -0.0002  0.0130  0.0795
> quantile(diff.GrPressureMed.Low, probs=c(0.025,0.975))
   2.5%   97.5%
-0.0399  0.0379
```

Unsurprisingly, the partial slopes for Prod = Medium and Prod = Low are virtually identical. These trivial calculations are just to highlight the wealth of possibilities available to obtain true probability distributions from any set of operations carried out on the posterior distribution of a Bayesian GLM. Nothing comparable is available in the frequentist framework.

We can also obtain the conditional plots that are very relevant to interpreting the model results. First, we obtain the predicted values of the response variable for each variable and interaction used in the model m1.brms using the command conditional_effects from the brms package:

```
> m1.brms.cond.eff=conditional_effects(m1.brms)
> names(m1.brms.cond.eff)
[1] "Gr.pressure"      "Prod"             "patch.size"       "inter.patch.dist"
[5] "Gr.pressure:Prod" "patch.size:Prod"
```

The resulting object m1.brms.cond.eff is a list containing each of the six predicted conditional effects. Given that the two interactions Gr.pressure:Prod and patch.size:Prod were relevant as the partial slopes differed depending on Prod, the single effects of each of these variables (i.e., the first three elements of the list) are irrelevant and should not be considered. Figure 12.10 shows the conditional plots for each combination of explanatory variables, keeping the other variables at zero, i.e., at their mean values. We can see that changes of one standard deviation in the structure of the vegetation (inter.patch.dist and patch.size) lead to larger changes in Totalcover than similar changes in Gr.Pressure, regardless of Prod.

Analysis of the goodness of fit of model m1.brms requires first obtaining the randomized quantile residuals (Chapter 8) using the package DHARMa based on the posterior predictive distribution of the model (cf. Chapters 9–11): `post.dist.m1.brms = posterior_predict(m1.brms, nsamples = 1000)`. Next, the randomized quantile residuals are calculated:

```
qres.m1.brms=createDHARMa(simulatedResponse = t(post.dist.m1.brms),
    observedResponse = cover.s$Totalcover,
 fittedPredictedResponse=apply(post.dist.m1.brms,2,median),integerResponse=T)
```

and then converted to have a normal distribution: `res.m1.brms=data.frame(res=qnorm(residuals(qres.m1.brms)))`. The next step is to put all the components required for the residual analysis in a data frame:

```
res.m1.brms=cbind(res.m1.brms,
            cover.s[,c("Gr.pressure","patch.size","inter.patch.dist","Prod")],
            fitted=fitted(m1.brms, nsamples=1000)[,1],
            pareto=loo(m1.brms, pointwise=T)$diagnostics$pareto_k)
```

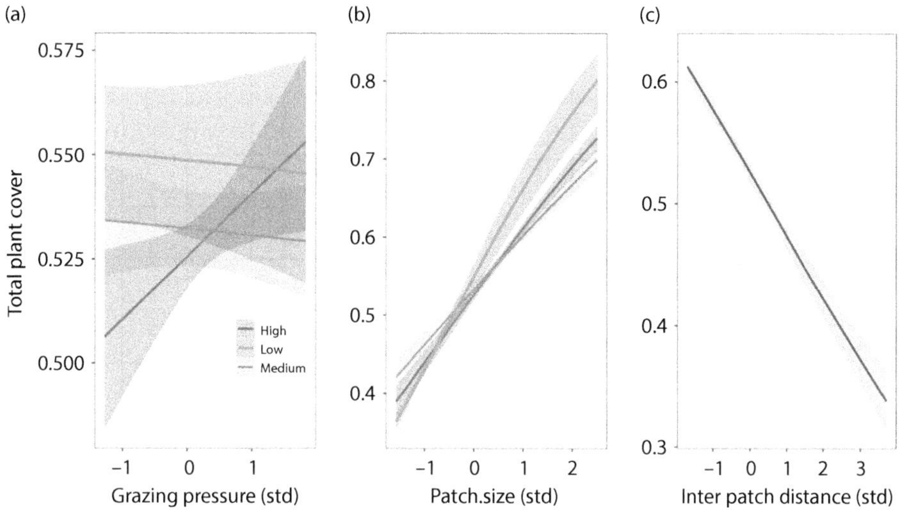

Fig. 12.10 Conditional plots showing the relation between the response variable total plant cover and the explanatory variables of the Bayesian proportion GLM `m1.brms`. The lines show the predicted mean and the 95% credible intervals.

Figure 12.11 suggests that model `m1.brms` has acceptable goodness of fit to the data: the residuals appear to have random scatter and homogeneous variation, and to lack discernible trends (Fig. 12.11a,b,d,e), most of them are very close to a normal distribution (Fig. 12.11c), and all but two data points are in the "OK" range (Vehtari et al. 2017).

Finally, we can use posterior predictive checks based on the posterior predictive distribution (Chapter 7) to assess the generative capacity of the model to produce the data in hand. There are a wealth of possible posterior predictive checks that may be implemented with a family of commands starting with `ppc_` in the package `bayesplot`. All these commands operate on the posterior predictive distribution as `m1.brms.post = posterior_predict(m1.brms, nsamples = 1000)`, and then use a function or a descriptive statistic (i.e., mean, variance, max, 41st percentile, etc.) to compare them with the actual data. A model with good generative capacity should be able to produce the observed metrics or functions in the actual data as a "typical" result of the ensemble of predictions.

Figure 12.12 shows that `m1.brms` has pretty good generative capacity, in complete correspondence to its adequate goodness of fit (Fig. 12.11). There are, of course, statistics such as the maximum and minimum that are always harder to predict for any statistical model. Any realistic model will break or perform badly if we push it far enough. Thus, we can view these harder-to-predict metrics as "stress tests" of the typical predictive capacity of statistical models.

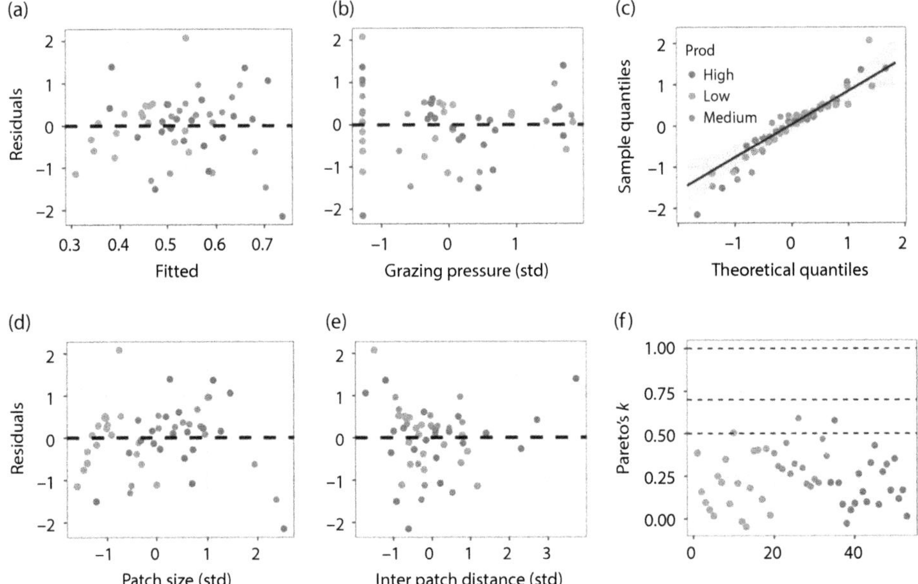

Fig. 12.11 Residual analysis of the Bayesian proportion GLM `m1.brms` based on randomized quantile residuals converted to normal residuals, showing the mean of the Pearson residuals vs. the mean of the fitted value (a), the mean of the Pearson residuals vs. the explanatory variables in the model (b,d,e), the quantile–quantile plot of the randomized quantile residuals residuals (c), and the Pareto *k* cross-validation statistic for each data point (f).

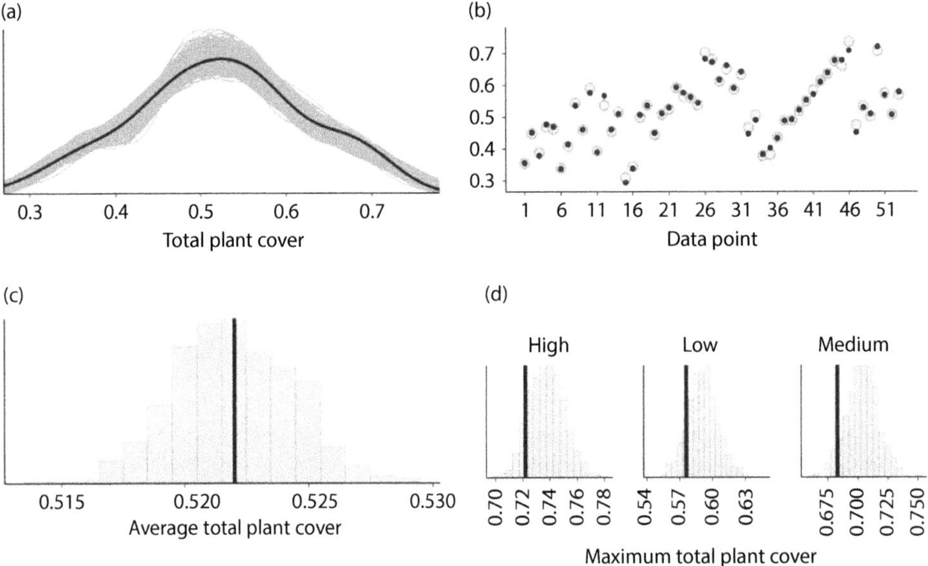

Fig. 12.12 Posterior predictive distribution of the Bayesian proportion GLM `m1.brms` showing (a) density curves for the response variable and observed total plant cover (thick line); (b) mean and 95% credible intervals (that are too precise to be seen) for each observed value of the response variable (dots); (c) frequency distribution of the predicted average plant cover and the observed value; and (d) maximum predicted total plant cover for each level of the primary productivity, showing the observed values as thick lines.

12.6 Modeling positive, real-valued response variables

Many examples of positive, real-valued response variables in the biological and social sciences quickly spring to mind: measures of mass, volume, length, money or wealth, etc. The distributions of these response variables are often right-skewed (i.e., with a longer upper tail), meaning that there are many fewer large than small values. The lognormal distribution (Chapter 4) is defined for real, positive values, and for most combinations of its parameter values has a right-skewed shape. Scientists typically deal with this type of response variable by applying a logarithmic transformation to render its distribution more symmetric and closer to a normal distribution. It is well known that if $Y \sim$ Lognormal, then $\log(Y) \sim$ Normal, and the transformed variable is used as the response variable in the general linear model (Chapters 4 to 6).

Nevertheless, the lognormal is not the only probability distribution that could describe positive, real-valued response variables. The gamma distribution is part of the exponential family of distributions (Chapter 8). Its origins are unclear: it might have been used by Laplace in 1836, and it was obtained by Karl Pearson and known as a Pearson type III distribution until it got its current name some time in the 1940s (Dagpunar 2019). The gamma distribution can be expressed in terms of its mean μ_Y and a dispersion or shape parameter ϕ that can produce a variety of shapes (Fig. 12.13), including the exponential distribution as a special case:

$$\Pr(Y = y) = \frac{\left(\frac{Y}{\phi\mu_Y}\right)^{\frac{1}{\phi}}}{y\Gamma(1/\phi)} \exp\left(-\frac{y}{\phi\mu_Y}\right), \tag{12.2}$$

where $\Gamma()$ is the gamma function that gives its name to the probability distribution. The mean of the gamma distribution is, of course, μ_Y, and its variance increases proportionally with its mean as $\mu_Y\phi$. Aside from the few parameter combinations for which the gamma distribution renders an exponential shape (Fig. 12.13), it is typically difficult to discriminate it from the lognormal distribution in most data sets (Dick 2004).

Although the gamma distribution has a canonical link function ($-1/\mu_Y$; Table 8.1), the log link function appears to be more commonly used in practice. Nevertheless, the fitting of saturating response functions such as Michaelis–Menten enzyme kinetics or the Holling type II functional response of predators (it is the same function) constitutes an interesting case where the inverse link function can be applied. The function describing the speed of an enzymatic reaction V as a function of the substrate concentration S is $V(S) = \frac{V_{\text{MAX}}S}{K_{\text{M}}+S}$, where V_{MAX} is the maximum rate of enzymatic reaction, and K_{M} is the substrate concentration yielding half of V_{MAX}. This non-linear function $V(S)$ can be linearized using the Lineweaver–Burk transformation that consists of applying the inverse function to both sides such that $\frac{1}{V(S)} = \frac{K_{\text{M}}}{V_{\text{MAX}}} + \left(\frac{1}{S}\right)\frac{1}{K_{\text{M}}}$. This is a linear equation for $1/V$ in terms of $1/S$, with a slope of $1/K_{\text{M}}$ and intercept $K_{\text{M}}/V_{\text{MAX}}$. If we were to fit this relation as a gamma GLM, the statistical model would be V $\sim$ Gamma(μ_Y, ϕ) and $(-1 / \mu_Y) = \beta_0 + \beta_1(1 / S)$, where β_0 and β_1 are the intercept and the slope of the linearized relation. In R we would write `glm(V ~ I(1/S), family = Gamma(link = "inverse")`, where I is the "as is" function that literally interprets `I(1/S)` as we write it. Having the estimates of slope and intercept, it is easy to obtain the parameters of the Michaelis–Menten equation.

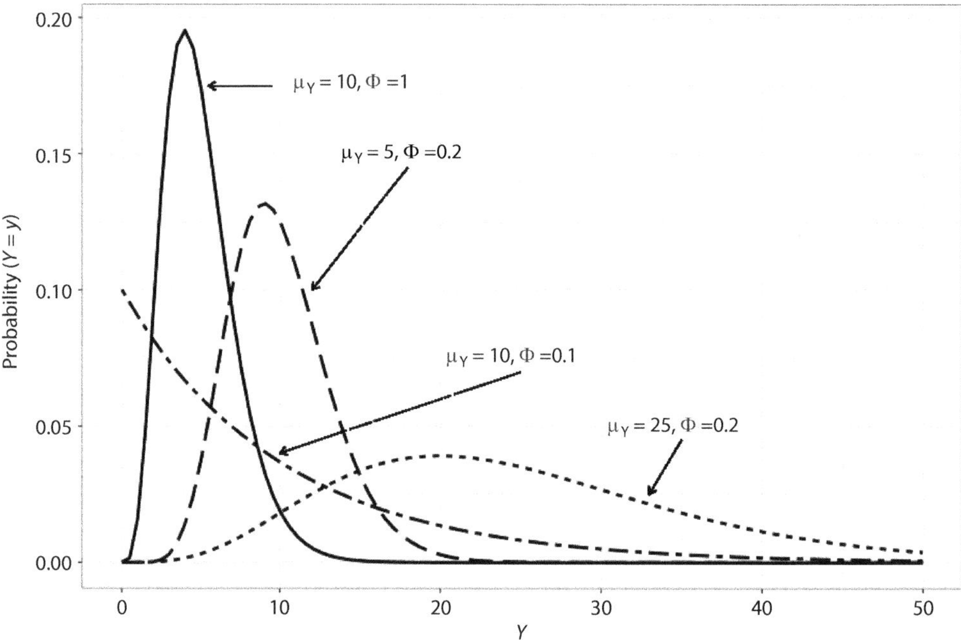

Fig. 12.13 Probability density for different combinations of parameters of the gamma distribution.

12.7 Predicting tree seedling biomass

THE DATA IN CONTEXT: Philipson et al. (2012) investigated the growth of seedlings of 21 dipterocarp tree species in Malaysia by growing them in a greenhouse under standardized conditions. In order to estimate the above-ground seedling biomass prior to harvesting, they wished to estimate the relations between this response variable and the height, diameter, and number of leaves (the explanatory variables). They used an average of 20 additional seedlings of each species (range: 11–40) that were destructively harvested to obtain the data from which to estimate the desired empirical relation. The response variable, Biomass, takes positive, real values that might be modeled with the Gamma distribution $Y \sim \text{Gamma}(\mu_Y, \phi)$ with $\log(\mu_Y) = X\beta$, where X are the numerical explanatory variables and β the parameters to be estimated. Using the log link function will make sure that its inverse will render strictly positive values of μ_Y, as the response variable Biomass should be. Of course, stating that the response variable follows a gamma distribution is just a conjecture at this point.

EXPLORATORY DATA ANALYSIS: We start by importing and summarizing the data:

```
> trees=(read.csv("Tree biomass Chapter 12.csv", header=T))
> str(trees)
'data.frame':   290 obs. of  9 variables:
 $ Genus       : Factor w/ 5 levels "Dipterocarpus",..: 1 1 1 1 1 1 1 1 1 1 ...
 $ Species     : Factor w/ 21 levels "argentifolia",..: 4 4 4 4 4 4 4 4 4 4 ...
 $ Sp.name     : Factor w/ 19 levels "Dip.con","Dry.bec",..: 1 1 1 1 1 1 1 1 1 1 ..
 $ Height      : int  245 263 256 386 402 568 540 551 599 561 ...
 $ leaves      : int  5 5 6 7 7 9 7 11 7 11 ...
 $ Diam        : num  4.95 3.95 7.05 5.5 4.7 7.55 5.2 5.25 6.15 5.5 ...
 $ leaf.area.1: num  38.2 40 17.6 53.1 57.5 ...
 $ leaf.area.2: num  63.3 42.5 37.9 50.6 31.5 ...
 $ Biomass     : num  2.84 1.85 2.55 3.83 2.97 7.12 4.62 5.32 8.27 6.09 ...
```

The data set also contains the variable `Sp.name` that abbreviates the genus and species name, and two measures of leaf area that we are not going to use. We are not going to fit a model per species, but a general model for all species. In truth, this data set should be analyzed as a generalized linear mixed model (Part III) with species as a random intercept, but let's not burn through the stages in our progressive understanding of GLMs.

We next explore the best probability distribution for the response variable using the package `fitdistrplus`. We consider the gamma, lognormal, and normal distributions:

```
lognor.biom=fitdist(trees$Biomass,"lnorm")
normal.biom=fitdist(trees$Biomass,"norm")
gamma.biom=fitdist(trees$Biomass,"gamma")
```

Figure 12.14 shows that the gamma distribution is clearly the best-fitting distribution, and that the lognormal and gamma distributions only differ in the upper tail.

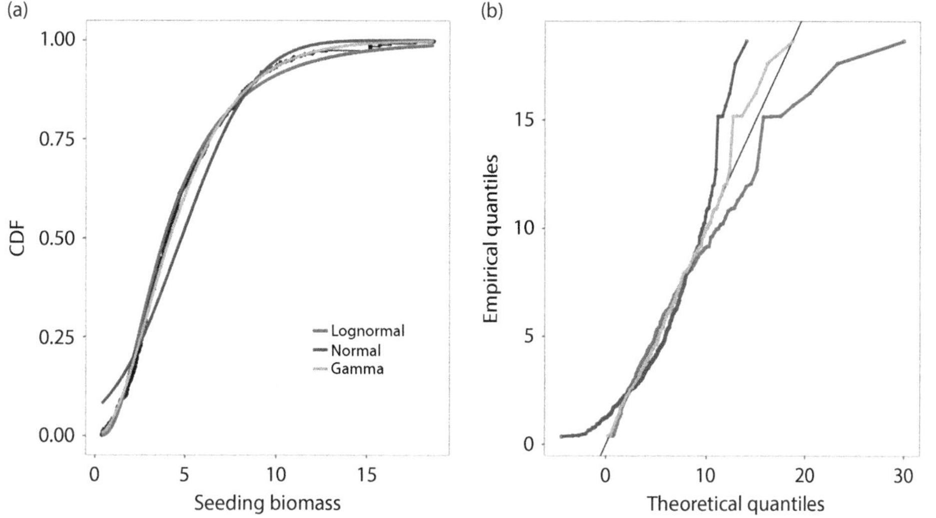

Fig. 12.14 Goodness-of-fit assessment of the response variable seedling biomass showing the cumulative distribution function (a) and the quantile–quantile plot (b) comparing the lognormal, normal, and gamma distributions using the package `fitdistrplus`.

The command `ggpairs` from the package `GGally` allows us to explore the relations between the response and explanatory variables, and the correlations among the latter. Figure 12.15 shows that there appear to be positive relations between the response and explanatory variables (bottom row), and that the latter are not strongly correlated with each other (upper triangular part). Hence, all the explanatory variables should enter into the analyses. In case you were wondering whether there are grounds to imagine that there might be interactive effects between pairs of explanatory variables on the response variable, there are not. You can check this as follows:

```
ggplot(data=trees, aes(x=Diam, y=Biomass))+
  geom_point()+
  theme_bw()+
  geom_smooth(method="glm")+
  facet_wrap(~ cut_number(Height, n=4))+
  theme(axis.title=element_text(size=18),
        axis.text = element_text(size=16),
        strip.background=element_rect(colour = "black", fill = "white"),
        strip.text = element_text(size=12))
```

this will plot `Biomass` vs. `Diam` for four different ranges of values of `Height`. Should there be an interaction between the explanatory variables, the relation between `Biomass` and `Diam` would differ among the ranges of `Height` (figure not shown).

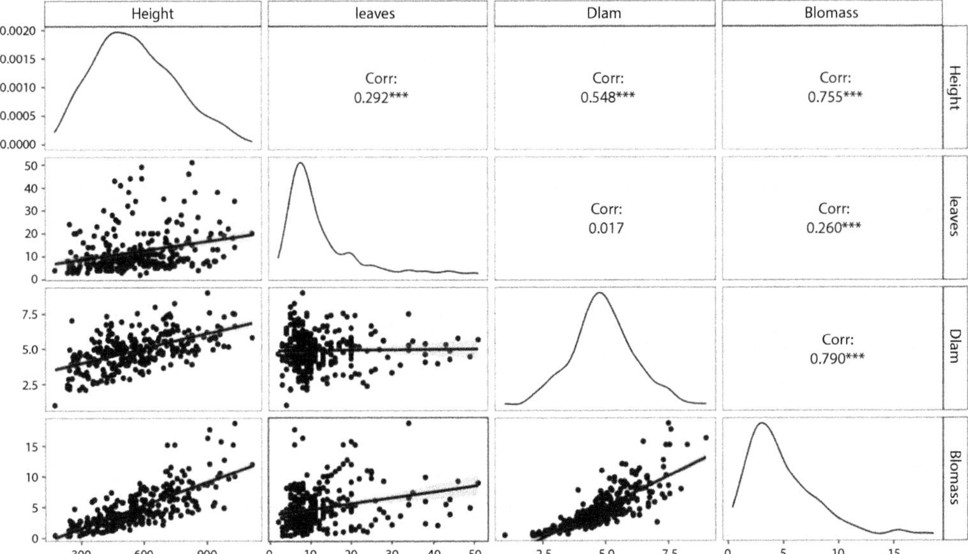

Fig. 12.15 Covariation between the explanatory variables in the gamma GLM. The diagonal shows the density plots of each explanatory variable for the entire data set and depending on the value of the response variable. Below the diagonal are shown all pairwise *x*–*y* plots, with presences colored in red and absences in green. Above the diagonal are the pairwise correlation of the explanatory variables for the entire data set and for each value of the response variable. The figure was produced with the `GGally` package.

As usual, we standardize the numerical explanatory variables prior to the analysis:

```
trees.s=data.frame(scale(trees[,c("Height","leaves","Diam")],scale=T, center=T),
                   Biomass=trees$Biomass, sp=trees$Sp.name)
```

12.8 Frequentist fitting of a gamma GLM

We are going to fit the model `m2 = glm(Biomass ~ Diam + Height + leaves, data = trees.s, family = Gamma(link=log")`, whose summary is:

```
> summary(m2)

Call:
glm(formula = Biomass ~ Diam + Height + leaves, family =
    data = trees.s)

Deviance Residuals:
     Min        1Q    Median        3Q       Max
-0.77356  -0.18785  -0.02421   0.14158   1.12760

Coefficients:
            Estimate Std. Error t value Pr(>|t|)
(Intercept)  1.38696    0.01600  86.681  < 2e-16 ***
Diam         0.43827    0.01947  22.507  < 2e-16 ***
Height       0.26085    0.02036  12.815  < 2e-16 ***
leaves       0.08290    0.01703   4.868 1.87e-06 ***
---
Signif. codes:  0 '***' 0.001 '**' 0.01 '*' 0.05 '.' 0.1

(Dispersion parameter for Gamma family taken to be 0.074

    Null deviance: 126.546  on 289  degrees of freedom
Residual deviance:  20.481  on 286  degrees of freedom
AIC: 851.74
```

We can see that all the explanatory variables have strong effects on log(E(Biomass)). Expressing their effects as a percentage change in the mean of the response variable (Chapter 10), we can see that changes of one standard deviation in `Diam`, `Height`, and `leaves` lead to 53.7%, 29.7%, and 8.3% changes in the mean of `Biomass`. However, the effect of `Diam` is (0.43 / 0.26 =) 1.65 times more important than that of `Height`, and (0.43 / 0.08 =) 5.37 times stronger than `leaves`. The point estimate of the dispersion or shape parameter ϕ of the gamma distribution is 1 / 0.074 = 13.5. The fitted model accounts for a large proportion of the deviance in the data, using McFadden's (1974) metric.

```
> 1-(m2$deviance/m2$null.deviance)
[1] 0.838
```

We use randomized quantile residuals (Chapter 8) obtained with package `DHARMa` to assess the goodness of fit of model `m2`. We first simulate these residuals, `res.m2 = simulateResiduals(fittedModel = m2, n = 500, refit = F, plot = F)`, then gather the data needed for the plots using the package `broom`, `resid.m2 = augment(m2)`, and finally put the simulated residuals converted to have a normal distribution into the data frame: `resid.m2$.std.resid = residuals(res.m2, quantileFunction = qnorm)`. Figure 12.16 shows that model `m2` has a reasonable goodness of fit to the data. The residuals have a seemingly random scatter and homogeneous variation when plotted against the fitted values and explanatory variables, and they are not too far from having a normal distribution. The three data points with larger Cook distances can be identified with:

```
> trees[resid.m2$.cooksd>0.04,c("Height","leaves","Diam","Biomass","Sp.name")]
    Height leaves Diam Biomass Sp.name
3      256      6 7.05    2.55 Dip.con
75     267     20 4.60    4.71 Hop.ner
136    605     12 3.90    7.75 Sho.mac
```

When compared to `summary(trees)` shown above, we see that the values seem neither extreme nor extraordinary since they are in the bulk of the data.

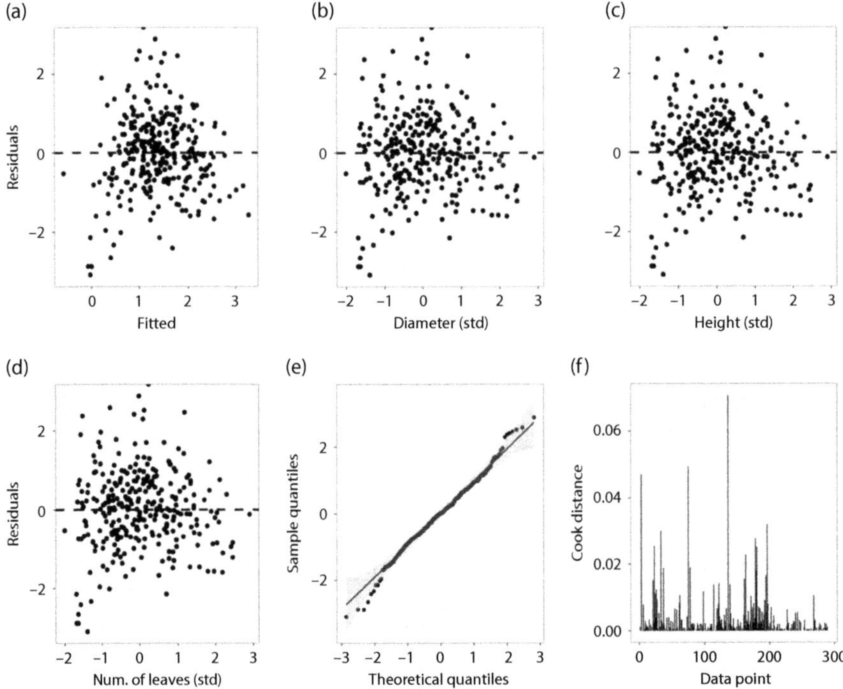

Fig. 12.16 Residual analysis of the frequentist gamma GLM `m1` based on simulated randomized quantile residuals converted to normal residuals, showing the relation between these residuals and the fitted values and to the explanatory variables (a to d), the quantile–quantile plot and the 95% confidence band of the residuals (e), and the Cook distances for each data point (f).

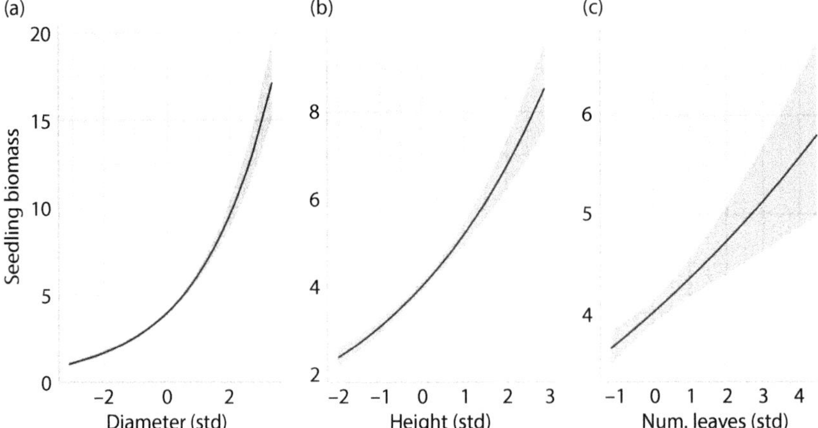

Fig. 12.17 Conditional plots of the frequentist proportion GLM `m1` showing the predicted relations between the standardized explanatory variables (mean of zero and unit standard deviation) and the response variable. The bands or intervals in all the plots correspond to the 95% confidence interval around each predicted relation. The plots show the predicted relation for each variable, holding all other variables at zero (i.e., their mean values)

We finally obtain the plots of marginal predictions of model m2 using the package ggeffects. For that, we first need to obtain a set of predicted values for each explanatory variable:

```
pred.m2a=ggpredict(m2, terms = c("Diam [all]"))
pred.m2b=ggpredict(m2, terms = c("Height[all]","Prod"))
pred.m2c=ggpredict(m2, terms = c("leaves [all]"))
```

Model m2 requires values of the three explanatory variables to generate the predicted conditional curves. Each plot in Fig. 12.17 renders the conditional mean and 95% confidence interval, setting the values of all other variables not included to zero (i.e., to their means). Notice that the predictions are in the actual scale of the data after inverting the log link function used in the fitting of model m2.

12.9 Bayesian fitting of a gamma GLM

We use the package brms to fit the Bayesian analogue of the gamma GLM for seedling biomass. We start by finding out the priors that need to be defined:

```
> get_prior(formula=Biomass~Diam+Height+leaves,data=trees.s,family=Gamma(link="log"))
                prior     class   coef group resp dpar nlpar bound        source
               (flat)         b                                           default
               (flat)         b   Diam                                (vectorized)
               (flat)         b Height                                (vectorized)
               (flat)         b leaves                                (vectorized)
 student_t(3, 1.4, 2.5) Intercept                                         default
     gamma(0.01, 0.01)     shape                                          default
```

We define the following priors:

```
prior.m2brms = c(set_prior("normal(0,0.5)",class = "b"),
          set_prior("normal(log(70),(log(160)-log(20))/4", class = "Intercept"))
```

The priors for the partial slopes in the log link scale indicate that we expect 95% of the slopes for the standardized numerical explanatory variables to be in [–1, 1]. The prior for the intercept implies the the average weight of seedlings would be 70 g, and we set its standard deviation using the approach of Wan et al. (2014) based on expected ranges. We point out that standardizing the numerical explanatory variables greatly simplifies the setting of priors for the partial slopes. We leave the default weakly informative prior for the dispersal parameter ϕ of the gamma distribution. The prior for ϕ could be set by using the ratio var / mean of the seedling biomass of other tree species to define a range of plausible values for this parameter.

We fit the following model:

```
m2.brms=brm(formula=Biomass~Diam+Height+leaves, data=trees.s,prior=prior.m2brms,
         family=Gamma(link="log"),warmup = 1000,chains=3, iter=3000, future=T,
         control = list(adapt_delta =0.99))
```

whose summary is:

```
> summary(m2.brms)
 Family: gamma
  Links: mu = log; shape = identity
Formula: Biomass ~ Diam + Height + leaves
   Data: trees.s (Number of observations: 290)
Samples: 3 chains, each with iter = 3000; warmup = 1000; thin = 1;
         total post-warmup samples = 6000

Population-Level Effects:
          Estimate Est.Error l-95% CI u-95% CI Rhat Bulk_ESS Tail_ESS
Intercept     1.39      0.02     1.36     1.42 1.00     5281     4134
Diam          0.44      0.02     0.40     0.48 1.00     3841     4051
Height        0.26      0.02     0.22     0.30 1.00     3516     3897
leaves        0.08      0.02     0.05     0.12 1.00     5041     4267

Family Specific Parameters:
      Estimate Est.Error l-95% CI u-95% CI Rhat Bulk_ESS Tail_ESS
shape    14.13      1.17    11.94    16.54 1.00     5089     3986
```

We see that `Rhat` and the effective sample size metrics all suggest the convergence of the Hamiltonian Monte Carlo algorithm to a stationary posterior distribution, with many sampled parameter estimates. We will spare you from the usual plots assessing these two features (but you can find the code for the two plots on the companion website). The marginal posterior distributions for the partial slopes are comparatively narrow, meaning that their means are very precise, and their 95% credible intervals are very narrow. The three explanatory variables are clearly important in determining the mean of $\log(\mu_Y)$, with `Diam` being very clearly the most important variable. As for every GLM with a log link function, we can again express the mean of the posterior distributions of the partial slopes as a percentage change in the mean of the response variable as 100 × (exp(mean slope) − 1) (see Chapter 10). The mean of the dispersal or shape parameter's posterior distribution is also very similar to the point estimate in the frequentist model `m2` (note that the identity link scale was used for this parameter). Agreement does not imply correctness, but at least it is reassuring. The number of parameters is also sufficiently small to dispense with the need for a separate plot to depict the summary table of `m2.brms`.

We use Gelman et al.'s (2018) metric to quantify the percentage of variance of the response variable accounted for by the model `m2.brms`:

```
> bayes_R2(m2.brms)
   Estimate Est.Error Q2.5 Q97.5
R2    0.851   0.00474 0.84 0.858
```

We next obtain the conditional effects predicted by `m2.brms`: `m2.brms.cond.eff = conditional_effects(m2.brms)`, and with `names(m2.brms.cond.eff)` we can detect the order of the predicted effects that matches the order of entry of variables in `m2.brms`. The resulting figure is really the identical to Fig. 12.17 produced after the frequentist analogue and is not worth repeating (you can again find the code to generate the conditional plots on the companion website).

Finally, we simulate the randomized quantile residuals with the package `DHARMa`, for which we first obtain the posterior predictive distribution of `m2.brms`, `post.dist.m2.brms = posterior_predict(m2.brms, nsamples = 1000)`. We then simulate the residuals:

```
qres.m2.brms=createDHARMa(simulatedResponse = t(post.dist.m2.brms),
        observedResponse = trees.s$Biomass,
        fittedPredictedResponse=apply(post.dist.m2.brms, 2, median))
```

which are later converted to have a normal distribution with `res.m2.brms= data.frame(res = qnorm(residuals(qres.m2.brms)))`. We next create a data frame to store all the elements required for the plots shown in Fig. 12.18:

```
res.m2.brms=cbind(res.m2.brms,
                  trees.s[,c("Diam","Height","leaves")],
                  fitted=fitted(m2.brms, nsamples=1000)[,1],
                  pareto=loo(m2.brms, pointwise=T)$diagnostics$pareto_k)
```

We can see that the overall goodness of fit of model `m2.brms` is acceptably good. Figures 12.18a, b, d, and e are slightly worrisome, suggesting some heterogeneity in the variation of the residuals, but we should not over-interpret their right-hand sides, which are based on very few points. The Q–Q plot and Pareto *k* values are excellent.

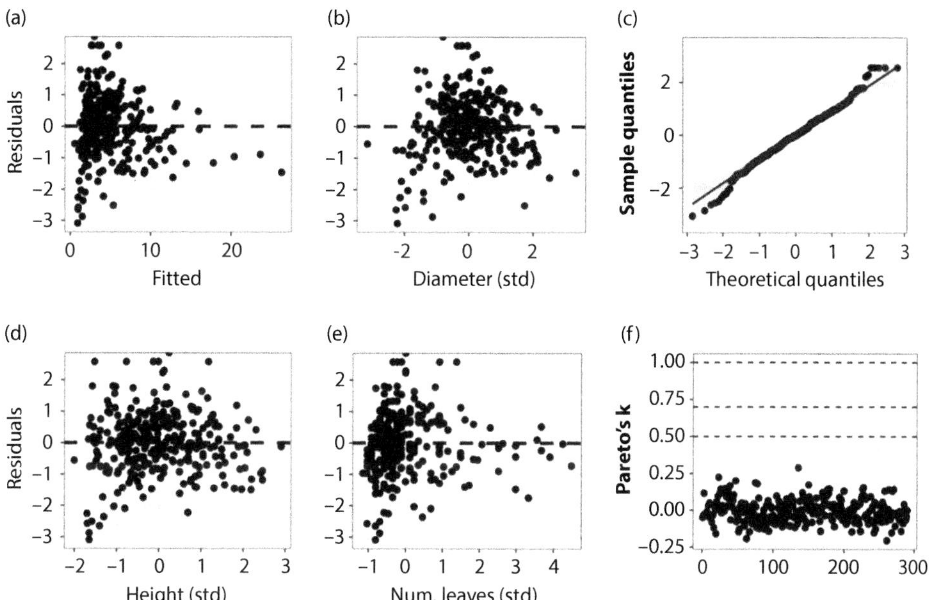

Fig. 12.18 Residual analysis of the Bayesian gamma model `m2.brms` based on randomized quantile residuals converted to normal residuals, showing the mean of the Pearson residuals vs. the mean of the fitted value (a), the mean of the Pearson residuals vs. the explanatory variables in the model (b,d,e), the quantile–quantile plot of the randomized quantile residuals (c), and the Pareto *k* cross-validation statistic for each data point (f).

12.10 Other related yet important cases of positive, real-valued response variables

There are a few extra cases involving positive, real-valued response variables that are close to the GLMs covered in this chapter and deserve to be mentioned. First, there are cases where the response variable has some proportion of zeros, and its remaining real values may have a gamma or beta distribution. For instance, when we work with fishery or rainfall data, there are days with no catch or no rain, and other days with a real number of kilos of fish landed or cubic millimeters of rainfall. Some researchers divide these data sets into one part to be modeled as a binary GLM, and the other as a gamma GLM. This is a pretty bad idea: the two parts are not easily stitched together as a single (Frankenstein, we may add) model allowing an integrated interpretation of the evidence. Also, we lose statistical power by splitting the precious data into two parts. These cases cannot be modeled with the gamma distribution as it does not admit zeros, nor with the beta distribution whose link function admits neither zeros nor ones. The solution for these cases is to use mixture models (Chapter 11). There are zero-augmented beta models (Ospina and Ferrari 2012) and zero-augmented gamma (also called hurdle gamma models) that are two- or three-part models with separate link functions for different values. For example, the zero–one-augmented beta models use one logit link function to model the relations of the explanatory variables for proportions in (0, 1), and two other logit link functions for the zero and one values. We could put explanatory variables in one or all parts of these mixture models. The `brms` package allows the fitting of these continuous mixture models in the Bayesian framework, and `glmmTMB` fits zero-augmented gamma models in the frequentist framework.

Another case concerns response variables having arbitrary relations between their mean and their variance. While the gamma and lognormal distributions allow the variance to increase with the mean, there are some cases where this is not enough. Jorgensen's (1997) exponential dispersion model generalized the exponential family that is the basis of GLMs (Chapter 8), and it includes the Tweedie distribution as a special case. This distribution includes the normal, Poisson, gamma, and other distributions as special cases depending on certain parameter values (Dunn and Smyth 2018; Ye et al. 2021). When applied to real-valued response variables, the Tweedie distribution allows us to consider flexible relations because $\text{var}(Y) = \text{mean}(Y)^{\varepsilon}$, where the parameter ε is also estimated from the data. The Tweedie distribution also allows fitting positive, real-valued response variables with a large number of zeros. Because the package `brms` does not include the Tweedie distribution, we direct you to the package `cplm` (Zhang 2013) for frequentist and Bayesian fittings of the Tweedie distribution, and also to Ye et al. (2021) for Stan code that could be used for this distribution.

12.11 Problems

Visit the companion website at https://www.oup.com/companion/InchaustiSMWR to obtain the data sets for these problems.

12.1 Aranda et al. (2021) investigated the main determinants of the establishment and growth of the invasive honey locust population in the intensive agricultural plains of central Argentina. They marked and followed seedlings in maize and soybean crops during one wet and another dry growing season. Their experiment involved two main crops (maize, soybean), the occurrence of the typical intensive management of main crops (yes, no), and the crop cover (yes, no), and they measured the percentage of starch reserves in seedlings. File: Pr12-1.csv

12.2 Gabler and Siemann (2013) studied the effect of fertilization (0, F), competition (0, N), and stress (D, F) on the above-ground biomass of the invasive Chinese tallow tree in greenhouse experiments. They also established five window duration treatments by exposing pots to identical well-drained and well-watered conditions for 4, 6, 8, 10, or 12 weeks before imposing water stress. Hints: just consider the single effect of explanatory variables, and pay attention to the distribution of the response variable. File: Pr12-2.csv

References

Agresti, A. (2010). *Analysis of Ordinal Categorical Data*, 2nd edn. John Wiley and Sons, New York.

Aranda, M., Tognetti P., and Mazía N. (2021). Are field crops refuge for woody invaders? Rainfall, crop type and management shaped tree invasion in croplands. *Agriculture, Ecosystems and Environment*, 319, 07564.

Dagpunar, J. (2019). The gamma distribution. *Significance*, 16, 10–11.

Dick, P. (2004). Beyond 'lognormal versus gamma': Discrimination among error distributions for generalized linear models. *Fisheries Research*, 70, 351–366.

Dormann, C., Elith, J., Bacher, S. et al. (2013). Collinearity: A review of methods to deal with it and a simulation study evaluating their performance. *Ecography*, 36, 27–46.

Douma, J. and Weedon, J. (2019). Analysing continuous proportions in ecology and evolution: A practical introduction to beta and Dirichlet regression. *Methods in Ecology and Evolution*, 10, 1412–1430.

Dunn, P. and Smyth, G. (2018). *Generalized Linear Models With Examples in R*. Springer, New York.

Ferrari, S., and Cribari-Neto, F. (2004). Beta regression for modelling rates and proportions. *Journal of Applied Statistics*, 31, 799–815.

Fox, J. (2015). *Applied Regression Analysis and Generalized Linear Models*, 3rd edn. Sage, New York.

Gabler, C., and Siemann, E. (2013). Timing of favorable conditions, competition and fertility interact to govern recruitment of invasive chinese tallow tree in stressful environments. *PLoS ONE*, 8, e71446.

Gelman, A., Carlin, J., Stern, H. et al. (2014). *Bayesian Data Analysis*, 3rd edn. Chapman and Hall, New York.

Gelman, A., Goodrich, B., Gabry, J. et al. (2018). R-squared for Bayesian regression models. *American Statistician*, 73, 307–309.

Jorgensen B. (1997). *The Theory of Dispersion Models*. Chapman and Hall, London.

Long, J. (2007). *Regression Models for Categorical and Limited Dependent Variables*. Sage, New York.

McCullaugh, P. and Nelder, J. (1990). *Generalized Linear Models*. Chapman and Hall, London.

McFadden, D. (1974). Conditional logit analysis of qualitative choice behavior, in P. Zarembka (ed.) *Frontiers in Econometrics*. pp. 105–142. Academic Press, New York.

Oñatibia, G., Boyero, L., and Aguiar, M. (2018) Regional productivity mediates the effects of grazing disturbance on plant cover and patch size distribution in arid and semi-arid communities. *Oikos*, 127, 1205–1215.

Ospina, R. and Ferrari, S. (2012). A general class of zero-or-one inflated beta regression models. *Computational Statistics and Data Analysis*, 56, 1609–1623.

Philipson, C., Saner, P., Marthews, T. et al. (2012). Light-based regeneration niches: Evidence from 21 Dipterocarp species using size-specific RGRs. *Biotropica*, 44, 627–636.

Smithson, M. and Merkle, E. (2014). *Generalized Linear Models for Categorical and Continuous Limited Dependent Variables*. CRC Press, New York.

Smithson, M. and Verkuilen, J. (2006). A better lemon squeezer? Maximum likelihood regression with beta-distributed dependent variables. *Psychological Methods*, 11, 54–71.

Smyth, G. (1989). Generalized linear models with varying dispersion. *Journal of the Royal Statistical Society B*, 51, 47–60.

Vehtari, A., Gelman, A., and Gabry, J. (2017). Practical Bayesian model evaluation using leave-one-out cross validation and WAIC. *Statistical Computing*, 27, 1413–1432.

Wan, X., Wang, W., Liu, J. et al. (2014). Estimating the sample mean and standard deviation from the sample size, median, range and/or interquartile range. *BMC Medical Research Methodology*, 14, 135–147.

Warton, D. and Hui, C. (2011). The arcsine is asinine: The analysis of proportions in ecology. *Ecology*, 92, 3–10.

Ye, T., Lachos, V., Wang, X. et al. (2021). Comparisons of zero-augmented continuous regressions models from a Bayesian perspective. *Statistics in Medicine*, 40, 1073–1100.

Zhang, Y. (2013). Likelihood-based and Bayesian methods for Tweedie compound Poisson linear mixed models. *Statistical Computing*, 23, 743–757.

Approaches to Defining Priors

Defining priors remains an issue of perennial debate in the Bayesian framework. There is no principled way of defining priors that has gained universal or even widespread acceptance. While this might well be an elusive goal, one would hope that, at least for some large class of commonly used models (e.g., generalized linear mixed models), there ought to be at least a conventional approach to setting priors to guide the data analysts. Those working within the frequentist framework do not have such problems, as we operate in a state of almost complete amnesia as if the world starts again with every data analysis. We may criticize Bayesian statistics all we want, but it is the explicit specification of the assumed priors what allows us to debate their influence on the results.

We have pointed out that priors tend to have a larger influence on the results of data analyses when we are trying to detect smaller magnitudes of effects of the explanatory variables with data sets of modest size. Unfortunately, this is the most common situation with experimental and survey data in the life sciences. Effect sizes of large magnitude are likely to have been detected in previous studies. The amount of information contained in large (meaning many hundreds of thousands of data points) data sets is likely to overwhelm most reasonable priors. Priors can also be used to regularize results by reducing the chances of obtaining unrealistically large effect sizes due to chance, as may occur with data sets of modest size.

This book lacks a single chapter devoted to the definition of priors and its related debates. This was a deliberate choice. The book was written for a reader with a moderate amount of background statistical knowledge. In this regard, we preferred to progressively introduce several features involved in the definition of priors in different chapters. Here, we summarize them in a single list:

- Chapter 4: using "guesstimates" of priors and plotting the prior distributions to gauge their implied ranges of feasible parameters. The role of hyperparameters.
- Chapter 5: discussion of weakly informative priors, and the difficulty of precisely defining them in the context of the data at hand.
- Chapter 6: using prior sensitivity analysis to assess the implications of different priors in a specific data analysis. Joint plotting of prior and posterior distributions to visualize the impact of the data on our beliefs about ranges of parameter values.
- Chapter 8: using the prior predictive distributions to discard impossible or implausible ranges of parameter values.
- Chapter 11: using prior predictive checks (introduced in Chapter 7) to examine the effect of priors in reproducing notable features of the available data.

We wish to stress that these features are by no means mutually exclusive: we may combine them as we please in any data analysis. Only space constraints have prevented us from making more extensive use of them in the data analysis of every chapter. Data analysts in the Bayesian framework do have a considerable arsenal of recent tools and approaches at their disposal to address the perennial debate of the influence of priors.

PART III

Incorporating Experimental and Survey Design Using Mixed Models

CHAPTER 13

Accounting for Structure in Mixed/Hierarchical Models

Packages needed in this chapter:

```
packages.needed<-c("ggplot2","fitdistrplus","brms","future","broom",
"GGally","lme4","broom.mixed","glmmTMB","ggeffects", "performance",
"cowplot","HLMdiag","arm","pbkrtest","DHARMa","qqplotr","reshape2",
"bayesplot","parallel","doBy","ggbreak","ggdist")
lapply(packages.needed, FUN = require, character.only = T)
```

13.1 Introduction

The most fundamental assumption of all statistical models is that, unless stated otherwise, the values of the response variable are independent and identically distributed. This is called the "iid assumption." Independence denotes that the probability of the response variable attaining a given value does not affect the probability of attaining any other. Identically distributed means that there is a common set of parameters valid to predict the probability of observing all possible values of the response variable. In its basic form, the iid assumption in a statistical model without explanatory variables implies a constant and homogeneous (not to be confused with uniform) world over space and time. We build statistical models to actually ascertain the extent to which our observations differ from this idealized, causality-free worldview embodied by the iid assumption. The iid assumption implies that anything generating structure or relatedness in the values of the response variable is to be included in the model as explanatory variables. Thus, after accounting for the relations between parameters of the response variable and the explanatory variables, a well-fitting statistical model must have structure-less remaining residual variation. This is why we examine the randomness and absence of trends when plotting the residuals against the fitted values and those of the explanatory variable.

Part III of the book is devoted to mixed/hierarchical models. The title of this chapter alludes to "structure," a term that is as general and all-encompassing as it is non-specific. By structure we mean the arrangement of, and relations between, parts of the data that need to be explicitly accounted for in a statistical model. A structured data set neither occurs by accident nor does it result from sheer bad luck. Rather, the structure arises from either the experimental or sampling design (Chapter 14). Our statistical models need to explicitly include the reasons for structure (i.e., differences, heterogeneity, regularities, variation) so as to adequately model the variation of the response variable, and obtain adequate estimates of model parameters in different subsets of the data.

Statistical Modeling With R. Pablo Inchausti, Oxford University Press. © Pablo Inchausti (2023).
DOI: 10.1093/oso/9780192859013.003.0013

There are at least four non-mutually-exclusive reasons for structure in a data set. They are completely general as they apply regardless of the type and nature of the response variable. First, most scientific studies involve an experimental design involving a set of categorical variables that denote the purposeful manipulation (or at least the opportunistic use) of different conditions. These categorical explanatory variables may have simple or interactive effects on the response variable (see examples in Chapters 6 and 11). In the simplest terms, they correspond to specific sets of comparisons whose outcomes are related to specific scientific hypotheses. By dividing the data set into different subsets for which we estimate and contrast the parameters of our statistical models, the manipulated categorical variables impose a structure on our data set.

Second, our observations can have a clustered structure, thereby violating the iid assumption. For example, we may measure several fledglings from the same nest and several nests from the same tree, pose several questions or tests to the same individuals, make several measurements in a few seemingly distinct locations, etc. The clustering may result from the practicalities and constraints involved in the execution of the sampling design. For instance, we might wish to randomly select k sites for sampling, but economic considerations do not permit sampling all k sites out of a much larger population of sites due to the effort involved in traveling between sites. We thus select the k sites but notice afterwards that some sampled sites are similar in some respect so as to be considered a cluster. Measurements obtained in a cluster are unlikely to be independent from each other because they tend to be more similar to each other than to measurements from other clusters. Therefore, related observations from the same cluster do not contribute independent information to the fitting of the statistical model.

Third, besides the explanatory variables stemming from the scientific hypotheses, we often measure other explanatory variables known or suspected to have an effect on the response variable. These are often called "confounding variables." They are included in statistical models with the goal of controlling or discounting their influence, despite not being our main focus of interest. Of course, declaring a variable "confounding" is entirely contingent on the research goals and hypotheses, and on the existence of previous knowledge of the putative effect of the confounding variable(s). Controlling for the effect of confounding variables so as to uncover the effect of manipulated variables is a main feature in the design of experiments (Chapter 14). In the case of numerical confounding variables, they are included as another explanatory variable. However, when confounding variables are categorical they define groups, and their putative effect on the response variable would be identical for all sampling units in each group. There are conceptual differences in the way in which main and confounding categorical variables are modeled in the frequentist (but not in the Bayesian) framework that we need to understand.

The final reason for structure in the data is the sampling of the response variable across space or time because of either the experimental or the sampling design employed. For instance, we may repeatedly sample the same experimental units over time, or there might be a known spatially clustered distribution of a key variable affecting the response variable that we need to account for during the data sampling. The values of the response variable that are closer in space or time are bound to be more similar than those separated by large distances or time spans. Provided we register the spatial coordinates and the time at which each sample was obtained, we model the decay in similarity with either the distance or time elapsed between pairs of values known as spatial or temporal autocorrelation, respectively. When the values of the response variable are structured over space or time, they are not statistically independent precisely because their degree of non-independence or similarity changes with the distance or time separating pairs of samples (see Chapter 15 for one example). Another important reason for non-independence in the life sciences is that values of the response variable are related by either a pedigree or a phylogenetic tree. Again, there will be a gradient of resemblance depending on the degree of closeness to common ancestors or species that is reflected as a distance or variance–covariance matrix. There is extensive literature in quantitative genetics (e.g., Wilson et al. 2010) and comparative biology (e.g., Ives and Helmus 2011) on these issues that will not be covered here.

13.2 Fixed effects and random effects in the frequentist framework

Frequentist statistics makes a fundamental distinction for modeling categorical explanatory variables based on Eisenhart's (1947) distinction between fixed and random effects.

On the one hand, the *fixed effects* refer to categorical explanatory variables denoting a purposeful experimental manipulation in the general sense discussed above. They are considered to correspond to a complete and exhaustive set that is deliberately chosen in relation to the scientific hypotheses. In (quasi-)experimental settings, the fixed effects are the primary focus of interest. They represent the overall or population-level average effect of a specific term in the model, be it a main effect or an interaction between explanatory variables. Since the levels of a categorical variable modeled as fixed effects were deliberately chosen in relation to our specific hypotheses, what we learnt from their effects cannot be extended to any other plausible or actual level of the explanatory variable. The inference for variables modeled as fixed effects is therefore strictly limited to the levels considered. In this sense, when we compare treatments and control to ascertain the effect of the imposed experimental perturbation, it is relevant to make a posteriori comparisons to discern which levels of a categorical explanatory variable (or variables, when we seek to evaluate their interaction) differ from each other (Chapters 6 and 11). These sampling estimates of model parameters are the maximum likelihood estimates, for which we also estimate their precision in the standard errors. In Chapters 4 to 12 we have modeled all categorical explanatory variables as fixed effects without telling you. For completeness, any numerical explanatory variable is modeled as a fixed effect in the frequentist framework.

On the other hand, we have *random effects*. These are categorical variables whose levels are considered to be randomly selected from a much larger super-population of levels. The random effects denote the stochastic variability associated with group-specific effects. They capture the extent to which the group-level response differs from the overall or global mean of the response variable reflected as a fixed effect. To make a simple example, suppose we evaluate whether a drug has a curative effect on a disease. We define a categorical variable with two levels, drug and placebo, that is to be modeled as a fixed effect. But we know in advance that individual genetic makeup also affects the incidence of the disease and perhaps the putative effects of the drug. We then obtain a random sample of individuals that happen to correspond to 17 different genotypes (there is more than one individual per genotype). Comparing the effect of treatment between these 17 genotypes is not the goal of our experiment. In fact, we have advance knowledge that genotype is likely to affect what we really want to measure, i.e., the effect of the drug compared to the placebo. This advance knowledge is precisely the reason why we measured the genotype in the first place. In the context of this experiment, genotype would be considered a confounding variable to be modeled as a random effect. Its stochastic, group-specific effects on the response variable can be quantified to help make a more reliable inference about the effect of the drug.

It makes little sense to carry out a posteriori tests to contrast the 17 genotypes because these happen to be just chance sampling events: if we were to repeat the experiment, it would be very unlikely that we sample precisely the same 17 genotypes again. However, what we learnt about the importance of genotype modeled as a random effect can be extended to the super-population of genotypes from which we randomly obtained 17 random samples.

A *mixed model* in the frequentist framework contains categorical (and perhaps numerical) explanatory variables modeled as fixed effects, and others as random effects. Other names for mixed models are multilevel models in social sciences and economics (e.g., Hox et al. 2018), and hierarchical models (e.g., Cressie et al. 2009). In the frequentist framework, deciding whether to model a categorical variable as either a fixed or random effect is thought to mostly depend on the goals of the analysis. The general but debatable wisdom (see Bolker 2015 for a discussion) is to model as fixed effects those categorical explanatory variables that reflect the main goals of the research related to the hypotheses, and as random effects the confounding variables that result from experimental or sampling design (Chapter 14). Associating a random effect with a confounding variable would seem to be akin to considering it a "nuisance variable" whose effects we wish to estimate and discount from the inference focused on fixed effects. While superficially correct from the extended viewpoint of just focusing on the fixed effects in significance testing, viewing random effects as nuisance variables would be incorrect. Doing so would miss the importance of understanding the relative importance of all explanatory variables in statistical models. Many examples in this chapter and in Chapter 15 will show that random effects may account for (or explain, if you wish) a substantial fraction of the variation of the response variable, and that they are essential to making accurate model predictions.

In Chapters 4 to 6 we introduced the general linear model as $Y \sim \text{Normal}(\mu_Y = X\beta, \sigma_Y^2)$ or, in its equivalent form, $Y = X\beta + \varepsilon$ where $\varepsilon \sim \text{Normal}(0, \sigma_Y^2)$. Here, Y is the response variable, X is design matrix with a first column of ones (for the estimation of the model intercept) and the categorical and/or numerical explanatory variables, and ε the residuals reflecting the random scatter around its mean explained by the explanatory variables. Including a categorical explanatory variable modeled as a random effect transforms a general linear model into a linear mixed model. Recalling the hypothetical assay of the effects of a drug, and the 17 genotypes, the equation for the linear mixed model is

$$Y = X\beta + Zb + \varepsilon, \tag{13.1}$$

where Z is a second design matrix for the random effects, with zeros and ones denoting whether each value of the response variable is associated with each of the 17 genotypes; b is a vector for the random effects, with b ~ Normal(0, G); and ε are still the random residuals with $\varepsilon \sim$ Normal(0, R). R and G are variance–covariance matrices that will allow us to specify different random effects (matrix R), and the temporal or spatial structured variation of residuals (matrix G).

To give just one example of a specific structure of the G matrix, let's consider the simplest model in quantitative genetics (i.e., the genetics of complex traits influenced by a large number of genes) for a continuous trait that has a normal distribution. The main parameter of interest here is the heritability h^2 of the trait that denotes how a population would respond to selection pressures; it is estimated as $V_A/\text{Var}(Y)$, where V_A is the additive genetic variance, and Var(Y) is the phenotypic variance of the trait. V_A is estimated with the linear mixed model $z = \mu + Ga + \varepsilon$, where the random effects a correspond to the additive genetic effects (also called breeding values) that are distributed as

Normal(0, AV_A), where A is a matrix of relatedness of individuals in the population, and $\varepsilon \sim$ Normal(0, σ^2_{res}), where σ^2_{res} is the residual variance (de Villemereuil 2018). In a more general sense, we will at this point leave both matrices unspecified, and we will return to them with specific examples in this and the following chapters.

Equation (13.1) states that the expected value E(Y) or mean μ_Y of the response variable in the linear mixed model is E(Y) = Xβ + Zb. This mean is actually a conditional mean, E(Y | b) = Xβ: it reflects the effect of the fixed effects conditional on the random effects. The fixed effects (Xβ) represent the expected relation between the measured categorical explanatory variable (drug in our example) irrespective of the genotype an individual has.

The random effects (Zb) in Eq. (13.1) capture how much the estimated mean effects related to the explanatory variable drug differ for each genotype. The group-specific shifts from the overall mean of fixed effects Xβ are denoted by random coefficients whose magnitudes may differ between groups. We can view random effects as zero-centered, group-specific offsets that are added to the fixed effects to account for the lack of statistical independence of individuals of a genotype. Random effects are sometimes called "latent variables" precisely because their effects cannot be directly measured.

In the drug–genotype example, an appropriate design matrix G would reflect that individuals of the same genotype would be highly similar to each other in their responses to the drug, and that they would be unrelated to individuals of other genotypes. This matrix is called "compound symmetry by group" (e.g., Pinheiro and Bates 2004). We do not need to go into the mathematical equations designating compound symmetry by group, as the G matrix is set by our definition of the random effects in the statistical model. Compound symmetry by group implies that the variance of the response variable, $\sigma^2_Y = \sigma^2_{\text{genotypes}} + \sigma^2_\varepsilon$, can be partitioned into the "shared" (within a genotype) component, $\sigma^2_{\text{genotypes}}$, and the non-shared component or residual variation, σ^2_ε. The parameter $\sigma^2_{\text{genotypes}}$ quantifies the heterogeneity of the genotype-specific random shifts from the conditional mean estimated from the fixed effects. Had we not included random effects in our model, all variation around the mean response would have been absorbed in σ^2_ε. Now, having $\sigma^2_\varepsilon > 0$ leads to a smaller residual variance σ^2_ε and to greater precision (i.e., smaller standard errors) of the estimates of model parameters (see Chapter 14).

The generalized linear models (Chapter 8) expressed the relation between the mean of the response variable and the explanatory variables in the scale of the link function g such that g(E(Y)) = Xβ. The link function g could be logit (for binary and proportion data; Chapters 9 and 12), log (for count data; Chapters 10 and 11), or inverse (for certain cases of positive, real-valued data modeled with the gamma distribution; Chapter 12). We can also include categorical variables modeled as random effects, and by so doing the GLM turns into a generalized linear mixed model (GLMM). If the response variable of the drug experiment with the 17 genotypes were non-Gaussian, the equation for the resulting GLMM would be

$$g(\text{E}\,(Y|\text{b}) = \text{X}\beta + \text{Zb}. \tag{13.2}$$

Equation (13.2) states that the effect of the explanatory variable (drug) is modeled as a fixed effect, and that of genotype is modeled as a random effect. Their effects are additive on the scale of the link function g. As in the linear mixed model, the mean in the GLMM is conditional on the estimated random effects: g(E(Y | b)) = Xβ. As in Eq. (13.1), Z is a second design matrix denoting whether each value of the response variable is associated with each group or cluster, and b is a vector of coefficients for the random effects with b ~ Normal(0, G). The model parameters (fixed effects and variance associated with the genotypes) are estimated in the scale of the link function g, which is later inverted to provide model predictions comparable with the observed data (see Chapters 9–12). Nothing new here. The residual variation term ε of Eq. (13.1) does not appear in Eq. (13.2) because the mean and the variance of the response variable are not independent parameters in non-Gaussian GLMs (see Chapter 8).

Equations (13.1) and (13.2) state that the coefficients b reflecting the magnitude of the random effects follow a normal distribution with mean zero and a variance to be estimated. Needless to say, this is a model assumption, not necessarily a true state of affairs. Accordingly, we should verify whether the normality assumption of the random effects holds for each data set. Why then assume a normal distribution for the random effects? One reason is that the assumption greatly simplifies the algorithms of parameter estimation in linear mixed models that were developed before GLMMs. A second non-decisive reason is that the assumption of normality for random effects seems to work reasonably well with many data sets. Besides, simulating the multivariate normal distribution that is needed in mixed models (see below) is easier than other plausible probability distributions such as gamma or lognormal (McElreath 2019). If needed, the hierarchical generalized linear models consider random effects with distributions other than normal (Jin and Lee 2020, Lee et al. 2010). It is thought that mixed models are relatively robust to misspecifying the distributional shape of random effects (e.g., McCulloch and Neuhaus 2011, Schielzeth et al. 2020, but see Hui et al. 2021).

13.3 Defining mixed effects models

We are going to formulate several linear mixed models, leaving their solution in the frequentist and Bayesian frameworks for later in the chapter. The context is as follows. Hill et al. (2005) studied students' gain in math achievement score from the spring of kindergarten to the spring of first grade (mathgain, the response variable) in relation to a series of explanatory variables, of which we are only going to consider sex (0 = boys, 1 = girls) and the student math score in the spring of their kindergarten year (mathkind). The data comprise 1,190 children from 288 classes in 99 schools in Michigan, USA. In this observational data set, classes had different numbers of students and schools had different numbers of classes. Neither the individual students of each class nor the classes in each school may be considered independent sampling units. In both cases, it is easy to envisage several common variables affecting either all children in a class or all classes in a school, leading to greater within-cluster (compared to between-cluster) similarity and the breaking of the iid assumption. Hill et al.'s (2005) sampling scheme was hierarchical, much like a Russian doll: children nested in a classroom, the latter nested in a school, and (if we wish to pursue the logic) schools nested within Michigan school districts, etc. We are going to model the break-up of the iid assumption at the children level (the elementary unit of replication in this observational study) by formulating a linear mixed model.

Before doing so, let's recall that our main goal is to relate the mean of mathgain with mathkind, and to study the variation of this relationship between the sexes. We are going to model these explanatory variables as fixed effects: `mathgain ~ mathkind*sex`. Our main goal should strike you as an analysis of covariance (Chapter 6), assuming that the response variable follows a normal distribution.

We can now formulate our first linear mixed model by considering the categorical variable `schoolID`. This model will evaluate whether there is a school effect (equal for both sexes) on the mean of the response variable. Some schools will be better and some worse

Table 13.1 R syntax for the random effects in mixed/hierarchical models. `S` and `T` are two categorical explanatory variables modeled as random or group-level effects, and `X` and `W` are either numerical or categorical explanatory variables that are included as fixed or population-level effects in the statistical model. This is a non-exhaustive list, as other combinations such as `(1|S) + (X|T)`, `(1|S) + (X||T)`, `(1|S) + (0 + X|T)`, etc. are plausible.

Notation	Meaning of the fitted random effects
`(1|S)`	Random intercepts for each level of `S`
`(0 + X|S)`	Only random intercepts for each level of `S`; equivalent to `(-1 + X|S)`
`(1 + X||S)`	Uncorrelated random intercepts and random slope for the explanatory variable `X` for each level of `S`; equivalent to `(1|S) + (0 + X|S)`
`(X*W|S)`	Random intercepts and slopes for `X*W` for each level of `S`, including the single effect of `X` and `W` and their interaction `X:W`; equivalent to `(1 + X + W + X:W|S)` and to `(X + W + X:W|S)`
`(1|S) + (1|T)`	Independent (crossed) random intercepts for each level of `S` and `T`
`(1 + X|S/T)`	Nested random effects estimating random intercepts and slopes of all levels of `T` measured within each level of `S`; equivalent to `(1 + X|S) + (1 + X|S:T)`

than the "average school." This is a *random intercept* model that is written as mathgain ~ $\beta_0 + \beta_1$mathkind + $\beta_2$sex + $\beta_3$mathkind:sex + $b_{0,S}$, with the school-varying effect $b_{0,S}$ assumed to be Normal(0, $\sigma^2_{b_{0,S}}$). Assuming the sex of reference to be boys (because it was coded as 0), β_0 is the mean of mathgain for boys when mathkind = 0, β_1 is the differential in the mean of mathgain for girls when mathkind = 0, β_2 is the slope for the sex of reference, and β_3 is the differential of the slopes between the sex of reference and girls. Nothing new thus far. A small value of $\sigma^2_{b_{0,S}}$ would imply that the random differentials for each school are small, and thus that schools are largely homogeneous in the value of the mean of mathgain. In R, the model is written as `m1 = mathgain ~ mathkind*sex + (1 | schoolID)` (see Table 13.1). More formally, the statistical model for `m1` is

$$\begin{aligned}&\text{mathgain} \sim \text{Normal}(\mu_Y, \sigma^2{}_Y)),\\ &\text{Fixed part:} \quad \mu_Y = \beta_0 + \beta_1\text{mathkind} + \beta_2\text{sex} + \beta_3\text{mathkind:sex},\\ &\text{Random part:} \quad b_{0,S} + \varepsilon, \text{ where } b_{0,S} \sim \text{Normal}\left(0, \sigma^2{}_{b0,S}\right) \text{ and } \varepsilon \sim \text{Normal}\left(0, \sigma^2_{\text{residual}}\right).\end{aligned} \tag{13.3}$$

`m1` then has six parameters to be estimated from the data. The random effects do not alter the interpretation of the fixed or population effects.

Besides group-level changes in the mean of the response variable for both sexes, we can envision that the slope β_1 governing the linear relation between `mathgain` and `mathkind` for the sex of reference (boys) differs among schools. That is, schools may differ not only in the average mathgain but also in the extent to which there are changes of the response variable associated with mathkind. This model is called a *random slopes and intercept or* just *random slopes* model, and it is written as mathgain ~ $\beta_0 + \beta_1$mathkind + $\beta_2$sex + $\beta_3$mathkind:sex + $b_{0,S}$ + $b_{1,S}$mathkind. The fixed and the random parts are denoted Greek and Latin letters, respectively. After collecting terms the model is mathgain ~ (β_0 + $\sigma^2_{b_{0,S}}$) + (β_1 + $b_{1,S}$)mathkind + $\beta_2$sex + $\beta_3$mathkind:sex, with the school-varying effects $b_{0,S}$ ~Normal(0, $\sigma^2{}_{b0,S}$) and $b_{1,S}$ ~ Normal(0, $\sigma^2_{b_{1,S}}$). These are sometimes called vectorial random effects (Bolker 2015), whose two components are typically correlated. The statistical model in this case can be formally written as

mathgain ~ Normal(μ_Y,$\sigma^2{}_Y$)),
Fixed part: $\mu_Y = \beta_0 + \beta_1$mathkind + $\beta_2$sex + $\beta_3$mathkind: sex,
Random part: random effects + ε, where the vectorial random effects are

$$\begin{pmatrix} b_{0,S} \\ b_{1,S} \end{pmatrix} = \text{MultivariateNormal}\left(\begin{pmatrix} 0 \\ 0 \end{pmatrix}, \begin{pmatrix} \text{Var}\,(b_{0,S}) & \text{Covar}\,(b_{0,S}, b_{1,S}) \\ \text{Covar}\,(b_{0,S}, b_{1,S}) & \text{Var}\,(b_{1,S}) \end{pmatrix}\right) \tag{13.4}$$

and the residual errors are $\varepsilon \sim \text{Normal}(0, \sigma^2_{\text{residual}})$ as before. The random slope and intercept model needs a multivariate normal distribution (mentioned in the previous section) with means (0, 0) and a variance–covariance matrix containing the variances of the random intercept ($b_{0,S}$) and of the random slopes ($b_{1,S}$), and their covariance. From the latter, we obtain their correlation from Covar($b_{0,S}$, $b_{1,S}$) = correlation($b_{0,S}$, $b_{1,S}$)SD($b_{0,S}$)SD($b_{1,S}$). In R, this new model `m2` would be written as `m2 = mathgain ~ mathkind*sex + (1 + mathkind|schoolID)` (see Table 13.1).

We can also conceive that the differentials between sexes (indicated by β_2 in the fixed part) may also differ among schools. This would mean schools are not homogeneous in the extent to which the mean of mathgain differs between sexes. You may not be shocked by now if we write the model in R as `mathgain ~ mathkind*sex + (1 + sex|schoolID)`. Again, this model will estimate the fixed effects (β_0, β_1, β_2, β_3) denoting the overall effects of the explanatory variables at the population level irrespective of the variable school-level effects, and the variances of each random effect ($b_{0,S}$ and $b_{3,S}$), and their correlation at the school level. We can fit a random slopes model for a categorical explanatory variable such as `sex`. The random slopes model actually corresponds to an interaction between a variable modeled as a fixed effect (`sex`) and another modeled as a random effect (`schoolID`). The explanatory variable modeled as random effects can be recorded either at the lowest level (the individual child in our example) or at a higher level such as schools. An example of this would be the percentage of poor households in a school's catchment area. In economics and social sciences these variables are called "context effects" (e.g., Heisig and Schaeffer 2019) where a contextual characteristic mediates the strength of lower-level relationships. Note that it is also possible to consider random slopes and intercept models where the two random effect parameters are uncorrelated (see Table 13.1). Considering uncorrelated random slopes and intercepts may greatly reduce the number of parameters and help algorithmic convergence (see below) when the categorical variable modeled as a random effect has a large number of levels. However, the choice between correlated and uncorrelated random effects should be decided based on the relative support that each model receives from the data.

Finally, a statistical model can also have more than one random effect. These effects can be either independent (nested) or interactive (crossed). Phrased in the context of our example, the categorical explanatory variables schools and classes are said to be *crossed* when each level of schools is observed combined with all levels of classes (and vice versa). In contrast, the categorical variable classes is *nested* in schools when some classes are only observed combined with some schools. In the latter case, there is a hierarchy between the categorical variables schools and classes. Because classes are nested in schools, each level of classes is observed only in one school, and there cannot be interactive effects with nested random effects. Whether we should considered crossed or nested random effects is a property of the data. An R model considering crossed random effects for intercepts would be `mathkind*sex + (1|schoolID) + (1|classID)`, and a model with nested random effects would be `mathkind*sex + (1|schoolID/classID)`. The random effects structure of the latter model can also be written as `(1|schoolID) + (1|schoolID:classID)` (see Table 13.1).

It may be tempting to formulate mixed models with very complex random effects. The temptation is probably strongest in exploratory (i.e., not hypothesis-driven or confirmatory) data analysis. But the limit of what is feasible is often lower than our fertile imagination. Models with complex random effects require a large amount of detailed data, and very often have difficulties in achieving convergence. The "adequate" level of complexity of random effects in mixed models remains an unresolved issue. On the one hand, Barr et al. (2013) took the position "keep it maximal," suggesting that only complex random effects have acceptable type I error rates for the fixed effects. On the other, Matuschek et al. (2017) showed that overly complex random effects lead to steep reductions in statistical power, and sensibly advocated using model selection to decide the adequate complexity in the random effects for each data set. We cover model selection in the context of mixed models in Section 13.7.

13.4 Problems and inconsistencies with the definition of random effects

We admit to having provided somewhat superficial definitions of fixed and random effects in Section 13.2. With very slight variations, these are the definitions repeated in many textbooks in the frequentist framework (e.g., Crawley 2002). Gelman (2005, p. 20) contrasted several definitions of random effects in the frequentist framework and found them imprecise and mutually contradictory. Bolker (2015) also discussed the inconsistencies of the random effects definition regarding whether the levels of the categorical variable were actually randomly sampled from a larger population. If so, can we model species as a random effect when we have sampled all species in a site, or the 26 countries in the European Union, or all years in a three-year experiment? There seems to be a fair amount of leeway and personal choice in modeling categorical variables as either fixed or random effects.

Going further, Gelman (2005, p. 21) sensibly stated his strong preference for abandoning the overloaded terms of "fixed" and "random" effects for constant or population-level effects and varying or group-level effects, respectively. In Gelman's (2005) view, the fixed or population-level effects are constant if they are identical for all groups in a population, whereas group-varying effects may differ among clusters or groups defined by a categorical variable, regardless of whether they are interesting in themselves. In the Bayesian framework, random effects are sets of variables whose parameters are drawn from, and therefore constrained by, a probability distribution (Bolker 2015). The variance of the normal distribution of the categorical variable modeled as random effect functions is a regularizing parameter (Chapter 5) that constrains the values of the group-level coefficients based on the shared information between groups (see Section 13.8).

Bolker (2015, p. 314) summarized the main attributes of a categorical variable to be modeled as a random effect in the frequentist framework. They include: (i) focusing on quantifying variability rather than testing for differences among levels; (ii) wishing to make a prediction about unobserved levels of the categorical variable, provided that the observed levels were randomly sampled from a larger super-population of possible levels; and (iii) having a categorical variable that is a nuisance variable, possibly resulting from the experimental or sampling design, whose effect is to be controlled.

An unresolved aspect related to modeling a categorical variable as a random effect involves the minimal number of levels of the grouping variable. The core of the issue is the bias and low precision of the variance of the random effect variable when the number of levels is very low (Bolker 2015). In practical terms, we need to know how many levels is actually "very low." Some textbooks (e.g., Crawley 2002) state that less than five levels is "very low" without any justification. The determination is complicated since the number of replicates per level of the random effect variable also affects the bias and precision of the parameter estimates. The approximate answer can only be established through extensive simulations that cannot possibly cover all possible cases. For instance, McNeish (2016) found that having less than 15 clusters with sample sizes smaller than 10 per group leads to non-convergence, large biases of the random effect variances, and low coverage of the confidence intervals in binary GLMMs. The situation is similar but somewhat less stringent for count GLMMs (McNeish 2019), and slightly less so for linear mixed models (McNeish and Stapleton 2016). For the latter, having at least five levels of the random effect variable and with sample sizes of more than 10 replicates per level provides largely unbiased estimates of the fixed effects and of the variance of the random effects, and adequate coverage of the confidence intervals. We have glossed over many details in McNeish's simulations that also compared different algorithms and approximations used in the estimation of mixed models. In terms of the severity of the very "low threshold" for the number of levels of random effects, it seems clear that both the number of levels and the sample sizes per level in binary GLMMs are required to be larger than for other GLMMs, and that, unsurprisingly, the (co)variance terms of the random effects are by far the worst estimated parameter in all cases examined.

Generalized estimating equations (Liang and Zeger 1986) is an algorithmic method applied to estimate parameters of GLMMs without explicitly specifying the data structure as random effects. This method aims to correct the bias in the standard errors of the fixed effects arising from the structured nature of data, and essentially treats random effects as a nuisance variable by imposing a statistical adjustment for the sake of adequacy the statistical significance tests. Generalized estimating equations then give the equivalent of "corrected" population-level or fixed effect parameter estimates that can neither make any inference about specific clusters, nor attempt any partitioning of variation between levels. If you were to be treated for a disease, and there is a large variation in the success rate among hospitals, would you be content with just knowing the overall effect of the treatment, or would you also want to know which hospital to select/avoid? What if the variation among hospitals is comparable with the magnitude of the difference between treatment and placebo? Compared to GLMMs, generalized estimating equations often have biased parameter estimates and confidence intervals with poor coverage unless the number of groups or clusters is very large (> 50; e.g., McNeish and Stapleton 2016). They will not be discussed further in this book.

13.5 Population-level and group-level effects in Bayesian hierarchical models

Mixed or multilevel models are called hierarchical models in Bayesian statistics. While every mixed model is a special case of a Bayesian hierarchical model, the latter comprises a larger domain of models (more on this below). A Bayesian model is hierarchical whenever we can decompose the joint posterior distribution into a series of valid conditional distributions (Berliner 1986). Let's recall the fundamental equation used for parameter estimation in the Bayesian framework (cf. Chapter 3):

$$[\text{parameters} \mid \text{data}] \propto [\text{data} \mid \text{parameters}]\,[\text{parameters}]\,, \tag{13.5}$$

where the joint posterior distribution is proportional to the product of the likelihood of the data and the joint prior distribution of parameters. In the Bayesian framework there are two sources of uncertainty that are modeled using probability: the sampling variation and our degree of knowledge about the model parameter values. The observation of any biological process postulated to link the response and explanatory variables is always limited or imperfect in some fashion. Reasons for this uncertainty include imperfect knowledge about causal processes, spatial or temporal variation, structured data due to experimental or sampling designs, limitations in the quantification of variables, and others. This suggests that randomness should play a role in process modeling. We can then associate a probability distribution to reflect process uncertainty (e.g., Cressie et al. 2009, Hobbs and Hooten 2015) and thus the middle (the likelihood) term in Eq. (13.5) now becomes [data | process, parameters][process | parameters].

Bayesian hierarchical models distinguish (at least) three layers or stages in the statistical model, indicated here as conditional probability functions:

Data model: [observed data | process, parameters]

Process model: [process | parameters]

Parameter model: [parameters]

where the data model includes the response and explanatory variables, the process model indicates how these variables are functionally related by a parametric relation (for instance, a straight line) and our uncertainty or variation about the process, and the parameter model denotes our previous knowledge about all the parameter values (Wikle 2003). We can now modify Eq. (13.5) as:

$$\underbrace{[\text{process, parameters} \mid \text{data}]}_{\text{Posterior distribution}} \propto \underbrace{[\text{data} \mid \text{process, parameters}]\,[\text{process} \mid \text{parameters}]}_{\text{(Conditional) likelihood}}\,\underbrace{[\text{parameters}]}_{\text{Prior distribution}} \tag{13.6}$$

Equation (13.6) says that the data model is conditional on the process and on the sampling variability of the data as reflected by other parameters. Let us give an example based on a simplified version of model `m1` for just one sex (say, boys):

Data model: [observed data | process, parameters] mathgain ~ Normal(μ_Y, σ_Y^2). The sampling variation is reflected in σ_Y^2.

Process model: [process | parameters] $\mu_Y = (\beta_0 + \mathrm{b}_0) + \beta_1$mathkind.s, and the process variation is reflected as heterogeneity of intercepts among schools: b_0 ~ Normal (0, $\sigma_{\mathrm{b}_0}^2$). The random intercepts reflect the variation in the process for reasons other than sampling variation (see Section 13.2) that arise from the structured nature of the data set.

Parameter model: [parameters]. These are the joint prior probability distributions of all model parameters: $[\beta_0, \beta_1, \sigma_{\mathrm{b}_0}^2, \sigma_Y^2]$. This is generally simplified by assuming that all parameters are independent such that the joint prior distribution is the product of the priors: $[\beta_0][\beta_1][\sigma_{\mathrm{b}_0}^2][\sigma_Y^2]$. It turns out that inference is rarely sensitive to this commonly used simplifying assumption (Hobbs and Hooten 2015).

$$\underbrace{\left[\beta_0, \beta_1, \sigma_Y^2, \mathrm{b}_0 \mid \text{data}\right]}_{\text{Posterior distribution}} \propto \underbrace{\left[\text{data} \mid \beta_0, \beta_1, \sigma_Y^2, \mathrm{b}_0\right]}_{\text{(Conditional) likelihood}} \underbrace{\left[\mathrm{b}_0 \mid \beta_0, \beta_1\right][\beta_0][\beta_1]\left[\sigma_Y^2\right]\left[\sigma_{\mathrm{b}_0}^2\right]}_{\text{Prior distribution}} \quad (13.7)$$

Equation (13.7) is a Bayesian hierarchical model because the unobserved quantity b_0 (the random intercepts) appears on both sides of the conditioning sign "|" of the conditional likelihood (Hobbs and Hooten 2015). In general, a model is hierarchical if the probability of one parameter can be conceived to depend on the value of another parameter. Bayesian hierarchical models are written modularly, or in terms of sub-models that can be factored as a chain of dependencies among parameters as in Eq. (13.7).

Hierarchical models can be represented as Bayesian networks or directed acyclic graphs to represent the relations between the parameters and variables. These diagrams are often useful to depict parameter relations in more complex hierarchical models involving latent (unobservable) variables, and models involving the joint use of many sources of data (see Hobbs and Hooten 2015 for a readable introduction). An example of the former may illustrate a key difference between mixed and hierarchical models. For example, ecologists rarely estimate the true population abundance of a wild species at a site. The reasons for this include the measurement errors of large counts, and the non-observability of individual animals at the particular time and site that mainly (but not only) affects low counts. Similar issues occur when we estimate the numbers of drug addicts or tax evaders: what we count are just a shadow of the actual numbers. Say we want to model the effect of a numerical explanatory variable X on the population abundance. The problem can be framed as a hierarchical model as follows:

Data model: [observed data | process, parameters] $Y = \text{Count}_{\text{obs}} \sim$ Binomial $(\pi,\ \mu_{\text{Count\_true}})$, where π is the probability of observing an animal given that it is present. As in every non-Gaussian GLM, the sampling variation is implicitly reflected by the relation between $\mu_{\text{Count\_true}}$ and $\sigma^2_{\text{Count\_true}}$ (see Chapter 8).

Process model: [process | parameters] $\log(\mu_{\text{Count\_true}}) = \beta_0 + \beta_1 X$, and the process variation is reflected in the uncertainty of the observed counts. The relation between the population abundance and the explanatory variable occurs for the true but unobservable counts that we only observe as a latent variable. Note that we could also introduce explanatory variables (say, weather conditions) for the probability of observation as $\text{logit}(\pi) = \beta_3 + \beta_4\text{weather}$.

Parameter model: [parameters]. These are the product of the prior probability distributions of all model parameters: $[\beta_0][\beta_1][\beta_3][\beta_4][\pi]$.

The resulting hierarchical model is a composite of two nested GLMs. There is considerable literature dealing with this class of hierarchical models (e.g., Kéry and Royle 2016) that lack a direct correspondence in the frequentist framework. This is why we stated at the start of the chapter that while every mixed model is a hierarchical model, the converse is not always true.

In Bayesian statistics there are no fixed effects: all effects are modeled as random variables with associated probability distributions. The fixed effects in the frequentist framework correspond here to uncertain constants denoting population-level effects. Random effects are similarly modeled in both frameworks, i.e., with a common probability distribution constraining the variation of the group-varying effects through the shrinking or borrowing strength effect (see Section 13.8). In Bayesian estimation, priors have a regularizing effect on the likelihood that constrains how the information contained in the data through the likelihood determines parameter estimates (Hooten and Hefley 2019). Thus, Bayesian hierarchical models contain two components that regularize or shrink parameter estimates toward the population-level values, and lead to low variance estimates (Hobbs and Hooten 2015). While the extent of the shrinkage depends on the degree of informativeness of the parameter priors, weakly informative priors are often enough to induce this desirable shrinkage effect in most cases (e.g., Hobbs and Hooten 2015, Chapter 5). However, be advised that strongly informative priors in models fitted with small data sets may induce excessive shrinkage and biased parameter estimates. In a sense, in the Bayesian framework we essentially make a bet on the choice of the priors being correct, or at least not harmful. Prior sensitivity analysis (Chapter 5) and prior predictive distributions (Chapter 7) are two useful procedures to gauge the appropriateness of the chosen priors.

13.6 Fitting mixed models in the frequentist framework

Frequentist mixed models aim to obtain the maximum likelihood estimates of the parameters β for the fixed effects, and the (co)variances involved in the random effects. Assuming that the samples within each group (i.e., the level of the variable defining the random effect) are independent after conditioning on the random effects b (i.e., after accounting for the structure in the data), the conditional probability of observing the entire data set given the fixed and random effects is $\mathrm{Pr}\,(\text{all data} \mid \beta, \mathrm{b}, \text{other params}) = \prod_i \mathrm{Pr}\,(Y_i \mid \beta, \mathrm{b}, \text{other params})$, where $\prod_i$ is the sequence of products of the probability of observing each ith value of the response variable Y_i, Pr is the probability mass or density function adequately describing its variability (i.e., normal, binomial, Poisson, etc.), and "other params" refers to the other parameters required in Pr to generate the probabilities such as the variance or dispersion parameter (see Chapters 4, 10–12 for examples).

Because the random effects b are unobserved, inferences for the fixed effects β and its covariance matrix G are obtained by integrating over the random effects b to obtain what is called the marginal likelihood function:

$$\text{Likelihood}\left(\beta, \sigma_{\mathrm{b}}^{2} \mid \text{all data}\right) = \prod_i \int \mathrm{Pr}\,(Y_i \mid \beta, \mathrm{b}, \text{other params})\,\text{Normal}\left(0, \sigma_{\mathrm{b}}^{2}\right) db. \quad (13.8)$$

Equation (13.8) looks terribly complicated. First, let's recall that the likelihood on the left-hand side is proportional to the probability of observing the data given the parameter values (Chapter 3) that is shown on the right-hand side. The product of terms inside the integral is just an application of the well-known equation for conditional probabilities: $\mathrm{Pr}(A \text{ and } B) = \mathrm{Pr}(A \mid B)\mathrm{Pr}(B)$. "Normal" in Eq. (13.8) refers to the distribution used to model the random effects. db is the variable of integration or infinitesimal sum, such that we are summing the contribution of each real value of b drawn from its normal distribution to produce the observed data. We can apply logarithms to both sides of Eq. (13.8) to transform the products into sums, which are easier to handle (Chapter 3). Equation (13.8) was written assuming the simplest case of only random intercepts. When we contemplate vectorial random effects or random effects for more than one categorical variable (be they nested or crossed), the integral in Eq. (13.8) becomes multidimensional, considerably uglier, and much harder computationally because we need to account for the plausible values of each component of the random effects.

Equation (13.8) was intended neither to belabor the mathematical details nor "to look smart." We need Eq. (13.8) to explain the methods and approximations employed to fit mixed models in the frequentist framework. Recall that the fitting of any statistical model in this framework aims to obtain the maximum likelihood estimates, i.e., the values of the model parameters that maximize the likelihood of generating the data in hand. This is done by maximizing the log-transformed Eq. (13.8). Two main cases can be distinguished depending on whether or not $\Pr(Y \mid \beta, \mathrm{b}, \text{other params})$ is normal.

In the first case, we would obtain a closed-form solution (i.e., a formula) for the maximum likelihood expression. This analytical solution is known to yield biased parameter estimates when the number of clusters is smaller than 30 (e.g., McNeish and Stapleton 2016). The reason for the bias is akin to the need to divide by $N - 1$ rather than by N to correct the bias of the maximum likelihood estimate of the variance when $Y \sim$ Normal. The solution to the biased estimates is to use the restricted maximum likelihood algorithm (REML; Patterson and Thompson 1971) implemented in virtually every piece of software that fits frequentist linear mixed models. REML starts by first obtaining initial estimates of the fixed effects parameters disregarding the random effects. It then estimates the random effects and their variance components on the residuals (i.e., the difference between Y_{obs} and Y_{pred} after the fitted fixed effects). Finally, REML repeats the estimation of both the fixed and random effects until they finally converge to unbiased parameter estimates (McCulloch and Searle 2001, Pinheiro and Bates 2004). REML has become the default option for fitting mixed models in the frequentist framework.

In the second case, when Y is not normal, Eq. (13.8) does not have a closed-form solution. In these cases the term inside the integral (known as the integrand) of Eq. (13.8) needs to be approximated, and two techniques have become the standard approaches. Adaptive Gaussian quadrature partitions the integrand into multiple components around a large number (often > 20) of nodes, and evaluates the integral by a weighted sum of the partitions. The higher the number of nodes, the better the precision of the approximation at the expense of a higher computational burden. Rather than computing the integral of Eq. (13.8) with numerical methods, the Laplace approximation uses a Taylor series expansion of the integrand to obtain a closed-form expression that can be evaluated and maximized to obtain the parameter estimates (see McNeish 2017 for a not-too-technical explanation). Should you have a hard-to-fit mixed model, it is useful to have at least an intuitive understanding of these methods in order to read the help in the R packages and troubleshoot your possible fitting problems. Needless to say, these are the most commonly used approximations in mixed models, but far from being the only ones available. The Laplace method corresponds to adaptive Gaussian quadrature with one node; it is a lot faster but less precise, and is implemented by default in the R packages `lme4` and `glmmTMB` (Bolker 2015) that we are going to use to fit frequentist mixed models. Without further ado, let's fit the models `m1` and `m2` outlined in Section 13.3 using `lme4`, which has become the R standard for this type of model. As, always, we start by importing the data and undertaking an exploratory data analysis. We read the data, `DF = read.csv("schools Chapter 13.csv", header = T)`, and then convert the variables to factors: `DF$schoolid = as.factor(DF$schoolid)`, `DF$classid = as.factor(DF$classid)`, `DF$sex = as.factor(DF$sex)`, `DF$childid=as.factor(DF$childid)`. We also standardize the numerical explanatory variable: `DF$mathkind.s = as.vector(scale(DF$mathkind, scale = T, center = T))`. The overall summary of the data is:

```
> summary(DF)
 sex        minority          mathkind         mathgain             ses              yearstea
 0:588   Min.   :0.0000   Min.   :290.0   Min.   :-110.00   Min.   :-1.61000   Min.   : 0.00
 1:602   1st Qu.:0.0000   1st Qu.:439.2   1st Qu.:  35.00   1st Qu.:-0.49000   1st Qu.: 4.00
         Median :1.0000   Median :466.0   Median :  56.00   Median :-0.03000   Median :10.00
         Mean   :0.6773   Mean   :466.7   Mean   :  57.57   Mean   :-0.01298   Mean   :12.21
         3rd Qu.:1.0000   3rd Qu.:495.0   3rd Qu.:  77.00   3rd Qu.: 0.39750   3rd Qu.:20.00
         Max.   :1.0000   Max.   :629.0   Max.   : 253.00   Max.   : 3.21000   Max.   :40.00

    mathknow           housepov          mathprep          classid         schoolid        childid
 Min.   :-2.5000   Min.   :0.0120   Min.   :1.000   26     :  10   11     :  31   1      :   1
 1st Qu.:-0.7200   1st Qu.:0.0850   1st Qu.:2.000   13     :   9   12     :  27   2      :   1
 Median :-0.1300   Median :0.1270   Median :2.300   42     :   9   71     :  27   3      :   1
 Mean   : 0.0312   Mean   :0.1782   Mean   :2.612   189    :   9   76     :  27   4      :   1
 3rd Qu.: 0.8500   3rd Qu.:0.2550   3rd Qu.:3.000   205    :   9   77     :  24   5      :   1
 Max.   : 2.6100   Max.   :0.5640   Max.   :6.000   253    :   9   31     :  22   6      :   1
 NA's   :109                                        (Other):1135   (Other):1032   (Other):1184
   mathkind.s
 Min.   :-4.26110
 1st Qu.:-0.66111
 Median :-0.01589
 Mean   : 0.00000
 3rd Qu.: 0.68360
 Max.   : 3.91575
```

Let's examine the distribution of the response variable (`mathgain`) using the package `fitdistrplus`. The values of the response variable are integers, ranging from −110 to +253, and there are a few commonly used probability distributions that could be used. The normal distribution appears to fit the bulk of the data set well, except for a small percentage of values in both tails (Fig. 13.1).

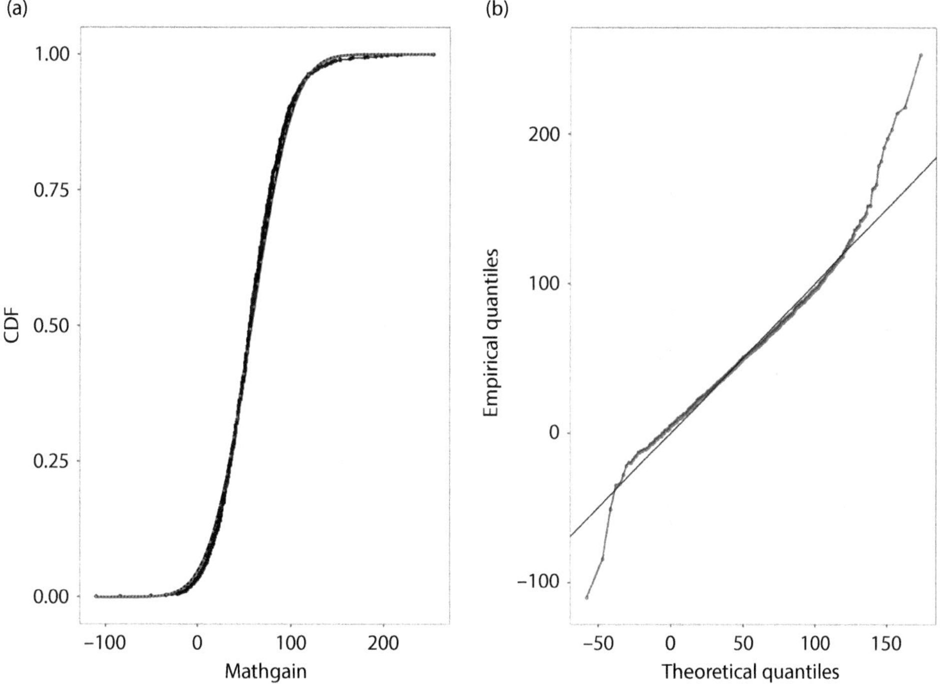

Fig. 13.1 Goodness-of-fit assessment of the response variable `mathgain` showing the cumulative distribution function (a) and the quantile–quantile plot (b) for the normal distribution using the package `fitdistrplus`.

It is also important to understand the hierarchical structure of the data, with children nested within classes, and the latter within schools. Given that each of the 1,190 lines corresponds to a child (and that there is only one data item per child, which you can verify with `table(DF$childid)`), the following command shows the number of children for whom we have data in the 99 schools:

```
> table(DF$schoolid)

  1   2   3   4   6   7   8   9  10  11  12  13  14  15  16  17  19  20  21  23  24  25  26  27
 11  10  14  12  12  14  16   6  18  31  27   9  15  13   6  13  11  12  17   8   9   7  15  21
 28  29  31  32  33  34  35  36  37  38  39  40  41  42  43  44  46  47  48  49  50  51  52  53
 10  11  22   9  22   7  11   8  11   9  12   7   9  18   4  22  13  10   6   8  12   4   9   8
 54  55  56  57  58  60  61  62  64  65  66  67  68  69  70  71  72  73  74  75  76  77  78  79
  7  16  10   8   9  13  11  16  10   9   8  12  16   5  19  27  11  10   6  12  27  24  15  12
 80  81  82  83  84  85  86  87  88  90  91  92  93  94  95  96  97  98  99 100 101 102 103 104
  7   9  19  12  14  20   7  10  10   9  10   9  21  15   5   8   6   2  19  13  16  11   8   6
105 106 107
 10   2  10
```

Now we can count the number of schools for which we have data for a given number of children:

```
> table(table(DF$schoolid))

 2  4  5  6  7  8  9 10 11 12 13 14 15 16 17 18 19 20 21 22 24 27 31
 2  2  2  6  6  8 11 11  8  9  5  3  4  5  1  2  3  1  2  3  1  3  1
```

This shows that there are two schools for which we have data for 2, 4, and 5 children, and one school for which we have data for 31 children. For instance, the schools for which we have data for two children are:

```
> table(DF$schoolid)[table(DF$schoolid)==2]

 98 106
  2   2
```

If we convert `table(table(DF$schoolid))` into a data frame, we could make a histogram with `ggplot` to depict the uneven amount of information that we have across schools. We could also calculate the number of children per class with `table(DF$classid)`, and we will see that there are data between two and five children per class, and that most often we have information for three or four children per class. There is no need to spend a figure on it. Another, perhaps simpler, way of proceeding is to store the unique identifiers of schools and classes, `class.school = unique(DF[,c("classid", "schoolid")])`, and then use it to count the number of schools for which there is data for a given number of classes:

```
> table(table(class.school$schoolid))

 1  2  3  4  5  8
 6 40 30 15  7  1
```

The table shows that there are 40 schools for which we have data for two classes, and so on. In sum, for most schools we have data for two or three classes, and for most classes we have data for three or four children.

Let's recall that the goal of the problem is to relate the response variable `mathgain` to the explanatory variables of the problem, `mathkind` and `sex`, and their variation across schools (and perhaps classes). Making either a succession of plots for the 99 schools or a single plot with 99 panels would be pointless overkill. We wish to know if these variables are related and if the relation changes with sex and across schools. One compromise would be to make a few plots for the schools for which we have some "reasonable" amount of data. We are going to select data for schools with at least 18 students: `schools18 = as.data.frame(table(DF$schoolid)[table(DF$schoolid) > 17])[,1]`. The command `table(DF$schoolid)[table(DF$schoolid) > 17]` gives the table of schools with more than 17 students. This is converted into a data frame whose first column (the `schoolsID` satisfying the condition) is stored as `schools18`. The threshold of

18 students is completely arbitrary, or, if you prefer, just bigger than the prime number 13. Figure 13.2 shows that `mathgain` is related to `mathkind` in different ways for boys and girls (i.e., there might be an interaction between these variables to be modeled as fixed effects), and that these relationships are noticeably heterogeneous across (at least some) schools.

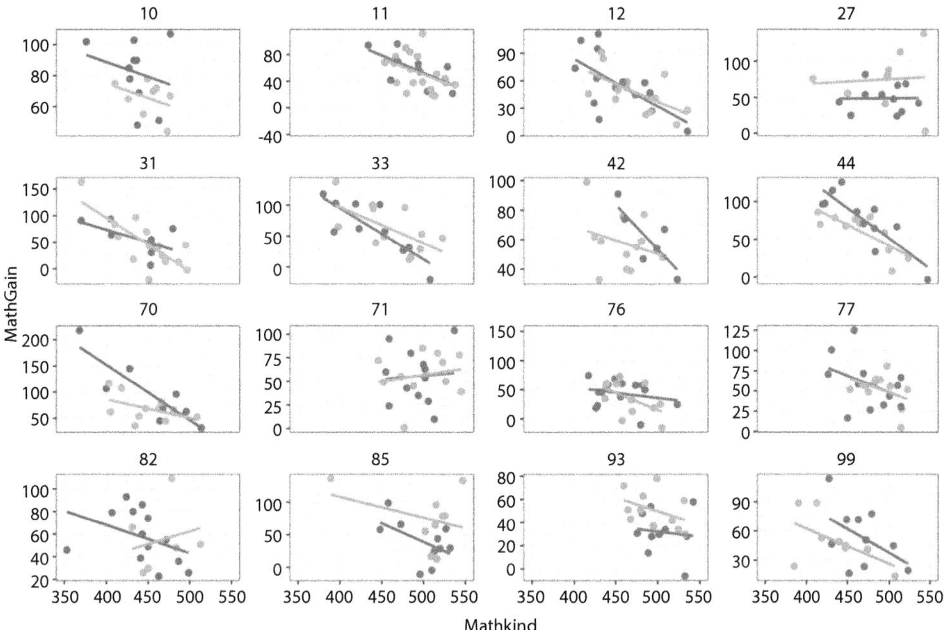

Fig. 13.2 Relation between `mathgain` and the explanatory variables `sex` (just to be contrarian, pink: boys) and `mathkind` for the schools with data for at least 18 children. The threshold of 18 children is entirely arbitrary, and it aims to show the variability of the data without showing as many plots as the 99 schools.

Let's fit the statistical model `m1` given above, `m1 = lmer(mathgain ~ mathkind.s*sex + (1|schoolid), REML = T, data = DF)`, whose summary is:

```
> summary(m1)
Linear mixed model fit by REML. t-tests use Satterthwaite's m
Formula: mathgain ~ mathkind.s * sex + (1 | schoolid)
   Data: DF

REML criterion at convergence: 11430

Scaled residuals:
   Min      1Q Median     3Q    Max
-6.078 -0.606 -0.030  0.564  4.354

Random effects:
 Groups   Name        Variance Std.Dev.
 schoolid (Intercept) 101      10.0
 Residual             818      28.6
Number of obs: 1190, groups:  schoolid, 99

Fixed effects:
                Estimate Std. Error      df t value Pr(>|t|)
(Intercept)        57.86       1.59  176.18   36.38   <2e-16
mathkind.s        -18.84       1.23 1185.65  -15.37   <2e-16
sex1               -1.31       1.70 1157.00   -0.77     0.44
mathkind.s:sex1     1.84       1.70 1147.37    1.08     0.28
```

The summary is divided in two parts corresponding to the fixed and random effects. We already know how to interpret the fixed effects part as this is what we did for all the models in Part II. Nothing new here. The fixed effects are the point estimates and the standard errors for four parameters: the intercept, or mean of the response variable for the sex of reference (boys, coded as 0) for the average value of `mathgain`, the slope of `mathkind.s` for the sex of reference, the differential of the intercept between sexes (the intercept for girls is 18.84 smaller than the boys'), and the differential of slopes between the sexes (the one for girls is 1.84 units higher than for the boys). These parameter values denote the overall relation between the variables irrespective of the school where the students learnt. The relative magnitude of the difference in intercepts between sexes (1.31 / 57.86 = 2.2%) was much smaller than that of the slopes (1.84 / 18.84 = 9.7%). As before, the *t*-values are the ratio between the coefficient and its partial slope. As always, the *p*-values denote the statistical significance of the terms modeled as fixed effects. There are debatable issues in the testing for statistical significance of mixed models (including the Satterthwaite method) that will be discussed in the next section. The random effects part of the summary provides the variances for the random intercepts, and the residual standard deviation for the mixed model. The standard deviation of the random intercepts indicates the magnitude of the heterogeneity of the intercepts (i.e., the mean of `mathgain` for the average value of `mathkind.s`) across schools.

We can visualize the parameter estimates for the 99 schools as:

```
> coef(m1)
$schoolid
   (Intercept) mathkind.s  sex1 mathkind.s:sex1
1         57.1      -18.8 -1.31            1.84
2         67.1      -18.8 -1.31            1.84
3         78.2      -18.8 -1.31            1.84
4         48.9      -18.8 -1.31            1.84
6         68.3      -18.8 -1.31            1.84
7         51.6      -18.8 -1.31            1.84
8         60.5      -18.8 -1.31            1.84
9         49.9      -18.8 -1.31            1.84
10        62.3      -18.8 -1.31            1.84
```

The intercepts for boys (we only show the first 10 of them) for each school differ from the value reported in the fixed effects. We may also use these 99 equations to obtain school-specific predictions of the effects of the explanatory variables. The random intercepts per school actually translate into a set of 99 parallel lines for each sex, some of which will be above and some below the average relationship denoted as the fixed effects. We will make such a display for model `m2`.

The command `ranef(m1)` returns a list with the school-specific coefficients reflecting the extent to which a school's intercept differs from the global intercept of 57.86 given as `fixef(m1)[1]`. We can store these estimates of random intercepts into a data frame, `rand.eff.m1 = data.frame(school = unique(DF$schoolid), rand.int = unlist(ranef(m1)))`, and get rid of the ugly row names with `row.names(rand.eff.m1) = NULL`. Here are the first four random intercepts (out of the 99):

```
rand.eff.m1
 school  rand.int
      1  -0.78788
      2   9.20642
      3  20.39139
      4  -8.91057
```

The summary statistics for the 99 estimated random intercepts is:

```
> summary(rand.eff.m1$rand.int)
   Min. 1st Qu.  Median    Mean 3rd Qu.    Max.
 -16.62   -6.03    0.01    0.00    5.85   20.39
```

Their mean equals zero because the random intercepts are the deviations from the same "average intercept" for boys, equal to 57.86. We can also visualize a density curve of the distribution of the estimated random school intercepts. Figure 13.3 depicts the density plot of the estimated random effects together with a superimposed normal distribution with mean zero and standard deviation 10.0, as shown in the model summary. We can now use the probability distribution Normal($\mu_{\text{rand.int}}$ = 0, $\sigma_{\text{rand.int}}$ = 10.0) to predict the possible random intercepts for non-sampled schools from the super-population of schools in Michigan.

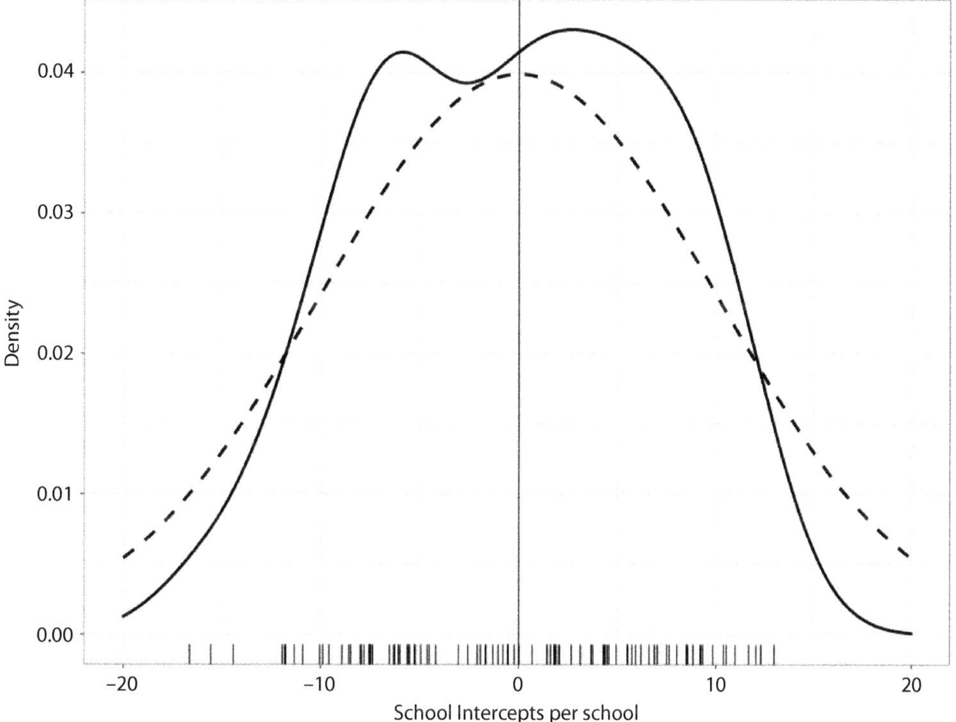

Fig. 13.3 Density plots for the estimated random intercepts for the 99 schools with the frequentist linear mixed model `m1`, with the actual values shown on the *x*-axis. The dotted line is the normal probability distribution implied in the modeling of random intercepts with a mean of zero and a variance equal to the estimated variance of the random intercepts.

The inclusion of random intercepts accounted for the presumably greater similarity (and hence lack of statistical independence) of the sampling units (the children) within groups (schools) than among schools. The intra-class correlation coefficient (ICC) quantifies "the proportion of the variance explained by the grouping structure in the population," and it is calculated as ICC = $\frac{\sigma_{\text{rand.int}}}{\sigma_{\text{rand.int}}+\sigma_{\text{resid}}}$ = 0.11 (the actual formula for the ICC changes with the sampling design; see Lohr 2019). The ICC is also called repeatability (Schielzeth and Nakagawa 2010). To an extent, our model takes into account and quantifies the structured nature of the data in a rather descriptive sense that cannot attribute any putative reasons for the degree of intra-school similarity among the children.

We can extract the fitted parameters with the package `broom` and store them with:

```
pars.m1=tidy(m1)
pars.m1
A tibble: 6 × 6
effect   group    term                estimate std.error statistic
<chr>    <chr>    <chr>                  <dbl>     <dbl>     <dbl>
fixed    NA       (Intercept)             57.9      1.59    36.4
fixed    NA       mathkind.s             -18.8      1.23   -15.4
fixed    NA       sex1                   -1.31      1.70    -0.769
fixed    NA       mathkind.s:sex1         1.84      1.70     1.08
ran_pars schoolid sd__(Intercept)         10.0      NA       NA
ran_pars Residual sd__Observation         28.6      NA       NA
```

We can also obtain the 95% confidence intervals and store them in another data frame: `CI.m1 = data.frame(confint(m1))`.

```
> CI.m1
                 X2.5.. X97.5..
.sig01             7.60   12.58
.sigma            27.40   29.80
(Intercept)       54.73   60.97
mathkind.s       -21.24  -16.43
sex1              -4.64    2.02
mathkind.s:sex1  -1.48     5.16
```

Putting the 95% confidence intervals inside `pars.m1` requires some juggling because the order of the parameters does not coincide between them. It can be done in three lines of code:

```
CI.m1=rbind(CI.m1[3:6,],CI.m1[1:2,])
pars.m1[,7:8]=CI.m1
names(pars.m1)[7:8]=c("l2.5", "u97.5")
```

We can now plot the parameter estimates and their 95% confidence intervals in Fig. 13.4. As for the random effects that range in [–16.62, 20.39] (or, better, expressed as a percentage of the fixed effect intercept: [–28.7%, +35.2%]), Fig. 13.5 displays their values.

We use the package `ggeffects` to generate the conditional curves predicted by the model `m1`. These curves are the population-level values of the response variable predicted either by only the fixed effects and their uncertainty (Fig. 13.6a), or by including the random effects by making the fixed effects conditional on the estimates of the random-effect variances.

We can also obtain R^2 statistics (Nakagawa and Schielzeth 2012, Nakagawa et al. 2017) for mixed models using the package `performance`. The conditional R^2 is the proportion of the variation of the response variable explained by both the fixed and random effects. The marginal R^2 is the proportion explained only by the fixed effects, and their difference is, of course, the proportion explained by the random effects.

```
> r2(m1)
# R2 for Mixed Models

  Conditional R2: 0.342
     Marginal R2: 0.261
```

As always, our final step is assessment of the goodness of fit of model `m1` through residual analysis. Because linear mixed models involve a normally distributed response

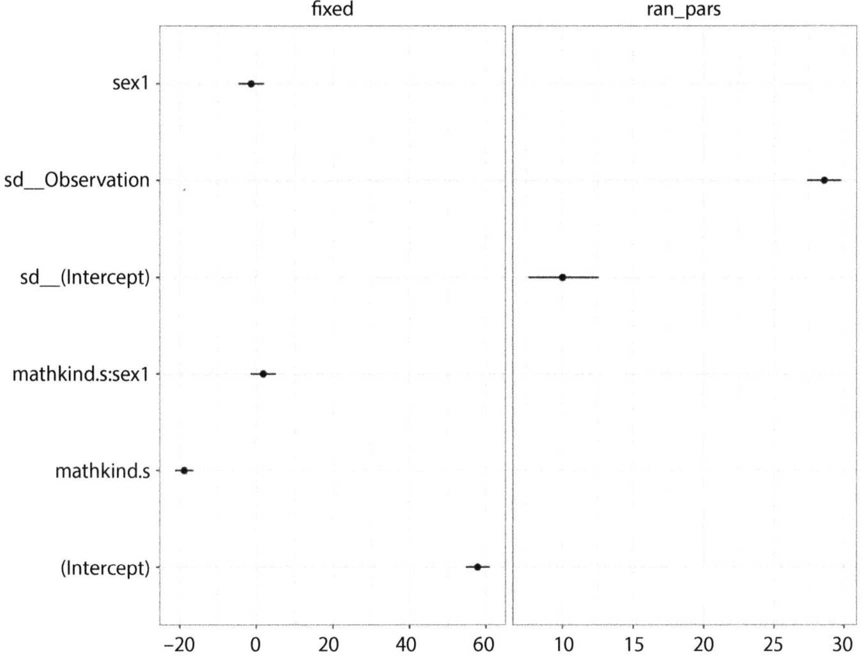

Fig. 13.4 Parameter estimates and 95% confidence intervals for the frequentist linear mixed model `m1`.

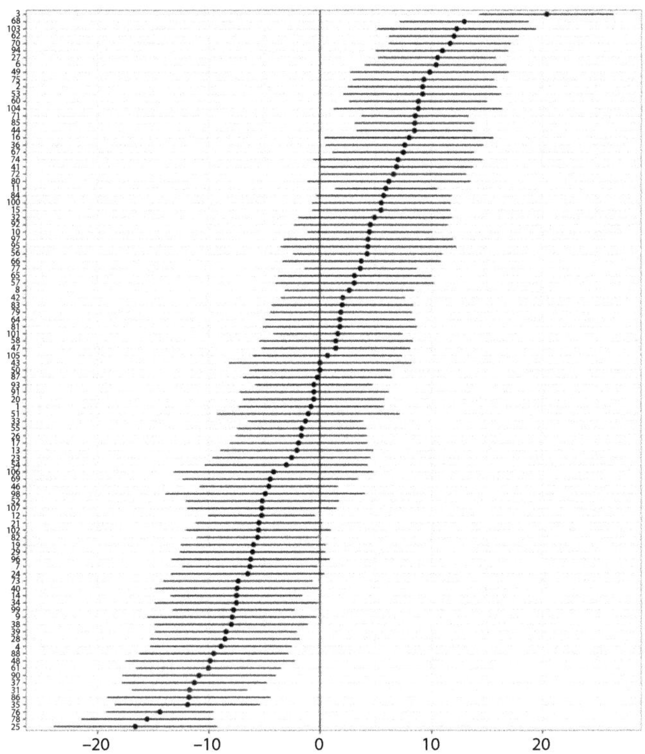

Fig. 13.5 Estimated random intercepts and their standard errors for the 99 schools for the frequentist linear mixed model `m1`, shown in increasing order.

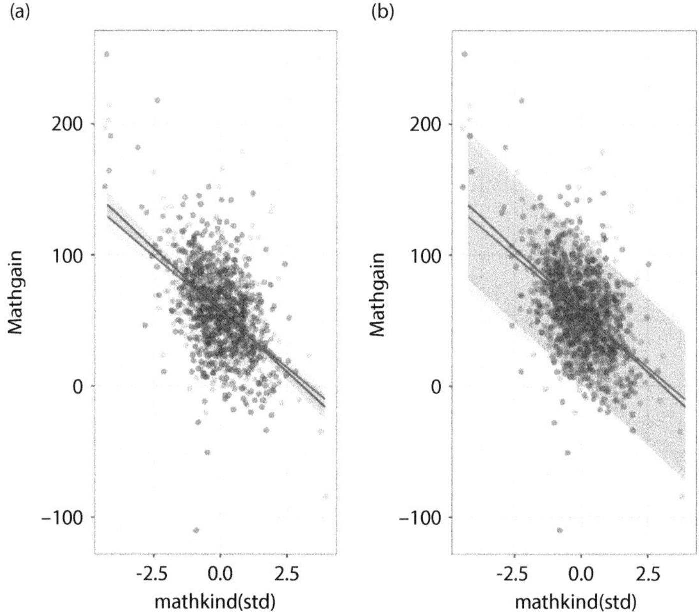

Fig. 13.6 Conditional plots for the frequentist linear mixed model `m1` showing the predicted relations between the standardized explanatory variable `mathkind.s` and `sex` (pink: boys) and the mean of the response variable `mathgain`. Plot (a) shows the population-level predicted curves using only the fixed effects and their standard errors. Plot (b) also shows the population-level predicted curves but taking into account both fixed and random effects.

variable, we could employ either the raw or standardized residuals in our evaluation of their goodness of fit (Chapter 8). We gather nearly all the information needed for the assessment using the command `resid.m1 = augment(m1)` from the package `broom`. In mixed models, we also need to check the normality of the random effects. We can also use a separate Cook distance calculated with the package `HLMdiag` to assess the sensitivity of the parameter estimates to deletion of all data from the grouping variable `schoolid` that was modeled as a random effect. This is done with `cd.school = as.data.frame(cooks.distance(m1, level = "schoolid"))` and `names(cd.school) = "Cook.school"` just to simplify the labeling.

Figure 13.7 shows that model `m1` had overall an adequate fit to the data. Figure 13.7a and b show a random scatter of residuals, there was a very small proportion of Cook distances with large values (Fig. 13.7d and e), and the random intercepts were normally distributed (Fig. 13.7f). Figure 13.7c shows that a fair number of data points deviated from the expected normal distribution at both tails; this is not too worrisome since linear mixed models are known to be robust to mild violations of the assumption of normal distribution of residuals (cf. Fig. 13.1).

We now quickly fit model `m2` involving both random intercepts and slopes: `m2 = lmer(mathgain ~ mathkind.s*sex + (1 + mathkind.s|schoolid), REML = T, data = DF)`. Its summary is:

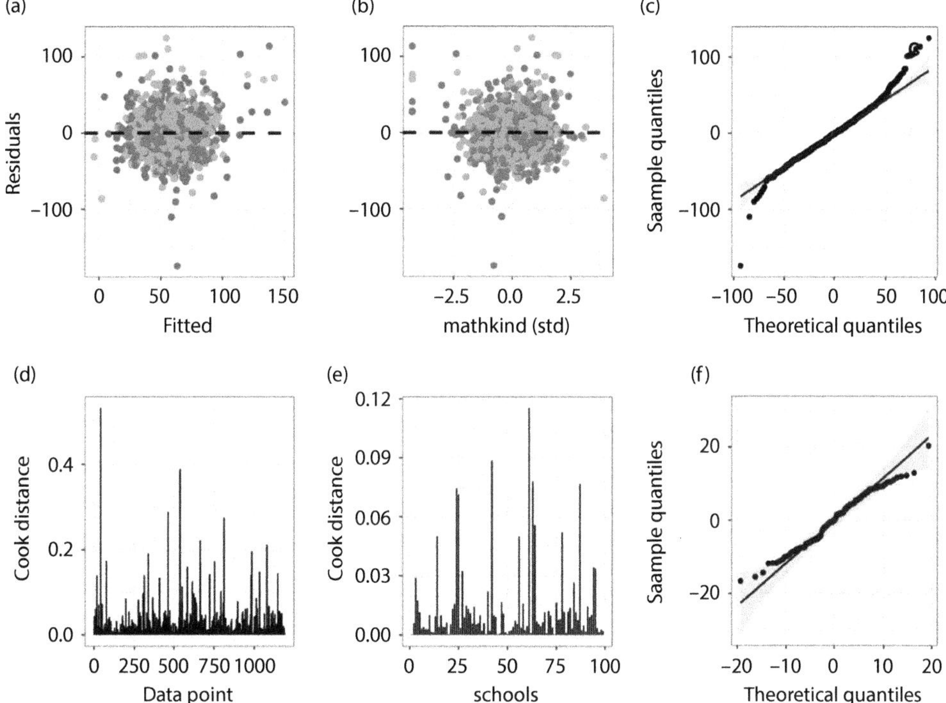

Fig. 13.7 Residual analysis for the frequentist linear mixed model `m1` based on raw residuals. Plots (a) and (b) show the relation between the residuals and the fitted values and the explanatory variable respectively (pink: boys). Plots (d) and (e) show the Cook distances from each data point, and for all the data of each school, respectively. Plots (c) and (f) are quantile–quantile plots (and the 95% confidence band) for the residuals and the estimates of the random intercepts, respectively.

```
> summary(m2)
Linear mixed model fit by REML. t-tests use Satterthwaite's method
Formula: mathgain ~ mathkind.s * sex + (1 + mathkind.s | schoolid)
   Data: DF

REML criterion at convergence: 11412

Scaled residuals:
   Min     1Q Median     3Q    Max
-6.015 -0.599 -0.033  0.562  4.460

Random effects:
 Groups   Name        Variance Std.Dev. Corr
 schoolid (Intercept)  99.0     9.95
          mathkind.s   44.6     6.68    -0.28
 Residual             777.5    27.88
Number of obs: 1190, groups:  schoolid, 99

Fixed effects:
                Estimate Std. Error      df t value Pr(>|t|)
(Intercept)        57.50       1.59  176.02   36.11   <2e-16 ***
mathkind.s        -17.66       1.44  198.32  -12.29   <2e-16 ***
sex1               -1.24       1.68 1142.68   -0.74     0.46
mathkind.s:sex1     1.90       1.74 1072.90    1.09     0.28
```

The interpretation of the fixed effects part of the summary was given above for model `m1`.

The random effects now contain the estimated variances of the random intercepts and slopes, and their correlation. You may recall that the intercept and slope in a linear regression are not independent. Now we have 99 estimates of each parameter whose correlation across schools equals –0.28. Again, the fixed effects are the population-level parameter estimates (irrespective of the random effects) that give the overall view of the effects of the explanatory variables. We also have the estimated coefficients at the school level (we only show the first 4 out of 99) incorporating the random differentials in intercept and slope for the sex of reference (boys):

```
> coef(m2)
$schoolid
    (Intercept) mathkind.s  sex1 mathkind.s:sex1
1          57.0     -16.66 -1.24             1.9
2          68.7     -22.74 -1.24             1.9
3          78.4     -21.42 -1.24             1.9
4          49.2     -10.85 -1.24             1.9
```

Note that the differentials between sexes in the intercept and slope are identical among schools. As a side note, we should point out that it is important to at least center the explanatory variable modeled as a random slope to avoid having a spuriously high correlation between random slopes and intercepts, which is often a sign of a poorly fitting mixed model.

Let's display the estimated parameters of model `m2` and its estimated random effects. We first gather the parameter estimates with `pars.m2 = tidy(m2)`. We then obtain the 95% confidence intervals with `CI.m2 = as.data.frame(confint(m2))` and, yet again, change the order of the parameters with `CI.m2 = rbind(CI.m2[5:8,], CI.m2[1:4,])`. We can now place the 95% confidence intervals into `pars.m2[,7:8] = CI.m2` and give names to the last two columns added with `names(pars.m2)[7:8] = c("l2.5", "u97.5")`. All this work is needed because we lack a simple way of having all the parameter estimates and their 95% confidence intervals in a single data frame that could be the input to a `ggplot` (Fig. 13.8a).

As for the random effects of `m2`, we first gather the estimates using the command `melt` from package `reshape2`: `rand.eff.m2=melt(ranef(m2)$schoolid, id=c("(Intercept)", "mathkind.s"))`. Then, we obtain their standard errors with a command from the package `arm`, `rand.eff.m2[,3:4]=se.ranef(m2)$schoolid`, and make a few changes with `names(rand.eff.m2)=c("rand.int","rand.slope", "SE.rand.int","SE.rand.slope")` and `rand.eff.m2$schools = row.names (ranef(m2)$schoolid)` to get a readable data frame for `ggplot` (Fig. 13.8b). The proportion of variation explained by the fixed and random effects of model `m2` obtained with package `performance` is actually very similar to what we obtained for model `m1`, and needs no further comment.

```
> r2(m2)
# R2 for Mixed Models

  Conditional R2: 0.354
     Marginal R2: 0.234
```

We can use the package `ggpredict` to obtain the conditional plots based on the predictions obtained as `pred.m2.r=ggpredict(m2, type="random", terms = c("mathkind.s [all]", "sex"))`. Plotting the predicted relations for each sex per school requires a few lines of code to gather the random intercept and slopes for each school:

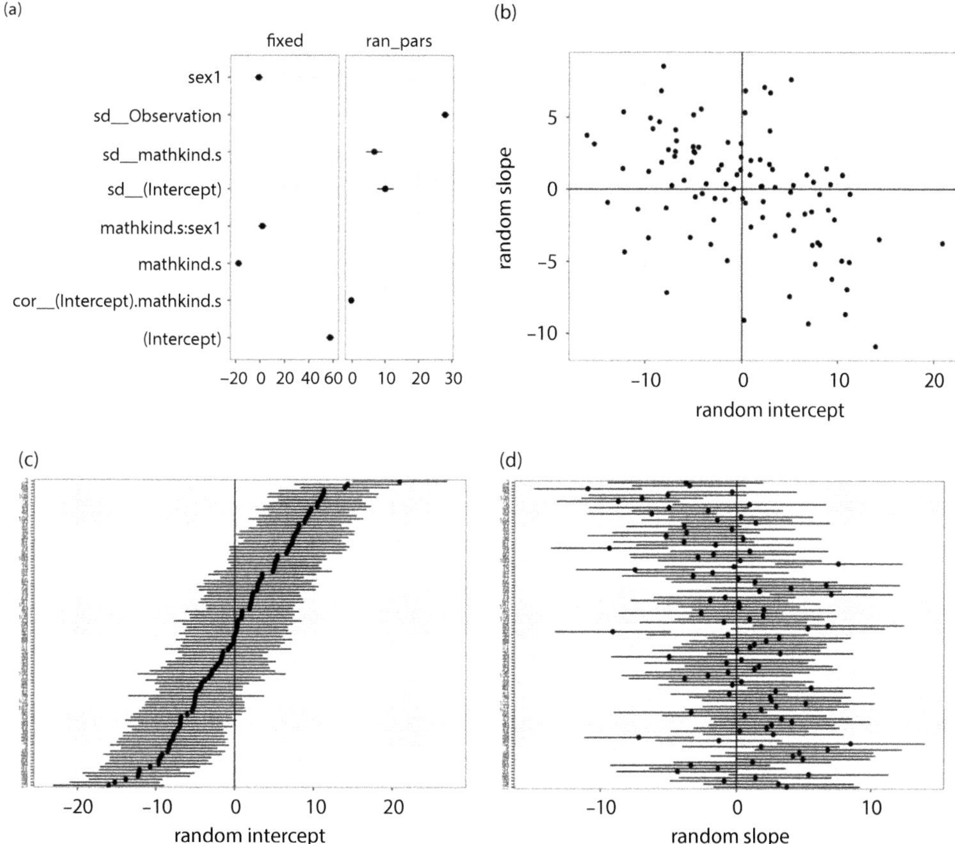

Fig. 13.8 Estimates of the fixed and random effects for the frequentist linear mixed model m2. Plot (a) shows the parameter estimates and their 95% confidence intervals. Plot (b) shows the weak correlation between random slopes and intercepts. Plots (c) and (d) show the estimated random intercepts and slopes and their standard errors for each school; the schools are ordered according to the value of the random intercept.

```
int.slopes.m2.boys=coef(m2)$schoolid[,c("(Intercept)", "mathkind.s")]
names(int.slopes.m2.boys)=c("intercept", "slope")
int.slopes.m2.girls=as.data.frame(cbind(
   coef(m2)$schoolid[,"(Intercept)"]+coef(m2)$schoolid[,"sex1"],
   coef(m2)$schoolid[,"mathkind.s"]+coef(m2)$schoolid[,"mathkind.s:sex1"]))
names(int.slopes.m2.girls)=c("intercept", "slope")
int.slopes.m2=as.data.frame(rbind(int.slopes.m2.boys, int.slopes.m2.girls))
int.slopes.m2$school=as.factor(rep(rownames(coef(m2)$schoolid),times=2))
int.slopes.m2$sex=as.factor(rep(c(0,1), times=99))
```

The first line extracts the parameters for boys (the sex of reference in the analysis) from the list `coef(m2)` that contains the data frame `schoolid`. Then we name the columns of the data frame `int.slopes.m2.boys`. We then gather the intercepts and slopes for girls for each school, knowing that they can be obtained by adding the differentials of the intercept and slope to the values of the sex of reference. After naming the columns of `int.slopes.m2.boys`, we merge both data frames into a single data frame `int.slopes.m2`, to which we add the labels school and sex. Now we finally have all that is needed to produce Fig. 13.9. Figure 13.9b shows that at the school level there is a much smaller variation between the slopes than between the intercepts, consistent with the variances of the random effects shown in `summary(m2)`. We can also appreciate that, viewed at the school level, the predicted relations for girls (blue) appear to be systematically above those of boys (pink), despite the difference between the intercepts at the population level (Fig. 13.9a) not being statistically significant ($p = 0.46$ in `summary(m2)`). The point here is showing that we can improve our understanding and the presentation of modeling results with just a little effort without leading to R code indigestion.

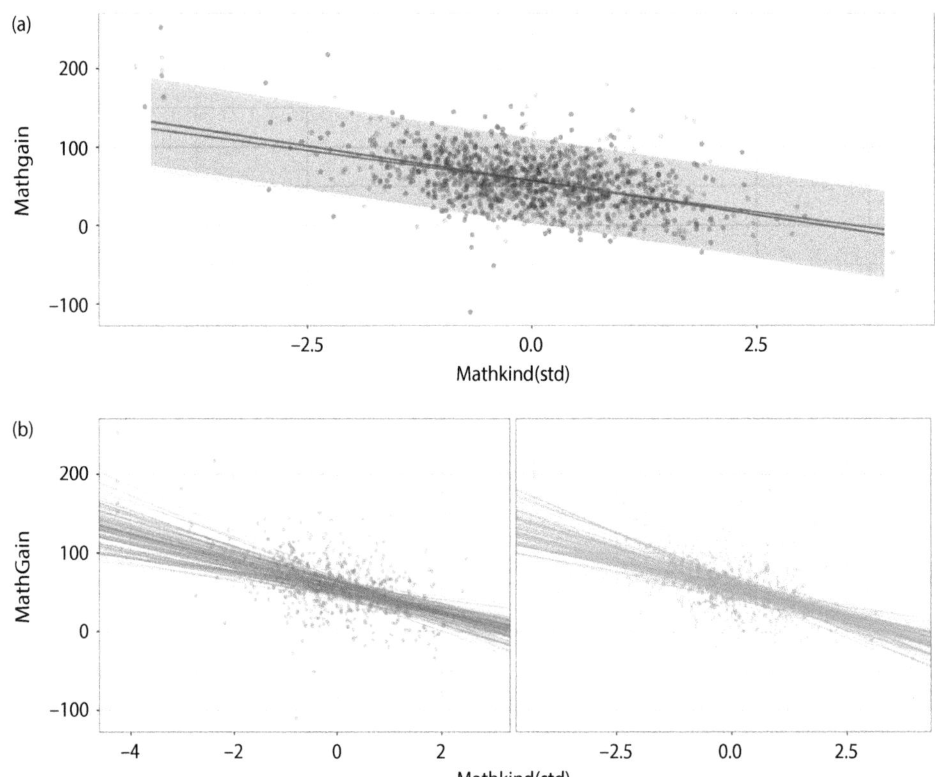

Fig. 13.9 Conditional plots for the frequentist linear mixed model `m2` showing the predicted relations between the standardized explanatory variable `mathkind.s` and `sex` (pink: boys) and the mean of the response variable `mathgain`. (a) The population-level predicted curves using only the fixed effects, and their standard errors taking into account both fixed and random effects. (b) The group-level predicted curves for each of the 99 schools separating the two sexes (pink: boys).

We should evaluate the goodness of fit of model `m2` using exactly the same procedure followed for the linear mixed model `m1` using the packages `broom` and `HLMdiag`. The only difference is that we now also need to assess whether the random slopes also have a normal distribution. The actual plots are rather similar to those shown in Fig. 13.7, and hence are not shown here; we refer you to the R code on the companion website.

13.7 Statistical significance and model selection in frequentist mixed models

Assessing the statistical significance of fixed effects in mixed models remains a polemic issue that is often treated in different ways depending on the book you read or the software you use. The issue is as follows: it is VERY IMPORTANT to realize that *the difficulties involved with the number of parameters related to the random effects are common to all mixed models regardless of the distribution of the response variable.* The root of the problem stems from disagreements on how to count the number of parameters associated with the random effects. Using our previous example, if there were 999 or just 9 schools, we would still estimate a single variance of the random intercepts in model `m1`. To state the obvious, an adequate number of parameters in the mixed model `m1` would be some number between one (the variance of the random intercepts) and the number of schools minus one (Bolker 2015). The actual number of parameters of a statistical model is fundamental to calculating the degrees of freedom necessary to obtain the asymptotic sampling distributions that are used to calculate the *p*-values in the frequentist framework. You should not be shocked to learn that the number of degrees of freedom of mixed models does vary a great deal depending on the complexity of the random effect structure (i.e., vectorial random effects, nested and crossed random effects) and the size of the data set. In mixed models, the degrees of freedom of fixed effects depend not only on the overall size of the data set but, crucially, on the structure of the data modeled as random effects. There are a few expressions to calculate the degrees of freedom of fixed effects only for cases with balanced, or at least proportional, sample sizes across the levels of the variables modeled as random effect(s). Needless to say, these simple and ideal cases are rare exceptions in most scientific data.

The options for assessing the statistical significance in fixed effects can be ordered from worst to best (Bolker 2015) as follows:

(i) Wald test: This is the simplest and most commonly used approach. The Wald statistic is the ratio between each fixed effect coefficient in a mixed model and its standard error. The use of Wald tests in GLMs relies on a quadratic approximation of the likelihood profile (Chapter 3) to make a large-sample approximation from which we obtain the *p*-values. This approximation in GLMs remains valid even for moderate sample sizes. In contrast, in mixed models, Wald tests are known to be anti-conservative, i.e., they systematically over-estimate the *p*-values, particularly for small and moderate sample sizes (Luo et al. 2021). Software packages that rely on large-sample, asymptotic approximations to derive *p*-values for fixed effects using Wald tests are "at best ad hoc solutions" (Bates et al. 2015) of questionable reliability and generality for small to moderate sample sizes.

(ii) Likelihood ratio test: We encountered the likelihood ratio test (LRT) in Chapter 3 to compare two nested models. The larger or more encompassing model contains the reduced or simpler model in which a single effect being tested equals zero. For large sample sizes, the LRT asymptotically follows a chi-square distribution with degrees of freedom equal to the difference in the number of parameters between the more encompassing and the reduced models. Unlike the Wald test, the LRT requires fitting the encompassing model, and a series of reduced statistical models obtain the *p*-values of each term in the former. When applied to mixed models, the LRT is also anti-conservative because it over-estimates the true *p*-values (Pinheiro and Bates 2004).

(iii) Small sample corrections and approximations: There are to main approaches to obtain *p*-values of mixed models for small to moderate sample sizes. First, Satterthwaite's (1946) approximation, developed only for linear mixed models, corrects the degrees of freedom related to random effects and provides *p*-values for fixed effects. The summaries of models `m1` and `m2` above used this approximation implemented in the package `LmerTest`. Second, Kenward and Roger's (1997) approximation both corrects the standard error of the fixed effects and makes an adequate approximation of the degrees of freedom for all mixed models. It is based on a sequence of two Taylor series approximations during the estimation of the variances of the random effects and of the fixed effects parameters, followed by a moment-matching approximation of the degrees of freedom (McNeish 2017). While the Kenward and Roger (1997) approximation requires the mixed model to be fitted with REML, Satterthwaite's works with both maximum likelihood and REML (Luke 2017). While both methods yield comparable and acceptable type I error rates for linear mixed models across different sample sizes (e.g., Luke 2017), Kenward and Roger's (1997) approximation has much greater generality than Satterthwaite's, and should generally be preferred (Stroup 2013, Singmann and Keller 2019). We will use the Kenward and Roger method to compare two nested mixed models implemented in package `pbkrtest`.

(iv) Parametric bootstrap: The bootstrap (Efron and Tibshirani 1993) is one of the great advances in the frequentist framework of the second half of the twentieth century. This computer-intensive method allows us to obtain sampling distributions of statistics in cases where we lack an expression (e.g., the standard error of the median) or cannot rely on asymptotic results (our current case). There are two main varieties of the method: non-parametric and parametric bootstrap; we are only going to deal with the latter here. After fitting the statistical model of interest, the parametric bootstrap uses this as a generative model to simulate many (say 1,000) synthetic data sets of the same size as the original data. Then, it re-estimates the model parameters in each synthetic data set to obtain sampling distributions for each parameter. Rather than assuming an asymptotic chi-square distribution as in the LRT, or a small-sample correction as for the Kenward–Roger approximation, the parametric bootstrap generates a reference distribution from which we obtain the *p*-values. To obtain the *p*-value for each term of the fixed effects, we must again consider the encompassing and the reduced model, from which we calculate an empirical log-likelihood ratio. The procedure is to simulate many synthetic data sets from the reduced model (equivalent to assuming that one parameter of the encompassing model equals zero), and then fit both models to each synthetic data set to produce a reference distribution of likelihood ratio values under the null hypothesis. The parametric bootstrap *p*-value for each model term is the percentage of likelihood ratio values of the reference distribution that are larger than the observed likelihood ratio value. The main problem with the parametric bootstrap is the time and computational effort required to obtain the reference distributions. Bolker (2015) and Luke (2017) compared the parametric bootstrap very favorably with other methods employed to estimate the statistical significance of fixed effects in mixed models. In later chapters we use the package `afex` for a simple implementation of the parametric bootstrap function of the package `pbkrtest` to estimate the *p*-values of mixed models.

It may be thought necessary to assess the statistical significance of random effects, namely to test whether the variance of random effect terms significantly differs from zero. There are important technical problems in making such a statistical test at the boundary of the admissible values of a parameter (here variance = 0) that remain unresolved (Bolker 2015). Beyond any statistical test, we need to bear in mind that the random effects account for the structure of the data set. Eliminating a random effect would amount to pretending that the data were generated with an experimental or sampling design different from the one actually employed. We should really think very hard and possess very good reasons to argue for the elimination or simplification of random effects in mixed models.

We can perform model selection in frequentist mixed models through very careful use of the LRT combined with parametric bootstrap. The use of parametric bootstrap in LRT model selection probably sidesteps the need to refit the models being compared using `REML = F` to avoid using ("unrestricted") maximum likelihood parameter estimates (e.g., Pinheiro and Bates 2004). With the possible exception of linear mixed models, the use of the AIC in the model selection of mixed models seems to rest on dodgy theoretical grounds and, in practical terms, is better avoided (Bolker 2015) until further work clarifies many lingering issues. Those (e.g., Stroup 2013, Hox et al. 2018) who use the AIC for model selection with mixed models rely on specific estimators of the number of parameters that are not general for different random effect structures, and whose validity remains disputed. There is a clear need for further work in this area. Using the AIC to perform selection of models with different structures of random effects is subject to the same concerns just discussed for testing at the boundary of admissible values of the variance (Bolker 2015). At the time writing, the combined use of LRT and parametric bootstrap seems the only uncontroversial and general means to carry out model selection in the frequentist framework.

Let's illustrate model selection using parametric bootstrap with models `m1` (random intercepts) and `m2` (random intercepts and slopes). We recall that while random intercepts are essential to represent the structured nature of the Michigan school data, the use of random slopes largely remains an open and debatable option (see Section 13.3). We first obtain the actual LRT value from comparing the two models as `LRT.obs = 2 * (as.numeric(logLik(m2)) - as.numeric(logLik(m1)))` (`as.numeric` is needed just to store the actual value). We then use the parametric bootstrap of package `pbkrtest` to obtain the reference distribution that will be used to assess the statistical significance of `LRT.obs`: `RefDistPB = PBrefdist(largeModel = m2, smallModel = m1, nsim = 1000, cl = detectCores())`. We need to identify the encompassing or large model (`m2`) that contains the reduced or small model (`m1`) as a special case, the number of bootstrap simulations, and the number of clusters to be used to parallelize the calculations (from package `parallel`). We need some R juggling to convert `RefDistPB` into a data frame, `RefDistPB = as.data.frame(as.vector(RefDistPB))`, that is the input for Fig. 13.10. We can observe in Fig. 13.10 that `LRT.obs` is at the upper tail of `RefDistPB`, suggesting that there is a significant difference between the two models. We can be more precise and use the simulated reference distribution to calculate the p-value as:

```
> PBmodcomp(largeModel=m2,smallModel=m1,ref=RefDistPB)
Bootstrap test;  samples: 1; extremes: 0;
large : mathgain ~ mathkind.s * sex + (1 + mathkind.s | schoolid)
mathgain ~ mathkind.s * sex + (1 | schoolid)
        stat df p.value
LRT     17.4  2 0.00016 ***
PBtest 17.4     0.00100 **
```

We showed the reference distribution obtained with parametric bootstrap for illustrative purposes. You could have obtained just the same result with `PBmodcomp(largeModel = m2, smallModel = m1, nsim = 1000, cl = detectCores())`.

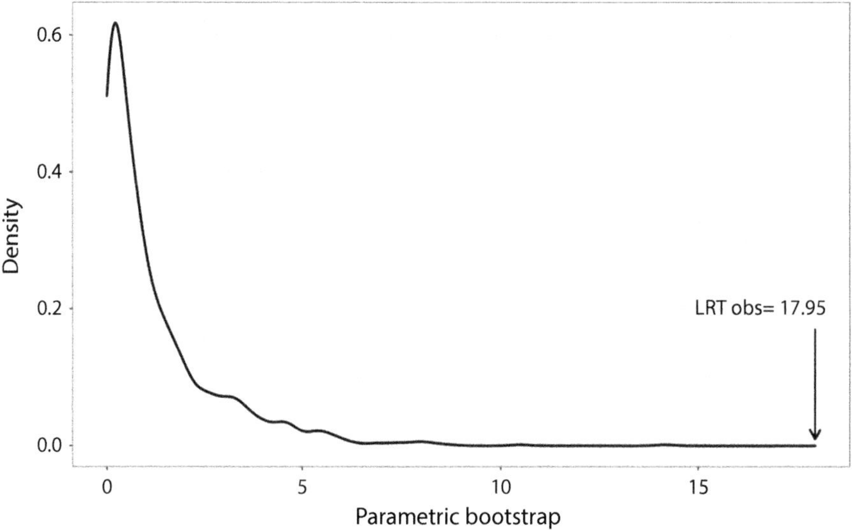

Fig. 13.10 Parametric bootstrap reference distribution for the model selection between the linear mixed models `m1` (only random intercepts) and `m2` (random intercepts and slopes) based on the likelihood ratio test comparison. The distribution was obtained with the package `pbkrtest`. The actual value of the likelihood ratio comparison between the models is shown in the graph.

We illustrate the use of the Kenward–Roger correction implemented in package `pbkrtest` to compare two nested mixed models. We first formulate `m1.1 = lmer(mathgain ~ mathkind + sex + (1|schoolid), data = DF)`, which differs from `m1` by not including the interaction between `mathkind` and `sex`. Note that these two models have the same structure of random effects. Their comparison would then amount to testing the statistical significance of the `mathkind` and `sex` interaction. We can see that the interaction between `mathkind` and `sex` is relatively unimportant to explain the data.

```
> KRmodcomp(largeModel = m1, smallModel=m1.1)
large : mathgain ~ mathkind.s * sex + (1 | schoolid)
small : mathgain ~ mathkind + sex + (1 | schoolid)
         stat     ndf     ddf F.scaling p.value
Ftest    1.17    1.00 1148.19         1    0.28
```

Finally, when appropriate, we could also perform a posteriori tests to account for the multiplicity effect (Chapter 5) in fixed effect comparisons using the package `multcomp`.

13.8 The shrinkage or borrowing strength effect in mixed models

Whenever we have a hierarchical data set such as the Michigan schools, we have three options for analyzing it. These options appear in the statistical literature under the names no pooling, complete pooling, and partial pooling.

At one extreme, no pooling means that we perform as many `mathgain ~ mathkind.s*sex` analyses as we have schools. Each of the 99 analyses occurs as if the other schools did not exist. We afterwards compile the 99 sets of parameter estimates, and perhaps calculate their means and standard deviations to get an overall summary. No pooling effectively considers that the variances among schools are very large (or infinite if you wish) so that each requires a separate analysis. We saw above that for 48 out of 99 schools we have data for 10 children or less, and presumably not an even representation of the two sexes. Would you bother to fit an analysis of covariance at the school level for such small data sets? Even if you do, the standard errors of the parameter estimates would be so large as to render the point estimates worthless.

At the other extreme, the complete pooling option amounts to ignoring the structured nature of the Michigan data by fitting a single model, `mathgain ~ mathkind.s*sex`, to the entire data set. The standard error of the parameter estimates would now be very small because we have data for 1,190 children. But this would be very wrong because the 1,190 children cannot be considered to be statistically independent replicates: children from the same school (and even more, from the same class) share many attributes that make them more similar to each other than to children from other schools. The complete pooling option effectively considers that the variance among schools is so small (or zero, if you prefer) as to be worth ignoring.

The final (and, as always, the correct) option is partial pooling. This is precisely what mixed models do. They estimate a global relation, `mathgain ~ mathkind.s*sex`, across schools as fixed effects, and they also characterize as random effects the extent to which each school deviates from the fixed effects. But there is more involved. We know that mixed models are made of population-level or fixed effects, and group-varying or random effects. The main distinction between them is that the latter have an associated normal probability distribution centered at zero with a variance to be estimated. That is, random effect coefficients are not just fortuitous departures from population-level effects as they are modeled as coming from a single probability distribution.

The main practical implication of varying parameters is to induce the shrinkage of group-level coefficients towards the population-level or fixed effects. The shrinkage effect reflects the regression to the mean effect: deviations need to compensate each other in sign and magnitude so that the random effect values conform to a normal distribution with given parameter values. *Shrinkage is a fundamental property of mixed or hierarchical models that occurs in both frequentist and Bayesian frameworks.* In practical terms, it means that group-level estimates are "shrunk" toward the fixed effect mean, and the strength of the shrinkage is inversely proportional to the sample size of each level of the categorical variable modeled as a random effect. That is, the groups having smaller sample sizes (and hence a smaller amount of information) are shrunk more towards the fixed effect mean than the clusters with larger sample sizes. This is why the shrinkage effect is also called the borrowing strength effect: levels of the random effects with more information "lend their strength" to allow the estimation of coefficients for clusters with small sample sizes. Thus, each estimation of a level of the variable modeled as a random effect "borrows strength" from others via their assumed global probability distribution. The shrinkage effect is particularly important when the amount of information (in terms of their sample sizes and precision) is very disparate across the groups. It is because of the partial pooling of information across groups that the mixed effects can obtain predicting equations not only at the population level but also at the group level, even for those schools for which there was little data.

We can illustrate the workings of the shrinkage effect by comparing the intercepts and slopes at the school level obtained from no pooling with those from the mixed model `m2` (partial pooling). To keep it simple, we are just going to do it for boys. We first make a data frame to hold the no pooling parameter estimates: `no.pool = data.frame(school = unique(DF$schoolid), id = 1:length(unique(DF$schoolid)), intercept =`

`NA, slope = NA)`. We next calculate the parameter estimates for boys at each school and put the estimate in the newly created data frame:

```
for (i in 1:length(DF$schoolid)) {
                    no.pool[i,3:4]=as.vector(coef(lm(mathgain~mathkind.s,
                        data=DF[DF$sex==0 & DF$schoolid==no.pool$school[i],])))}
```

This code requires some explanation. First, `lm(mathgain ~ mathkind.s, data = DF[DF$sex == 0 & DF$schoolid == no.pool$school[i],])` estimates a linear regression for a subset of the data frame `DF` for boys (`DF$sex == 0`) at the `i`th school. Next, `as.vector(as.coef())` gathers the estimated slope and intercept for each school, and converts them into a vector to be stored in the ith row of the data frame `no.pool`. We make the data frame `all.int.slopes` with `all.int.slopes = no.pool[,c("intercept", "slope")] - int.slopes.m2.boys` to calculate the differences between the intercepts and slopes for boys just obtained with no pooling, and those from the mixed model `m2` (partial pooling). And finally, we use the package `doBy` to obtain the number of boys at each school, and place the results in the same data frame: `all.int.slopes$boys = summaryBy(mathgain ~ schoolid, data = DF[DF$sex == 0,], FUN = length)[,2]`. We can now generate Fig. 13.11, which shows that the magnitude of the difference between the two sets of parameter estimates decreases with the number of boys at each school. The difference between the sets is very noticeable, even for the small range of values in the number of boys per school.

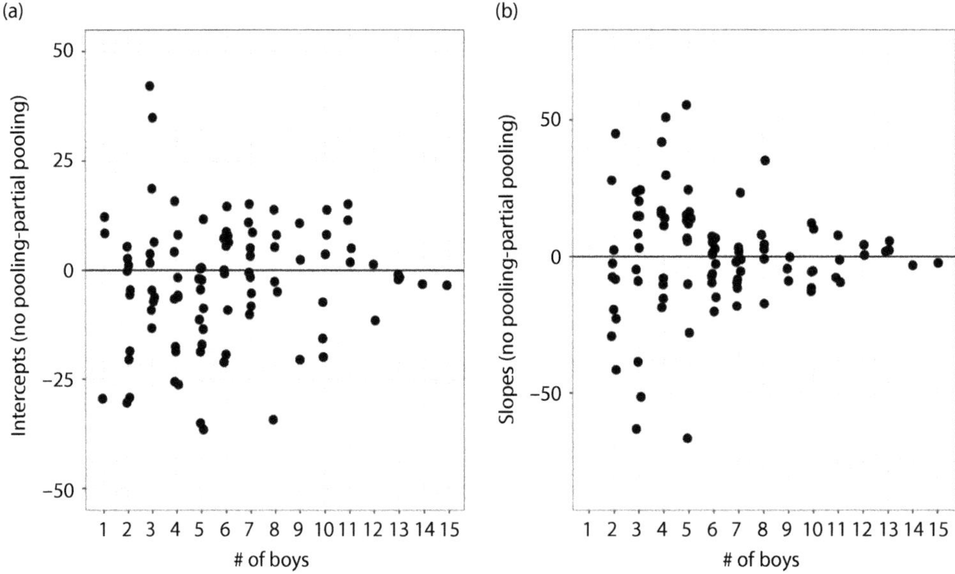

Fig. 13.11 Differences in the intercepts and slopes for each school between estimates obtained with (a) no pooling (each school analyzed separately) and (b) partial pooling (from the mixed model `m2`) for boys in relation to the number of boys at each school. The magnitude of the difference reflects the strength of the shrinking effect due to partial pooling.

13.9 Fitting mixed models in the Bayesian framework

Mixed models in the Bayesian framework are fitted using the MCMC algorithm. The fitting is not based on the approximations and small sample–size corrections that are required in their frequentist analogues to obtain the statistical significance of the fixed effects. We will fit the Bayesian analogues of models `m1` and `m2` using the package `brms`. The equations for these two models are Eqs. (13.3) and (13.4).

We start by defining the formula, `formula.m1.brms = bf(mathgain ~ mathkind.s*sex + (1|schoolid))`, which we use to obtain the default model priors:

```
> get_prior(formula.m1.brms, data=DF,family='Normal')
                 prior     class          coef    group resp dpar nlpar bound        source
                (flat)         b                                                    default
                (flat)         b    mathkind.s                                 (vectorized)
                (flat)         b mathkind.s:sex1                               (vectorized)
                (flat)         b          sex1                                 (vectorized)
 student_t(3, 56, 31.1) Intercept                                                   default
  student_t(3, 0, 31.1)       sd                                                    default
  student_t(3, 0, 31.1)       sd                  schoolid                     (vectorized)
  student_t(3, 0, 31.1)       sd     Intercept schoolid                        (vectorized)
  student_t(3, 0, 31.1)    sigma                                                    default
```

We have the default weakly informative priors for the population-level (fixed) effects grouped in the top bracket, and the standard deviation of the group-level intercepts (group-level effects), and for the standard deviation of the response variable (or "family-specific parameters") grouped in the lower bracket. The parameters for the population-level effects are the `Intercept` (the mean of the response variable for the reference sex (boys, or sex = 0)), `sex1` (the differential of intercepts between sexes), `mathkind.s` (the slope of the standardized explanatory variable for the sex defined as the group of reference), and `mathkind.s:sex1` (the differential of slopes between sexes). Again, it is only by being able to write the statistical model as Eq. (13.3), and by knowing how this model is implemented in R, that we can define sensible priors (when possible) and correctly interpret the fitted models. In `brms`, `student_t(3, 56, 31.1)` refers to 3 degrees of freedom, mean = 56 (i.e., the median of the response variable), and sd = 31.1. The standard deviation was obtained using Rousseeuw and Croux's (1993) estimator involving the median absolute difference, defined as MAD = median $(|Y - \text{median}(Y)|)$ and calculated as `1.4826 * median(abs(DF$mathgain - median(DF$mathgain)))`. We recall that a t-distribution with zero degrees of freedom becomes a Cauchy distribution (thicker tails), and one with infinite (or a very large number of) degrees of freedom becomes a normal distribution (thinner tails). Therefore, the default t-distribution with df = 3 is closer in shape to a Cauchy than to a normal distribution.

Lacking any direct knowledge on the performance of children in standardized tests in early primary school, let's make a rough guess that the difference in the mean of the response variable between sexes (`sex1`) is likely to be at most ±20%, i.e., 56 × 1.2 = 67.5 and 56 × 0.8 = 44.8, and likewise for the difference in their slopes (`mathkind.s:sex1`). In Chapter 5 we used Wan et al.'s (2014) approximation to obtain the standard deviation of a prior from its range / 4. Thus, based on this plausible range [44.8, 67.5], we could set the prior for `sex1` as Normal(mean = 0, sd = 5.55) so that 95% of its values would be in [–10.9, 10.9], where 10.9 = 1.96 × 5.5. We do likewise for `mathkind.s:sex1`. Having even less intuition for the heterogeneity of the mean of `mathgain` for the reference sex (boys) among schools (i.e., the random intercept) and for the variance of the response variable, we leave untouched the default weakly informative `brms` priors. We then modify the default priors, `prior.m1 = c(prior(normal(0, 5.55), class = b, coef = "sex1"), prior(normal(0, 5.55), class = b, coef = "mathkind.s:sex1"))`, and run the model: `m1.brms = brm(formula = mathgain ~ mathkind.s*sex + (1|schoolid), data = DF, chains = 3, prior = prior.m1, control = list(adapt_delta = 0.99), warmup = 1000, iter = 3000)`. The summary is:

```
> summary(m1.brms)
 Family: gaussian
  Links: mu = identity; sigma = identity
Formula: mathgain ~ mathkind.s * sex + (1 | schoolid)
   Data: DF (Number of observations: 1190)
  Draws: 3 chains, each with iter = 3000; warmup = 1000; thin = 1;
         total post-warmup draws = 6000

Group-Level Effects:
~schoolid (Number of levels: 99)
              Estimate Est.Error l-95% CI u-95% CI Rhat Bulk_ESS Tail_ESS
sd(Intercept)     9.63      1.30     7.22    12.36 1.00     2263     3256

Population-Level Effects:
                Estimate Est.Error l-95% CI u-95% CI Rhat Bulk_ESS Tail_ESS
Intercept          57.70      1.46    54.80    60.55 1.00     3267     4099
mathkind.s        -14.08      0.98   -16.01   -12.16 1.00     5539     4697
sex1               -0.81      1.29    -3.37     1.75 1.00     7354     4277
mathkind.s:sex1    -1.79      1.26    -4.29     0.66 1.00     6283     4891

Family Specific Parameters:
      Estimate Est.Error l-95% CI u-95% CI Rhat Bulk_ESS Tail_ESS
sigma    28.86      0.62    27.71    30.10 1.00     6300     4729
```

The output is divided into three parts: the group- and population-level effects, and the family-specific parameters, all of which are related to the statistical model (Eq. (13.3)). The first is the standard deviation of the group-level intercepts (or random effects in the frequentist framework). The second denotes the overall effects of the explanatory variables on the mean of the response variable `mathgain`, irrespective of school-induced heterogeneity (or fixed effects in the frequentist framework). The final part is the standard deviation of the response variable (cf. Eq. (13.3)). As usual, we have the mean ("Estimate") and the standard deviation ("Est. Error"), and the lower and upper limits of the 95% credible intervals for each model parameter. The "Rhat" and ESS metrics indicate that the chains of the MCMC algorithm converged to a single stationary posterior distribution, and that we have large effective sample sizes of estimates for each parameter (Fig. 13.12).

`summary(m1.brms)` shows that the differences in intercepts and slopes between sexes were small to moderate: 1.4% (= 0.81 / 57.70) and 12.7% (= 1.79 / 14.08), respectively. This suggests that the important effect of `mathkind.s` (that was standardized before the analysis) is very similar for both sexes but always slightly smaller for girls. We will not delve further into the interpretation of the population-level effects, as they are discussed at length for the analysis of covariance in Chapter 6.

We can also evaluate the autocorrelation of the sampled parameter estimates of model `m1.brms`. We have marginal posterior distributions not only for the population-level effects, but also for the group-level effects for the 99 schools, and for the standard deviations of the random intercepts and of the response variable (the family-specific parameter). Clearly, space constraints do not allow plotting a graphical examination of the the convergence and degree of autocorrelation of all parameters for the three chains. We therefore just choose to examine the autocorrelation of the sampled estimates of the population-level effects in Fig. 13.13. We can observe that the sampled parameter estimates of the population-level effects can, by and large, be considered to be statistically independent.

The posterior distributions of the population-level parameters of model `m1.brms` are shown in Fig. 13.14 and provide a much better graphical summary than the table of coefficients obtained with `summary(m1.brms)`.

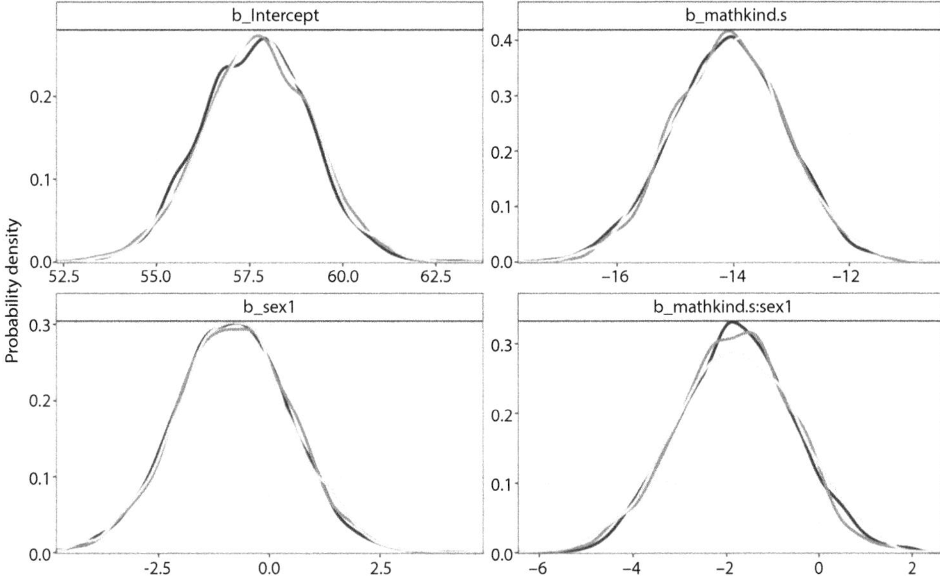

Fig. 13.12 Convergence of the three chains of the Bayesian mixed model `m1.brms` to a common stationary posterior distribution for the population-level parameters (which correspond to the fixed effects in the frequentist framework).

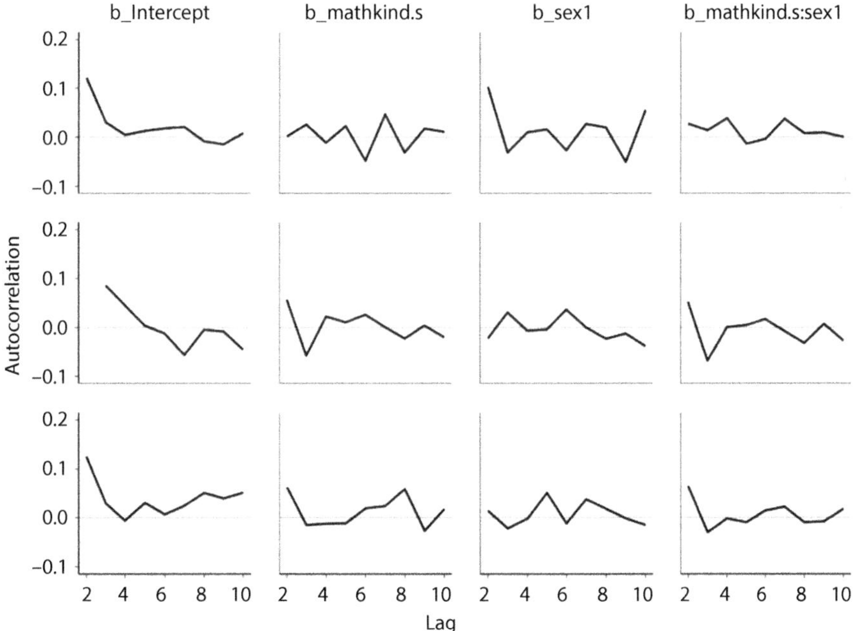

Fig. 13.13 Autocorrelation plots of population-level parameters for the three chains used to fit the Bayesian mixed model `m1.brms`. The very low autocorrelation for all parameters and chains suggest that the sampled values of the posterior distributions can be considered largely statistically independent.

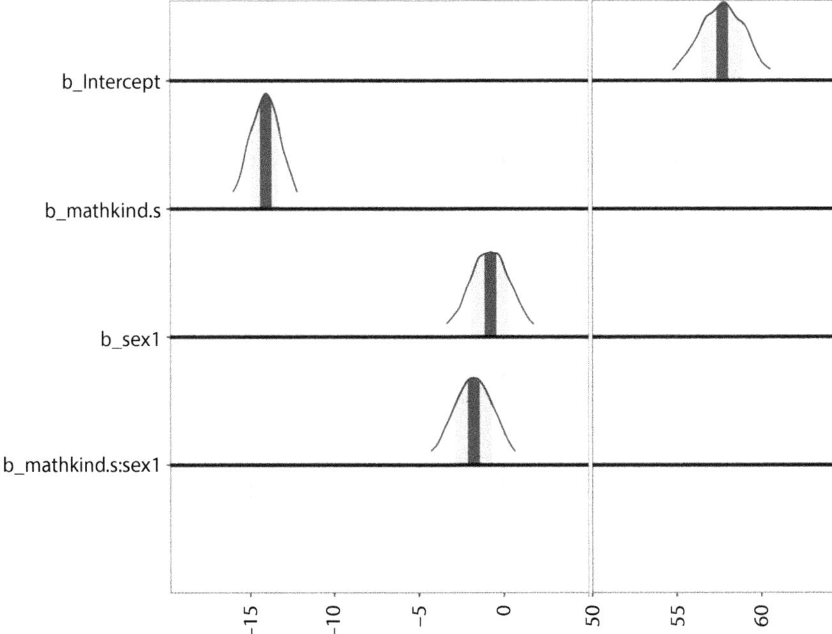

Fig. 13.14 Posterior distributions of the population-level parameters of the Bayesian mixed model `m1.brms` showing their 62% (dark blue) and 95% credible intervals.

The visualization of the posterior distributions of the group-level effects for the 99 schools represents a challenge because of the number of plots. We first use the package `posterior` to gather the posterior distributions with `gr.eff.m1.brms = as_draws_df (m1.brms, variable = "^r_", regex = TRUE)`. The resulting data frame has 6,000 rows (= 3 chains × (3,000 iterations – 1,000 warm-ups)) and 102 columns (the 99 schools + 3 extra columns for ".chain," ".iteration," and ".draw" that we are going to delete): `gr.eff.m1.brms = gr.eff.m1.brms[,-c(100,101,102)]`. The column names of the data frame `gr.eff.m1.brms` are very awkward (e.g., `r_schoolid[1, Intercept]`), and we shorten them to just `schoolid: names(gr.eff.m1.brms) = unique(DF$schoolid)`. Finally, we use the package `reshape2` to produce a data frame in the "long format" that simplifies the `ggplots: gr.eff.m1.brms = melt(gr.eff.m1.brms, value.name = "gr.eff.school")`.

Figure 13.15 shows the posterior distributions for just the first six schools. Each posterior probability distribution depicts both the range and the likelihood of effects in each school. The group-level effects show the deviations of each school intercept from the population-level intercept with mean of 57.70. Note that the means of the group-level effects range from –20 to +20 (Fig. 13.15c), and thus the school-level intercepts differ by almost 35% from the population-level intercept. The data for Fig. 13.15c are just descriptive statistics obtained from the posterior distributions plotted in Fig. 13.15a. They were obtained as `stats.ranef.m1.brms = as.data.frame(ranef(m1.brms))`, then the variable names were changed with `names(stats.ranef.m1.brms) = c("mean", "se", "Q2.5", "Q97.5")`, and finally the `schoolid` labels were changed with `stats.ranef.m1.brms$school = unique(DF$schoolid)`. We used Gelman et al.'s (2018) to obtain the proportion of the variation in mathgain explained by the model.

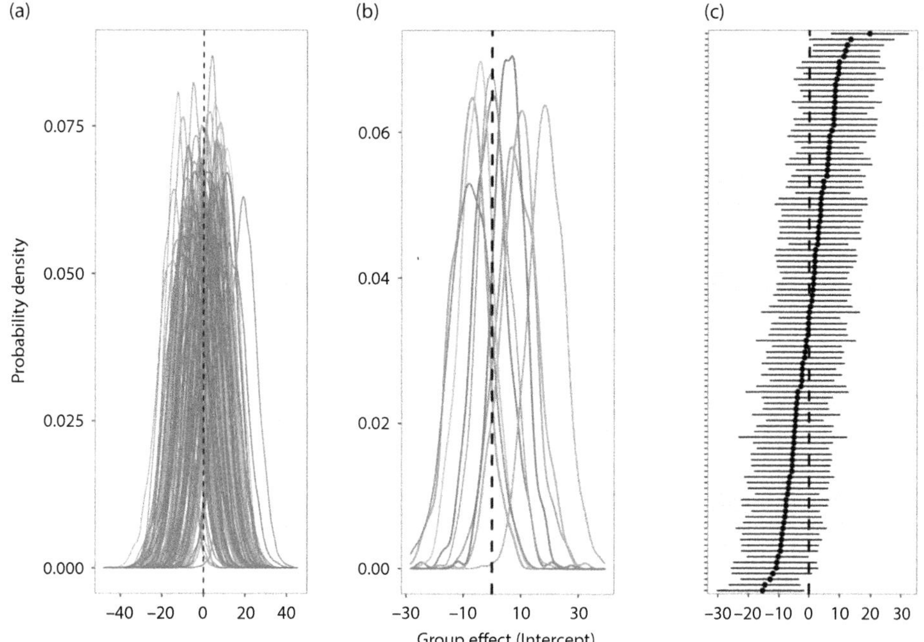

Fig. 13.15 Posterior distributions of the group-level effects in the intercepts (mean of `mathgain` for the sex of reference: boys) of the Bayesian mixed model `m1.brms` for (a) all 99 schools, (b) only the first 10, and (c) the mean and 95% credible intervals for all schools in increasing order of their means.

```
> bayes_R2(m1.brms)
   Estimate Est.Error  Q2.5 Q97.5
R2    0.263    0.0209 0.222 0.303
```

We now gather all data needed to assess of the goodness of fit of model `m1.brms`. We start by obtaining the posterior predictive distribution (Chapter 7) as `post.dist.m1.brms = posterior_predict(m1.brms, ndraws = 1000)`, and simulate the randomized quantile residuals (Chapter 8) with `qres.m1.brms = createDHARMa (simulatedResponse = t(post.dist.m1.brms), observedResponse = DF$mathgain, fittedPredictedResponse = apply(post.dist.m1.brms, 2, median))`, which are then converted to a normal distribution: `res.m1.brms = data.frame(res = qnorm(residuals(qres.m1.brms)))`. We are following the same procedure previously used for all Bayesian GLMs in Part II. While Pearson residuals would also have been appropriate for a linear mixed model, randomized quantile residuals are adequate as well. We finally calculate the mean of the fitted values and the Pareto *k* values, and also put the explanatory variables involved in `m1.brms` into a single data frame: `res.m1.brms = cbind(res.m1.brms, DF[,c("mathkind.s", "sex", "schoolid")], fitted = fitted(m1.brms, draws = 1000)[,1], pareto = loo(m1.brms, pointwise = T)$diagnostics$pareto_k)`. We also need to verify whether the (means of the) group-level effects follow a normal distribution, for which we use `stats.ranef.m1.brms$mean`.

Figure 13.16 shows that `m1.brms` had an excellent goodness of fit to the data. Figure 13.16a and b had the seemingly random scatter of points of a well-fitting model,

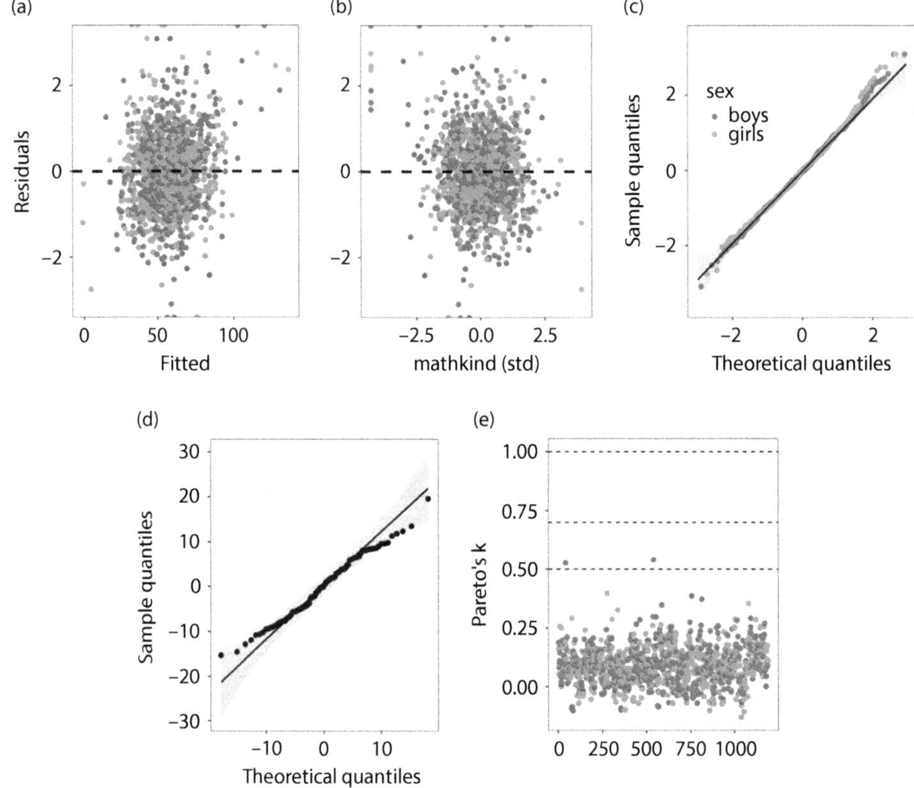

Fig. 13.16 Residual analysis of Bayesian mixed model `m1.brms` based on randomized quantile residuals converted to normal residuals, showing the mean of the Pearson residuals vs. the mean of the fitted value (a), the mean of the Pearson residuals vs the explanatory variables in the model (b), the quantile–quantile plot of the randomized quantile residuals (c), the Pareto *k* cross-validation statistic for each data point (e), and the quantile–quantile plot of mean group effects for each school (d).

the randomized quantiles residuals were largely normally distributed (with a slight deviation for girls in the upper tail), the Pareto *k* statistics were all "excellent" (except for two values in the "very good" range), and the group-level effects also followed a normal distribution.

We can produce plots showing the conditional relations between the response and explanatory variables predicted by model `m1.brms`. We use the package `ggpredict` to obtain data frames with the predicted values only for population-level effects, for both population- and group-level effects, and for three arbitrarily chosen schools but only for the sex of reference of the analysis (= boys). The data frames are:

```
cond.pred.m1brms.fixef=ggpredict(m1.brms,type="fixed", terms =
c("mathkind.s [all]", "sex"))
cond.pred.m1brms.ranef=ggpredict(m1.brms, type="random", terms =
c("mathkind.s [all]", "sex"))
cond.pred.m1brms.ranef3=ggpredict(m1.brms, terms = c("mathkind.s [all]",
"schoolid [11,33,93]"), type =."random")
```

Figure 13.17 shows the plethora of possibilities. We see again that the conditional curves help visualize the general relations between the response and the explanatory variables, and there is still both large variability across schools and variation in the data not explained by the model.

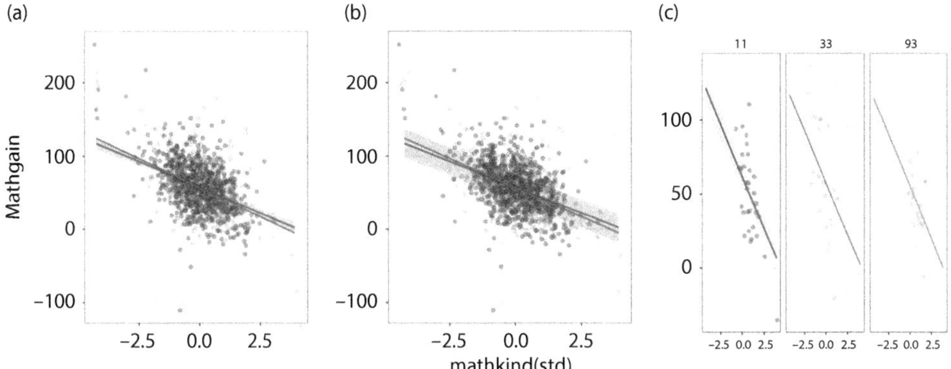

Fig. 13.17 Conditional predictive curves of Bayesian mixed model `m1.brms` for the population-value effects (a), population- and group-level effects (b), and for three arbitrarily chosen schools (11, 33, 93) for the sex of reference in the analysis (boys, corresponding to pink). The explanatory variable `mathkind` was standardized prior to the analysis.

Our next and final model is `m2.brms` that contemplates group-level variation in both the intercepts and slopes (Eq. (13.4)): `formula.m2.brms = bf(mathgain ~ mathkind.s*sex + (1 + mathkind.s|schoolid))`. Let's examine the priors that need to be defined:

```
> get_prior(formula.m2.brms, data=DF,family='Normal')
                prior     class           coef    group resp dpar nlpar bound        source
               (flat)         b                                                     default
               (flat)         b     mathkind.s                                 (vectorized)
               (flat)         b mathkind.s:sex1                                (vectorized)
               (flat)         b           sex1                                 (vectorized)
               lkj(1)       cor                                                     default
               lkj(1)       cor                schoolid                        (vectorized)
 student_t(3, 56, 31.1) Intercept                                                   default
  student_t(3, 0, 31.1)        sd                                                   default
  student_t(3, 0, 31.1)        sd                schoolid                      (vectorized)
  student_t(3, 0, 31.1)        sd      Intercept schoolid                      (vectorized)
  student_t(3, 0, 31.1)        sd     mathkind.s schoolid                      (vectorized)
  student_t(3, 0, 31.1)     sigma                                                   default
```

Compared with model `m1.brms`, we can see that we must also now define priors for the standard deviation of the slopes (`student_t(3, 0, 31, 1 sd mathkind.s schoolid`), and for the correlation between intercepts and slopes at the school level (`lkj(1) cor schoolid`). The latter prior demands an explanation. The package `brms` uses the Lewandowski et al. (2009) (LKJ) distribution to define the priors for a correlation matrix. What correlation matrix? `m2.brms` contains vectorial group-level effects that are modeled using a multivariate normal distribution with means (0, 0) and a variance–covariance matrix (Eq. (13.4)). It turns out that it is more convenient to transform (a Cholesky decomposition to diagonalize, for those who like matrix algebra) this variance–covariance matrix as:

$$\begin{pmatrix} \text{Var}\,(b_{0,S}) & \text{Covar}\,(b_{0,S}, b_{1,S}) \\ \text{Covar}\,(b_{0,S}, b_{1,S}) & \text{Var}\,(b_{1,S}) \end{pmatrix}$$
$$= \begin{pmatrix} \text{Var}\,(b_{0,S}) & 0 \\ 0 & \text{Var}\,(b_{1,S}) \end{pmatrix} \begin{pmatrix} 1 & \text{Corr}\,(b_{0,S}, b_{1,S}) \\ \text{Corr}\,(b_{0,S}, b_{1,S}) & 1 \end{pmatrix} \begin{pmatrix} \text{Var}\,(b_{0,S}) & 0 \\ 0 & \text{Var}\,(b_{1,S}) \end{pmatrix}. \tag{13.9}$$

In this way we can separate the priors for the group-level effects between the variance (or squared standard deviations) of each parameter and their correlation. To give at least some intuition, the LKJ distribution that defines the prior of the correlation between random slopes and intercepts is essentially a stretched a beta distribution (Chapter 12) in the range [–1, 1] whose shape is defined by a single parameter η (Fig. 13.18).

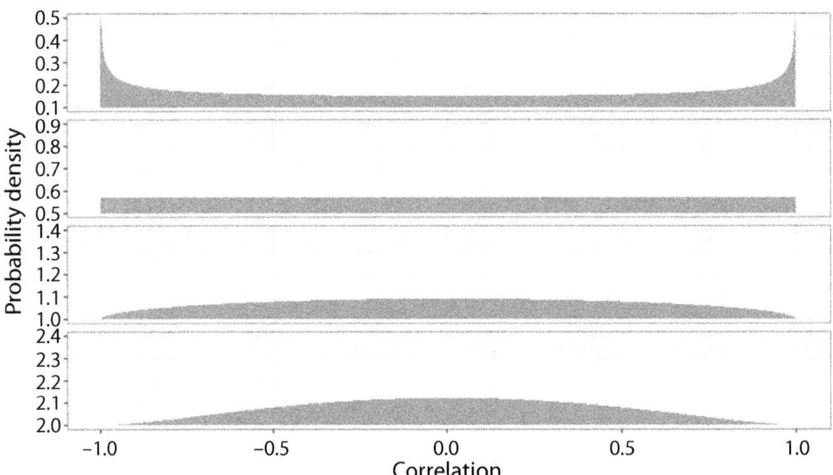

Fig. 13.18 Shapes of Lewandowski et al.'s (2009) distribution used to define the priors for the correlation matrix of group-level intercepts and slopes of Bayesian mixed model `m2.brms`. The plots correspond to η values of the LKJ distribution of 0.1, 0.5, 1.0, and 2.0 (top to bottom) for a 2 × 2 matrix. The figure was generated using the package `ggdist`.

We use the same priors as `m1.brms`, to which we add the prior for the LKJ distribution:

```
prior.m2=c(prior(normal(0, 5.55), class = b, coef="sex1"),
prior(normal(0, 5.55), class = b, coef="mathkind.s:sex1"),
prior(lkj(0.5), class = cor))
```

We then run the model, `m2.brms = brm(formula = formula.m2.brms, data = DF, prior = prior.m2, chains = 3, control = list(adapt_delta = 0.99), warmup = 1000, iter = 3000, future = T)`, and the summary is:

```
> summary(m2.brms)
 Family: gaussian
  Links: mu = identity; sigma = identity
Formula: mathgain ~ mathkind.s * sex + (1 + mathkind.s | schoolid)
   Data: DF (Number of observations: 1190)
  Draws: 3 chains, each with iter = 3000; warmup = 1000; thin = 1;
         total post-warmup draws = 6000

Group-Level Effects:
~schoolid (Number of levels: 99)
                            Estimate Est.Error l-95% CI u-95% CI Rhat Bulk_ESS Tail_ESS
sd(Intercept)                  10.10      1.33     7.61    12.82 1.00     2172     3636
sd(mathkind.s)                  6.70      1.34     4.07     9.34 1.00     1962     2030
cor(Intercept,mathkind.s)      -0.29      0.23    -0.74     0.18 1.00     1510     1545

Population-Level Effects:
                 Estimate Est.Error l-95% CI u-95% CI Rhat Bulk_ESS Tail_ESS
Intercept           57.44      1.59    54.28    60.45 1.00     3892     3907
mathkind.s         -17.58      1.47   -20.49   -14.67 1.00     5142     3927
sex1                -1.15      1.58    -4.29     1.96 1.00    10714     4639
mathkind.s:sex1      1.71      1.65    -1.47     4.93 1.00     6755     4530

Family Specific Parameters:
      Estimate Est.Error l-95% CI u-95% CI Rhat Bulk_ESS Tail_ESS
sigma    27.95      0.62    26.77    29.21 1.00     5418     4191
```

The only difference from the summary of `m1.brms` discussed above is that we now also estimated the standard deviation of the group-level slopes, and the correlation between them and the group-level intercepts. The means of the marginal posterior distributions for the parameters common to `m1.brms` and `m2.brms` would naturally be different, but not by much in the case of the population-level effects. In other words, the same interpretation given for `m1.brms` would also apply here. We also see that the `Rhat` values of all the parameters is essentially equal to one, and the effective sample sizes are all very large (although smaller for the group-level effects, which is to be expected). The latter two aspects suggest (but do not prove) that the MCMC algorithm has converged to a common stationary posterior distribution. We are going to spare you a visual verification of the convergence, and of the degree of autocorrelation of the sampled estimates of the three MCMC chains: they are pretty good (the code for these two figures is on the companion website).

Displaying the posterior distributions of the population- and group-level effects in `m2.brms` requires first gathering the relevant data from the model output. We follow the same procedure used for `m1.brms` above. We first obtain the posterior distributions of all population-level effects with `gr.eff.m2.brms = as_draws_df(m2.brms, variable = "^r_", regex = T)`. We then delete the last three columns with `gr.eff.m2.brms = gr.eff.m2.brms[,-c(199,200,201)]`. Next, we give the columns simpler and handier names: `names(gr.eff.m2.brms) = paste(rep(c("int", "slope"), each = length (unique(DF$schoolid))), rep(unique(DF$schoolid), times = 2))`. The last instruction will generate as many labels `int1`...`int107` as there are values of the posterior distribution for group-level intercepts, and will do likewise for group-level slopes. We then reformat `gr.eff.m2.brms` as "long format" by stacking columns using the package `reshape2: gr.eff.m2.brms = melt(gr.eff.m2.brms, value.name = "gr.eff.school")`. The data frame `gr.eff.m2.brms` has 1,188,000 rows (= 99 intercepts + 99 slopes) × 6,000 values for each posterior distribution resulting from 3 chains × 2,000 MCMC iterations). To keep things simple, we split it into two separate data frames for each group-level parameter: `int.eff.m2.brms = gr.eff.m2.brms[grep(pattern = "^int", gr.eff.m2.brms$variable),]`, `slo.eff.m2.brms = gr.eff.m2.brms[grep(pattern = "^slope",` `gr.eff.m2.brms$variable),]`. We could have prepared the data frames for the plot with less R code, but probably at the expense of confusing you some more. We also gather the means and 95% credible intervals for each group-level parameter: `stats.ranef.m2.brms = as.data.frame(ranef(m2.brms))`. We then create shorter and more meaningful variable names with `names(stats.ranef.m2.brms = c("mean.int", "se.int", "Q2.5.int", "Q97.5.int", "mean.slope", "se.slope", "Q2.5.slope", "Q97.5.slope")`, and add the school identifiers: `stats.ranef.m2.brms$school = unique(DF$schoolid)`. We now have all the data ready for the long-awaited Fig. 13.19.

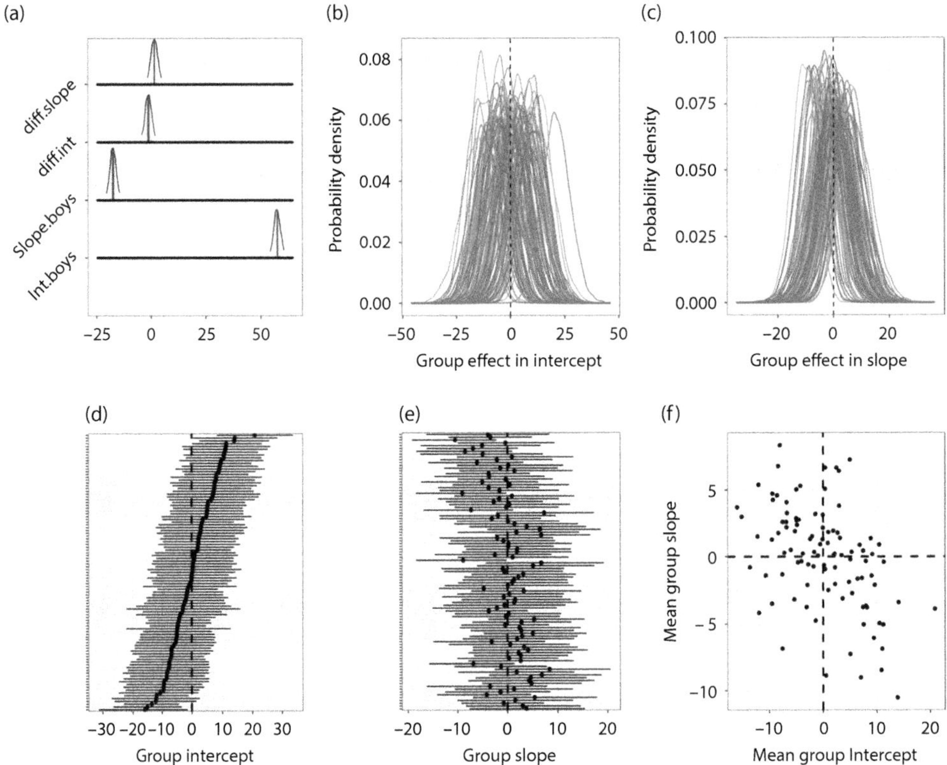

Fig. 13.19 Posterior distributions of parameter estimates of the Bayesian mixed model `m2.brms`. (a) Population-level means and 62% (dark blue) and 95% credible intervals. (b) and (c) Posterior distributions of the group-level effects in the intercepts and slopes for the 99 schools. (d) Means and 95% credible intervals of the group-level effects in the intercepts, with schools shown in ascending order of their means. (e) same as (d) but for group-level effects in slopes; schools shown in ascending order of their means for the intercepts. (f) Covariation between the mean group-level effects in the intercepts and slopes for the 99 schools.

This figure provides a visual depiction of the posterior distributions of the model parameters obtained when fitting a mixed model in the Bayesian framework. In the frequentist framework, we obtain just a point estimate (not a true probability distribution) of the group-level or random-effect parameters. The plots in Fig. 13.19d and e are just a simple summary of the posterior distributions shown in Fig. 13.19b and c, respectively. We can also see in Fig. 13.19f a characterization of the weak correlation between group-level intercept and slopes obtained in `summary(m2.brms)`. The weak negative correlation between slope and intercept differentials at the school level suggests that better than average schools (the vertical dashed line in Fig. 13.19f) tend to show negative effects of the explanatory variable `mathkind.s` (and vice versa for worse than the average schools), in a sort of "regression to the mean" effect explained in Section 13.8. Provided that the 99 schools were randomly chosen, we can use the fitted group-level effects whose standard deviations were shown in `summary(m2.brms)` to predict the magnitude of the group-level differentials in intercepts and slopes for other non-sampled schools of the super-population of schools in Michigan. We could use `rnorm(n = 10, mean = 0, sd = 10.17)` to predict 10 possible values of the school-level differentials of intercepts (i.e., the mean level of `mathgain` for the reference sex, boys). The same can, of course, be done with the frequentist mixed model `m2`.

Using Gelman et al.'s (2018) approach, we can obtain the proportion of variation explained by model `m2.brms`. This is slightly better than for `m1.brms`; it cannot be otherwise given that the latter is a special case of the former.

```
> bayes_R2(m2.brms)
   Estimate Est.Error  Q2.5 Q97.5
R2    0.351    0.0212 0.309 0.392
```

Model selection can also be performed with Bayesian mixed models using the LOO-CV criterion (Chapter 7). To that end, we must first add the criterion to each model:

```
m1.brms=add_criterion(m1.brms, criterion= "loo")
m2.brms=add_criterion(m2.brms, criterion= "loo")
```

Then:

```
> loo_compare(m1.brms,m2.brms)
        elpd_diff se_diff
m2.brms   0.0       0.0
m1.brms -13.5       7.9
```

`m2.brms` fits better (has the lower expected log pointwise predictive density (ELPD); see Chapter 7) than `m1.brms` and hence is ranked first. Because the difference in ELPD is much larger than its standard error (there is no clear-cut criterion here), we should prefer `m2.brms`.

Posterior predictive checks (Chapters 7, 11, and 12) can be used to assess the capacity of a Bayesian model to predict features of the observed data. We arbitrarily choose three features: the overall density of the response variable, its standard deviation, and the maximum value per sex. We must first obtain the posterior predictive distribution of the model: `post.dist.m2.brms = posterior_predict(m2.brms, ndraws = 1000)`. Figure 13.20 shows that `m2.brms` can reproduce the overall distribution of the response variable (a), but has increasing difficulty in predicting the standard deviation per sex (b), and largely under-predicts its maximum value for boys. Predicting features related to variability and to the tails of a distribution per group is always harder than the central tendencies of the entire data set. To an extent, Fig. 13.20b and c are very stringent predictive tests. Every realistic statistical model should fail at some point. It is worth highlighting that posterior predictive checks offer an open-ended array of possibilities to assess the predictive abilities of Bayesian models. There is no true counterpart of posterior predictive checks in the frequentist framework.

The final two tasks of this chapter would be to assess the goodness of fit of `m2.brms` with residual analysis, and to generate the conditional plots for the response variables. The procedures are nearly identical to those used to generate Figs. 13.16 and 13.17. However, the chapter is already much too long, and we choose to spare you the near reiteration of material. The figures are as good as or better than Figs. 13.16 and 13.17. If interested, you can find the code to generate these two figures on the companion website.

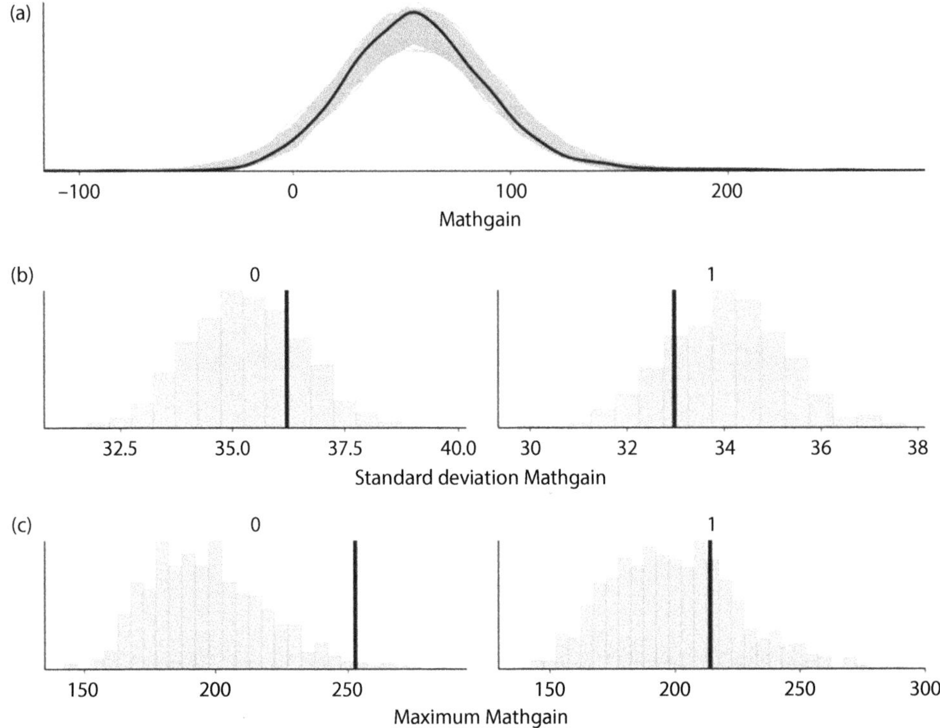

Fig. 13.20 Posterior predictive checks of the Bayesian mixed model `m2.brms` for different features of the data based on the posterior predictive distribution (pale blue) and the observed values (dark blue). (a) Density distribution of the response variable `mathgain`. (b) and (c) Standard deviation and maximum value per sex (boys = 0, girls = 1).

This chapter has covered a large amount of complex material. Our aim was to introduce mixed models showing the frequentist and Bayesian fittings side by side. We shall put to use the material explained here in several study cases involving (quasi-)experimental data for different types of response in Chapter 15. However, prior to that, we need to provide a short introduction in the next chapter to experimental designs and their representation as mixed models.

13.10 Problems

In this chapter we fitted two linear mixed models to the Michigan schools data: one with only random or group-level intercepts, the other with random or group-level intercepts and slopes. These models used `schoolid` as the categorical variable modeled as a random or group-level effect.

13.1 The linear mixed models just discussed implicitly assumed that the effect of schools on the children performance was unrelated to sex. This implies that the differential on the mean of the response variable associated with the school identity was the same for both boys and girls. How would you model a school effect that differs between sexes? How would you compare this model with the ones fitted in this chapter?

13.2 Another implicit assumption of the models fitted in the chapter was that the categorical variable `schoolid` allowed us to form "homogeneous sets" of children regarding the response variable `mathgain`. However, within each school, children attended different classes (`classid`) and were taught by different teachers. How heterogeneous is the performance of children among classes? Is it worth including the random or group-level effect `classid`? Should we include it as either a nested or a crossed effect?

References

Barr, D., Levy, R., Scheepers, C. et al. (2013). Random effects structure for confirmatory hypothesis testing: Keep it maximal. *Journal of Memory and Language*, 68, 255–278.

Bates D., Mächler, M., Bolker, B. et al. (2015). Fitting linear mixed-effects models using lme4. *Journal of Statistical Software*, 67, 1–48.

Berliner, M. (1986). Hierarchical Bayesian time series models. In K. Hanson and R. Silber (eds), *Maximum Entropy and Bayesian Methods*, pp. 15–22. Kluwer, Dordrecht.

Bolker, B. (2015). Generalized linear mixed models. In G. Fox, A. Negrete-Yankelevich, and V. Sosa (eds), *Ecological Statistics: Contemporary Theory and Applications*, pp. 309–333. Oxford University Press, Oxford.

Crawley, M. (2002). *The R Book*. John Wiley & Sons, London.

Cressie, N., Calder, C., Clark, J., et al. (2009) Accounting for uncertainty in ecological analysis: The strengths and limitations of hierarchical statistical modeling. *Ecological Applications*, 19, 553–570.

de Villemereuil, P. (2018). Quantitative genetic methods depending on the nature of the phenotypic trait. *Annals of the New York Academy of Sciences*, 1422, 29–47.

Efron, B. and Tibshirani R. (1993). *An Introduction to the Bootstrap*. CRC Press, New York.

Eisenhart, C. (1947). The assumptions underlying the analysis of variance. *Biometrics*, 3, 1–21.

Gelman A. (2005). Analysis of variance: Why it is more important than ever. *Annals of Statistics*, 33, 1–33.

Gelman, A., Goodrich, B., Gabry, J. et al. (2018). R-squared for Bayesian regression models. *American Statistician*, 73, 307–309.

Heisig, J. and Schaeffer, M. (2019). Why you should *always* include a random slope for the lower-level variable involved in a cross-level interaction. *European Sociological Review*, 35, 258–279.

Hill, H., Rowan, B., and Loewenberg, D. (2005). Effects of teachers' mathematical knowledge for teaching on student achievement. *American Educational Research Journal*, 42, 371–406.

Hobbs, N. and Hooten, M. (2005). *Bayesian Models: A Statistical Primer for Ecologists*. Princeton University Press, Princeton.

Hooten, M. and Hefley, T. (2019). *Bringing Bayesian Models to Life*. Routledge, New York.

Hox, J., Moerbeek, M., and van de Schoot, R. (2018) *Multilevel Analysis: Techniques and Applications*, 3rd edn. Routledge, New York.

Hui, F., Müller, S., and Welsh A. (2021). Random effects misspecification can have severe consequences for random effects inference in linear mixed models. *International Statistical Review*, 89, 186–206.

Ives, A. and Helmus, M. (2011). Generalized linear mixed models for phylogenetic analyses of community structure. *Ecological Monographs*, 81, 511–525.

Jin, S. and Lee, Y. (2020). A review of h-likelihood and hierarchical generalized linear model. *WIREs Computational Statistics*, 14, e1527.

Kenward, M. and Roger, J. (1997). Small-sample inference for fixed effects from restricted maximum likelihood. *Biometrics*, 53, 983–997.

Kéry, M. and Royle, A. (2016). *Applied Hierarchical Modeling in Ecology: Analysis of Distribution, Abundance and Species Richness in R and BUGS*, vol. 1. Academic Press, New York.

Lee, Y., Shen, X., and Noh, M. (2010). *Data Analysis Using Hierarchical Generalized Linear Models with R*. CRC Press, New York.

Lewandowski, D., Kurowicka, D., and Joe, H. (2009). Generating random correlation matrices based on vines and extended onion method. *Journal of Multivariate Analysis*, 100, 1989–2001.

Liang, K. and Zeger, S. (1986). Longitudinal data analysis using generalized linear models. *Biometrika*, 73, 13–22.

Lohr, S. (2019). *Sampling Design And Analysis*, 2nd edn. CRC Press, New York.

Luke, S. (2017). Evaluating significance in linear mixed models in R. *Behavioral Research*, 49, 1494–1502.

Luo, W., Li, H., Baek, E. et al. (2021). Reporting multilevel modeling: A revisit after 10 years. *Review of Educational Research*, 91, 311–355.

Matuschek, H., Kliegl, R., Vasishth, S. et al. (2017). Balancing type I error and power in linear mixed models. *Journal of Memory and Language*, 94, 305–315.

McCulloch, C. and Neuhaus, J. (2011). Misspecifying the shape of a random effects distribution: Why getting it wrong may not matter. *Statistical Science*, 26, 388–402.

McCulloch, C., and Searle, S. (2001). *Generalized Linear and Mixed Models*. John Wiley & Sons, New York.

McElreath, R. (2019). *Statistical Rethinking: A Bayesian Course with Examples in R and Stan*, 2nd edn. CRC Press, New York.

McNeish, D. (2016). Estimation methods for mixed logistic models with few clusters. *Multivariate Behavioral Research*, 51, 790–804.

McNeish, D. (2017). Small-sample methods or multilevel modeling: A colloquial elucidation of REML and the Kenward–Roger correction. *Multivariate Behavioral Research*, 52, 661–670.

McNeish, D. (2019). Poisson multilevel models with small samples. *Multivariate Behavioral Research*, 54, 444–455.

McNeish, D. and Stapleton, L. (2016). Modeling clustered data with very few clusters. *Multivariate Behavioral Research*, 51, 495–518.

Nakagawa, S., Johnson, P., and Schielzeth, H. (2017). The coefficient of determination R^2 and intra-class correlation coefficient from generalized linear mixed-effect models revisited and expanded. *Journal of the Royal Society Interface*, 14, 20170213.

Nakagawa, S. and Schielzeth, H. (2010). A general and simple method for obtaining R^2 from generalized linear mixed models. *Methods in Ecology and Evolution*, 4, 133–142.

Patterson, H. and Thompson, R. (1971). Recovery of inter-block information when block sizes are unequal. *Biometrika*, 58, 545–554.

Pinheiro, J. and Bates, D. (2004). *Mixed-Effects Models in S and S-plus*. Springer, New York.

Rousseeuw, P. and Croux, C. (1993). Alternatives to median absolute deviation. *Journal of the American Statistical Association*, 88, 1273–1283.

Satterthwaite, F. (1946) An approximate distribution of estimates of variance components. *Biometrics*, 2, 110–114.

Schielzeth, H., Dingemanse, N., Nakagawa, S. et al. (2020). Robustness of linear mixed-effects models to violations of distributional assumptions. *Methods in Ecology and Evolution*, 11, 1141–1152.

Schielzeth, H. and Nakagawa, S. (2010). Repeatability for Gaussian and non-Gaussian data: A practical guide for biologists. *Biological Reviews*, 85, 935–956.

Singmann, H. and Keller, D. (2019). An introduction to mixed models for experimental psychology. In D. Speier and E. Schumacher (eds) *New Methods in Cognitive Psychology*, pp. 4–31. Psychology Press, New York.

Stroup, W. (2013). *Generalized Linear Mixed Models: Modern Concepts, Methods and Applications*. CRC Press, New York.

Wan, X., Wang, W., Liu, J. et al. (2014). Estimating the sample mean and standard deviation from the sample size, median, range and/or interquartile range. *BMC Medical Research Methodology*, 14, 135–147.

Wilke, C. (2003). Hierarchical models in environmental science. *International Statistical Review*, 71, 181–199.

Wilson, A., Réale, D., Clements, M. et al. (2010). An ecologist's guide to the animal model. *Journal of Animal Ecology*, 79, 13–26.

CHAPTER 14

Experimental Design in the Life Sciences

The basics

14.1 Introduction

The goal of this chapter is to provide a very broad introduction to experimental designs in the life sciences. A single chapter cannot do justice to a fundamental topic that already has book-length treatments such as Mead et al. (2012), Lawson (2014), Montgomery (2017), and Dean et al. (2017), among many others. Our aim here is simpler and more modest. We explained the need for mixed models as means of accounting for structure in the data mostly (but not always) associated with categorical explanatory variables (Chapter 13). We already knew that these variables divide the data into disjoint subsets whose parameter values are compared or contrasted in relation to the statistical hypotheses. This chapter aims to show how the main types of experimental designs most often used in the life sciences can be implemented as mixed models through the definition of suitable random or group-level effects. The presentation will be largely conceptual, and we defer the solving of actual study cases to Chapter 15.

We use a generic definition of "experiment" as a systematic inquiry aiming to understand the relation between manipulated input variables on the observed values of a response variable in a clearly defined study system. The need to perform experiments stems from the inherent variability of experimental results. Practical experience shows that this variability arises for several non-mutually-exclusive reasons. They include the vagaries of the measurement procedures and the inherent intrinsic variability of the experimental material, chance effects that differentially affect some experimental units, and others (Chapter 2). Because variables other than those involved in the experimental manipulation may also affect the response variable, experiments aim to control for the putative effect of confounding variables through either the design and execution of the experiment, or through the measurement of confounding variables (or both). The goal of designed experiments is to maximize the chances of quantifying the sole effect of the manipulated explanatory variables on the response variable with the aim of either testing specific hypotheses (confirmatory research) or looking for general patterns in the data to generate new hypotheses (exploratory research).

At the outset, any experiment or observational study requires defining a response variable reflecting the behavior of the study system, and a set of suitable independent variables or conditions amenable to experimental manipulation. It also involves defining which other variables or conditions must be held constant to prevent them from affecting the

Statistical Modeling With R. Pablo Inchausti, Oxford University Press.
DOI: 10.1093/oso/9780192859013.003.0014

inference. The choice of these confounding variables largely stems from our conceptual and background understanding of the study system. Therefore, the choice of these variables is entirely dictated by the scientific hypotheses, by our understanding of the study system, and by the available empirical evidence from previous studies. These are strictly scientific issues, and statistics cannot help here. Experimental design can only start once we have decided the identity of the response variable, which experimental factors are to be manipulated, and which confounding variables need to be measured to quantify and/or control their influence on the response variable.

Once we have chosen an adequate experimental design for the questions at hand, we need to decide the total sample size, and the number of samples for each level of the categorical explanatory variables involved in the experiment or observational design. The aim is to determine the minimally sufficient sample size to achieve the goals of the planned study.

14.2 The basic principles of experimental design

Comparisons and experiments with purposely collected data have been carried out for centuries by both scientists and lay people. For instance, Dr. James Lind on board HMS *Salisbury* during the Seven Year War performed in 1747 what we would now call a clinical trial (including a control group) to evaluate the effectiveness of citrus fruit as a cure for the scurvy that decimated the crews during long maritime voyages. Hinkelman (2015) provides a concise historical background on experimental design. Fisher's (1935) landmark book *The Design of Experiments* provided this topic with firm underpinnings based on the burgeoning statistical theory built on his maximum likelihood method (Chapter 3). This book provided a principled basis for the acquisition of experimental data, and synthesized a set of fundamental papers that Fisher published while working as Chief Statistician at the Rothamsted Agronomical Research Station, 25 km north of London, in 1919–1933 (Chapter 2). During this period, Fisher created analysis of variance for one factor (Chapter 5), factorial analysis of variance and analysis of covariance (Chapter 6) based on his maximum likelihood theory, and was later led to the principles of experimental design. In so doing, Fisher developed techniques for conducting experiments, and mathematical and arithmetical procedures for making sense of the results (Fisher Box 1978). During his period at Rothamsted, Fisher also wrote his landmark book *The Genetical Theory of Natural Selection* in 1930, and developed the conceptual and statistical basis of quantitative genetics that is still used in the selective breeding of animals and crops.

The three basic techniques fundamental to every experimental design are replication, blocking, and randomization (Fisher 1935). The first two help to increase precision in the experiment, and the last one aims to decrease bias in the parameter estimates. These techniques enhance the chance of obtaining meaningful inference from experiments in the face of variability. High replicability can be obtained by narrowing the scope (e.g., only studying one sex at one site) of the biological variability and by tightly controlling the experimental conditions to enhance the internal validity of results. However, high internal validity (essentially, low bias) is generally accomplished at the expense of low external validity, or the ability to generalize the experimental results to a larger realm of existing circumstances. Actual experiments achieve an acceptable compromise between internal and external validity through the use of suitable designs.

Replication is the repetition of parallel measurements of several experimental units (i.e., the entities to which we apply experimental conditions and measure the response variable) with the aim of both estimating reliable summary statistics and capturing their variation. Replication leads to greater reliability in the assessment of the effects of interest, and to better characterization of the variability of the response variable. Here we need to distinguish between "biological" and "technical" replicates. The former are replicates aiming to characterize the intrinsic biological variability by measuring several independent experimental units. The latter are independent measurements of the same experimental unit aiming to characterize the uncertainty or vagaries of the methods or measurements. We also need to differentiate between replication of experimental units and repeated measurements of the same experimental unit (see Section 14.4.4). When five subjects or experimental units are assigned to receive a drug and a measurement is taken on each subject, we have five independent observations on the effect of the drug. When we measure the effect of a drug five times for one experimental subject over time, these are repeated measurements that reflect the variation of a subject over time to the drug; they are not statistically independent and hence do not constitute replicates.

Blocking is a strategy to isolate part of the variability of the response variable by associating it with a categorical explanatory variable identifiable with a nuisance or confounding variable whose influence we wish to control and adjust in our statistical model. The goal of blocking is to remove as much variation of the response variable as possible to make the differences due to the experimental conditions more evident and precise. For example, the Michigan schools (Chapter 13) can be considered as a block or set of homogeneous experimental units subjected to a shared set of effects that are bound to be similar to each other, but different to other schools. Using schools as blocks arises from the clustered nature of the sampling or gathering of information of the experimental units. Mathematically, the blocking variable is treated as another categorical explanatory variable that is modeled as a random or group-level effect in mixed models. If we have batches of cell cultures, samples from each batch are bound to be more similar than those from different batches. By keeping track of the batch identity, we can isolate and quantify the variation among groups when subjecting the samples from different sets of experimental conditions. When the confounding variable whose effect we wish to control is a numerical variable, we need to record its values in each experimental unit and just introduce it as another explanatory variable in the statistical models.

The purpose of *randomization* is to prevent or minimize the effect of systematic and personal biases introduced into the experiment by the experimenter. The idea is that by randomly assigning experimental units to different experimental conditions we can avoid, minimize, or even out the effect of confounding variables. A random assignment means that each experimental unit has the same probability of being assigned to a set of experimental conditions. Old-fashioned statistical textbooks contained tables of pseudo-random numbers that were used for this purpose; today we can use computers and calculators (or even cell phones) to generate pseudo-random numbers for the assignment. The randomization can occur at the scale of the entire experiment (a completely randomized design) or at the scale of the blocks (randomized block design; see Section 14.4.2). While blocking controls the allocation of experimental conditions to experimental units that are similar in some respect, randomization offers no control and allows all possible allocations to occur with equal probability, but it avoids subjective allocation with its inevitable and unspecified tendency towards particular patterns (Mead et al. 2012). For example, researchers may unconsciously tend to give the drug to patients they deem to be more obviously in need, thus biasing the estimated effect of the drug. And patients may exhibit the well-known "placebo effect," whereby they may exhibit a marked improvement despite not receiving the experimental treatment(s). This is why clinical trials typically use a double-blind approach to randomize the experimental factor assignment so that neither the researchers nor the patients know what the patients actually received.

14.3 Surveys and observational studies

Our definition of experiment in Section 14.1 is to be contrasted with surveys or observational (also known as quasi-experimental) studies. In observational studies, researchers look for sets of contrasting empirical circumstances that differ in the putative action of the experimental variables. In a sense, it is the opportunistic selection of contrasting sites or experimental units by the researcher that would describe and reflect the impact of levels of the experimental factors in observational studies, rather than the careful perturbation of the latter by the conditions imposed in the controlled setup of a designed experiment. In life sciences, surveys or observational studies are performed in lieu of experiments whenever imposing the experimental conditions is infeasible (e.g., adding controlled amounts of iron oxide in the open ocean), impractical (e.g., eliminating all parasites from a host population), too expensive (e.g., characterizing the influence of genetic variation), illegal (e.g., decreasing the abundance of an endangered species), or unethical (e.g., capturing and sacrificing individuals in order to sample tissues).

The main and fundamental difference between experimental and observational studies is that in the latter the researcher cannot randomly assign experimental units to the different levels (or combinations of levels) of the experimental variables. The researcher neither imposes conditions through the controlled manipulation of input variables, nor can randomly assign the experimental units (i.e., the entities on which we both impose the input variables and measure the outcome as the response variable) to different groups defined by the (combinations of) levels of the experimentally manipulated variables (Mead et al. 2012). Experimental units may not only differ in the alleged impact of experimental variables, but also in other variables capable of inducing the same measured effect(s). One solution is to also measure other confounding variables lurking in the background so as to discount their effect when making the inference. In the absence of random assignment of experimental units, any measured effect can be due to either a true effect due to the manipulation, or result from bias in the formation of the groups being compared that may induce the alleged effects of the experimental variables. While both problems are resolved by randomization, it is not feasible, practical, or even ethical to use it in many interesting and relevant (quasi-)experimental comparisons. Therefore, surveys or observational studies constitute a weaker inference than controlled experiments to ascertain the effects of the controlling variables of study systems.

The design of observational studies and surveys nonetheless incorporates all features of experimental designs that aim to minimize bias (e.g., simultaneous controls) and the impact of sampling error (e.g., replication, blocking). That is, experiments and observational studies are both designed using a common set of principles and ideas, and they are analyzed with the same statistical models in either the frequentist or the Bayesian framework.

14.4 The main types of experimental design used in the life sciences

We cover the following types of experimental designs: factorial, randomized blocks, repeated measures, crossover, split-plot, and nested designs. Their selection was based on our perceived frequency of their use in the life sciences. To be sure, there are many other types of experimental design used in the sciences and engineering, for which detail explanations can be found in Ryan (2007), Lawson (2014), and Dean et al. (2017).

Our discussion will be based on two categorical explanatory variables A and B, each with two levels (A_1, A_2 and B_1, B_2), that are experimentally manipulated and the effect measured in a response variable Y. We will call variables A and B *experimental factors*. Depending on the experimental design, we will introduce additional explanatory variables as needed. While for simplicity we will assume equal sample sizes across the explanatory variables, it should be clear that all the ideas also apply to the far more common situation of unequal sample sizes. For each experimental design we will present the associated statistical model assuming that Y is any of the probability distributions of the exponential family (Chapter 8) such that the effect of the explanatory variables is assessed in the scale of the link function: $g(\mathrm{E}(Y))$ = experimental factors. When $Y \sim$ Normal, the link function $g()$ is the identity function, and the statistical model also has an error term corresponding to the unexplained variation around the mean unrelated to the experimental factors (Chapters 4–6). In all other cases, the error term does not exist because the mean and the variance of non-Gaussian response variables are not independent parameters (Chapters 9–12). In a general sense, $Y \sim \mathrm{ExponentialFamily}(\mu_Y, \phi)$, where the exponential family includes at least all the probability distributions of Table 8.1, μ_Y is the mean or expected value $\mathrm{E}(Y)$ of the response variable Y, and ϕ is a dispersion parameter whose meaning varies with the probability distribution (see Chapter 8).

The main analytic method used by Fisher for the analysis of data from designed experiments was some type of analysis of variance (Chapters 5 and 6). This provides an arithmetic procedure to divide the total variation between experimental units into separate components that represent different sources of variation, whose relative importance provides a basis for inferences about the effects of the experimental manipulations (Mead et al. 2012). You will find that virtually every textbook on experimental design presents the statistical models associated with each experimental design assuming a Gaussian response variable, and employing notation with a plethora of Greek letters, sub-indices, and sub-sub-indices to denote the effect of the experimental variables, their nesting and interactions, etc. However, any experimental design can be applied to every conceivable type of response variable (Table 2.1). It is just that we would be using the generalized rather than the general linear model to write the appropriate statistical model. After wrestling for some time with the confusing (but perhaps necessary) notation employed in the traditional presentation of statistical models associated with experimental designs in most textbooks, we chose a somewhat more informal approach to present the experimental designs aiming for terms closer to the phraseology employed by life scientists.

14.4.1 *Factorial design*

This design consists of the simultaneous application of two (or more) experimental factors, each of which has at least two levels. The goal is to estimate the single effect of each explanatory factor and their interaction on the mean of the response variable. In terms of our example, $g(\mathrm{E}(Y)) = A{*}B$, which includes the single effects of each categorical explanatory variable A and B and their interactions. In R, the statistical model is written `glm(y ~ A*B, family = xxx)` and `brm(formula = A*B, family = xxx)` in the frequentist and Bayesian frameworks, where `xxx` stands for the adequate probability distribution in the likelihood part of the statistical model. (Of course, we need to specify the priors for the Bayesian model.) We have presented, solved, and discussed the interpretation of factorial designs in Chapter 6 for $Y \sim$ Normal, and in Chapters 11 and 12 for response variables where Y has negative binomial or beta distributions. Note that, strictly speaking, there is no need to use random or group-level effects in factorial designs.

The key feature of factorial designs is to have replicates in every combination of the levels of the explanatory factors A and B in order to estimate the magnitude of their interaction. The factors A and B are said to being crossed. Subject to the replication constraint just explained, every experimental unit is to have the same chance of being assigned to the four groups A_1B_1, A_1B_2, A_2B_1, and A_2B_2 defined by the pairwise combinations of all levels of the two experimental factors. Because there are no restrictions in the random allocation of experimental units to the above-mentioned four groups (Fig. 14.1), the design is called a completely randomized factorial design. The replication and the randomization are essential here to minimize any sampling bias and to obtain precise estimates of model parameters. A set of experimental units having the same levels of the experimental factors are considered statistical replicates. This means that these experimental units provide independent information to estimate the parameters of the statistical model. In this regard, the first (and second, and so on) pair of rows in the data table are to be considered statistical replicates in the sketched factorial design.

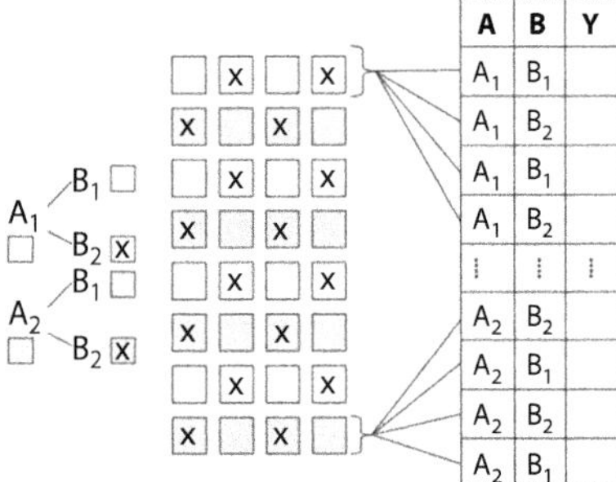

Fig. 14.1 Schematic representation of a completely factorial design. Each square represents an experimental unit to which the two experimental factors *A* and *B* were applied and the response variable was measured. The experimental factors *A* and *B* had two levels, indicated by the color (white or gray for factor *A*) and by the mark (none or an X for factor *B*) to form four groups defined by their 2 × 2 combinations. The scheme shows a completely randomized experiment with replicates for each combination of levels of the experimental factors. The table only shows the data input for the first four and the last four experimental units.

One practical problem with factorial designs is that the number of combinations of levels increases very quickly with the number of experimental factors and the number of levels. This leads to larger and more costly experiments. Unavoidably, not all combinations of factor levels are relevant to evaluating the scientific hypothesis that motivated the research. The fractional factorial designs (e.g., Lawson 2014, Dean et. al. 2017) that are more frequently used in engineering constitute an alternative to ever larger factorial designs, including only specific combinations of levels of the experimental factors. However, the restricted crossing of levels of the experimental factors implies that the estimates of some model parameters will be confounded, and we should be careful in the analysis and interpretation of parameter estimates.

14.4.2 *Randomized block design*

A block is a set of experimental units sharing a common set of effects that would amount to a potentially large source of variation in the response variable that is unrelated to the experimental factors. Defining a categorical variable as a block requires having pre-existing knowledge about the putative effect of the blocking variable. For instance, we may know that the area where we are designing an experiment can be partitioned into smaller subsets depending on the presumed, but not necessarily measured, blocking variable soil fertility. Experimental units in each subset would be similar to each other, and different from units in other subsets. By taking into account this loosely defined soil fertility as a categorical explanatory variable, we would be accounting for part of the variation of the response variable, and be able to obtain more precise estimates of the experimental factors.

When performing the actual experiment, we would first need to define the blocks, and then in each block use randomization to apply the different levels of the experimental factors (and combinations thereof). Figure 14.2 defines the blocks as dashed lines, with a given number of experimental units inside each block. To prevent cluttering, Fig. 14.2 does not include replication of the combinations A_1B_1, A_1B_2, A_2B_1, and A_2B_2 of the experimental factors. In some sense, using blocks would amount to making a number of "mini-experiments," each of which would contain experimental units to which we randomly assigned the combinations of levels of the experimental factors. While most researchers often have the same number of experimental units per block, this is not strictly required. Any set of experimental units having the same levels of experimental factors within a block (e.g., the first and third rows the data table in Fig. 14.2) can be considered statistical replicates.

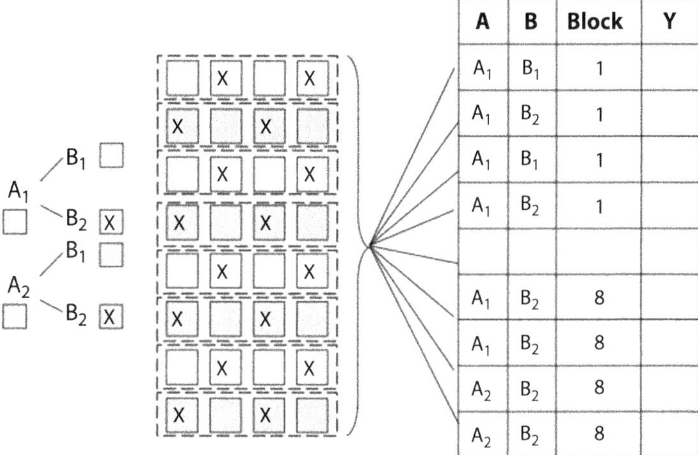

Fig. 14.2 Schematic representation of the randomized block design in which all combinations of levels of the experimental factors were applied to each experimental block (denoted by a dashed line). The symbols and colors are as in Fig. 14.1. The table only shows the data input for the first and eighth experimental block, each of which has two replicates of combinations of the experimental factors *A* and *B*.

The statistical model for a randomized block design is a mixed model consisting of the added effects of the fixed or population-level effects, and the random or group-level effects on the scale of the link function: $g(E(Y)) = A*B$ + random effects. The categorical variable `blocking` is typically modeled as a random or group-level effect in statistical models (Chapter 13). The statistical models are `glmmTMB(y ~ A*B + (1|blocking), family = xxx)` and `brm(formula = A*B + (1|blocking), family = xxx)` in the frequentist and Bayesian frameworks, respectively. The command `glmmTMB` is from the homonymous package; we could also have used `glmer` from `lme4`. The random or group-level intercept is modeled as Normal (0, $variance_{\text{blocking}}$). As we saw in the examples of Chapter 13, the fitted models would yield a coefficient (i.e., the random or group-level intercept) for each level of the explanatory variable `blocking`, quantifying the difference between the global mean of the response variable (i.e., the fixed or population-level value) and the group mean. These models would also estimate a variance for the the random or group-level associated with `blocking`. Provided that the actual instances of `blocking` were randomly sampled from a larger super-population of levels, we could use the estimated variance of the random or group-level intercept to predict the possible differentials for non-sampled levels of the `blocking` variable. Formally speaking, there are two randomization events involved in the execution of a randomized block design: first, the random selection of the realized levels of the blocking variable, and second, the random allocation of experimental units inside each level of the blocking variable.

A question arises in the frequentist framework when the number of levels of the blocking variable is small, meaning less than 5 or 10, depending on the author you consult. The issue here is that the precision in the estimated block variance is poor whenever the number of levels of `blocking` is small. In this case, it is possible to include it as a fixed effect as if it were an experimentally manipulated variable. In these cases, we can take advantage of an interesting default feature in R: the order of evaluation of the variables rests in your hands. We would want to evaluate (and thus isolate and discount) the effect of the blocking variables *before* evaluating the effects of the experimental factors. Our frequentist fixed-effects model would then be `glm(y ~ blocking + A*B, family = xxx)`. There are no such subtleties in the Bayesian framework, where there are no "fixed effects" as all parameters are modeled as random variables. Nevertheless, the amount of information to estimate the group-level variance would still be small, and we may need to use more informative priors (whose sensitivity we should definitely evaluate!) to attain convergence of the MCMC chains.

Two quick additional points. First, when a block does not contain all levels, or combinations of levels, of the experimental factors A and B, it is called an incomplete block design, but the statistical model remains the same as above. Second, it is possible to define blocks with respect to two variables whose combinations of levels are treated as unique, and the fixed or population-level effects are allocated to these unique levels. Names such as Latin, Graeco-Latin, and Youden squares describe these more complex blocking situations whose details can be found in Ryan (2007) and Lawson (2014).

14.4.3 *Split-plot design*

Split-plot designs are frequently applied in factorial experiments in industry and agronomy when one of the experimental factors is more difficult to vary than the others at the level of the experimental unit. The two previous designs implicitly made three assumptions. First, the experimental factors A and B were both equally important and relevant in the context of our hypothesis. Second, we considered that varying the levels of each factor and their application to the experimental units was equally feasible, and could be applied simultaneously with similar effort and costs. And third, we expected the differences between levels to have similar magnitudes in all experimental factors; more precisely, there were no "hard to change" or "easier to change" experimental factors. There are many cases where these three implicit assumptions do not hold in practice. For instance, it may be harder to apply irrigation, exclude herbivores, spray insecticides in smaller rather than in larger experimental areas; it may be easier to change food conditions to a bird colony than to selected nests; it may be more feasible to give an oral drug to change individual physiology than to administer it to specific tissues (Altman and Krzywinski 2014).

The underlying principle of the split-plot design is to apply or impose the experimental factors as a two-stage process. First, the experimental factor whose application is costlier, harder, and/or that we expect to have smaller variation between levels (say, factor *A* in Fig. 14.3) is applied in larger experimental areas. These large experimental areas (called whole-plot units) serve as a block that is split into smaller units (called the split plots) where the second experimental factor (say, *B* in Fig. 14.3) is then applied. In other words, the experimental factors *A* and *B* are sequentially applied or imposed to experimental areas of different size: larger for the hard-to-change, difficult, or costly-to-apply factor, smaller for the easier-to-apply factor for which we expect to find larger differences between levels. Just like in randomized designs, there are two levels of randomization in split-plot experiments: the first is conducted to determine the assignment of the block-level experimental factor to whole plots, and the second to randomly select the smaller subunits (the split plots) where the the second factor is applied (Jones and Nachtsheim 2009).

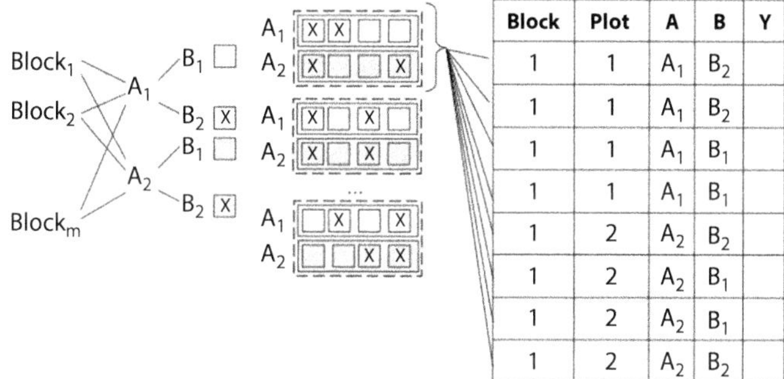

Block	Plot	A	B	Y
1	1	A_1	B_2	
1	1	A_1	B_2	
1	1	A_1	B_1	
1	1	A_1	B_1	
1	2	A_2	B_2	
1	2	A_2	B_1	
1	2	A_2	B_1	
1	2	A_2	B_2	

Fig. 14.3 Schematic representation of the split-plot design, showing a set of *m* experimental blocks (dashed lines) where the large-scale factor *A* was applied to whole plots, and the small-scale factor *B* was applied to parts of the whole plots within each experimental block. The symbols and colors are as in Fig. 14.1. The table only shows the data input for the first block, which is composed of two plots.

Let's consider again that `blocking` is a blocking variable defining a set of very large homogeneous areas, and that is to be modeled as a random or group-level effect. Then, assume that experimental factor *A* is the hard-to-change factor that is applied to large subsets (that would now become the whole plots) that are randomly selected within each block. Finally, the experimental factor *B* (the easier-to-change factor) is applied to randomly selected small subunits within each level of factor *A* in each block (Fig. 14.4). Using our example, the statistical model for a split-plot design is a mixed model with the fixed or population-level effects $A*B$ on the scale of the link function, and the random or group-level effects describing the nested structure of the split block: $g(\mathrm{E}(Y)) = A*B$ + random effects. In R, we would write `glmmTMB(y ~ A*B + (1|blocking/A), family = xxx)` and `brm(formula = A*B + (1|blocking/A), family = xxx)` in the frequentist (we could have used `glmer` from the `lme4` package) and Bayesian frameworks, respectively. The key to correctly specifying the statistical model is to keep track of the nested spatial structure of the experiment. It is relatively easy to mistake a split-plot for a randomized block design without a precise description of the application of the experimental factors.

The values of the response variable `Y` are taken at the level of the small experimental units or split plots. The smallest units to which we apply the experimental factor *B* in Fig. 14.3 are pseudo-replicates (i.e., they are not statistically independent) when it comes to testing hypotheses about the experimental factor *A* applied to the larger units. In split-plot designs, sub-plot experimental units are usually not independent because observations on pairs of sub-plots within the same whole plot are likely to be correlated. This is why it is fundamental to specify the correct nesting structure in the statistical models of split-plot designs. A set of experimental units having the same values of the experimental factors within the same block and plot are considered statistical replicates (see Fig. 14.3).

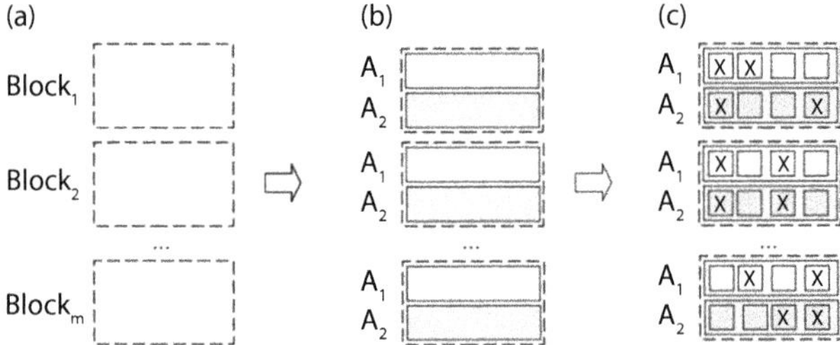

Fig. 14.4 The sequential implementation of a split-plot design (Fig. 14.3), showing the definition of large, homogeneous experimental blocks denoted with dashed lines (a), the application of the large-scale experimental factor *A* to whole plots (b), and the application of the small-scale experimental factor *B* to subsets of the whole plots where a level of the large-scale experimental factor *A* was applied (c).

A final word about the random or group-effect structure of the statistical models for split-plot designs: `(1|blocking/A)` indicates that the experimental factor *A* is nested within the blocking variable `blocking`. There is an equivalent and perhaps more transparent notation for these effects: `(1|blocking) + (1|blocking:A)`. The statistical models will estimate the group or random effect variance for `blocking` that will indicate how different or heterogeneous were the large areas defined as blocks (Fig. 14.4). It will also estimate the variance between levels of the experimental factor `A` within the blocks defined by `blocking` (Fig. 14.4). This would amount to an interaction between the experimental factor `A` and `blocking`. It will indicate the heterogeneity of the differences between the levels of the experimental factor *A* across the levels of `blocking` (see Lawson 2014 for details).

14.4.4 *Nested design*

The factorial design requires that the experimental factors are crossed to form all possible pairwise combinations of their levels. In our example, the same two levels B_1 and B_2 are used in combination with the two levels of factor *A*. If so, the two levels B_1 and B_2 are interchangeable between the levels of factor *A* (Fig. 14.1). The alternative to crossed factors is nested factors. In this case, the experimental factor *B* is nested in factor *A* if there is a completely different set of levels of *B* for every level of *A* (Fig. 14.5). Accordingly, in nested designs we cannot assess the interaction between the experimental factors *A* and *B*, and we are forced to assume that it does not exist. Moreover, the differences between the levels of the nested experimental factor (*B* in Fig. 14.5) are entirely contingent on each level of the outer experimental variable (*A* in Fig. 14.5).

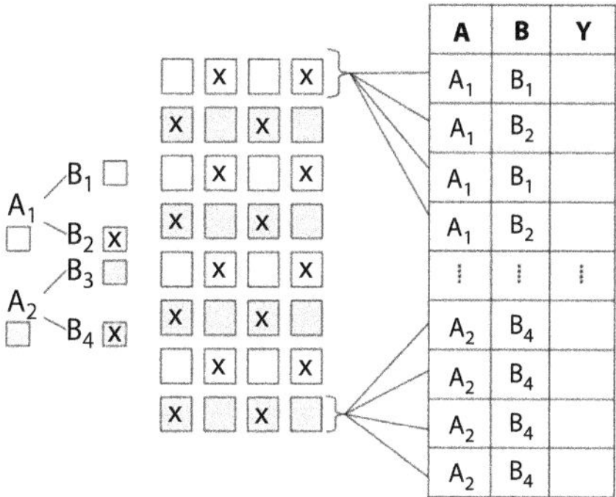

Fig. 14.5 Schematic representation of the nested design, showing the experimental factor *A*, whose two levels (A_1 and A_2) had different and specific levels of the experimental factor *B* (for which we had to create two extra levels). The table only shows the data input for the first and the last sets of values.

Nested designs are sometimes called hierarchical designs to denote the hierarchical structure between the experimental factors. In these cases we must build our statistical models to match the structure of the experimental factors. For instance, we may want to compare a phenotypic trait between biogeographical regions, for which you measure the traits in different species. Because some, but not all, species may be shared among biogeographical regions, we need to consider a nested design. In other cases, the crossing of levels of the experimental factors may be impractical, too expensive, or of minor relevance for the hypotheses being tested.

Although the nested factor(s) can be modeled as being either random or fixed, they are usually considered as random since they often arise from some kind of subsampling. Nested factors are usually, but not always, random effects, and they are usually, but not always, blocking factors (Oehlert 2010). We point out that the nested experimental factor should have many more than two levels (ideally, more than ten; Chapter 13), but we chose to use two levels (Fig. 14.5) in our example. The nesting structure need not be as simple as in Fig. 14.5, and it may include further categorical variables, as suggested for the Michigan school example in Chapter 13. This experimental design is commonly used in quantitative genetics to estimate the percentage of variation explained at different levels of the nesting hierarchy (Bate and Clark 2013). In another example, we could study the variation in egg size between two colonial bird species (the outer or fixed factor), among females of a colony, within females among years, and within females within years (i.e., clutches). Having an appropriate sampling design with a nested structure allows us to estimate the percentage of variation of the response variable at each level of the sampling hierarchy. This estimation will be extremely useful to plan the allocation of sampling effort at the level(s) associated with the largest percentage of variation.

Using our example, the statistical model for a nested design is a mixed model with the fixed or population-level effects (*A*) on the scale of the link function, and the random or group-level effect (*B*; see Fig. 14.5) describing the nesting structure: $g(E(Y)) = A + B$ + random effects. Note that we cannot fit the interaction `A*B` precisely because we do not have all combinations of levels of both experimental factors. In R, we would write `glmmTMB(y ~ A + B + (1|A:B), family = xxx)` and `brm(formula = A + B + (1|A:B), family = xxx)` in the frequentist (we could have used `glmer` from `lme4`) and Bayesian frameworks. The notation `(1|A:B)` indicates that the experimental factor *B* is nested within *A* (Fig. 14.5).

14.4.5 *Repeated measures design*

In this design we measure every experimental unit repeatedly over time. The primary purpose is to compare experimental units that were exposed to experimental factors when the response is measured repeatedly on each subject over time (Fig. 14.6). We now have "between-subject" (i.e., experimental subjects belonging to different groups defined by the levels of the experimental factors) and "within-subject" (i.e., temporal variation of the response variable for each experimental unit) effects, both of which are interesting in their own sake. Repeated measures designs are also called "longitudinal studies." Please be aware that you might find confusing jargon related to repeated measures design with considerable variation across the literature. This design is similar to a split-plot design with whole plots being the experimental subjects, and sub-plots being different observations over time on each experimental subject.

In the idealized execution of a repeated measures design, each experimental unit would be randomly assigned to "receive" one of the four combinations A_1B_1, A_1B_2, A_2B_1, and A_2B_2 of the experimental factors in our example. The set of measurements of every experimental unit are, of course, not statistically independent, and they form a block of values (Fig. 14.6). Of course, we need to have replicated sets of measurements in each of the four combinations in order to estimate the interaction between the experimental factors A and B. However, considering each set of values as a block does not completely model their lack of statistical independence. This is because the departure from statistical independence occurs in a very specific way: the sets of measurements of each experimental unit are ordered along a temporal axis. We must then model their temporal autocorrelation to describe how their degree of similarity declines with the time elapsed between pairs of measurements within experimental units. Measurements made over a short time span will likely be more similar (i.e., more correlated), but as the time between measurements increases, they will eventually become uncorrelated (i.e., statistically independent). There are several time-series models that can describe the decline of similarity over time. The simplest one is the autoregressive model of order 1, AR(1), that describes the time course of a random variable z as dependent on the previous values of a sequence: $z_{t+1} = \alpha z_t + v_t$, where $-1 < \alpha < 1$ and $v_t \sim \text{Normal}(0, \sigma_z^2)$. The previous equation (known as the Yule–Walker process) states that the next value in a temporal sequence is proportional to the previous one plus a random shift (otherwise it would be a straight line going through the origin). The autocorrelation function of the AR(1) process describing the correlation of pairs of values separated by t units of time is a decaying exponential function $\rho_t = \alpha^t$, where the parameter α is estimated as part of the fitting process of the statistical model. The AR(1) process is often enough for many repeated measures data sets, but there are other, more complex, autoregressive processes that could be used if necessary (Hedecker and Gibbons 2006).

Why would you wish to make repeated measurements of the response variable for each experimental unit? First, the temporal trend of effects, rather than a single end-point assessment, can be interesting in its own right. Second, they tend to be more efficient since subject variability is controlled better, which often increases power (i.e., the probability of detecting true effects) using fewer experimental units. Finally, repeated measure designs tend to cost less to implement than randomized designs (Hedecker and Gibbons 2006). In a sense, every experimental unit in a repeated measures design acts as its own control. Note that only sets of measurements in different experiments receiving the same values of the experimental factors can be considered statistical replicates (Fig. 14.6).

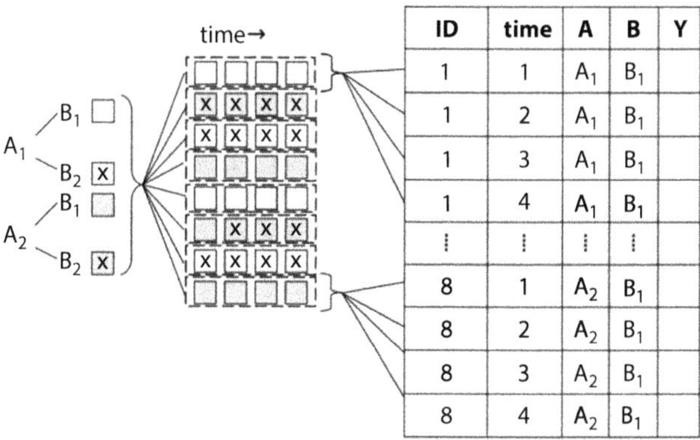

ID	time	A	B	Y
1	1	A_1	B_1	
1	2	A_1	B_1	
1	3	A_1	B_1	
1	4	A_1	B_1	
⋮	⋮	⋮	⋮	
8	1	A_2	B_1	
8	2	A_2	B_1	
8	3	A_2	B_1	
8	4	A_2	B_1	

Fig. 14.6 Schematic representation of repeated measures design, showing the application of combinations of levels of the experimental factors *A* and *B* to experimental units, whose set of measures over time constitute a block of values. Each experimental unit received the same combination of levels of factors *A* and *B*, and it is followed and measured over four time steps. The symbols and colors are as in Fig. 14.1. The table only shows the data input for the first and the last sets of four values of two experimental units.

Considering our example, the statistical model is $g(\mathrm{E}(Y)) = A*B$ + random effects + autocorrelation. The random or group-level effects are the set of measurements taken over time which, after verifying that one might reasonably envision a straight line fit, would be written as `(1 + time|exp.unit)`, where `exp.unit` is a categorical variable identifying the set of measurements over `time` associated with each experimental unit. These random or group-level effects would describe a different intercept and temporal slope for each experimental unit. In R, we would write `glmmTMB(y ~ A*B + (1 + time|exp.unit), family = xxx)` and `brm(formula = A*B + (1 + time|exp.unit), family = xxx)` in the frequentist and Bayesian frameworks. However, these models are incomplete since they do not model the temporal autocorrelation of the successive measurements at the level of the experimental units. To generate an AR(1) process for the temporal autocorrelation we would need to add `ar1(0 + time|exp.unit)` and `cor_ar(~ time|exp.unit, p = 1)` to the previous commands of the packages `glmmTMB` and `brms`, respectively. These commands generate a variance–covariance matrix that depends on the parameter α of the AR(1) process (Hedecker and Gibbons 2006). As always, in the Bayesian framework we should define suitable priors for the population- and group-level effects, and for the parameter(s) of the autocorrelation process. These final statistical models will now also model the non-independence of repeated measurements at the level of `exp.unit` through the random or group-level effect, and also the temporal dependence of the residuals through the AR(1) process.

14.4.6 *Crossover design*

The purpose of crossover designs is to increase the precision of the comparisons of levels of an experimental factor (or combinations thereof) by comparing them within each experimental unit. A crossover design differs from repeated measures and randomized designs in that each experimental unit receives a sequence of experimental factor(s), rather than a single level of an experimental factor (or combinations thereof; Fig. 14.7). What is actually randomized here is the allocation of the sequences of levels for each experimental unit, such that each has the same chance of receiving a given sequence of factor levels (Díaz-Uriarte 2002). The intrinsic heterogeneity between experimental units is not problematic because each of them that is sequentially measured over time serves as its own control. Needless to say, the set of measurements for each experimental unit are not statistically independent, and they are accordingly treated as a block. Crossover designs are used in medical trials, and in animal behavior, psychology, and pharmacology.

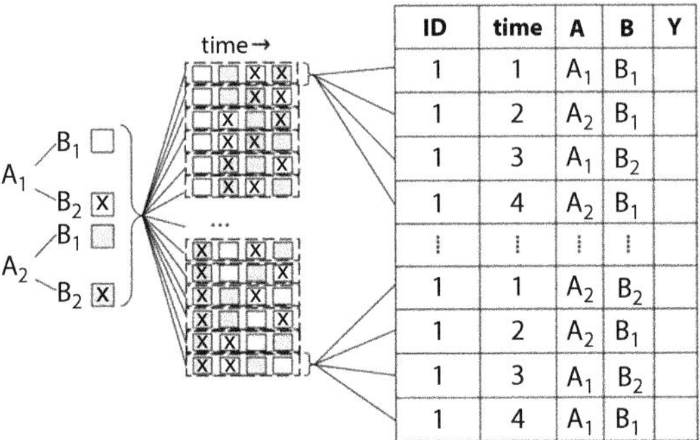

time→
A_1
B_1
B_2
A_2
B_1
B_2
ID | time | A | B | Y
1 | 1 | A_1 | B_1
1 | 2 | A_2 | B_1
1 | 3 | A_1 | B_2
1 | 4 | A_2 | B_1
1 | 1 | A_2 | B_2
1 | 2 | A_2 | B_1
1 | 3 | A_1 | B_2
1 | 4 | A_1 | B_1

Fig. 14.7 Schematic representation of crossover design, showing the successive application over four time units of combinations of levels of the experimental factors *A* and *B* to each experimental unit. Each experimental unit received a different sequence of combination of levels of factors *A* and *B*, and it is followed and measured over four time steps. Note that only a small subset of all possible sequences are shown with only one replicate. The table only shows the data input for the first and the last sets of four values of two experimental units.

The simplest crossover design is the two-period, two-level (either *A* or *B*) or 2 × 2 design. It is known as the *AB*/*BA* design. There are only two possible sequences: {*A*, rest, *B*} and {*B*, rest, *A*}. One problem with crossover designs is that the number of plausible comparisons or contrasts increases very quickly with the number of levels of an experimental factor. Consider an experimental factor with three levels {*A*, *B*, placebo} and three time periods without wash-out or resting periods. Here we have six possible sequences ranging from {*A*, *B*, placebo} to {placebo, *B*, *A*}. Hence, there will be 6 × (6 − 1) / 2 = 15 plausible pairwise comparisons of sequences of levels.

Three main issues arise in crossover designs that need to be considered in the planning and analysis of the experiment. They are: order effects, reversibility, and carry-over effects (Jones and Kenward 2015). Order effects describes how experimental units receiving *A* first and *B* later may not have the same values of the response variable if the experimental factors were applied in the opposite order. Reversibility refers to how an initial change in the mean of the response variable after the application of a level of an experimental factor can be annulled or compensated by the application of a second experimental factor. The relevance of these first two issues can be evaluated through the contrast of specific sequences of combinations of levels of the experimental factors. In turn, carry-over effects refer to the delayed effects of an experimental factor. Carry-over effects hinder the detection of differences between levels of the experimental factors in a sequence of application. The common approach to deal with carry-over effects is to introduce a "rest" or "wash-out" period between consecutive levels, and then to make the relevant comparisons to assess their importance.

In the context our example with two crossed experimental factors each with two levels, and assuming no wash-out period, each experimental unit would receive at each time period a combination of $\{A_1$ or $A_2\}$ and $\{B_1$ or $B_2\}$, thus forming the set of the four combinations $\{A_1B_1, A_1B_2, A_2B_1, A_2B_2\}$ (Fig. 14.7). Each experimental unit would need to be measured in four time periods if we are to evaluate the entire set on each of them. Including a wash-out period does increase the complexity in the execution and analysis of crossover designs. We would need to have a time explanatory variable indexed by {1,2,3,4} to keep track of the temporal sequence of levels of the experimental factors *A* and *B* applied to each experimental unit. Because the set of values of the response variable arising from the combinations of the experimental factors *A* and *B* are not statistically independent at the level of each experimental unit, we also need a categorical variable `exp.unit` to identify the block of measurements over `time` associated with each experimental unit. Needless to say, we would also need several replicates of each sequence of combinations of the experimental factors *A* and *B*. In a generic sense, the statistical model is $g(E(Y)) = A{*}B$ + random effects. In the crossover design, the random or group-level effects should denote the combinations of levels of the experimental factors *A* and *B* that are nested as a sequence of measurements over time for each experimental unit. In R, we would write `glmmTMB(y ~ A*B + time + (1|exp.unit:AB), family = xxx)` and `brm(formula = y ~ A*B + time + (1|exp.unit:AB), family = xxx)` in the frequentist and Bayesian frameworks, where `AB` is a composite categorical variable denoting the combination of levels of the experimental factors `A` and `B` given to an experimental unit at each time step. To create the composite variable `AB`, we can use `DF$AB = interaction(DF$A, DF$B, sep = ".")` to generate categories `A1.B1`, `A1.B2`, etc. for each row in the data frame `DF`. When the number of `exp.unit` is small to moderate (say, less than 10), we should not include it as a random or group-level effect (see Section 14.4.2), but as a fixed or population-level effect at the start of the model equation. It is possible that the set of measurements over time would exhibit temporal autocorrelation at the level of the experimental units. Nonetheless, the fact that we are applying different levels of the experimental factors at each time step is likely to decrease or eliminate the temporal autocorrelation. If needed, we should use the same approach described in the repeated measures design (Section 14.4.5). In general, defining adequate statistical models for crossover designs requires great care, particularly so when there are wash-out or rest periods involved. We suggest that you should at least consult Lawson (2014) and Jones and Kenward (2015) before embarking on the analysis.

14.5 How many samples should we take?

Once we define a suitable, or at least reasonable, experimental design for the goals of our study, the next questions are to define the total sample size, and how to distribute or allocate this total number of experimental units (and of measurements over time, in the case of the repeated measures and crossover designs) among the levels of the explanatory variables. These are neither simple nor trivial issues.

To understand why, let's consider the simplest possible experiment: the comparison of the means of two groups, i.e., the *t*-test in the frequentist framework. Its statistical model has a single categorical variable with two levels defining the means we wish to compare. This statistical model contains three parameters: the mean of the reference group, the magnitude of the difference in the means between the two groups, and the variance of the response variable (Chapter 4). In the simplest terms, the goal is to assess whether the estimated magnitude of the effect size (i.e., the difference between the two means) is large enough to be declared "relevant." Judging the scientific relevance of empirical findings based on statistical significance is rather foolish. For starters, we all know that a tiny effect size of irrelevant empirical implications may be declared statistically significant for a large enough sample size. We use this much decried and conventional approach to keep things concrete and simple in this discussion. Before collecting the data, we aspire to have a high probability of achieving our goals if our suspected underlying state of nature were true. In the frequentist framework, this achievement is quantified by the statistical power, i.e., the probability that the experiment would produce a statistically significant result assuming a given effect size whose true magnitude is unknown to us. (Had we known in advance the true magnitude of the difference between the two means, we would not have bothered collecting the data!) More formally, the statistical power is the complement of the probability of a type II error, which is the probability of wrongly rejecting the alternative hypothesis that the true effect size had a given magnitude.

To calculate the statistical power of the *t*-test, we need to provide the true magnitude of the effect size that we wish to detect, the sample size, the variance of the response variable (often guessed in advance), and α, the probability of incorrectly rejecting a hypothesis of zero mean difference, usually fixed at 0.05). We even have a formula for that calculation, from which we can also obtain the minimal sample size needed to attain a certain power for a given effect size α and variance of the response variable. Now, there is an oft-quoted standard that we must strive to have a power of at least 0.80. This comes from Cohen's (1988) view that "the mistaken rejection of the null hypothesis is considered four times as serious as mistaken acceptance," with no justification whatsoever. Thus, if $\alpha = 0.05$, and the probability of a type II error (incorrectly detecting a non-existing effect) $\beta = 4\alpha$, then the power is $1 - \beta = 0.8$. As you can see, the routine wish or requirement that power = 0.8 is entirely arbitrary and based on little more than thin air. There is no fundamental justification beyond unreasoned tradition to require that power = 0.8. By the way, it is hardly ever achieved in actual experiments for many scientific disciplines (e.g., Button et al. 2013, Smith et al. 2011, Smaldino and McElreath 2016) despite the recurrent calls to meet it that appear from time to time.

Going further than the *t*-test, there are a few general formulae to estimate the adequate sample size for a few experimental designs involving the general linear model (Chapters 4–6). And that's almost all of it. To the best of our knowledge, there is no R package capable of calculating the total sample size and the allocation of the sampling units for the experimental designs covered in this chapter, and for the types of response variables listed in Table 2.1 that would cover most of the analytical needs of life scientists. In all other cases (i.e., in virtually every other univariate statistical analysis) we need to resort to Monte Carlo simulations to answer the two fundamental questions mentioned at the start of this section. These simulations are typically designed to estimate power as a function of the sample size, the magnitude(s) of the effect sizes for each explanatory variable that we wish to estimate, the variance(s) of random or group-level effect(s) and of the error for Gaussian response variables, and a few other parameters depending on the experimental design and on the type of the response variable. Because we need to consider four or five values for each of these features, we would need to carry out many joint combinations of them in the numerical simulations (e.g., Arnold et al. 2011, McNeish 2016, McNeish and Stapleton 2019, Kain et al. 2015, Arend and Schäffer 2019). And, after carrying out the Monte Carlo simulations, we would need to do some "reverse engineering" to read out the sample size needed to attain the desired power for specific combinations of effect sizes, variances, and other GLMM parameters for a given experimental design. To best of our knowledge, the package `simr` (Green and MacLeod 2016) is the most general tool available to carry out these numerical simulations aiming to estimate power and thereafter the minimal sample sizes for GLMMs. Kain et al. (2015) and Arend and Schäffer (2019) are two very good starting points for setting up the Monte Carlo simulations necessary to estimate power, and from it the minimum sample sizes in GLMMs.

Our discussion about determining the sample size was entirely framed in the context of power and statistical significance that is only pertinent to the frequentist framework. There are no such things as statistical significance, *p*-values, or type II error (and its complement, statistical power) in the Bayesian framework. The calculation of the minimally adequate sample size(s) in the Bayesian framework uses a different approach. This may be framed in terms of the benefits of attaining the research goals, and the costs of pursuing them so as to make a cost–benefit analysis that may be evaluated in a decision-theoretic treatment (Chaloner and Verdinelli 1995; see also Kunzmann et al. 2021). This is a highly technical endeavor. Kruschke (2015) suggests an alternative working definition of power: the probability of attaining a set of research goals, including estimating an effect size with adequate precision, as a function of sample size and other model parameters. We may then use use this more general definition of power to estimate the minimal sample size necessary to attain the research goals. In analogy with the frequentist Monte Carlo simulations, we now need to define hypothetical distributions of parameter values to generate representative values from which we would generate random samples of plausible pseudo-data for the sampling design and different sample sizes. We would then use these pseudo-data and the prior distributions of the model parameters in a customary MCMC analysis to obtain posterior predictive distributions (Chapter 7). And we could then tally the attaining of the research goals (much like a posterior predictive check; Chapter 7) from these posterior distributions. Just like in the frequentist Monte Carlo simulations described above, Kruschke's (2015, pp. 360–376) suggested approach is far from a streamlined pipeline, as it does involve a minimum of programming to set up an automated approach to calculating the minimally adequate sample sizes.

While the seemingly innocent question of "How many samples should we take?" has a long, convoluted, and laborious tentative answer specific to each case, there are a few general messages emerging from the Monte Carlo simulations carried out thus far. First, in designs involving random or group-level effects (i.e., randomized blocks, nested, repeated measures, crossover), the power increases much faster with the number of levels of the blocking variable than with the number of replicates et each level of the blocking variable (e.g., Kain et al. 2015, Judd et al. 2017, Arend and Schäffer 2019). This is of particular importance when predicting group-level estimates is relevant for the study objectives. Second, estimates of variances of the random or group-level variables are notoriously imprecise for most types of experimental designs and response variables unless the number of levels of this explanatory variable gets close to 100, which is almost never attained in the life sciences (e.g., Westfall et al. 2014, McNeish 2016, McNeish and Stapleton 2019). Third, the dependence of power on the magnitude of the effect size that we wish to estimate for each explanatory variable poses several challenges in estimating power through simulations. On the one hand, while Cohen's (1988) three-level classification for the general linear model is sometimes used to conceptualize the effect sizes to be used as arguments in the power estimations, it does not automatically translate into the coefficients that need to be input into the simulations (Arend and Schäffer 2019). On the other, there are no metrics of effect size analogous to Cohen's (1988) for GLMs, and even less so for GLMMs. Thus, when our response variable is non-Gaussian, we even lack a simple-minded classification for the effect sizes associated with the explanatory variables. Unavoidably, the specific ways in which the effect sizes are defined in the Monte Carlo simulations affect the power estimations in non-Gaussian GLMMs, so beware. Fourth, details of numerical approximations and algorithms (i.e., Kenward–Roger, Satterwhaite, REML, Laplace or Gaussian quadrature; see Chapter 13) used in the Monte Carlo simulations do influence the power estimates obtained in numerical simulations. This is particularly so for the small sample sizes and weak effect sizes that are of the greatest interest to life scientists. The variation of these details greatly hinders the cross-study comparisons of Monte Carlo simulations of GLMMs aimed at estimating power (and minimum sample sizes). These messages are very likely to be valid for Kruschke's (2015) Bayesian power approach.

Finally, every meaningful calculation of frequentist or Bayesian power must be made before having the data. Let's not forget that the main and only goal of these simulations is to help design an experimental or observational study. Post hoc estimations of power are highly misleading and should definitely be avoided, because they are based on published effect size estimates with large and unknown downward bias (e.g., Goodman and Berlin 1994, Hoening and Helsey 2012, Gelman 2019). Any power analysis should thus be carried out before, not after, the experiment to serve as guidance when deciding the sample sizes and other features of the experimental design.

References

Altman, N. and Krzywinski, M. (2014). Split-plot designs. *Nature Methods*, 12, 165–166.

Arend, M. and Schäffer T. (2019). Statistical power in two-level models: A tutorial based on Monte Carlo simulation. *Psychological Methods*, 24, 1–19.

Arnold, B., Hogan, D., Colford, J. et al. (2011), Simulation models to estimate design power: An overview for applied research. *BMC Medical Research Methodology*, 11, 91–103.

Bate, S. and Clark, R. (2013). *The Design and Statistical Analysis of Animal Experiments*. Cambridge University Press, Cambridge.

Button, K., Ioannidis, J., Mokrysz, C. et al. (2013). Power failure: Why small sample size undermines the reliability of neuroscience. *Nature Reviews Neuroscience*, 14, 365–376.

Chaloner, K. and Verdinelli, I. (1995). Bayesian experimental design. *Statistical Science*, 10, 273–304.

Cohen, P. (1988). *Statistical Power Analysis for the Behavioral Sciences*, 2nd ed. Lawrence Erlbaum Associates, New York.

Dean, A., Voss, D., and Draguljic, D. (2017). *Design and Analysis of Experiments*. Springer, New York.

Díaz-Uriarte, R. (2002). Incorrect analysis of crossover trials in animal behaviour research. *Animal Behaviour*, 63, 815–822.

Fisher, R. (1935). *The Design of Experiments*. Oliver and Boyd, Edinburgh.

Fisher Box, J. (1978). *R. A. Fisher: The Life of a Scientist*. John Wiley and Sons, New York.

Gelman, A. (2019). Don't calculate post hoc power using observed estimate of effect size. *Annals of Surgery*, 269, e9–e10.

Goodman, S. and Berlin, J. (1994). The use of predicted confidence intervals when planning experiments and the misuse of power when interpreting results. *Annals of Internal Medicine*, 121, 200–206.

Green, P. and MacLeod, C. (2016). SIMR: An R package for power analysis of generalized linear mixed models by simulation. *Methods in Ecology and Evolution*, 7, 493–498.

Hedecker, D. and Gibbons, R. (2006). *Longitudinal Data Analysis*. John Wiley and Sons, New York.

Hinkelman, K. (2015). History and overview of design and analysis of experiments. In A. Dean, M. Morris, J. Stufken et al. (eds), *Handbook of Design and Analysis of Experiments*, pp. 3–62. CRC Press / Chapman & Hall, New York.

Hoening, J. and Helsey, G. (2012). The abuse of power: The pervasive fallacy of power calculations for data analysis. *American Statistician*, 55, 19–24.

Jones, B. and Kenward, M. (2015). *Design and Analysis of Cross-Over Trials*, 3rd edn. CRC Press / Chapman & Hall, New York.

Jones, C. and Nachtsheim, C. (2009). Split-plot designs: What, why, and how. *Journal of Quality Technology*, 41, 340–361.

Judd, C., Westfall, J., and Kenny, D. (2017). Experimenting with more than one random factor: Designs, analytical models and statistical power. *Annual Review of Psychology*, 68, 1–25.

Kain, M., Bolker, B., and McCoy, M. (2015). A practical guide and power analysis for GLMMs: Detecting among treatment variation in random effects. *PeerJ*, 4, peerj.1226.

Kruschke, J. (2015). *Doing Bayesian Data Analysis: A Tutorial with R, JAGS and Stan*, 2nd edn. Academic Press, New York.

Kunzmann, K., Grayling, M., Lee, K. et al. (2021). A review of Bayesian perspectives on sample size derivation for confirmatory trials. *The American Statistician*, 75, 424–432.

Lawson, J. (2014). *Design and Analysis of Experiments with R*. CRC Press / Chapman & Hall, New York.

McNeish, D. (2016). Estimation methods for mixed logistic models with few clusters. *Multivariate Behavioral Research*, 51, 790–804.

McNeish, D. and Stapleton, L. (2019), The effect of small sample size on two-level model estimates: A review and illustration. *Educational and Psychological Review*, 28, 295–314.

Mead, R., Gilmour, S., and Mead, A. (2012). *Statistical Principles for the Design of Experiments: Applications to Real Experiments*. John Wiley and Sons, New York.

Montgomery, D. (2017). *Design and Analysis of Experiments*, 9th edn. John Wiley and Sons, New York.

Oehlert, G. (2010). *First Course in Design and Analysis of Experiments*. W. H. Freeman, New York.

Ryan, T. (2007). *Model Experimental Design*. John Wiley and Sons, New York.

Smaldino, P. and McElreath, R. (2016), The natural selection of bad science. *Proceedings of the Royal Society Open Science*, 3, 160384.

Smith, D., Hardy, I., and Gammell, M. (2011). Power rangers: No improvement in the statistical power of analyses published in *Animal Behaviour*. *Animal Behaviour*, 81, 347–352.

Westfall, J., Kenny, D., and Judd, C. (2014). Statistical power and optimal design in experiments in which samples of participants respond to samples os stimuli. *Journal of Experimental Psychology: General*, 143, 2020–2045.

CHAPTER 15

Mixed Hierarchical Models and Experimental Design Data

Packages needed in this chapter:

```
packages.needed<-c("ggplot2","fitdistrplus","brms","future","broom",
"GGally","lme4","broom.mixed","glmmTMB","ggeffects","performance",
"cowplot","HLMdiag","arm","pbkrtest","DHARMa","qqplotr","reshape2",
"bayesplot","parallel","doBy","ggbreak","nlme","ggridges","ggdist",
"MuMIn","pROC","afex")
lapply(packages.needed, FUN = require, character.only = T)
```

15.1 Introduction

This chapter shows three examples of mixed/hierarchical models to analyze data from different experimental or observational designs. Our presentation will build on the basic concepts of the theory and fitting of mixed/hierarchical models (Chapter 13), and will include some of the experimental designs of Chapter 14. Table 2.1 lists the five main types of response variables depending on the scale and existence of boundaries of possible values, and the possible probability distributions that can be used to model them. It lists nine probability distributions, including mixture models for response variables having an excess of zeros (Chapters 10 and 12). Since each of the six experimental designs of Chapter 14 applies to every response variable in Table 2.1, this would make a total of 54 possible case studies, and 108 analyses given that we employ both the frequentist and Bayesian frameworks. This would clearly be too many combinations, and hence we need to make a selection. We therefore consider three experimental designs (randomized blocks, repeated measures, and split-plot) that are probably the most frequently used in the life sciences. We will not consider factorial design because it was implied in the study cases involving two categorical explanatory variables in the chapters of Part II (see Chapter 14).

The most relevant feature of these analyses will be the definition of the random or group-level effects that denote the structure of the data imposed by each experimental design. Therefore, should a specific case of interest for you (say, a response variable with a zero-augmented Poisson distribution involving a repeated measures design) not be considered in this chapter, we trust that by now you can write the adequate statistical model

Statistical Modeling With R. Pablo Inchausti, Oxford University Press. © Pablo Inchausti (2023).
DOI: 10.1093/oso/9780192859013.003.0015

by selecting a suitable probability distribution in the model likelihood, and select the random or group-level effects appropriate for the experimental design of your case study. In the case of models fitted in the Bayesian framework, we would, of course, need to define priors of all the parameters of the statistical model.

A very important point in this chapter is the following. When the random or group-level effects denote the structure of the data induced by the experimental design, we cannot possibly simplify these effects. We need to correctly define the random or group-level effects to account for the statistical non-independence of the experimental units (or set of measurements, in the case of repeated measures and crossover designs) imposed by the design. Changing the random or group-level effects through model selection would amount to pretending that the empirical evidence was obtained using an experimental or sampling design different from the one actually used. Details do matter and have consequences: parameter estimates and their measures of precision may change substantially if we meddle with the random or group-level effects imposed by the experimental design. We will see in this chapter that the random or group-level intercepts are the key feature to describing the experimental or sampling design. Besides the intercepts, we might consider including random or group-level slopes when suitable for the data at hand. The need for and importance of including random slopes remains debatable; we might consider their inclusion in the model a matter to be settled by means of model selection (see Chapter 13).

15.2 Binary GLMM with a randomized block design

THE DATA IN CONTEXT: Vicente et al. (2006) studied the prevalence of a nematode parasite in adult red deer of both sexes in 24 sites or farms in south central Spain. Each sampled animal was sexed and its body length measured. While the authors quantified parasite prevalence and other variables describing the vegetation and landscape at each site, we are going to focus on the presence/absence of parasites, and use the sampling site as an explanatory variable. This was not a designed but an observational experiment. That is, at each site, the experimental units did not have the same probability of being sampled, nor were they randomly assigned to each sampling site. It seems reasonable, however, to assume that all deer at a site were exposed to or experienced similar environmental conditions that were different from those at other sites. Thus, each level of the explanatory variable site would group sets of more similar (and hence non-independent) experimental units. The goal of the analysis was to detect whether the presence of the nematode varied among sexes and according to the individual body length.

STATISTICAL MODEL:

$$Y \sim \text{Binomial}(\pi).\ \mu_Y \equiv \text{E}(Y) = \pi \text{ (because we have ungrouped binary data)},$$
$$\text{logit}(\pi) = \beta_{0,\text{sex}} + \beta_{1,\text{sex}}\text{length},$$

where $\beta_{0,\text{sex}}$ and $\beta_{1,\text{sex}}$ are the fixed or population-level effects in the form of the intercept and slope of length for each sex in the scale of the logit link function (see Chapter 9). Thus written, the statistical model assumes that all experimental units are statistically independent. We are going to account for their lack of independence by modeling the sampling site as a random or group-level effect. Modeling the categorical explanatory variable farm as a random or group-level effect amounts to defining a randomized block design (see Chapter 14). Besides random or group-level intercepts for each farm, we will examine whether we should also include random (or group-level) slopes to make up vectorial random effects (Chapter 13). The corrected statistical model would then be:

$$Y \sim \text{Binomial}(\pi),$$
$$\text{logit}(\pi) = (\beta_{0,\text{sex}} + b_{0,\text{farm}}) + (\beta_{1,\text{sex}} + b_{1,\text{farm}})\text{length}, \quad \text{where the vectorial random effects are:}$$
$$\begin{pmatrix} b_{0,\text{farm}} \\ b_{1,\text{farm}} \end{pmatrix} = \text{MultivariateNormal}\left(\begin{pmatrix} 0 \\ 0 \end{pmatrix}, \begin{pmatrix} \text{Var}(b_{0,\text{farm}}) & \text{Covar}(b_{0,\text{farm}}, b_{1,\text{farm}}) \\ \text{Covar}(b_{0,\text{farm}}, b_{1,\text{farm}}) & \text{Var}(b_{1,\text{farm}}) \end{pmatrix}\right), \tag{15.1}$$

where $b_{0,\text{farm}}$ and $b_{1,\text{farm}}$ are the farm-level coefficients to be estimated that denote the extent to which each farm deviates from the fixed or population-level estimates of the intercept and slope, respectively. Note that the the fixed and random effects are additive on the scale of the logit link function where the parameters are to be estimated (Chapter 13).

Exploratory data analysis: Let's start by importing and examining the data for the 826 (= 378 + 448) individuals:

```
> DF1=read.csv("deer chapter15.csv", header=T)
> summary(DF1)
      farm          sex          length         pres.abs
 MO     :209   female:378   Min.   : 75   Min.   :0.000
 CB     : 85   male  :448   1st Qu.:151   1st Qu.:0.000
 QM     : 60                Median :163   Median :1.000
 BA     : 50                Mean   :162   Mean   :0.646
 PN     : 37                3rd Qu.:175   3rd Qu.:1.000
 MB     : 34                Max.   :216   Max.   :1.000
 (Other):351
```

Besides the descriptive statistics for `length`, and the number of individuals per `sex`, we also have the proportion of parasitized individuals (0.646). We can count the number of farms:

```
> length(unique(DF1$farm))
[1] 24
```

We use the package `doBy` to obtain descriptive statistics:

```
> summaryBy(length~pres.abs+sex, data=DF1, FUN=c(mean, sd, length))
  pres.abs    sex length.mean length.sd length.length
1        0 female         146      17.6           155
2        0   male         164      24.2           137
3        1 female         156      12.1           223
4        1   male         173      15.6           311
```

We observe that males are, on average, larger than females (no news), and that parasitized deer of each sex have larger bodies, on average, than non-parasitized ones. The entire data set contains large numbers of deer in every combination of `pres.abs` and `sex`. Binary GLMs often have difficulties in parameter estimation when one of the categories of the response variable is rare (say, < 5%), which is not the case here. This is not true, however, when we make the same examination at the farm level:

```
> ftable(pres.abs~farm, data=DF1)
        pres.abs   0   1
farm
AL                 9   6
AU                 4  28
BA                 9  41
BE                 0  13
CB                26  59
CRC                1   0
HB                15   2
LN                 3  30
MAN               16  11
MB                 4  30
MO               127  82
NC                14  13
NV                 9  11
PN                 3  34
QM                12  48
RF                 2  18
RN                16   7
RO                 2  28
SAU                3   0
SE                 6  20
TI                 1  18
TN                 7  18
VISO               2  11
VY                 1   6
```

We can see that in 11 of the 24 farms there were fewer than 5 presences or absences, thus making it impossible to envisage fitting separate binary GLMs in each of them. This would be even worse if we disaggregated the data by sex in each farm. Doing so would be a "no pooling" analysis (Chapter 13). At the other extreme, a "complete pooling" analysis would incorrectly assume that the sampling units are statistically independent of each other (Chapter 13). The binary GLMM that we will fit amounts to a "partial pooling" approach (see Section 13.8) that will obtain adequate parameter estimates of the global relation between the response and the explanatory variables, and also farm-level equations even for those farms for which there is little data. This is the borrowing strength effect in action: farms for which we have more data will allow us to obtain parameter estimates in those with little data. And this is not all. The estimated parameters of the farms for which there is less data will exhibit a larger shrinkage or shift towards the fixed or population-level estimates (Chapter 13).

Figure 15.1 shows that body length is related to the presence of parasites, and that the relationship seems to differ between sexes, thus suggesting a possible interaction between the explanatory variables. There are 5 farms for which there are at least 35 data points, with the threshold greater than 34 individuals being chosen arbitrarily. We use just these five farms to explore the variation of the relation between the response and explanatory variables among farms:

```
> table(table(DF1$farm)>34)

FALSE  TRUE
   19     5
```

Figure 15.2 shows that there is a fair amount of variation in the relation between the presence of the parasite and body length, as suggested by the lack of parallelism in the indicative fitted lines among farms. The uneven distribution of data by sex at the farm level is such that we could not visualize the same relation shown in Fig. 15.1 even for

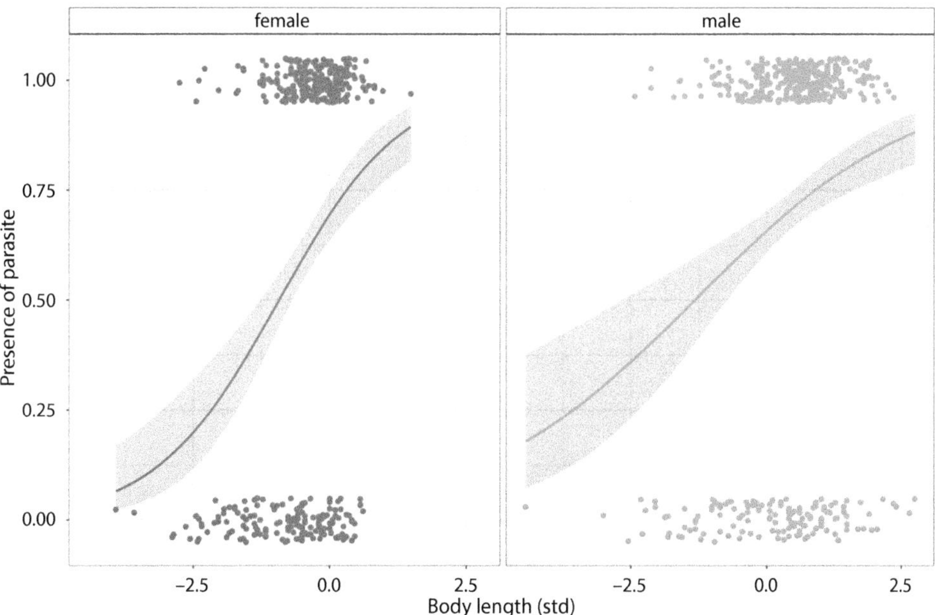

Fig. 15.1 Relation between the presence/absence of the nematode parasite and red deer body size of both sexes. All points are vertically jittered by adding a random value to allow better visualization. The curves correspond to a fitted univariate logistic regression with a 95% confidence band (gray).

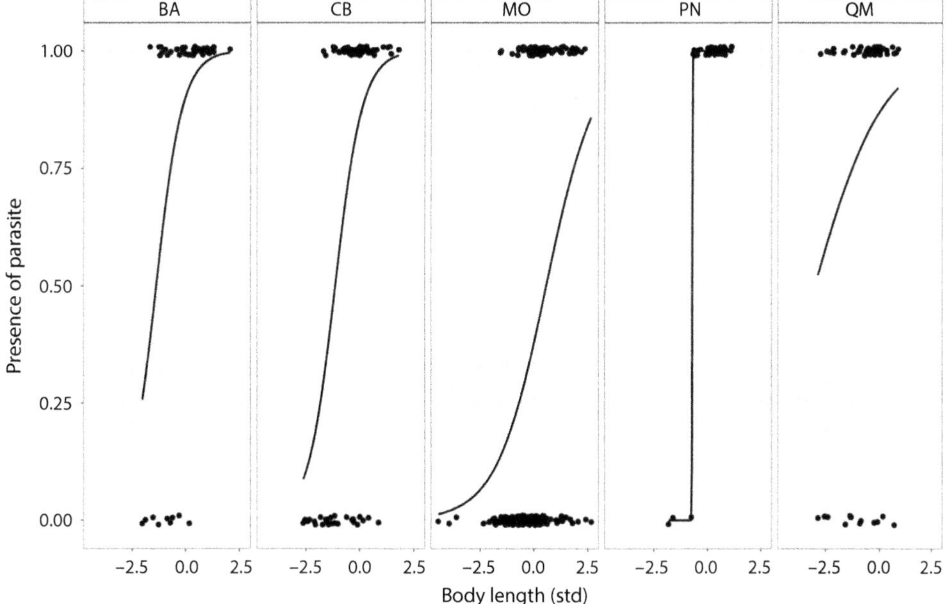

Fig. 15.2 Relation between the presence/absence of the nematode parasite and red deer body size for the five farms having at least 35 data points. All points are vertically jittered by adding a random value to allow better visualization. The curves correspond to a fitted univariate logistic regression for each farm.

these five farms for which there was more data. Prior to the analyses, we standardized the numerical explanatory variable to minimize the risk of obtaining a spuriously high correlation between random intercepts and slopes (Chapter 13), and to allow a more meaningful interpretation of the fixed or population-level intercept (Chapter 9):

```
> DF1$length.c=as.vector(scale(DF1$length, center=T, scale=T))
```

15.2.1 *Binary GLMM with a randomized block design: Frequentist models*

We use the package `glmmTMB` to fit the frequentist GLMMs of this chapter. Many, but not all, of these models can also be fitted with the package `lme4`. We start by comparing two models to decide whether to include random slopes:

```
m1=glmmTMB(pres.abs~sex*length.c +(1|farm), data=DF1, REML=T, family=binomial)
m2=glmmTMB(pres.abs~sex*length.c+(1+length.c|farm),REML=T,data=DF1,family=binomial)
```

Their summaries are:

```
> summary(m1)
 Family: binomial  ( logit )
Formula:          pres.abs ~ sex * length.c + (1 | farm)
Data: DF1

     AIC      BIC   logLik deviance df.resid
     838      862     -414      828      825

Random effects:

Conditional model:
 Groups Name        Variance Std.Dev.
 farm   (Intercept) 2.59     1.61
Number of obs: 826, groups:  farm, 24

Conditional model:
                 Estimate Std. Error z value Pr(>|z|)
(Intercept)         1.502      0.393    3.82  0.00013 ***
sexmale            -0.612      0.222   -2.75  0.00588 **
length.c            1.435      0.191    7.53    5e-14 ***
sexmale:length.c   -0.688      0.221   -3.11  0.00186 **
```

```
> summary(m2)
 Family: binomial  ( logit )
Formula:          pres.abs ~ sex * length.c + (1 + length.c | farm)
Data: DF1

     AIC      BIC   logLik deviance df.resid
     834      867     -410      820      823

Random effects:

Conditional model:
 Groups Name        Variance Std.Dev. Corr
 farm   (Intercept) 2.600    1.612
        length.c    0.389    0.624    0.63
Number of obs: 826, groups:  farm, 24

Conditional model:
                 Estimate Std. Error z value Pr(>|z|)
(Intercept)         1.564      0.399    3.92  8.9e-05 ***
sexmale            -0.548      0.227   -2.41    0.016 *
length.c            1.379      0.248    5.57  2.6e-08 ***
sexmale:length.c   -0.603      0.243   -2.49    0.013 *
```

Both models differ on the structure of the random effects, as the fixed (in this package called "conditional") effects have the same explanatory variable and effects. The AIC and the deviances of the two models are different too, of course.

Before discussing the output, we are going to use parametric bootstrap to select the most approriate model for these data (Chapter 13). Here, there are two options. First, had we used the `lme4` package to fit the models, we could have used the package `pbkrtest`: `res.boot = PBmodcomp(largeModel = m1b, smallModel = m1a, nsim = nboot, cl = cl)`, where `nboot` = 1,000 simulations, and `cl = makeCluster(rep("localhost", detectCores()))` from the package `parallel` makes a cluster of the available CPU cores in your computer to speed up the calculations. We gave one such example in Chapter 13. Second, models fitted with the `glmmTMB` package cannot be used in the command `PBmodcomp`. Thus, we need to write our own procedure for the model comparison with parametric bootstrap. The `lme4` and `glmmTMB` packages can fit GLMMs involving many probability distributions, but the latter includes many more probability distributions such as mixture distributions (Chapter 10), the beta distribution (Chapter 12), and a few others. This is why it is interesting to also consider the following code:

```
boot.res=data.frame(Dif.dev=rep(NA, nboot))
for (i in 1:nboot){
  sim.reduc=simulate(m1, nsim=1)
  mod.large=refit(m1, newresp =sim.reduc[[1]])
  mod.small =refit(m2, newresp =sim.reduc[[1]])
  boot.res$Dif.dev[i]=-2*(as.numeric(logLik(mod.large))-
                                    as.numeric(logLik(mod.small)))
}
dev.models=2*(as.numeric(logLik(m2))-as.numeric(logLik(m1)))
1-ecdf(na.omit(boot.res$Dif.dev))(dev.models)
```

We first create the data frame `boot.res` to store the results. Next, we simulate new values of the response variable using the parameter estimates from the small model `m1` and store them in `sim.reduc`. Next, we refit both the small (`m1`; a special case of `m2`) and large (`m2`) models using the newly simulated data `sim.reduc`, and store the results in `mod.small` and `mod.large`, respectively. Both models need to be refitted each time to the same newly simulated dataset. We then compute the differences in deviance between these two models, and store them in `boot.res$Dif.dev`. This is the reference distribution that will be used to obtain the p-value associated with the actual difference in deviance between the models that is stored in `dev.obs`. Let's recall that the whole point of doing parametric bootstrap is to obtain this reference sampling distribution that could not be generated otherwise because of the ongoing controversy about the number of degrees of freedom related to the random effects in mixed models (see Chapter 13). The p-value is calculated with the empirical cumulative distribution function of `boot.res$Dif.dev` after we eliminated the `NA` values occurring in the cases when one of the models did not achieve numerical convergence. The lack of convergence of binary GLMMs may occur by chance whenever the simulated values in `sim.reduc` contain a very small proportion of either zeros or ones. This is known as the problem of complete separation (Heinze and Schemper 2002). We sidestep the problem by simply discarding the non-converging simulations having a log-likelihood equal to `NA`. These are the summary statistics of the simulated distribution of the difference in deviances between models `m1` and `m2`, including the 134 `NA` values:

```
> summary(boot.res$Dif.dev)
   Min. 1st Qu.  Median    Mean 3rd Qu.    Max.    NA's
    0.0     0.2     0.7     1.4     1.8    12.5     134
```

We can also obtain the 95% confidence interval of this reference distribution:

```
> quantile(na.omit(boot.res$Dif.dev), probs=c(0.025,0.975))
   2.5%   97.5%
0.00296 6.55066
```

We note that the actual difference in deviance, `dev.models` = 7.93, is not included in the interval. More precisely, we can calculate the *p*-value of the actual difference in deviance as the proportion of the reference distribution having values equal to larger than `dev.models`:

```
> 1-ecdf(na.omit(boot.res$Dif.dev))(dev.models)
[1] 0.0127
```

This means that `dev.obs` leaves about 1.27% of the area of the cumulative distribution to its upper tail, which is smaller than the customary significance level of 5%. What we just did is equivalent to the log-likelihood Wilks test (command `anova`; see Chapter 3) without using the chi-square distribution, which is unreliable for GLMMs. Therefore, the small difference in deviance between the two models is statistically significant. Whenever two nested statistical models are significantly different, we should select the most general of the two (`m2` in this case), which includes the other as a special case. Should the two nested models being compared not be significantly different, we may select the simpler or more parsimonious one (`m1` in this case).

Having `m2` as our working model, we should now obtain the statistical significance of the model terms. Again, we should use parametric bootstrap because the *p*-values from the Wald statistics rely on large-sample approximations and other assumptions about the number of degrees of freedom of random effects that are always debatable for GLMMs (Bolker 2015). The package `afex` contains the command `mixed` that obtains the *p*-values of model terms by comparing the full model with all parameters with a reduced model in which each model term is in turn withheld or set to zero (Singmann and Keller 2019). The command is:

```
m2.boot=mixed(pres.abs~sex*length.c +(1+length.c|farm), method="PB", data=DF1,
family=binomial, progress=T, args_test = list(nsim = nboot, cl = cl))
```

and the model summary is:

```
> summary(m2.boot)
Generalized linear mixed model fit by maximum likelihood (Laplace Approximation) ['glmerMod']
 Family: binomial  ( logit )
Formula: pres.abs ~ sex * length.c + (1 + length.c | farm)
   Data: data

     AIC      BIC   logLik deviance df.resid
     830      863     -408      816      819

Scaled residuals:
   Min     1Q Median     3Q    Max
-6.783 -0.606  0.255  0.514  3.266

Random effects:
 Groups Name        Variance Std.Dev. Corr
 farm   (Intercept) 2.389    1.545
        length.c    0.329    0.573    0.65
Number of obs: 826, groups:  farm, 24

Fixed effects:
             Estimate Std. Error z value Pr(>|z|)
(Intercept)     1.367      0.360    3.80  0.00014 ***
sex1            0.279      0.113    2.46  0.01386 *
length.c        1.148      0.193    5.95  2.7e-09 ***
sex1:length.c   0.306      0.122    2.52  0.01188 *
---
Signif. codes:  0 '***' 0.001 '**' 0.01 '*' 0.05 '.' 0.1 ' ' 1

Correlation of Fixed Effects:
            (Intr) sex1  lngth.
sex1        0.100
length.c    0.483  0.290
sx1:lngth.c 0.130  0.274 0.154
```

The interpretation of fixed effects is as explained for binary GLMs in Chapter 9. `(Intercept)` is the logit of the sex of reference (0, or females) for an individual of average length (recall that this variable was standardized prior to the analysis); `sex1` is the differential of logit for `sex` = 1 (males) compared with the sex of reference. Therefore, the intercept for `sex1` in the logit scale is 1.367 + 0.279 = 1.596. Exponentiating the logits, we obtain the odds ratios for each sex: exp(1.367) = 3.92 for females and exp(1.596) = 4.93 for males. Females of average body length are 3.92 times more likely to be parasitized than not (and likewise for males). `length.c` is the slope in the logit scale for the sex of reference: when `length.c` increases by one standard deviation (`sd(DF1$length) = 19.6`), the logit increases by 1.148 units. `sex1.length.c` is the differential of slopes for `sex1` (males) compared with the sex of reference, which is again statistically significant. The slope for `sex1` in the logit scale is 1.148 + 0.306 = 1.454. We can use Gelman et al.'s (2014) "rule of four" (Chapter 9) to interpret the estimate slopes for each sex: 1.148 / 4 = 0.287 for females and 1.454 / 4 = 0.363 for males. Therefore, a change in one standard deviation of body length (= 19.6 cm) leads to an increase of 28.7% and 33.6% in the probability of being parasitized. Both the differences between sexes of the intercepts and slopes in the logit scale happen to be statistically significant. We can see that the variance of the random intercepts is 7.3 times larger than that of random slopes among farms. We can use the standard deviation of the random intercepts (1.545) to define a 95% confidence interval as [−1.545 × 1.95, +1.545 × 1.96] = [−3.03, +3.03], where 1.96 is the 95th percentile of a standard normal distribution. If the farms were a random sample from a large superpopulation of farms, the random intercept of a non-sampled farm would differ from the intercept of a "typical" farm shown in the fixed effects (1.367 for females, and 1.596 for males) by a differential in [−3.03, +3.03]. Thus, the random intercept at the farm level for females can be expected to be in [−3.03 + 1.367, +3.03 + 1.367] = [−4.4, +4.4] in the logit scale.

We should also obtain the 95% confidence intervals of the fixed effects using parametric bootstrap. This estimation may take a few minutes, depending on the size of the data set, the number of bootstrap simulations requested (500 should be a minimum), and the power of your computer. We use the command `bootMer` from `lme4`: `m2.CI.boot = bootMer(m2, FUN = function(x) fixef(x)$cond, nsim = nboot, parallel = "snow", cl = cl)`. This command uses the function `fixef(x)$cond`, which gives the fixed effects that are called "conditional model" in the output of the package `glmmTMB`. Had the model been fitted with `lme4`, we would have used `fixef(m2)`. If we wanted the 95% confidence intervals of the standard deviations of the random effects, we would need to replace `fixef(x)$cond` with `VarCorr(x)` in the function call (this is also valid in `lme4`). We need to process the object `m2.CI.boot` to obtain the 95% confidence intervals of the fixed effects. The simplest way is to write a small function to automate the process. The function `extract_ci` takes the output of `bootMer` stored in `m2.CI.boot` and calculates the 95% confidence intervals using the bootstrap percentile method (Efron and Tibshirani 1993) for a set of model parameters:

```
extract_ci <- function(b, conf = 0.95, ...) {
  p=length(b$t0) # number of parameters for which we want the 95% CI
  out=matrix(data=0,nrow=p,ncol=2) # matrix for the output
  colnames(out)=c("l2.5%", "u97.5%")
  rownames(out)=names(b$t0) # putting names of each bootstrapped parameter
  for (j in seq_len(p)) {
    out[j,]=boot.ci(b, conf, type = "perc", index = j, ...)$percent[4:5]
  }
  out=cbind(out, b$t0) # ads the parameter estimates
  colnames(out)[3]="estimate"
  return(out)
}
```

This is the actual function call: `m2.CI.res = data.frame(extract_ci(m2.CI.boot), se = summary(m2.boot)$bootSE)`.

```
> m2.CI.res
                     l2.5.   u97.5. estimate    se
(Intercept)          0.894  2.4630    1.649 0.390
sexmale             -1.033 -0.0768   -0.558 0.235
length.c             0.967  2.0056    1.456 0.258
sexmale:length.c -1.118 -0.1368   -0.611 0.249
```

Note that we also added the bootstrapped standard errors of the fixed effects, and transformed the output into a data frame so that it can be used as the input for a `ggplot`. But let's first examine `m2.CI.res`, and compare it with the standard 95% confidence intervals based on Wald's quadratic approximation of the likelihood function (Chapter 3):

```
> confint(m2, method="wald")[1:4,]
                    2.5 % 97.5 % Estimate
(Intercept)         0.782  2.346    1.564
sexmale            -0.994 -0.103   -0.548
length.c            0.894  1.865    1.379
sexmale:length.c -1.078 -0.127   -0.603
```

(the `[1:4,]` is to avoid displaying the confidence intervals of the random effects for which we did not obtain the bootstrapped equivalents). We can see that even for our very large sample size (n = 826), all bootstrapped confidence intervals differ from those based on the large-sample approximation. Their difference clearly becomes more important for smaller data sets. While we might use the command `tidy` from the package `broom.mixed` to gather the parameter estimates and their 95% asymptotic confidence intervals of model `m2` (see Chapter 13), we instead use the bootstrapped ones.

We need to work a little bit to put the random effects of model `m2` into a data frame suitable for plotting. We first extract the estimates and standard errors of the random intercepts and slopes with `rand.eff.m2 = as.data.frame(ranef(m2, condVar = T))`. The problem is that the random estimates and their variances are not in separate columns, but one on top of the other. One way to solve this problem is to split `rand.eff.m2` into a list `prov` of two data frames according to the variable `rand.eff.m2$term`, which indicates whether each coefficient is a random intercept or a random slope for each farm: `prov = split(rand.eff.m2, rand.eff.m2$term)`. Now we can reassemble the parts into a single data frame: `rand.eff.m2 = cbind(prov[["(Intercept)"]][,4:6], prov[["length.c"]][,5:6])`. This requires an explanation. `prov[["(Intercept)"]]` is the first half of the list `prov` involving the random intercepts, and `[4:6]` selects the columns of this data frame containing the farm label, the random intercepts, and their variances for each farm; `prov[["length.c"]][,5:6])` does likewise for the bottom part of the list `prov` containing the random slopes and their variances. We are going to display the standard errors as the square root of the variances at the moment of plotting each random effect. Then, both parts are joined as the data frame `rand.eff.m2`. We then provide meaningful names for the columns: `names(rand.eff.m2) = c("farms", "rand.int", "SE.rand.int", "rand.slope", "SE.rand.slope")`. Now we are finally ready to generate Fig. 15.3. Plots 15.3B and C show that those farms having a larger random intercept also tend to have a large random slope, as suggested by the positive correlation of the vectorial random effects in `summary(m2)`.

The R^2-like statistics (Nakagawa et al. 2017) obtained by the package `performance` show that the fixed and random effects of `m2` account for 54.1% of the variance of the response variable, and the fixed effects are associated with 12.5% of this variance.

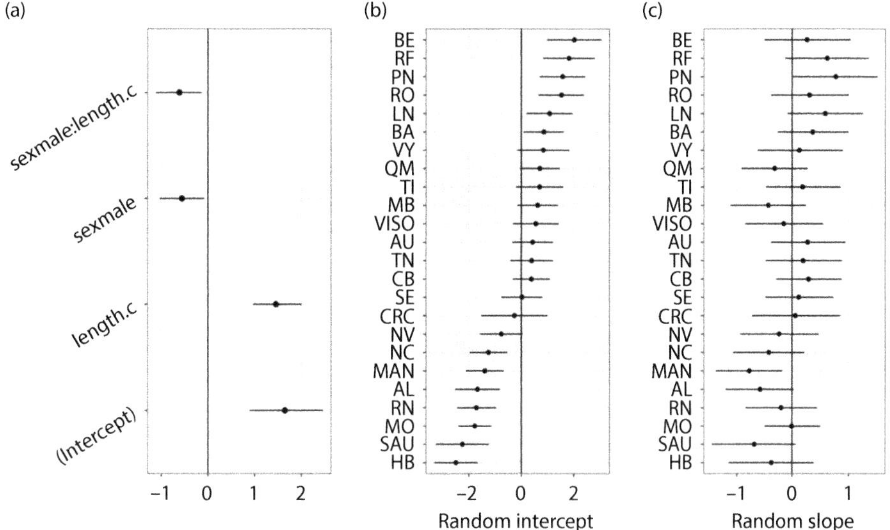

Fig. 15.3 Results of the frequentist binary GLMM `m2`. (a) Parameter estimates and 95% confidence intervals obtained by parametric bootstrap for the fixed effects. (b) Random intercepts and their standard errors for each farm, shown in increasing magnitude. (c) Random slopes for body length and their standard errors for each farm, shown in increasing order of the values of their intercepts.

```
> r2(m2)
# R2 for Mixed Models

  Conditional R2: 0.541
     Marginal R2: 0.125
```

We now need to assess the goodness of fit of `m2` with residual analysis based on randomized quantile residuals computed with the package `DHARMa` (Chapters 8 and 9). We start by gathering the information for the residual analysis using the command `augment` from the package `broom.mixed`: `resid.m2 = augment(m2)`. Next, we simulate the residuals, `res.m2 = simulateResiduals(fittedModel = m2, n = 1e3, integerResponse = T, refit = F, plot = F)`, that are converted from a uniform to a normal distribution (see Chapter 8) with `resid.m2$.std.resid = residuals(res.m2, quantileFunction = qnorm)`.

One feature that is missing are the Cook distances. Because we cannot obtain Cook distances with models fitted with the package `glmmTMB`, we need to calculate them ourselves. Had we fitted `m2` with `lme4`, we could have used the command `cooks.distance.estex` from the package `influence.ME` that works for all GLMMs fitted with `lme4` (see an example below). As with the parametric bootstrap, this is another small price to pay for using the `glmmTMB` package, which can model many more types of response variables than `lme4`. We need to write another function to automate the application of the Cook distance formula given in Section 4.5. The function `Cookdist.glmmTMB` takes any mixed model fitted with `glmmTMB` as input, and returns a vector of the Cook distances for each data point:

```
Cookdist.glmmTMB=function(model){
  coef.model=as.vector(unlist(fixef(model)$cond))
  mcov=solve(vcov(model)$cond)
  cook=vector(length = nrow(model$frame))
  for (i in 1:nrow(model$frame)){
    m=update(model, data=model$frame[-i,])
    refitted=as.vector(unlist(fixef(m)$cond))
    cook[i]= t(coef.model-refitted)%*% mcov %*% (coef.model-refitted)
  }
  return(cook)
}
```

Here, `coef.model` is the vector of the fixed-effect parameter estimates of the model, `mcov` is the inverse of the variance–covariance matrix of the fixed effects, and `cook` is a vector to store the Cook distances of each data point (see Chapter 4). The loop starting with `for` runs for as many data points as there are in the original data frame, fits a model `m` without the `i`th data point, stores the fixed-effect parameter estimates in the vector `refitted`, and then calculates the Cook distances for the `i`th data point. After loading the function in the working session, the actual function call is `resid.m2$Cook = Cookdist.glmmTMB(m2)`.

We need to write another function to calculate the Cook distances for each level of the random effect of model `m2`. This function is an adaptation of the previous one. It takes a GLMM (assuming that there is a single explanatory variable modeled as a random effect) fitted with package `glmmTMB` as input, and gives a vector of Cook distances as output.

```
Cookdist.glmmTMB.ranef=function(model){
  coef.model=as.vector(unlist(fixef(model)$cond))
  mcov=solve(vcov(model)$cond)
  ranef.name = model$modelInfo$grpVar
  levels.ranef=levels(model$frame[,ranef.name])
  cook=numeric(length(levels.ranef)) # vector to store results
  for (i in seq_along(levels.ranef)){
    keep.these = model$frame[ranef.name] != levels.ranef[i]
    m=update(model, data=model$frame[keep.these,])
    refitted=as.vector(unlist(fixef(m)$cond))
    cook[i]= t(coef.model-refitted)%*% mcov %*% (coef.model-refitted)
  }
  names(cook) <- levels.ranef
  return(cook)
}
```

The only changes with respect to the first function are `ranef.name = model$modelInfo$grpVar`, which gathers the name of the variable modeled as a random effect, and `keep.these = model$frame[ranef.name] != levels.ranef[i]`, which selects all but one levels of the random effect variable. After loading it in the working session, the actual call of the function `Cookdist.glmmTMB` is:

```
Cookdist.ranef.m2=as.data.frame(resid.m2$Cook=Cookdist.glmmTMB(m2))
names(Cookdist.ranef.m2)="Cook"
```

With `m2` being a binary GLMM, we can also calculate the the ROC curve, the AUC, and the Brier score (Chapter 9). We use the package `pROC` to obtain the ROC, `roc.m2 =`

`roc(response = DF1$pres.abs, predictor = fitted(m2))`, and its associated AUC:

```
> auc(roc.m2)
Area under the curve: 0.858
```

This means that `m2` can correctly predict 85.8% of a random sample of zeros and ones.

```
> BrierScore=mean((predict(m2,type='response')-DF1$pres.abs)^2)
> BrierScore
[1] 0.144
```

The closer the score is to zero, the better the predictive ability of the model. Now that we have all the components, we can show the residual analysis of model `m2` in Fig. 15.4.

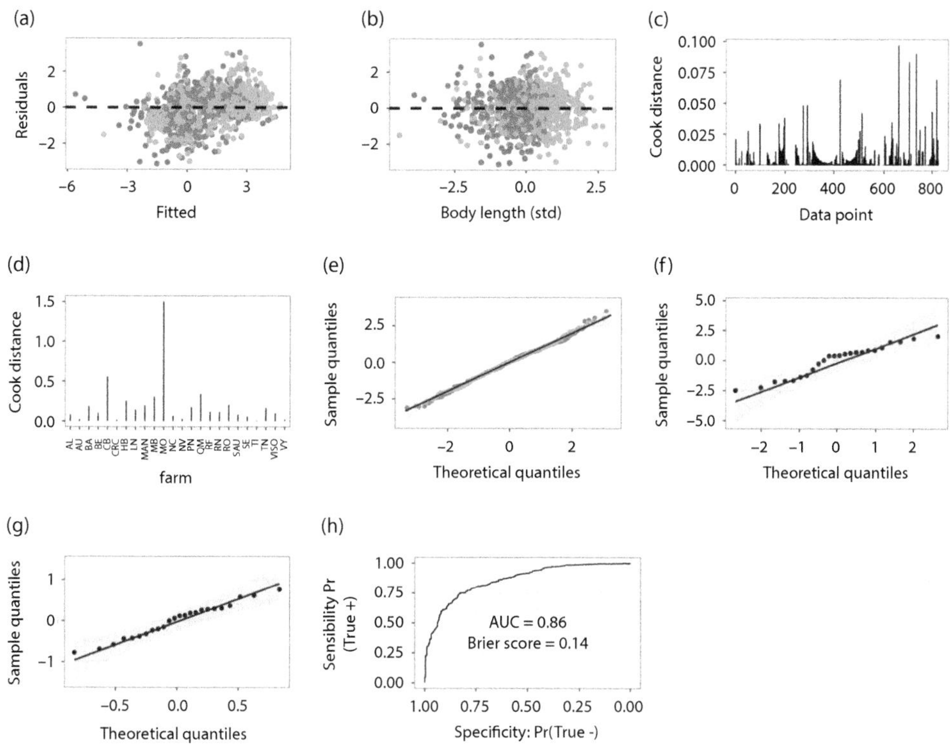

Fig. 15.4 Residual analysis of the frequentist binary GLMM `m2` based on randomized quantile residuals. Plots (a) and (b) show the relation between the residuals and the fitted values and to the explanatory variable (pink: girls). Plot (c) shows the Cook distances for each data point. Plots (e)–(g) are the quantile–quantile plots (and the 95% confidence band) for the residuals, and the estimates of the random intercepts and random slopes, respectively.

Figure 15.4a and b suggest that the simulated residuals seem to have a random scatter, and without any discernible trends for either sex. Figure 15.4c shows that there are four ($\approx$ 0.4%) data points with Cook distances larger than 0.05. We could examine the sensitivity of the parameter estimates of `m2` to these few points, but it is unlikely that omitting this very small percentage of points would have much effect on the parameter estimates of `m2`. Figure 15.4D, however, shows that farm MO had a much larger Cook distance than the other farms, something that requires further examination. Figures 15.4e, f, and g show that the residuals, and the random intercepts and slopes, are close to having normal distributions. Figure 15.4g displays the ROC curve and its related AUC. With the possible exception of Fig. 15.4d, we can conclude that `m2` has a very reasonable goodness of fit to the data.

Is there something special about farm MO?

```
> summary(DF1[DF1$farm=="MO",c("pres.abs", "sex")])
    pres.abs          sex
 Min.   :0.000   female:126
 1st Qu.:0.000   male  : 83
 Median :0.000
 Mean   :0.392
 3rd Qu.:1.000
 Max.   :1.000
```

```
> summary(DF1[DF1$farm!="MO",c("pres.abs", "sex")])
    pres.abs          sex
 Min.   :0.000   female:252
 1st Qu.:0.000   male  :365
 Median :1.000
 Mean   :0.733
 3rd Qu.:1.000
 Max.   :1.000
```

This farm contains about 25% of the data, and its parasitism prevalence is almost 50% less than in the other 23 farms. It should be obvious that omitting 25% of the data that is substantially different from the other 75% is bound to change the parameter estimates. There is obviously something that sets farm MO apart from the other farms. The random effect farm allows us to take into account the sampling design leading to a randomized blocks design, but it cannot suggest any of the proximate reasons for why farms differ from each other. For this, we need to make separate, additional analyses including other explanatory variables measured at the farm level.

Finally, we can display the relations predicted by `m2` as conditional plots in the scale of the observations. We use package `ggeffects` to obtain the predicted relations as `pred.m2.f = ggpredict(m2, type = "fixed", terms = c("length.c [all]", "sex"))` and `pred.m2.r = ggpredict(m2, type = "random", terms = c("length.c[all]", "sex"))`. While `pred.m2.f` includes only the fixed effects, `pred.m2.r` includes both fixed and random effects, leading to wider 95% confidence intervals, as shown in Fig. 15.5.

We can also obtain conditional plots at the farm level with `pred.m2.r.farms = ggpredict(m2, type = "random", ci.lvl = NA, terms = c("length.c [all]", "sex", "farm"))`. `ci.lvl = NA` prevents the computation of the 95% confidence intervals that would constitute a case of "over-plotting" or displaying too much information in Fig. 15.6.

Let's first appreciate in Fig. 15.6 a nearly "magical" aspect of GLMMs: we obtained a predictive equation for each sex at a global scale, even for those farms for which there was hardly any data for one sex or even both sexes (e.g., farms CRC, NV, SAU, and TI). This is the borrowing strength effect (Chapter 13) in action, the data from the other farms allow parameters to be estimated for those farms with few data. We can see the parameter estimates at the farm level with `coef(m2)`. While you should not bet your life on the parameter estimates of poorly sampled farms, having them is barely short of a "statistical miracle." There is obviously a large degree of heterogeneity in the predicted relations between farms, as suggested by the variance of the random intercepts shown and discussed above.

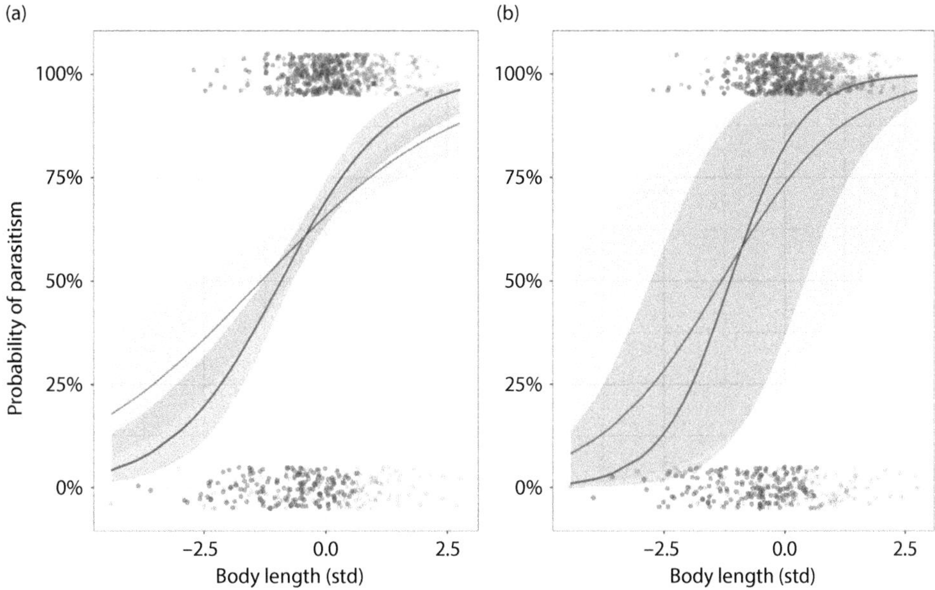

Fig. 15.5 Conditional plots of the frequentist binary GLMM `m2` including only the fixed effects (a) and both fixed and random effects (b), showing the predicted relation between the probability of parasitism in relation to (standardized) body length and sex. The colors red and blue correspond to females and males, respectively. All points are vertically jittered by adding a random value to allow better visualization. The bands correspond to the 95% confidence intervals.

15.2.2 *Binary GLMM with a randomized block design: Bayesian models*

We fit the Bayesian analogues of models `m1` and `m2` using the package `brms`. Equation 15.1 is the equation for the more general model with group-level intercept and slopes, whose formula is `formula.m2.brms = bf(pres.abs ~ length.c*sex + (1 + length.c|farm))`. Let's obtain its default model priors:

```
> get_prior(formula.m2.brms, data=DF1,family='Binomial')
                prior     class             coef group resp dpar nlpar bound       source
               (flat)         b                                                   default
               (flat)         b         length.c                             (vectorized)
               (flat)         b length.c:sexmale                             (vectorized)
               (flat)         b          sexmale                             (vectorized)
               lkj(1)       cor                                                   default
               lkj(1)       cor                   farm                       (vectorized)
 student_t(3, 0, 2.5) Intercept                                                   default
 student_t(3, 0, 2.5)        sd                                                   default
 student_t(3, 0, 2.5)        sd                   farm                       (vectorized)
 student_t(3, 0, 2.5)        sd        Intercept  farm                       (vectorized)
 student_t(3, 0, 2.5)        sd         length.c  farm                       (vectorized)
```

We encountered and discussed similar priors in our first mixed model in Section 13.9. We can see the default weakly informative priors for the population-level effects (top bracket, and the prior for the intercept), and for the group-level effects (lower bracket for the standard deviations of intercepts and slopes, and middle bracket for their correlation). The default priors for `formula.m1.brms` would be similar, but they would not include priors for the standard deviation of the group-level slopes, or for the correlation between group-level slopes and intercepts. For the group-level parameters, we are going to use the

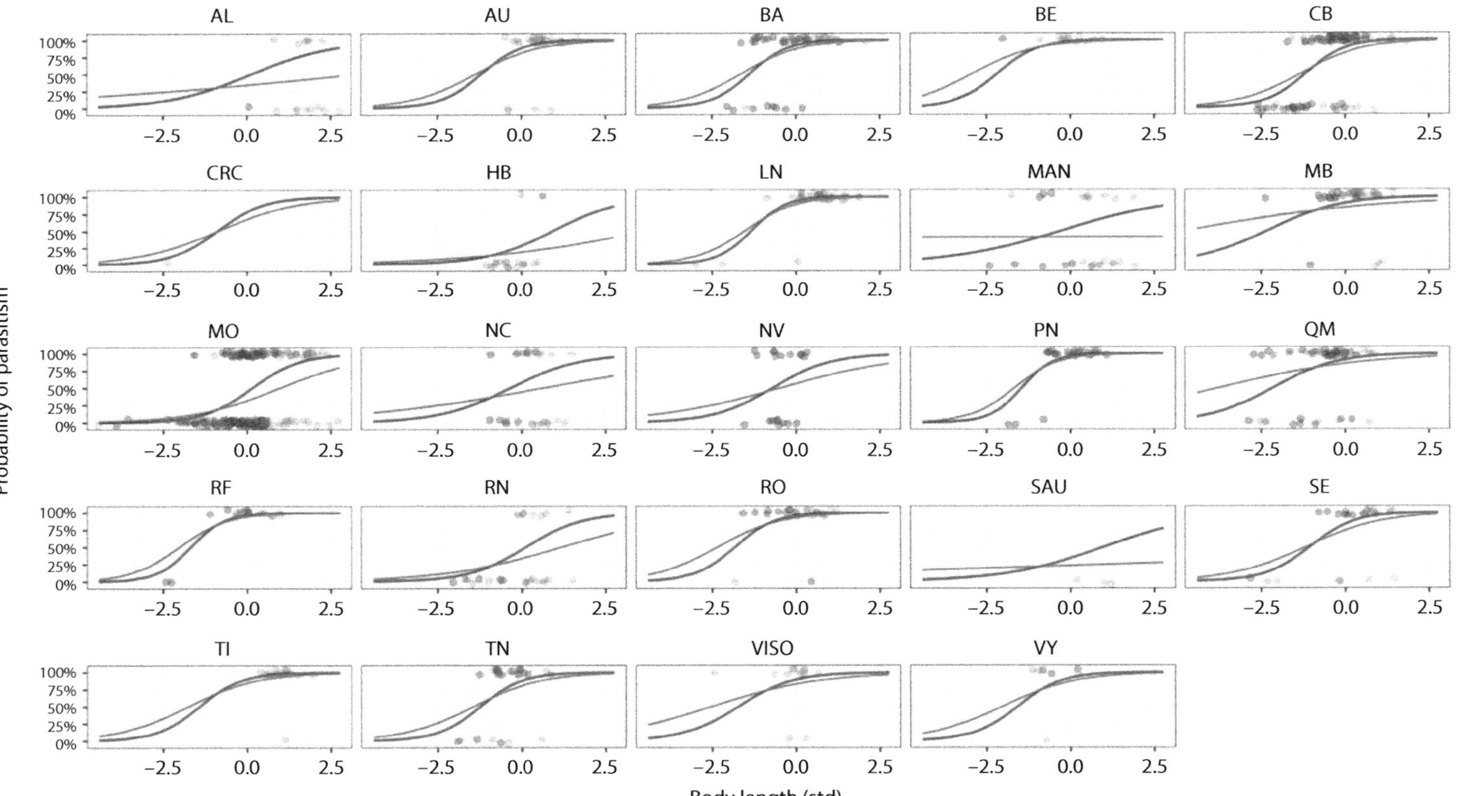

Fig. 15.6 Conditional plots of the frequentist binary GLMM `m2` at the farm level including both fixed and random effects, showing the predicted relation between the probability of parasitism in relation to (standardized) body length and sex. The colors red and blue correspond to females and males, respectively. All points are vertically jittered by adding a random value to allow better visualization.

default priors for the standard deviations of slopes and intercepts, and a weakly informative prior for the parameter of the LKJ distribution involved in their correlation (see Chapter 13 and Fig. 13.18). The meaning of the default prior `student_t(3,0,2,5)` was discussed in Chapter 13. For the population-level parameters, we are going to use the default priors for the intercept and slope of the reference group (females), and we are going to set weakly informative priors for the difference in intercepts (`sexmale`) and in slopes (`length.c:sexmale`). Counter-intuitive but true, recall that priors with a smaller standard deviation are less (rather than more) informative in the logit link scale (see Chapter 9). The priors for the two Bayesian binary GLMMs are:

```
prior.m2.brms = c(prior(normal(0, 1), class = b, coef="sexmale"),
                  prior(normal(0, 1), class = b, coef="length.c:sexmale"),
                  prior(lkj(0.5), class=cor))
prior.m1.brms = c(prior(normal(0, 1), class = b, coef="sexmale"),
                  prior(normal(0, 1), class = b, coef="length.c:sexmale"))
```

The models to be fitted are `m1.brms = brm(formula = formula.m1.brms, data = DF1, family = bernoulli, chains = 3, prior = prior.m1.brms, future = T, control = list(adapt_delta = 0.99), warmup = 1000, iter = 3000)` and `m2.brms = brm(formula = formula.m2.brms, data = DF1, family = bernoulli, chains = 3, prior = prior.m2.brms, future = T, control = list(adapt_delta = 0.99), warmup = 1000, iter = 3000)`. Let's first examine the summary of `m1.brms`:

```
> summary(m1.brms)
 Family: bernoulli
  Links: mu = logit
Formula: pres.abs ~ length.c * sex + (1 | farm)
   Data: DF1 (Number of observations: 826)
  Draws: 3 chains, each with iter = 3000; warmup = 1000; thin = 1;
         total post-warmup draws = 6000

Group-Level Effects:
~farm (Number of levels: 24)
              Estimate Est.Error l-95% CI u-95% CI Rhat Bulk_ESS Tail_ESS
sd(Intercept)     1.71      0.34     1.17     2.48 1.00     1226     1935

Population-Level Effects:
                 Estimate Est.Error l-95% CI u-95% CI Rhat Bulk_ESS Tail_ESS
Intercept            1.51      0.42     0.68     2.34 1.00      989     1730
length.c             1.45      0.19     1.09     1.81 1.00     2988     3792
sexmale             -0.59      0.21    -1.00    -0.17 1.00     4239     4519
length.c:sexmale    -0.67      0.22    -1.10    -0.24 1.00     3145     3681
```

Let's recall that we first need to decide if it is appropriate to also include group-level slopes (this would be `m2.brms`). At this point, we should just note that in `summary(m1.brms)` the `Rhat` values are all equal to 1.00, and that the effective sample sizes are very large for all parameters, suggesting that `m1.brms` has converged to a stationary posterior distribution. Once we select the appropriate model, we will make a better assessment of model convergence.

The summary for the other binary GLMM model is:

```
> summary(m2.brms)
 Family: bernoulli
  Links: mu = logit
Formula: pres.abs ~ length.c * sex + (1 + length.c | farm)
   Data: DF1 (Number of observations: 826)
  Draws: 3 chains, each with iter = 3000; warmup = 1000; thin = 1;
         total post-warmup draws = 6000

Group-Level Effects:
~farm (Number of levels: 24)
                          Estimate Est.Error l-95% CI u-95% CI Rhat Bulk_ESS Tail_ESS
sd(Intercept)                 1.77      0.38     1.15     2.66 1.00     1889     3205
sd(length.c)                  0.70      0.28     0.19     1.32 1.00     1340     1028
cor(Intercept,length.c)       0.52      0.31    -0.22     0.95 1.00     2210     1472

Population-Level Effects:
                 Estimate Est.Error l-95% CI u-95% CI Rhat Bulk_ESS Tail_ESS
Intercept            1.61      0.45     0.76     2.50 1.00     1319     2428
length.c             1.45      0.27     0.94     2.01 1.00     3321     3742
sexmale             -0.53      0.23    -0.97    -0.08 1.00     8614     4823
length.c:sexmale    -0.60      0.24    -1.07    -0.12 1.00     7262     4895
```

The summary of `m2.brms` suggests that this model also appears to have converged to a stationary posterior distribution. The interpretation of the means of the marginal posterior distributions of the model parameters shown in `summary(m2.brms)` is exactly the same as provided for the frequentist analogue `m2` (see Section 15.2.1), and it will not be repeated here.

We need to add the leave-one-out-criterion (`loo`; Chapter 7) to the fitted models (see examples in Chapter 10), and then perform the model selection:

```
> m1.brms=add_criterion(m1.brms, criterion="loo")
> m2.brms=add_criterion(m2.brms, criterion="loo")
> loo_compare(m1.brms,m2.brms)
        elpd_diff se_diff
m2.brms  0.0       0.0
m1.brms -3.3       3.3
```

The expected log-pointwise predictive difference (`elpd`) when multiplied by –2 equals the WAIC (Chapter 7). Model `m2.brms` has a larger `elpd` than `m1.brms`. However, the magnitude of their difference is similar to the standard error of this difference, even considering the warnings given in Vehtari et al. (2017) to interpret the latter. It seems unclear which model to select. Without any strong or decisive reasons, we select `m2.brms` because its summary appears to suggest slightly better model convergence than `m1.brms`. This is admittedly weak, but we have to make a choice at this point.

Let's examine the convergence of `m2.brms` using the package `bayesplot`. Figure 15.7 shows that the three chains of `m2.brms` converged to common marginal posterior distributions for each parameter, in agreement with the `Rhat` values shown in its summary. The command `variables(m2.brms)` gives the names of all the parameters for which we obtained posterior distributions. This includes posterior distributions of the group-level parameters (intercept and slope, and their correlations) estimated at the farm level. While there is no direct frequentist equivalent to these posterior distributions of group-level effects, we could find an analogue by replacing `fixef(m2)` with `ranef(m2)` in the command `m2.CI.boot` that obtains the parametric bootstrap sampling distributions.

Figure 15.8 shows that the sampled estimates in the chains of many (but not all) parameters of `m2.brms` have very small autocorrelation, and can be considered to be statistically independent estimates. It is not too worrisome for the quality of the inference that the

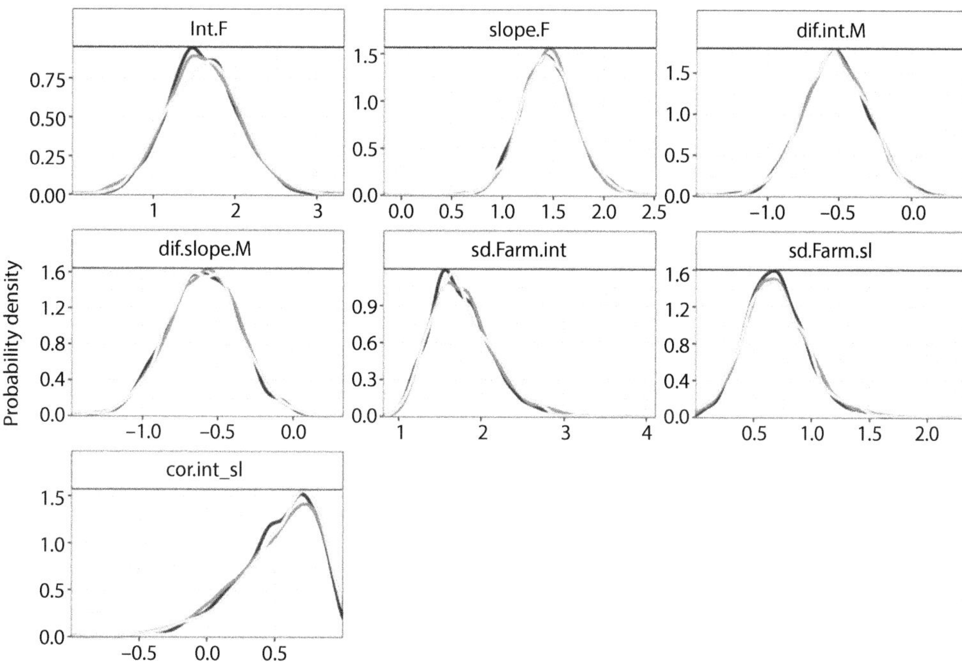

Fig. 15.7 Convergence of the three chains to a common stationary posterior distribution for the population- and group-level parameters of the Bayesian binary GLMM `m2.brms`. Int.F is the intercept of the reference sex (females), slope.F is the slope of the reference sex, dif.int.M is the differential of intercepts between sexes, dif.slope.M is the differential of slopes between sexes, sd.Farm.int and sd.Farm.sl are the standard deviations of the farm-level intercepts and slopes, and cor.int_sl is the correlations between farm-level slopes and intercepts.

population-level intercept for females and the standard deviations of the group-level slopes and intercepts have autocorrelations greater than 0.1 for small lags. This is because we still have large effective sample sizes (ESS). We could probably fix this small glitch by increasing the default value of `thin` when running the model from its default value of one to (say) four. This change would make the command longer to run as it would only accept one out of every four sampled parameter values.

We can finally show the posterior distributions of the model parameters shown in `summary(m2.brms)` in Fig. 15.9. The wiggly plots in Fig. 15.9 are not theoretical but MCMC-generated probability density distributions. The limits of the 95% credible intervals are shown in `summary (m2.brms)` above. We use Gelman et al.'s (2018) metric to obtain the percentage of variation of the response variable accounted for by the model, and its 95% credible interval:

```
> bayes_R2(m2.brms)
   Estimate Est.Error  Q2.5 Q97.5
R2    0.347    0.0198 0.306 0.384
```

We assess the goodness of fit of model `m2.brms` using randomized quantile residuals (Chapter 8). To obtain these residuals, we first obtain the posterior predictive distribution (Eq. 8.15): `post.pred.m2.brms = predict(m2.brms, nsamples = 1e3, summary = F)` (see the example in Chapter 9). We then use the package `DHARMa` to obtain the randomized quantile residuals: `qres.m2.brms = createDHARMa(simulatedResponse =`

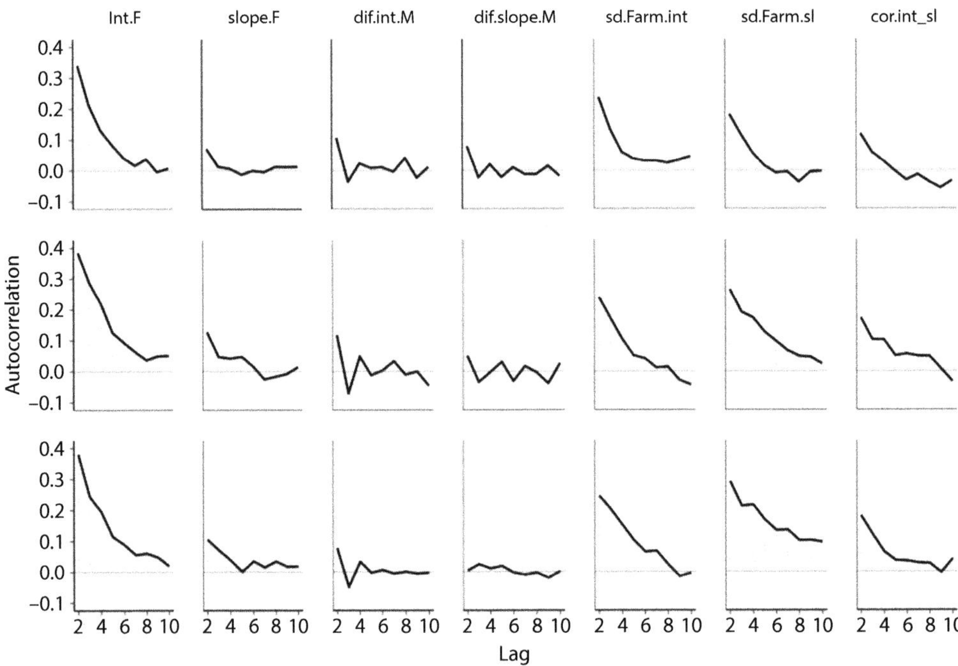

Fig. 15.8 Autocorrelation plots of the population- and group-level parameters for the three chains used to fit the Bayesian binary GLMM `m2.brms`. The very low autocorrelation for all parameters and chains suggest that the sampled values of the posterior distributions can be considered largely statistically independent. The names of the parameters are as in Fig. 15.7.

`t(post.pred.m2.brms), observedResponse = DF1$pres.abs, fittedPredictedResponse = apply(post.pred.m2.brms, 2, median), integerResponse = T)`. These have a uniform distribution (see Chapter 8), and are converted to a normal distribution with `res.m2.brms = data.frame(res = qnorm(residuals(qres.m2.brms)))`. We next create a data frame with median values of the residuals, the average fitted values, the Pareto *k* of LOO-CV (see Chapter 8), and the explanatory variables:

```
res.m2.brms=cbind(res.m2.brms,
                  DF1[,c("length.c", "sex", "farm")],
                  fitted=fitted(m2.brms, ndraws=1000)[,1],
                  pareto=loo(m2.brms, pointwise=T)$diagnostics$pareto_k)
```

To assess the normality of the group-level effects, we need to gather the mean farm-level estimates of intercept and slopes in a separate data frame, `stats.ranef.m2.brms = as.data.frame(ranef(m2.brms))`, whose variable names are shortened with `names(stats.ranef.m2.brms) = c("mean.int", "se.int", "Q2.5.int", "Q97.5.int", "mean.slope", "se.slope", "Q2.5.slope", "Q97.5.slope")`. We also add the farm names to the data frame: `stats.ranef.m2.brms$school = unique(DF1$farm)`. We now have all the components needed for the plots of Fig. 15.10.

We can see that the fit of model `m2.brms` to the data is excellent: the residuals appear random and without evident trends (Fig. 15.10a and b), and they have a normal distribution (Fig. 15.10d). There are only two data points outside the LOO-CV "good" range (Fig. 15.10c). Both the group-level intercepts and slopes appear to have normal distributions. We could hardly ask for more.

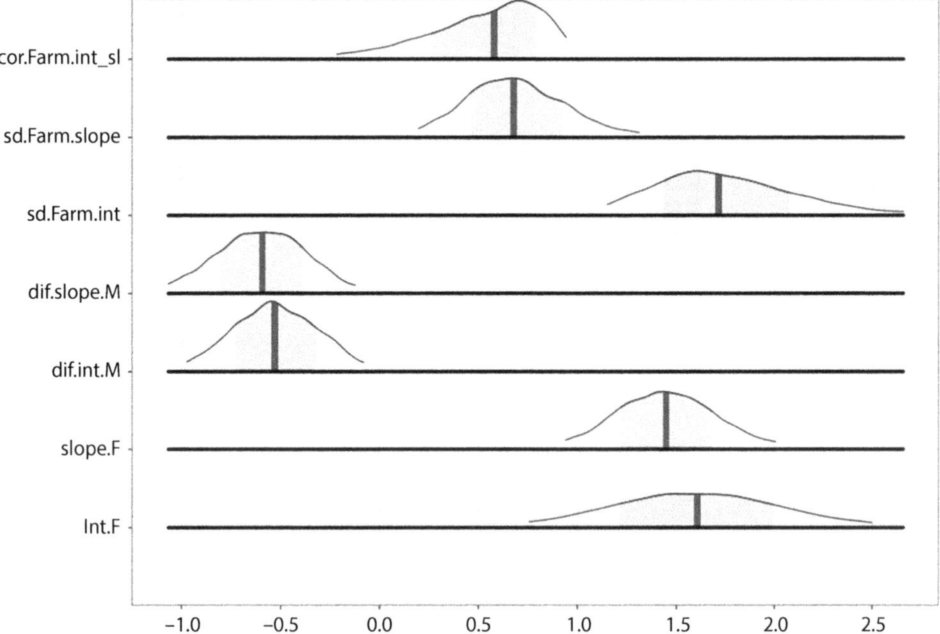

Fig. 15.9 Posterior distributions of the population- and group-level parameters of the Bayesian binary GLMM `m2.brms` showing their 62% (dark blue) and 95% credible intervals. The names of the parameters are as in Fig. 15.7.

We can use the posterior predictive distribution (Chapter 7) to assess the capacity of the model to reproduce features of our data. As in Chapter 9, we are going to apply the function `presences = function(x) mean(x == 1)` to the posterior predictive distribution to assess whether the model can reproduce the observed prevalence of parasitism as a typical prediction in a posterior predictive check. Of course, we could have applied many other functions to the posterior predictive distribution of `m2.brms` (see the examples in Chapter 13). We can also calculate the posterior distributions of the AUC and the Brier score using the code explained in Chapter 9:

```
nsamp=1e3
ROC.m2.brms=vector(length=nsamp)
Brier.m2.brms=vector(length=nsamp)
for (i in 1:nsamp){
  temp=auc(roc(predictor=post.pred.m2.brms[i,],response=DF1$pres.abs))
  ROC.m2.brms[i]=unlist(temp[grep("0", temp)])
  Brier.m2.brms[i]= mean((post.pred.m2.brms[i,]-DF1$pres.abs)^2)}
```

The posterior distributions of both metrics are then put in a data frame, `AUC.Brier = data.frame (AUC = ROC.m2.brms, Brier = Brier.m2.brms)`, to plot them in Fig. 15.11. The posterior predictive check (Fig. 15.11a) shows that the model can reproduce the density plot of the observed data as a "typical" prediction (i.e., it was not a lucky stroke) based on the posterior distributions of its parameter values. Unlike the single values without even a standard error obtained in the frequentist binary GLMM, we have here true posterior distributions of the AUC and Brier score (Fig. 15.11b and c) that we can use to obtain 95% credible intervals and any other summary statistics that we may wish.

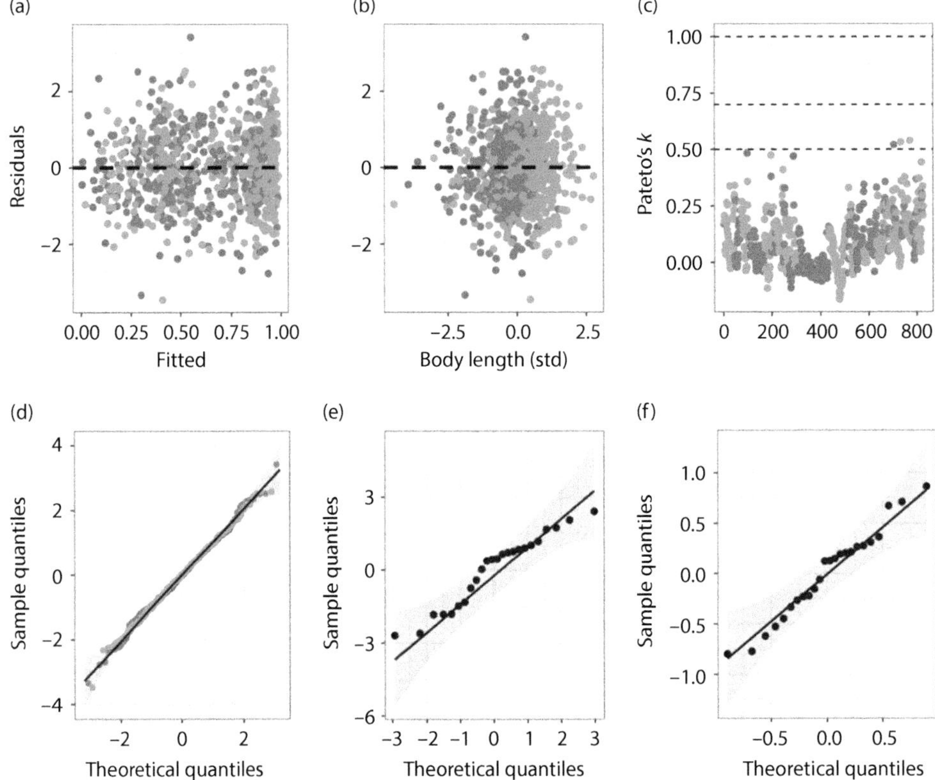

Fig. 15.10 Residual analysis of the Bayesian binary GLMM `m2.brms` based on randomized quantile residuals. Plots (a) and (b) show the relation between the residuals and the fitted values and to the explanatory variable (pink: girls). Plot (c) is the Pareto *k* cross-validation statistic for each data point. Plots (d)–(f) are the quantile–quantile plots (and their 95% credible band) for the residuals, and the estimates of the random intercepts and random slopes, respectively.

The final aspect is to visualize the conditional plots and the posterior distributions of the group-level effects. First, we need to obtain the conditional predictions using the package `ggffects`, `cond.pred.m2brms.fixef = ggpredict(m2.brms, type = "fixed", terms = c("length.c [all]", "sex"))`, including only the population-level effects, and use `cond.pred.m2brms.ranef = ggpredict(m2.brms, type = "random", terms = c("length.c [all]", "sex"))` to include both the population- and group-level effects. We then need to obtain the posterior distributions of the group-level effects with `post.gr_lev.m2.brms = posterior_samples(m2.brms, pars = "^r")`. This data frame (`str(post.gr_lev.m2.brms)`) will have 6,000 rows (= 3 chains × 2,000 MCMC draws) and 48 columns (= 24 farms × (intercept + slope)). We then need to reformat it to the the "long format" by stacking the 6,000 draws of each farm on top of each other. We start by separating the posterior distributions of slopes and intercepts with `post.gr_lev.m2.brms.INT = post.gr_lev.m2.brms[,1:length(unique(DF1$farm))]` and `post.gr_lev.m2.brms.SLOPES = post.gr_lev.m2.brms[,(length(unique(DF1$farm))+1): (2 * length(unique(DF1$farm)))]`. There are 24 farms (= `length(unique(DF1$farm))`), and hence the first command takes columns 1–24, and the second columns

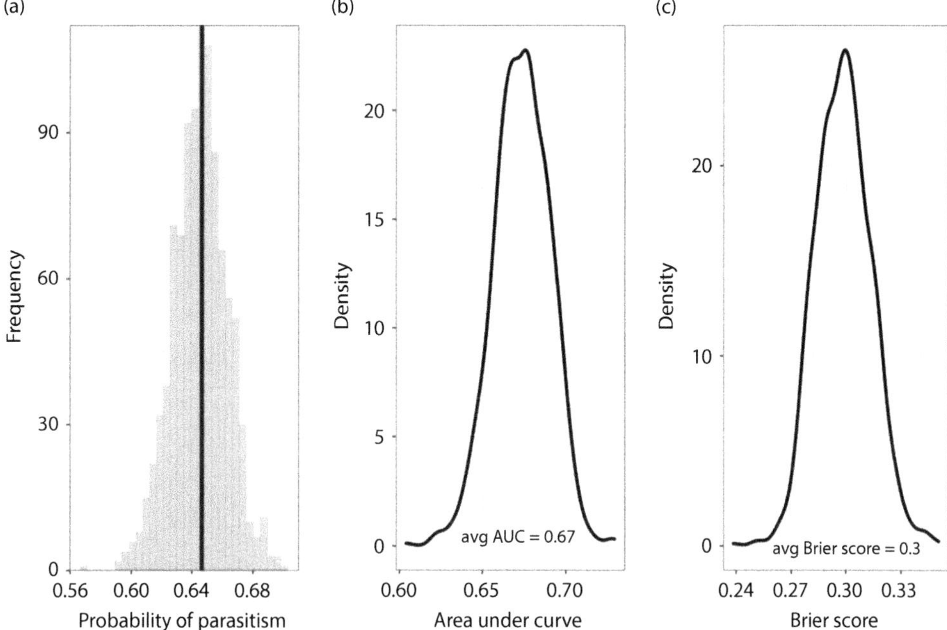

Fig. 15.11 (a) Posterior predictive distribution of the Bayesian binary GLMM `m2.brms` for the incidence of parasitism, with the observed value shown by the thick line. (b, c) Posterior distributions for the AUC and the Brier score for `m2.brms`.

25–28 of `post.gr_lev.m2.brms`. We then give better names to the columns of the two resulting data frames with `names(post.gr_lev.m2.brms.INT) = names(post.gr_lev.m2.brms.SLOPES) = unique(DF1$farm)`. Finally, we use the package `reshape2` to stack the 6,000 draws of the intercepts and slope of each farm to form a single column:

```
post.gr_lev.m2.brms.INT=melt(post.gr_lev.m2.brms.INT, value.name="intercept")
post.gr_lev.m2.brms.SLOPES=melt(post.gr_lev.m2.brms.SLOPES, value.name="slope")
```

Figure 15.12a and b shows the conditional curves predicting the relation between the incidence of parasitism and body length and sex. Note that the credible interval of the curves including both population- and group-level effects are so wide as to be nearly worthless. This shows that farm-level heterogeneity can easily blur the predicted effect of the population-level effects, which suggests that we would need to include other explanatory variables measured at the farm level to account for and explain the farm-level heterogeneity. Figure 15.12c and d display the posterior predicted distributions of farm-level differentials of intercepts and slopes, respectively. Recall that we can also obtain specific group-level posterior distributions as we did in Chapter 13. These are true posterior probability distributions, not just point estimates and standard error as in the frequentist binary GLMM. Note that the scales of Figure 15.12c and d differ greatly, mirroring the differences in the standard deviations of group-level intercepts and slopes shown in `summary(m2.brms)`.

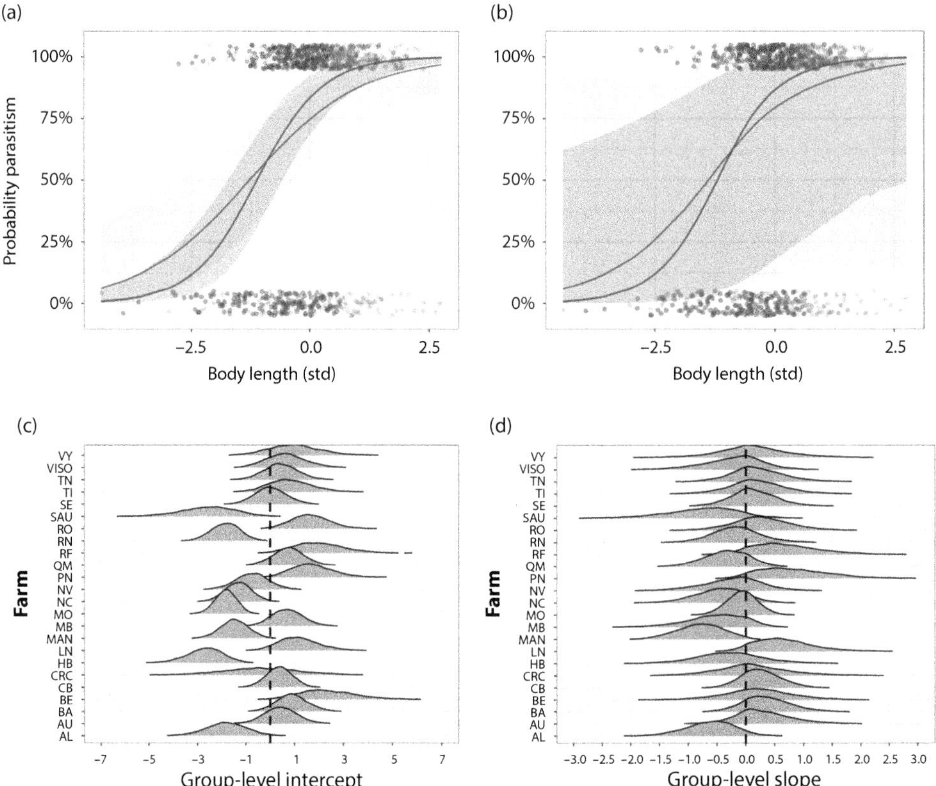

Fig. 15.12 Conditional plots of the Bayesian binary GLMM `m2.brms` including only the population-level effects (a) and both population- and group-level effects (b), showing the predicted relation between the probability of parasitism in relation to (standardized) body length and sex. The colors red and blue correspond to females and males, respectively. All points are vertically jittered by adding a random value to allow better visualization. The bands correspond to the 95% credible intervals. Plots (c) and (d) show the posterior distributions of group-level intercepts and slopes for each farm.

15.3 Gaussian GLMM with a repeated measures design

THE DATA IN CONTEXT: Franco-Trecu et al. (2014) studied the comparative trophic ecology of two sympatric marine mammals (South American fur seal and sea lion, referred to as SAFS and SASL) over two years on the coast of Uruguay. They measured the isotopic ratios of ^{13}C at the individual and population levels in the whiskers of individuals of both sexes and species. Being continually growing tissue, the amount of ^{13}C at different lengths from a whisker's base reflects the importance of carbon from land sources consumed at the time when the whisker's amino acids were synthesized. Franco-Trecu et al.'s (2014) sampling of each whisker at known lengths from its base amounted to obtaining individual profiles of ^{13}C over time. While not a designed experiment, the sampling design is equivalent to a repeated measures design (Chapter 14).

EXPLORATORY DATA ANALYSIS: Let's start by importing and examining the data:

```
> DF2=read.csv("whiskers chapter 15.csv", header=T)
> summary(DF2)
      ind           sp          sex          length            dC
 SASLf6 : 48   SAFS:210   f:256   Min.   :  2.0   Min.   :-16.6
 SAFSf9 : 44   SASL:204   m:158   1st Qu.: 26.0   1st Qu.:-15.2
 SASLm12: 28                      Median : 49.0   Median :-14.2
 SASLm16: 26                      Mean   : 54.0   Mean   :-14.1
 SAFSm3 : 24                      3rd Qu.: 76.8   3rd Qu.:-13.0
 SAFSf11: 23                      Max.   :164.0   Max.   :-12.2
 (Other):221
```

The categorical variable `id` shows that we have 48, 44, and 28 values for the first three individuals listed, and that we have a total of 210 + 204 = 414 values in the entire data set. `dC` is shown as negative values because of how chemists measure ^{13}C. We have two categorical explanatory variables, `species` and `sex`, and we have replicate individual profiles of ^{13}C for each combination of levels of these factors. We can easily count the number of females and males for which we have data:

```
> unique(DF2$ind)
 [1] SAFSf10 SAFSf11 SAFSf13 SAFSf14 SAFSf9  SAFSm17 SAFSm1  SAFSm2  SAFSm3  SASLf4  SASLf5  SASLf6
[13] SASLf7  SASLf8  SASLm12 SASLm15 SASLm16
```

We then have five profiles for females of both species, and four profiles for SAFS males and three for SASL. We use the package `doBy` to obtain the total lengths (in mm) of the 17 individual profiles:

```
> summaryBy(length~ ind, data=DF2, FUN=max)
       ind length.max
1  SAFSf10         96
2  SAFSf11        124
3  SAFSf13        133
4  SAFSf14        135
5   SAFSf9        164
6   SAFSm1         78
7  SAFSm17        145
8   SAFSm2        123
9   SAFSm3        145
10  SASLf4         99
11  SASLf5        105
12  SASLf6        137
13  SASLf7        104
14  SASLf8        105
15 SASLm12        139
16 SASLm15        102
17 SASLm16        144
```

We can obtain summary statistics for the combinations of sex and species as follows:

```
> summaryBy(dC~ sp+sex, data=DF2, FUN=c(mean, sd, length))
    sp sex dC.mean dC.sd dC.length
1 SAFS   f   -15.3 0.436       126
2 SAFS   m   -14.9 0.492        84
3 SASL   f   -12.9 0.353       130
4 SASL   m   -13.2 0.372        74
```

We can see that difference in the average `dC` between sexes changes sign depending on the species: females > males for SASL and the other way around for SAFS. This suggests that we should include an interaction between these categorical explanatory variables in our model. We should also examine the profiles of `dC` at the individual level in Fig. 15.13. We can see that the individual profiles of negative `dC` (to make things simpler) differ between individuals in their slopes and intercepts. Of course, the set of values of each individual profile are not replicates because they are not statistically independent of each other (Chapter 14).

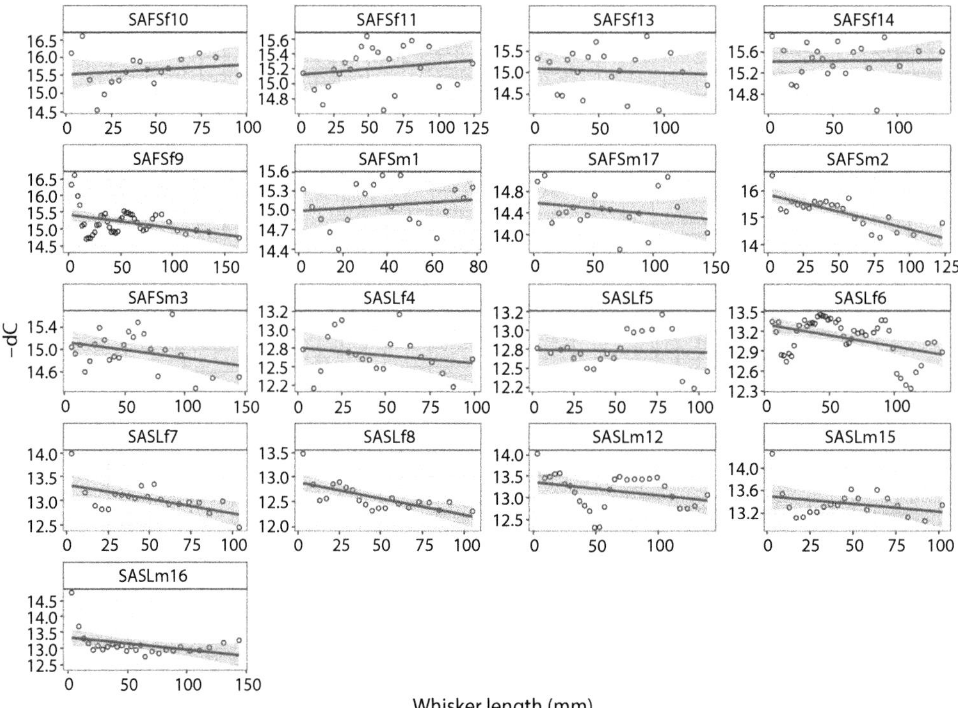

Fig. 15.13 Individual profiles of negative `dC` for different lengths from the whisker base as a sequence of samples along a temporal axis. The fitted lines and their 95% confidence intervals are shown to aid the interpretation.

We finally need to decide which probability distribution will be used in the likelihood part of the statistical models. We consider three candidate distributions to model negative `dC`: normal, lognormal and gamma (recall that the last two distributions are defined for strictly positive real values of the response variable). As before, the evaluation uses the package `fitdistrplus`, and starts by fitting each candidate distribution:

```
norm.whisker=fitdist(-DF2$dC,"norm")
lnormal.whisker=fitdist(-DF2$dC,"lnorm")
gamma.whisker=fitdist(-DF2$dC,"gamma")
```

and then comparing their CDF and Q–Q plots in Fig. 15.14. Figure 15.14a and b show that the three candidate probability distributions are very similar in their capacity to model the response variable (negative `dC`). The reason is that the response variable is plurimodal (Fig. 15.14c), with different modes for the four combinations of sex and species. The similarity of the three candidate probability distribution can be seen with:

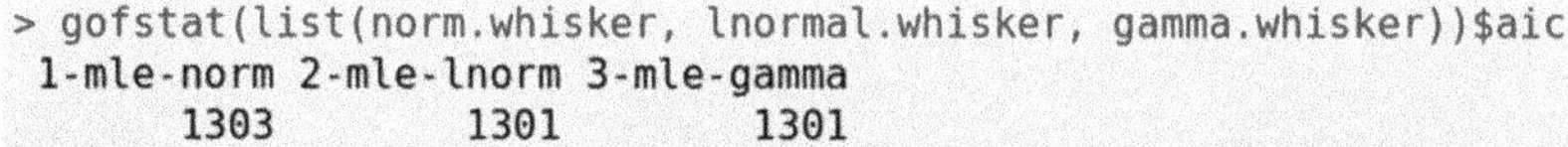

```
> gofstat(list(norm.whisker, lnormal.whisker, gamma.whisker))$aic
 1-mle-norm 2-mle-lnorm 3-mle-gamma
       1303        1301        1301
```

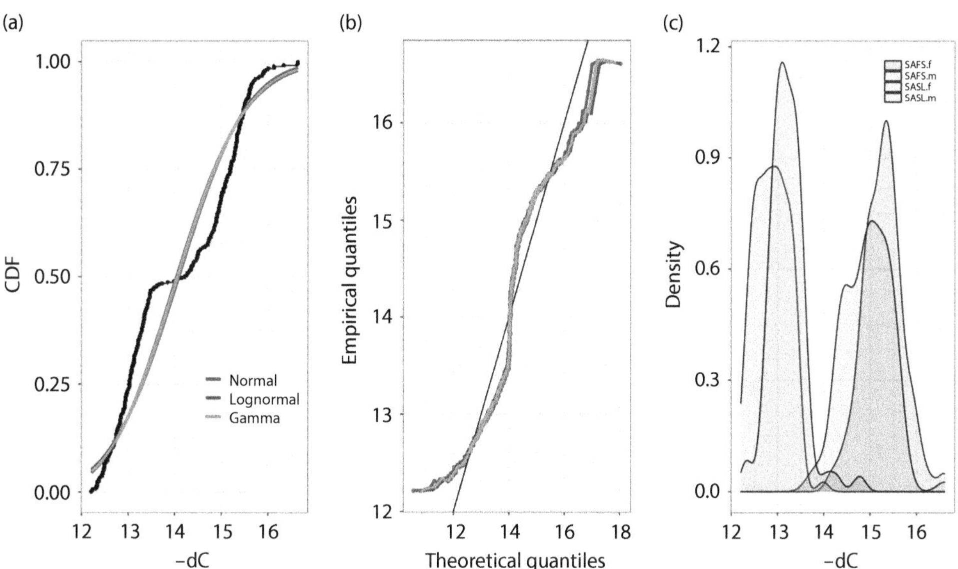

Fig. 15.14 Goodness-of-fit assessment of the response variable negative `dC` showing the cumulative distribution function (a) and the quantile–quantile plot comparing the normal, lognormal, and gamma distributions (b), and the density plots for the four groups formed by the 2 × 2 combinations of sex and species. Plots generated with the package `fitdistrplus`.

We therefore choose the normal distribution.

THE STATISTICAL MODEL:

$$Y \sim \text{Normal}(\mu_Y, \sigma_Y^2), \text{ with } E(Y) \equiv \mu_Y = \text{sex} + \text{species} + \text{sex:species}$$

and the vectorial random effects

$$\begin{pmatrix} \mathrm{b}_{0,\mathrm{ind}} \\ \mathrm{b}_{1,\mathrm{ind}} \end{pmatrix} = \text{MultivariateNormal}\left(\begin{pmatrix} 0 \\ 0 \end{pmatrix}, \begin{pmatrix} \mathrm{Var}(\mathrm{b}_{0,\mathrm{ind}}) & \mathrm{Covar}(\mathrm{b}_{0,\mathrm{ind}}, \mathrm{b}_{1,\mathrm{ind}}) \\ \mathrm{Covar}(\mathrm{b}_{0,\mathrm{ind}}, \mathrm{b}_{1,\mathrm{ind}}) & \mathrm{Var}(\mathrm{b}_{1,\mathrm{ind}}) \end{pmatrix}\right), \tag{15.2}$$

where Y is the response variable, negative `dC`, and μ_Y and σ_Y^2 are its mean and variance. The mean of the response variable is to be related to the fixed or population-level effects of sex and species. b_0 and b_1 are the random or group-level intercept and slope denoting the extent to which the sequence of values of negative `dC` change with `length`, which here represents a pseudo-temporal axis. We also need to model the autocorrelated nature of the set of values of negative `dC` at the individual level. We are going to use an AR(1) process (see Chapter 13 for details). The goals of the analysis are to estimate the magnitude of the differences in `dC` between sexes for each species, between species for each sex, and to determine whether the magnitudes of these differences change with the other explanatory variable. Prior to any analysis we standardize the numerical explanatory variable, `DF2$length.s = as.vector(scale(DF2$length, center = T, scale = T))`, to avoid any spurious correlation between the random or group-level intercepts and slopes.

15.3.1 *Gaussian GLMM with a repeated measures design: Frequentist models*

We might consider the package `glmmTMB` to fit our models. The `lme4` package does not allow modeling the autocorrelated nature of our data. The R model would be `m3 = glmmTMB(-dC ~ sex*sp + (1 + length.s|ind) + ar1(0 + as.factor (length.s)|ind), data = DF2, family=gaussian)`, where `ar1(0 + as.factor (length.s)|ind)` sets up the AR(1) process at the individual level. In `glmmTMB` package we need to define a "`0`" to generate a design matrix adequately linking the values of negative `dC` with the model random effects. An AR(1) process requires the values of the response variable to be equispaced over time. This is certainly not the case for our pseudo-temporal variable `length`. Here we have two options. First, to make an imperfect, approximate model with `glmmTMB` that will apply to many other probability distributions in the more frequent case of equally spaced observations over time. The `glmmTMB` package requires that we specify `length.s` as `as.factor(length.s)` in the `ar` structure. This would make `m3` an over-parameterized model because it fits the random effect coefficients for every value of `length.s`, which is incorrect from the point of view of Eq. (15.2). We have shown this use of `glmmTMB` because it allows fitting statistical models of repeated measures designs for a variety of non-Gaussian response variables, provided that the temporal axis is equispaced. The second option is to use the package `nlme` that allows modeling a continuous AR(1) process not requiring equally spaced observations, but which is only valid for Gaussian response variables. The syntax for the correct modeling with `nlme` is: `m3.1 = lme(-dC ~ sp*sex, random = ~length.s|ind, data = DF2, correlation = corCAR1(form = ~length.s|ind))`. The model summary is:

```
> summary(m3.1)
Linear mixed-effects model fit by REML
  Data: DF2
  AIC BIC logLik
  207 243  -94.4

Random effects:
 Formula: ~length.s | ind
 Structure: General positive-definite, Log-Cholesky parametrization
            StdDev Corr
(Intercept) 0.210  (Intr)
length.s    0.143  -0.492
Residual    0.384

Correlation Structure: Continuous AR(1)
 Formula: ~length.s | ind
 Parameter estimate(s):
   Phi
0.0362
Fixed effects:  -dC ~ sp * sex
            Value Std.Error  DF t-value p-value
(Intercept) 15.31     0.107 397   143.0  0.0000
spSASL      -2.53     0.155  13   -16.3  0.0000
sexm        -0.48     0.163  13    -3.0  0.0110
spSASL:sexm  0.94     0.241  13     3.9  0.0018
```

We have four groups defined by the 2 × 2 combinations of levels of `sp` and `sex`. As in Chapter 6 for the factorial analysis of variance, the reference levels of the categorical explanatory variables are those not appearing in the summary: `SAFS` for `sp`, and `f` for `sex`. The intercept (15.31) is the mean of the response variable for `SAFS` and `f`. `sexm` is the difference in the mean of the response variable for the reference level of species (SASL): male SASL have a smaller mean than females of the same species: 15.32 – 0.48 = 14.8.

Table 15.1 Predicted means of the (negative) content of ^{13}C in whiskers for the four groups representing male and female South American fur seals and sea lions.

	SAFS	SASL
Females	15.32	12.80
Males	14.80	13.20

`spSASL` is the difference in the mean of the response variable for the reference level of sex (females): female SAFS have a smaller mean of the response variable than females of the other species: 15.32 – 2.53 = 12.8. With `sexm:spSASL` we can calculate the predicted mean of the response variable for the remaining group (males of SASL) as 15.32 – 0.48 – 2.53 + 0.94 = 13.20 (see Chapter 6). All these contrasts happen to be statistically significant, and they respond to the main goals of the analysis. We can summarize the above simple calculations in Table 15.1.

The magnitude of the contrast (females – males) differs and also changes in sign depending on the species: +0.52 for SAFS and –0.50 for SASL. There is also a difference in the contrast (SAFS – SASL): +2.52 for females and +1.6 for males. This is precisely what an interaction means: the dependence of the magnitude of a contrast or an effect on the level or value of another explanatory variable. As a result, we cannot interpret the effect of each explanatory variable in isolation: we must interpret their effects together. The p-values of each term in `summary(m3.1)` obtained from the Wald statistics used the large-sample approximation that is adequate for linear mixed models. Performing parametric bootstrap in models such as `m3.1` to obtain better p-values and confidence intervals often requires careful programming to avoid destroying the (pseudo-)temporal structure. Doing so is not worth the effort for our purposes here. `summary(m3.1)` also contains the standard deviations of the random intercepts and slopes, and their correlation; it also shows the estimate of the parameter α in the equation describing the exponential decline in the autocorrelation of an AR(1) process (α^t; see Section 14.4.5). The estimate (0.0362) shows a weak temporal correlation that should not concern us much here.

We can use the package `broom` to gather the fixed effects parameter estimates and their 95% confidence intervals:

```
> tidy(m3.1, conf.int = T)
# A tibble: 8 × 10
  effect   group    term              estimate std.error    df statistic   p.value conf.low conf.high
  <chr>    <chr>    <chr>                <dbl>     <dbl> <dbl>     <dbl>     <dbl>    <dbl>     <dbl>
1 fixed    fixed    (Intercept)         15.3       0.107   397   143.     0          15.1      15.5
2 fixed    fixed    spSASL              -2.53      0.155    13   -16.3    4.80e-10  -2.87     -2.20
3 fixed    fixed    sexm                -0.483     0.163    13    -2.96   1.10e- 2  -0.835    -0.131
4 fixed    fixed    spSASL:sexm          0.940     0.241    13     3.90   1.82e- 3   0.420     1.46
5 ran_pars ind      sd_(Intercept)       0.210    NA        NA    NA     NA          0.105     0.419
6 ran_pars ind      cor_length.s.(I…    -0.492    NA        NA    NA     NA         -0.974     0.796
7 ran_pars ind      sd_length.s          0.143    NA        NA    NA     NA          0.0789    0.259
8 ran_pars Residual sd_Observation       0.384    NA        NA    NA     NA         NA        NA
```

We could make a figure corresponding to Table 15.1 with the command `ggcoef(m3.1)` from the `ggeffects` package, and add the usual changes to the background and font sizes (see the examples in Chapter 13). We use the command `ggeffect` from `ggeffects` to obtain the predicted values of model `m3.1` to make the conditional plot: `predm3.1f = ggeffect(m3.1, type = "fixed", terms = c("sp", "sex"))`. We also gather the random effect estimates with `ranef.m31 = ranef(m3.1)`, and then change the variable names with `names(ranef.m31) = c("int", "slope")`. We cannot obtain the standard errors of the estimated random effects using the command `se.ranef` from the `arm` package, because `m3.1` was fitted with package `nlme`, not with `lme4` as in Chapter 13. The problem is the compatibility of R objects between packages, not a conceptual one. We show the conditional plot and the plots of the fitted random effects of model `m3.1` in Fig. 15.15. The conditional plots shown in Fig. 15.15a are equivalent to, and in our view clearer than, the display of the fixed effects parameters of model `m3.1`.

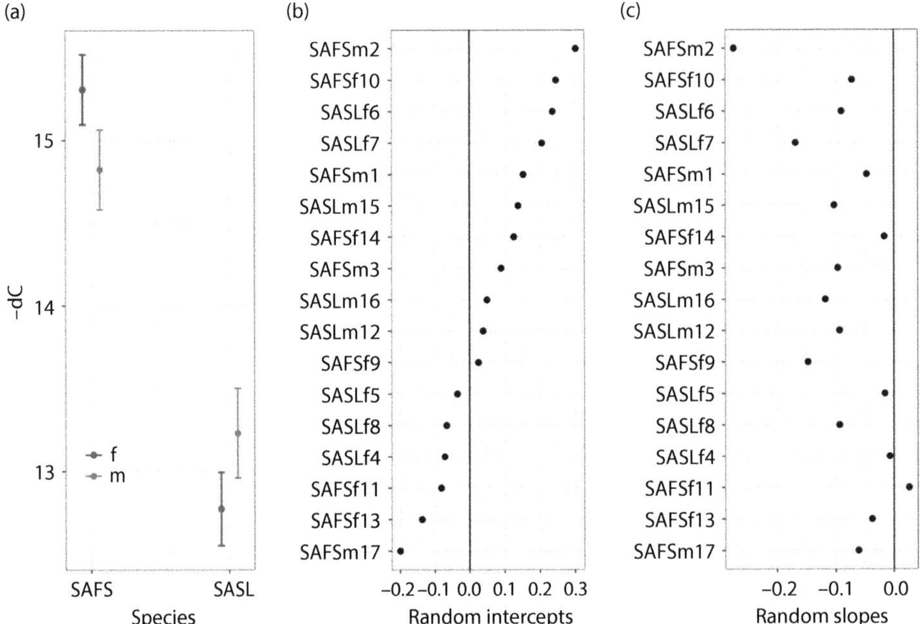

Fig. 15.15 Conditional plot of the frequentist GLMM `m3.1` including only the fixed effects (a), showing the predicted means of negative `dC` in their 95% confidence intervals for each species and sex (red: females). Plots (b) and (c) show the random intercepts and slopes for each individual, with the latter being shown in increasing magnitude of the random intercepts.

We use the package `MuMin` to estimate the proportion of the variation of the response variable explained by the fixed effects (`R2m`), and by the fixed and random effects:

```
> r.squaredGLMM(m3.1)
          R2m   R2c
[1,] 0.854 0.898
```

Both proportions are pretty high, suggesting that `m3.1` is a reasonable statistical model.

We can gather most of the inputs needed to assess the goodness of fit of model `m3.1` with the command `augment(m3.1)` from the package `tidy`. Given that `m3.1` is a linear mixed model, we can safely use the standard residuals without having to simulate the randomized quantile residuals that would give similar results. We also evaluate whether the random intercepts and slopes have a normal distribution, and whether the residuals are autocorrelated using the command `ACF` from the package `base`.

Figure 15.16a and b show that the residuals appear to have a random scatter and lack any clear tendencies. Figure 15.16c shows that only one data point has a large Cook distance; it is unlikely that omitting this single point would actually change the parameter estimates

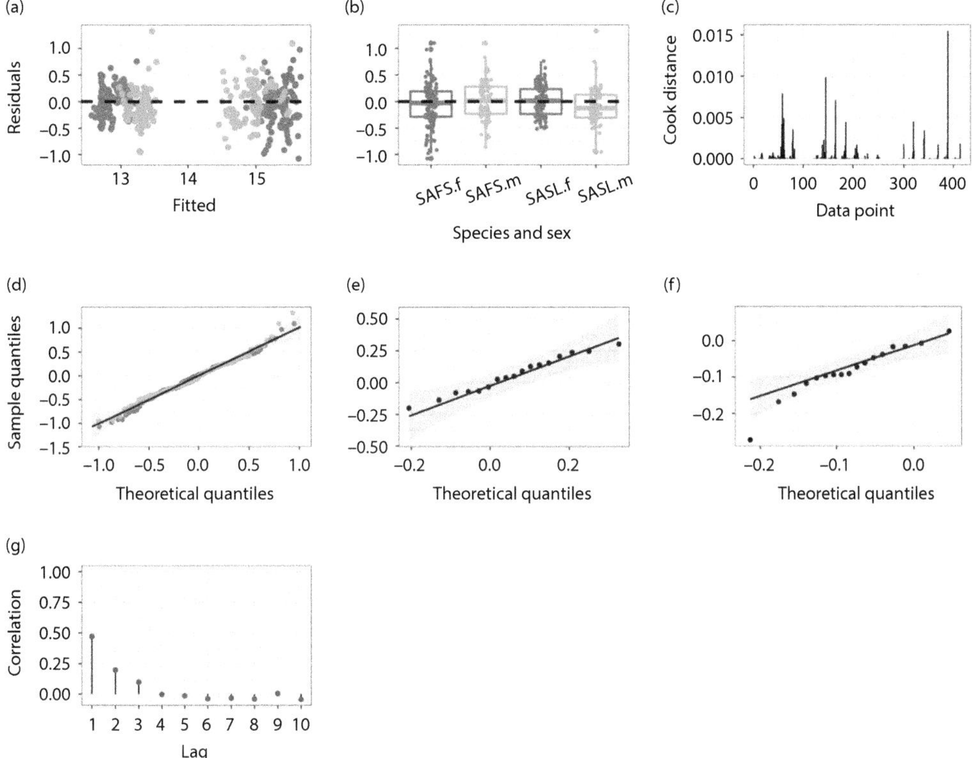

Fig. 15.16 Residual analysis of the the frequentist GLMM `m3.1` based on standard residuals. Plots (a) and (b) show the relation between the residuals and the fitted values and to the explanatory variables (red: females). Plot (c) shows the Cook distance for each data point. Plots (d)-(f) are the quantile-quantile plots (and the 95% confidence band) for the residuals, and the estimates of the random intercepts and random slopes, respectively. Plot (g) is the autocorrelation function of the residuals for the first 10 lags.

much. Figure 15.16d, e, and f show that the residuals and the random intercepts and slopes appear to have normal distributions. Figure 15.16g shows that only adjacent data in the temporal axis have a correlation close to 0.5. What can be done about this? One option would be to consider a more complex time series process than our chosen AR(1) to model the temporal autocorrelation of data at the individual level. While there are other options (see the help in `?corrClasses`) for equispaced data, we lack alternatives for non-equispaced data such as ours. On balance, we think that the goodness of fit of `m3.1` is very good.

15.3.2 *Gaussian GLMM with a repeated measures design: Bayesian models*

As always, we are going to use the package `brms` to fit this model. We start by defining the model formula: `formula.m3.brms = bf(-dC~sp*sex + (1 + length.s|ind))`. Note that this formula does not contain the autocorrelation component. We can then visualize the default weakly informative model priors:

```
> get_prior(formula.m3.brms, data=DF2,family='gaussian')
                     prior     class         coef group resp dpar nlpar bound        source
                    (flat)         b                                                default
                    (flat)         b         sexm                              (vectorized)
                    (flat)         b       spSASL                              (vectorized)
                    (flat)         b  spSASL:sexm                              (vectorized)
                    lkj(1)       cor                                                default
                    lkj(1)       cor                ind                        (vectorized)
 student_t(3, 14.2, 2.5) Intercept                                                default
    student_t(3, 0, 2.5)        sd                                                default
    student_t(3, 0, 2.5)        sd                ind                        (vectorized)
    student_t(3, 0, 2.5)        sd    Intercept   ind                        (vectorized)
    student_t(3, 0, 2.5)        sd     length.s   ind                        (vectorized)
    student_t(3, 0, 2.5)     sigma                                                default
```

We see yet again that it is not only very important to be able to write the statistical model, but also to understand how the model is implemented in R. We are going to define the following priors for the effect sizes that we aim to estimate:

```
prior.m3.brms = c(prior(normal(0, 2), class = b, coef="sexm"),
                  prior(normal(0, 2), class = b, coef="spSASL"),
                  prior(normal(0, 2), class = b, coef="spSASL:sexm"),
                  prior(lkj(0.5), class=cor))
```

These priors imply that we believe that the difference in means between sexes for the species of reference `SAFS` would be in [−2 × 1.96, +2 × 1.96] with probability 0.95, and likewise for the difference in means between species for the sex of reference `f` and for the interaction between these categorical explanatory variables. We keep the default weakly informative priors for the standard deviations of the group-level effects. The prior for the parameter of the LKJ distribution is also weakly informative (see Fig. 13.18).

We can now run the model: `m3.brms = brm(-dC ~ sp*sex + (1 + length.s|ind), cor_ar(~ length.s|ind, p = 1), data = DF2, prior = prior.m3.brms, family = "gaussian", control = list(adapt_delta = 0.99), warmup = 1000, iter = 5000, chains = 3, future = T)`. Note that here we specified the first-order autcorrelation structure. That is, we did not model a continuous, first-order autocorrelated process using the package `brms` as we should have done. We used a much larger number of iterations than in other models because of the warnings about a low effective sample size obtained in a first run with `iter = 2000`. The model summary is:

```
> summary(m3.brms)
 Family: gaussian
  Links: mu = identity; sigma = identity
Formula: -dC ~ sp * sex + (1 + length.s | ind)
         autocor ~ arma(time = length.s, gr = ind, p = 1, q = 0, cov = FALSE)
   Data: DF2 (Number of observations: 414)
  Draws: 3 chains, each with iter = 5000; warmup = 1000; thin = 1;
         total post-warmup draws = 12000

Correlation Structures:
      Estimate Est.Error l-95% CI u-95% CI Rhat Bulk_ESS Tail_ESS
ar[1]     0.56      0.06     0.44     0.68 1.00     5970     5337

Group-Level Effects:
~ind (Number of levels: 17)
                         Estimate Est.Error l-95% CI u-95% CI Rhat Bulk_ESS Tail_ESS
sd(Intercept)                0.25      0.09     0.11     0.45 1.00     2710     4150
sd(length.s)                 0.18      0.05     0.09     0.31 1.00     5250     6581
cor(Intercept,length.s)     -0.36      0.49    -0.99     0.68 1.00     1025     2149

Population-Level Effects:
            Estimate Est.Error l-95% CI u-95% CI Rhat Bulk_ESS Tail_ESS
Intercept      15.28      0.13    15.02    15.53 1.00     2749     5386
spSASL         -2.45      0.16    -2.78    -2.13 1.00     4626     6195
sexm           -0.37      0.19    -0.75     0.00 1.00     3729     5499
spSASL:sexm     0.80      0.26     0.29     1.32 1.00     4771     6118

Family Specific Parameters:
      Estimate Est.Error l-95% CI u-95% CI Rhat Bulk_ESS Tail_ESS
sigma     0.31      0.01     0.29     0.34 1.00    10293     8184
```

The model summary has three parts: the population- and group-level effects that we have encountered in other mixed models, and the correlation structure part that is specific to this model. For all model parameters, the effective sample sizes were larger than 1,000 and `Rhat` equal to one, suggesting that the Hamiltonian Monte Carlo algorithm converged to a stationary posterior distribution. We have the mean (`Estimate`), standard deviation (`Est.Error`), and the lower (`l-95%`) and upper (`u-95%`) limits of the 95% credible intervals calculated for the marginal posterior distributions of each model parameter. We explained the interpretation of the population-level parameters for the same statistical model in Section 15.3.1. We also discussed the interpretation of the standard deviations of the intercept, slope, and their correlations estimated for each group, here defined by the sampled individuals. Note that for these parameters, and for the correlation structure parameter, we now have adequate estimates of precision and credible intervals. Because of our inability to model the necessary continuous, first-order autocorrelation, we should take these results with a grain of salt. We can visually evaluate the convergence of the three chains to a posterior stationary distribution for the population- and group-level parameters in Figure 15.17, which does show the expected convergence for all parameters. While the marginal posterior distributions of the population-level parameters are very close to a Gaussian shape, those of the group-level parameters with hard boundaries (sd > 0, –1 < corr < 1) are clearly asymmetric. This is important for using the median rather than the mean to characterize the "typical" values of these posterior distributions. We do not show the figure for the autocorrelation profiles of the population- and group-level parameters for each chain, but the code to obtain it can be found on the companion website.

Figure 15.18 show the marginal posterior distributions of the population- and group-level parameters of model `m3.brms`. The command `variables(m3.brms)` gives all the parameter names in `m3.brms`. The command `grep` can provide the subset of the population-level parameters whose names start with "b," `variables(m3.brms)[grep("b", variables(m3.brms))]`, to obtain their marginal posterior distributions with `pop.pars.m3.brms = as_draws_df(m3.brms, variable = variables(m3.brms)[grep("b", variables(m3.brms))])`. We can now operate to obtain the marginal posterior distributions of the means of the response variable for the groups formed by the 2 × 2 combinations of the explanatory variables `sp` and `sex`. In fact, we can apply any operation or mathematical function to these marginal posterior distributions and the result would still be a posterior probability distribution. To obtain the marginal distributions for the means of the four groups, we just need to recall the meaning of the model parameters explained in Section 15.3.1 and in Chapter 6. It can all be done in one line of code:

```
means.m3.brms=data.frame(
            SAFS.f=pop.pars.m3.brms$b_Intercept,
            SASL.f=pop.pars.m3.brms$b_Intercept+pop.pars.m3.brms$b_spSASL,
            SAFS.f=pop.pars.m3.brms$b_Intercept+pop.pars.m3.brms$b_sexm,
            SAFS.m=pop.pars.m3.brms$b_Intercept+pop.pars.m3.brms$b_spSASL+
                     pop.pars.m3.brms$b_sexm+pop.pars.m3.brms$"b_spSASL:sexm")
```

We now use the package `reshape2` to obtain a data frame in the "long format" with `means.m3.brms = melt(data = means.m3.brms, variable.name = "species.sex", value.name = "mean")` that is adequate for plotting. We also obtain the posterior distributions of the group-level intercepts and slopes with:

```
int.m3.brms=as_draws_df(m3.brms,
            variable =variables(m3.brms)[grep("Intercept]",variables(m3.brms))])
            sl.m3.brms=as_draws_df(m3.brms,
            variable =variables(m3.brms)[grep("length.s]",variables(m3.brms))])
```

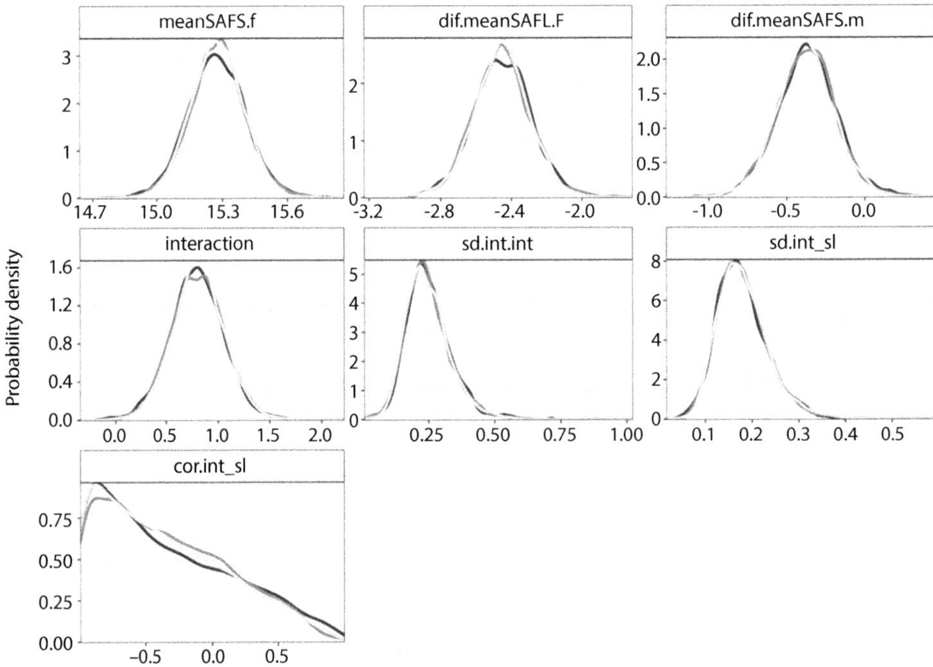

Fig. 15.17 Convergence of the three chains to a common stationary posterior distribution for the population- and group-level parameters of the Bayesian mixed model `m3.brms`. meanSAFS.f is the mean of the response variable for the reference levels (SAFS and f) that is shown as the model intercept; dif.mean.SASL.F is the differential in the mean of the response variable between species for the reference sex (f); dif.mean.SAFS.m is the differential in the mean of the response variable between sexes for the reference species (SAFS); interaction denotes the interaction between species and sex; sd.ind.int and sd.ind.sl are the standard deviations of group-level intercepts and slopes, and cor.int_sl is their correlation at the individual level.

The resulting data frames contain three last columns (`.chain`, `.iteration`, `.draw`) that we are going to delete to keep things simple: `int.m3.brms = int.m3.brms[,-((length(int.m3.brms)-2):length(int.m3.brms))]`, `sl.m3.brms = sl.m3.brms[,-((length(sl.m3.brms)-2):length(sl.m3.brms))]`. `length(sl.m3.brms)` counts the number of columns in a data frame, and allows the selection of the columns to be deleted from each data frame. We then rename the variables with `names(int.m3.brms)[1:length(unique(DF2$ind))] = unique(DF2$ind)` and `names(sl.m3.brms)[1:length(unique(DF2$ind))] = unique(DF2$ind)`, and put both data frames into the "long format" for plotting: `int.m3.brms = melt(data = int.m3.brms, variable.name = "ind", value.name = "int")`, `sl.m3.brms = melt(data = sl.m3.brms, variable.name = "ind", value.name = "slope")`.

Figure 15.19a shows the posterior distributions of the estimated means of negative `dC` for each sex and species. Figure 15.19b and c does likewise for the individual-level intercepts and slopes. Note that the spread of the posterior distributions of group-level intercepts is much wider than those of the slopes, in agreement with their standard deviations shown in `summary(m3.brms)`.

We use Gelman et al.'s (2018) metric to calculate the proportion of variation of the response variable accounted for by the model `m3.brms`:

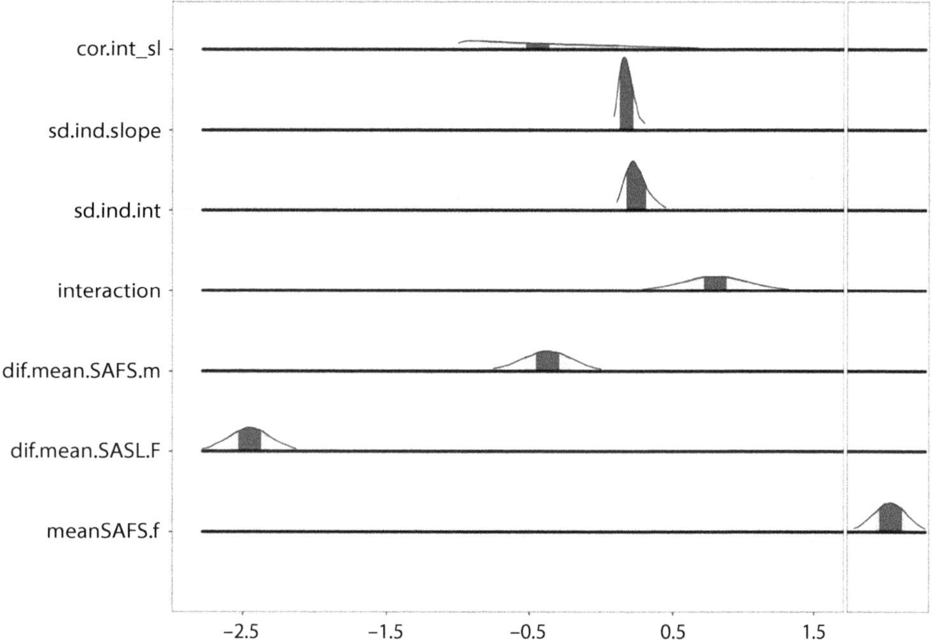

Fig. 15.18 Posterior distributions of the population- and group-level parameters of the Bayesian mixed model `m3.brms` showing their 62% (dark blue) and 95% credible intervals. The names of the parameters are as in Fig. 15.17. The posterior distributions of the group-level correlation between slopes and intercepts is rather spread out (see Fig. 15.17), and that of the population-level intercept (meanSAFS.f) is centered around its mean of 15.28 (see `summary m3.brms`), and hence off scale compared to those of the other parameters.

```
> bayes_R2(m3.brms)
   Estimate Est.Error  Q2.5 Q97.5
R2    0.905   0.00391 0.897 0.912
```

To assess the goodness of fit of `m3.brms` to the data, we first need to obtain the randomized quantile residuals using the package `DHARMa`. We could actually have used the Pearson residuals that we employed in Chapters 4–6, with identical results. We first obtain the posterior predictive distribution with `post.pred.m3.brms=predict(m3.brms, ndraws=1e3, summary=F)`, and then calculate the residuals with `qres.m3.brms = createDHARMa(simulatedResponse = t(post.pred.m3.brms), observedResponse = -DF2$dC, fittedPredicte-dResponse = apply(post.pred.m3.brms, 2, median), integerResponse = T)`; should the model fit well, these are converted to a normal distribution with `res.m3.brms = data.frame(res = qnorm(residuals(qres.m3.brms)))`. These are the same instructions we have already used many times, only changing the name of the fitted statistical model. We then create a data frame with all the variables needed for plotting: `res.m3.brms = cbind(res.m3.brms, DF2[,c("sp", "sex", "Sp.Sex", "ind")], fitted = fitted(m3.brms, ndraws = 1000)[,1])`. We note that the LOO_CV Pareto k estimates are not yet available for models with temporal autocorrelation. We also need the summary statistics of the group-level intercepts and slopes whose posterior distributions were showed in Fig. 15.19 to assess whether they have a normal distribution. These statistics are obtained

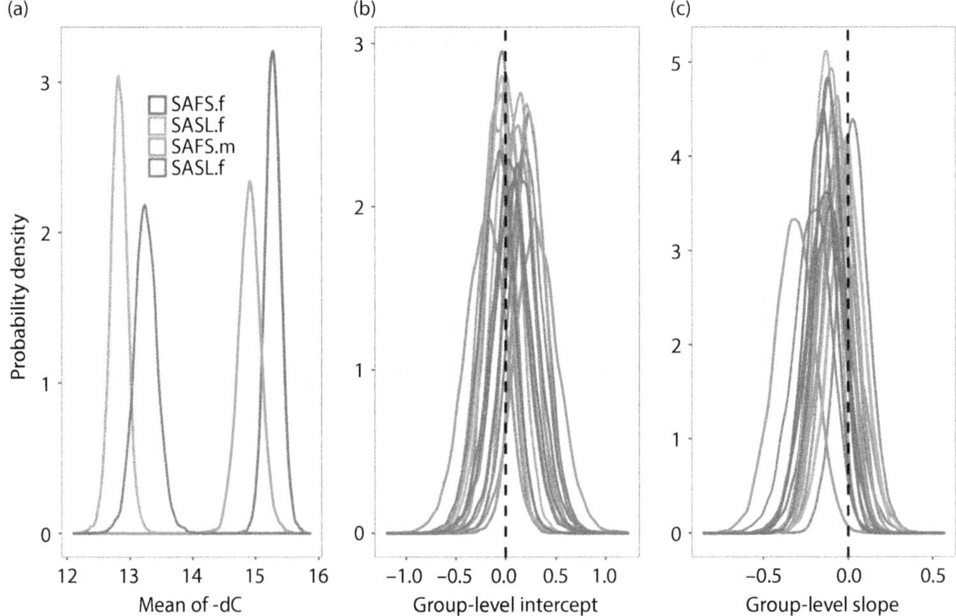

Fig. 15.19 Posterior distributions of the means of negative `dC` for each species and sex (a), and the group-level intercepts (b) and slopes (c) at the individual level from the Bayesian mixed model `m3.brms`.

with `stats.ranef.m3.brms = as.data.frame(ranef(m3.brms))` and the variables renamed: `names(stats.ranef.m3.brms) = c("mean.int", "se.int", "Q2.5.int", "Q97.5.int", "mean.slope", "se.slope", "Q2.5.slope", "Q97.5.slope")`.

Figure 15.20a and b show that the residuals appear to be randomly distributed, with a slightly larger variation for those of SAFS. Figure 15.20c suggests that the residuals have thicker tails than a normal distribution (perhaps a *t*-distribution?), which might be associated with the heterogeneity of variation of the residuals. Figure 15.20d and e show that the group-level intercepts and slopes are close to normal distributions. On balance, the goodness of fit of model `m3.brms` is not fantastic, perhaps only acceptable if we are generous. Options to improve the model, though unavoidably making it more complex, include the explicit modeling of the heterogeneity of variance as a function of the explanatory variables. But the main conclusions extracted are likely to be safe from the somewhat minor violations of assumptions detected here.

Finally, Figure 15.21 shows the conditional plot of model `m3.brms` including only the population-level effects (left), and both the the population- and group-level effects. The plots only show a slight difference in the 95% credible intervals on the group means.

15.4 Beta GLMM with a split-plot design

THE DATA IN CONTEXT: Aranda et al. (2021) studied the determinants of success of an invasive woody plant (honey locust) in the agricultural plains of central Argentina. They wanted to know whether the main `crop` (maize, soybean), its `management` (yes/no, with the latter meaning that no pesticides were used), and competition from the main crop (`crop_cover`: yes/no, with the latter meaning that all soybean and maize seedlings were

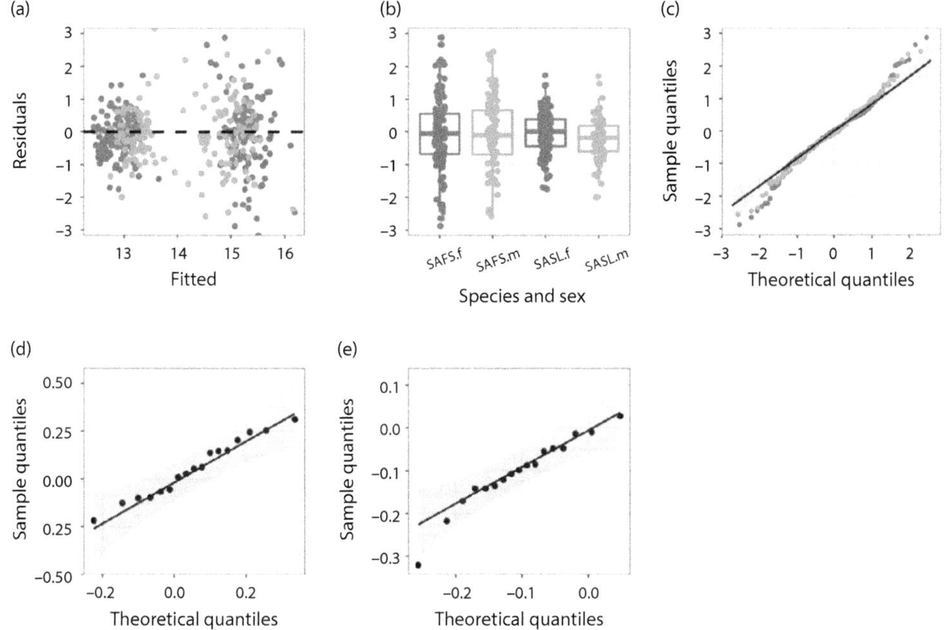

Fig. 15.20 Residual analysis of the Bayesian mixed model m3.brms based on randomized quantile residuals. Plots (a) and (b) show the relation between the residuals and the fitted values and to the explanatory variable. Plots (c)–(e) are the quantile–quantile plots (and their 95% credible band) for the residuals, and the mean estimates of the random intercepts and random slopes, respectively.

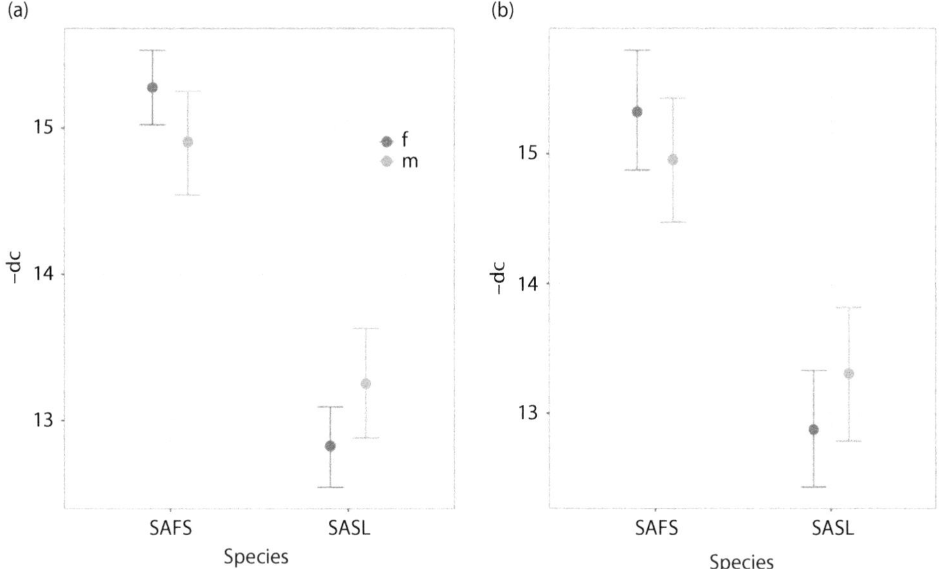

Fig. 15.21 Conditional plots of the Bayesian binary GLMM m2.brms including only the population-level effects (a) and both population- and group-level effects (b).

manually removed) affected the proportion of root biomass of the honey locust over two years (wet, dry). The authors defined three and six blocks for each crop in the wet and dry year that defined homogeneous crop management units, with respect to both soil heterogeneity and crop handling practices in each year. In each block, they defined two large plots where they randomly chose to apply one of the two levels of management. Within each plot, they defined smaller sub-plots to which they randomly assigned two crop presence levels (with, without). Their design was a split plot (Chapter 14). After the main crop emergence, they transplanted seedlings of the honey locust generated from seeds in the university greenhouse, let them grow in the field, and collected them after the harvesting of the main crop. The authors measured a series of response variables for each seedling growing in the sub-plots, but here we only focus on the proportion of above-ground biomass calculated from their measurements.

EXPLORATORY DATA ANALYSIS: Let's start by importing and examining the 666 (= 358 + 308) data points:

```
> DF3=read.csv("plant biomass chapt 15.csv", header=T)
> summary(DF3)
  year          crop          block         management crop_cover prop_above_biom
 dry: 83    maize  :358   Min.   : 1.00   no :377    no :298    Min.   :0.040
 wet:583    soybean:308   1st Qu.: 4.00   yes:289    yes:368    1st Qu.:0.200
                          Median : 7.00                         Median :0.264
                          Mean   : 7.38                         Mean   :0.269
                          3rd Qu.:10.00                         3rd Qu.:0.323
                          Max.   :18.00                         Max.   :0.581
```

We first examine the structure of the data set using the command `table` that just counts the number of rows (sampling units for the sub-plots, in our case) in the data frame:

```
> ftable(block~crop+year, data=DF3)
              block  1  2  3  4  5  6  7  8  9 10 11 12 13 14 15 16 17 18
crop    year
maize   dry          0  0  0  0  0  0  0  0  0  0  0  0 13 20 18  0  0  0
        wet         53 50 47 55 50 52  0  0  0  0  0  0  0  0  0  0  0  0
soybean dry          0  0  0  0  0  0  0  0  0  0  0  0  0  0  0 13 11  8
        wet          0  0  0  0  0  0 43 62 45 45 44 37  0  0  0  0  0  0
```

We verify that different blocks were defined for each type of year and type of main crop. Because block identity can be uniquely associated with homogeneous conditions of type of year and main crop, the estimated block variance depicts both the variation between years and between main crops.

Leaving out the blocks, the next table shows that there are unequal numbers of data points for all levels of the main explanatory variables measured at the sub-plot level. These values of the response variable measured for the sub-plots are statistical replicates only when they are in the same block. Thus, we can study the interactions between `management`, `year`, and `crop_cover`, provided that we correctly specify the spatial nesting of plots and sub-plots within blocks.

```
> ftable(crop~crop_cover+management +year, data=DF3)
                            crop maize soybean
crop_cover management year
no         no         dry           16      10
                      wet           82      77
           yes        dry            7       3
                      wet           65      38
yes        no         dry           13      14
                      wet           81      84
           yes        dry           15       5
                      wet           79      77
```

The exploratory graphical analysis of `DF3` is complicated because we have four explanatory categorical variables: `crop`, `management`, `year`, and `crop_cover`. Each variable has two levels, making $2^4 = 16$ groups to be compared. Before embarking on a brute force graphical exploration, let's recall that Aranda et al.'s research goals implied that they were at most interested in the third-order interaction between `management`, `year`, and `crop_cover`. A statistical model having a third-order interaction would also need to include all pairwise interactions, and the single-level effects of each explanatory variable. This is known as the "principle of marginality" that has become standard practice in statistical modeling. Of course, we could also choose to consider the most complex model possible, including up to the fourth-order interaction. There is nothing formally stopping us considering such a complex model. There are two main ways to avoid the brute force approach of "putting everything, including the kitchen sink" into the statistical modeling: having clear and specific hypotheses about only certain effects, and using exploratory data analysis to discard a few effects or interactions. The second approach is clearly not a principled way of proceeding as your visual conclusions might well differ from ours. Even more, in the Bayesian framework we are neither supposed to decide the structure of the statistical model nor set the priors after examining the data to be analyzed.

Figure 15.22 show how the median `prop_above_biom` (the horizontal lines in the boxes) varies for three-way combinations of `management`, `year`, and `crop_cover` for each `crop`. We also observe the number of points and their scatter for each of the $2^3 = 8$ three-way combinations of these explanatory variables. Consider Fig. 15.22a for maize. Comparing the two levels of `crop_cover`, the median `prop_above_biom` was always higher when the main crop was removed than when present, regardless of the year and management. Note also the wide scatter of data points for the eight groups, and the shortage of data in the dry year. These two features would suggest that the third-order interaction of `crop`, `management`, and `crop_cover` is unlikely to be very relevant. The same patterns are not visible in Fig. 15.22b for soybean, suggesting that here the third-order interaction might be relevant. We note, however, that there are few data for the two most dissimilar groups. Comparing the same groups in Fig. 15.22a and b, we note that the median `prop_above_biom` for soybean tended to be smaller than for maize.

The response variable `prop_above_biom` takes real values in [0, 1]. We can only use the beta distribution in the likelihood part of the statistical models (Chapter 12). We will use the logit link function to make sure of obtaining predictions in the scale of the data (Chapter 12).

STATISTICAL MODEL: $Y \sim \text{Beta}(\mu_Y, \phi)$, with $\text{logit}(\mu_Y)$ = `crop` + `management*year*crop_cover` + $b_{\text{blocks}} + b_{\text{plots w.block}} + b_{\text{subplots w.plots}}$.

The fixed or population-level effects are `crop` + `management*year*crop_cover`. The term involving the third-order interaction also includes all pairwise interactions and single effects of the explanatory variables. ϕ is the dispersion parameter of the beta distribution that governs the relation between its mean and variance (Chapter 12). The random or group-level effects in split-plot designs need to indicate the spatial structuring of subplots (the elementary sampling unit, where `crop_cover` was applied) within larger plots (where `management` was applied), and of the plots within the larger blocks. Each b coefficient involved in the random or group-level effects has a normal distribution with zero mean and its own specific variance. As in every GLMM, the random or group-level effects are additive with respect to the fixed or population-level effects in the scale of the logit link function.

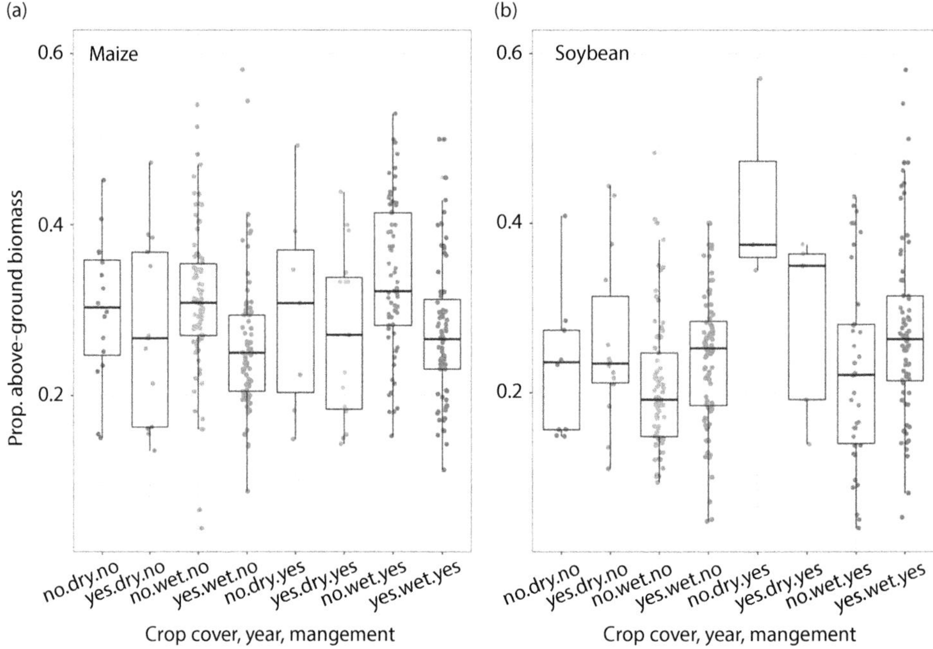

Fig. 15.22 Boxplots showing the relation between `prop_above_biom` for all combinations of the categorical explanatory variables `management`, `crop_cover`, and `year` for the main crops maize (a) and soybean (b) of the data frame `DF3`. All points are horizontally jittered by adding a random value to allow better visualization.

In R, we would write the random or group-level effects for this analysis as `(1|block/management/crop_cover)`. The notation means that there are three random or group-level effects, and that they are crossed effects (Chapter 13). Therefore, starting from the bottom of the hierarchy, there might be heterogeneity between levels of `crop_cover` in the sub-plots within each level of management applied to the larger plots, and that the plots might differ across the levels of `block`. It is almost like a set of Russian dolls. These random or group-level effects denote differentials or random shifts in the intercept, i.e., in logit(μ_Y), at the different levels of the hierarchy of effects above and beyond the fixed or population-level effects. The notation `(1|block/management/crop_cover)` makes R generate a sparse design matrix denoting which component of the random or group-level effects is associated with each data point (Chapter 13). Figure 15.23 shows the design matrix. The code to obtain it, melt it into a data frame with long format, and plot it are explained on the companion website.

15.4.1 *Beta GLMM with a split-plot design: Frequentist model*

We use the package `glmmTMB` to fit the model: `m4 = glmmTMB(prop_above_biom ~ crop + year*crop_cover*management + (1|block/management/crop_cover), data = DF3, REML = T, family = beta_family)`. The logit link function is assumed by default. The model summary is:

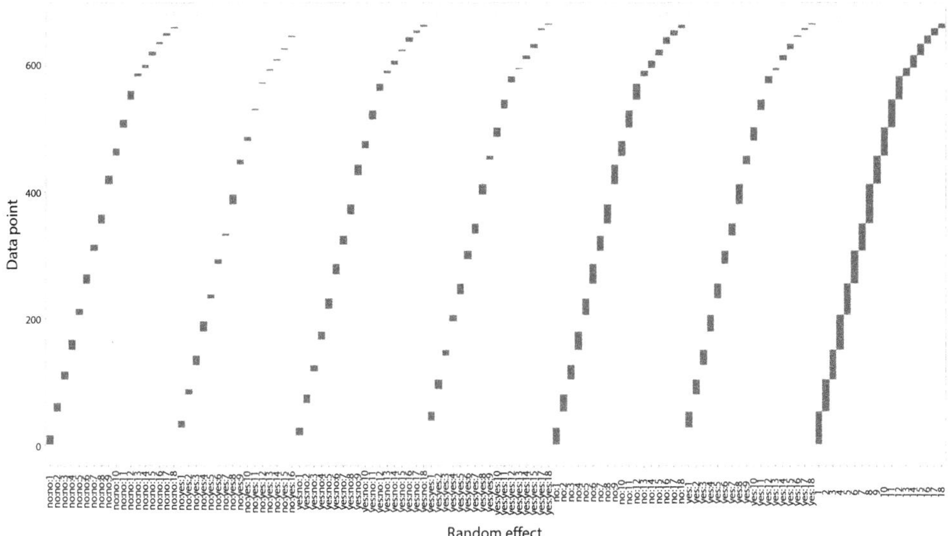

Fig. 15.23 Design matrix for the random or group-level effects for the beta GLMM with a split-plot design. The design matrix was transposed for better visualization. The plot shows the correspondence between each data point and the specific components of the random or group-level effects. These effects are ordered from the lowest (left) to the highest (right) level of the hierarchy of nested effects. The size of the rectangle denotes the number of data points associated with each component of the random or group-level effects.

```
> summary(m4)
 Family: beta  ( logit )
Formula:          prop_above_biom ~ crop + year * crop_cover * management + (1 |
    block/management/crop_cover)
Data: DF3

     AIC      BIC   logLik deviance df.resid
   -1320    -1261      673    -1346      662

Random effects:

Conditional model:
 Groups                         Name        Variance Std.Dev.
 crop_cover:management:block (Intercept) 5.49e-02 0.23439
 management:block            (Intercept) 3.25e-07 0.00057
 block                       (Intercept) 4.57e-03 0.06764
Number of obs: 666, groups:  crop_cover:management:block, 69; management:block, 36; block, 18

Dispersion parameter for beta family (): 27.3

Conditional model:
                                     Estimate Std. Error z value Pr(>|z|)
(Intercept)                           -0.8995     0.1346   -6.68  2.4e-11 ***
cropsoybean                           -0.2087     0.0760   -2.74   0.0061 **
yearwet                               -0.0535     0.1521   -0.35   0.7251
crop_coveryes                          0.0229     0.1798    0.13   0.8988
managementyes                          0.2569     0.2149    1.19   0.2321
yearwet:crop_coveryes                 -0.0717     0.2092   -0.34   0.7319
yearwet:managementyes                 -0.1570     0.2420   -0.65   0.5166
crop_coveryes:managementyes           -0.2764     0.2892   -0.96   0.3392
yearwet:crop_coveryes:managementyes    0.3308     0.3281    1.01   0.3133
```

As before, the model summary is divided in two parts: the random and the fixed effects. The dispersion parameter ϕ of the beta distribution is also estimated. Using the standard deviations, we see that there is much larger variation among sub-plots within plots and blocks than at other levels of the hierarchy of random effects. There was a modest variation among blocks, which is somewhat surprising given that the blocks were both defined to reflect homogeneous crop management units and associated with climatic variation.

To interpret the fixed effects, we must recall which are the reference levels of each categorical explanatory variable. They are those levels for each explanatory variable not appearing in the model summary: dry for `year`, maize for `crop`, no for `crop_cover`, and no for `management`. Therefore, the `(Intercept)` is the estimate of logit(μ_Y) for these reference levels. The other parameter estimates of the fixed effects are differences of logit(μ_Y) with respect to the reference level or to combinations of levels of the explanatory variables. For instance, logit(μ_Y) for the reference level maize is –0.8995, and for soybean is –0.8995 – 0.2087 = –1.1100. The difficulty in interpreting the remaining coefficients is that they reflect effects of variables that are involved in pairwise and third-order interactions. In the case of factorial analysis of variance (Chapter 6) with two explanatory variables and their interaction, we used the coefficients to calculate the mean of the response variable of a 2 × 2 table, and interpreted them as specific differences of their means. Now we have a 2 × 2 × 2 hypertable or cube, whose vertices are each of the eight combinations of levels of the explanatory variables (Fig. 15.24a).

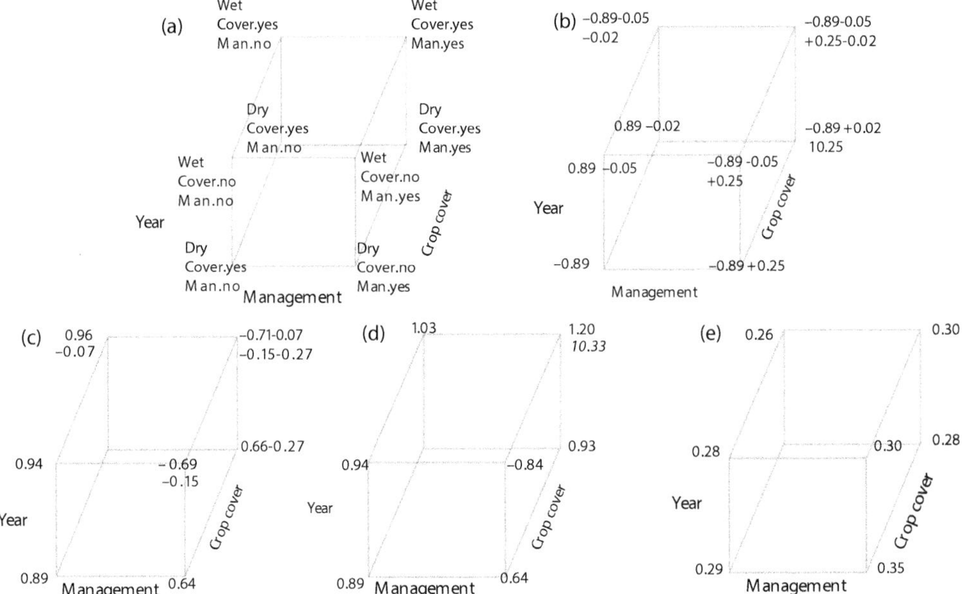

Fig. 15.24 Visualization of the fixed effects of the beta frequentist GLM `m4`. (a) Basic scheme showing the 2 × 2 × 2 combinations formed by the three categorical explanatory variables `year`, `crop_level`, and `management` as indicated at each vertex. The reference level of each variable is shown in bold. The `(Intercept)` of `summary(m4)` corresponds to the reference levels of the three variables at the bottom left vertex. (b) Inclusion of the single-variable effects (i.e., without interactions) estimated in `summary(m4)` at each vertex. (c) After totaling the single-variable effects, the estimated pairwise interactions (in bold) are included at each vertex. (d) Inclusion of the estimated third-order interaction (top, right in bold and italics) in the totals after plot C. (e) Means of the response variable estimated by model `m4`. obtained by applying the inverse logit to the values shown in plot (d).

Figure 15.24b contains the single effects (i.e., without any interaction) of each explanatory variable, shown in the top bracket next to `summary(m4)`. Figure 15.24c adds the effect of the three pairwise interactions (the second bracket next to `summary(m4)`) to the single effects on the logit scale. Figure 15.24d includes the effect of the third-order interaction (last bracket) and, after applying the inverse logit function, the last panel shows the predicted means of `prop_above_biom` for the eight groups formed by the 2 × 2 × 2 combinations of levels of the explanatory variables. We dwell on Fig. 15.24 because it is very difficult to put into words the meaning of the coefficients of a third-order interaction. We can appreciate in Fig. 15.22e that the model predicts small differences (ranging from 0.26 to 0.35) in the means of the response variable for the eight groups formed by the combinations of levels of the explanatory variables. This agrees with the absence of statistically significant effects in `summary(m4)`. Had any of the terms involved in interactions been statistically significant in `summary(m4)`, we would carry out a posteriori tests using the package `multcomp`, after creating a single variable combining pairs of them using `paste(DF3$crop_cover, DF3$management, sep = ".")` as we did in Chapter 6.

The command `tidy` from package `broom.mixed` allows us to gather the parameter estimates, their standard errors, and their 95% confidence intervals based on Wald's large-sample approximation of model `m4`: `m4.pars = tidy(m4, conf.int = T)`. We select the fixed effects only with `m4.pars = m4.pars[m4.pars$effect == "fixed",]`. We could also have obtained the 95% confidence intervals with parametric bootstrap as we did for the binary GLMM in Section 15.2. It is probably not worth the computing effort because `DF3` has 666 datapoints and with such a large sample size, bootstrapped confidence intervals are likely to differ little from shown here:

```
> m4.pars[,c("term", "estimate", "conf.low", "conf.high")]
# A tibble: 9 x 4
  term                              estimate conf.low conf.high
  <chr>                                <dbl>    <dbl>     <dbl>
1 (Intercept)                        -0.899    -1.16    -0.636
2 cropsoybean                        -0.209    -0.358   -0.0597
3 yearwet                            -0.0535   -0.352    0.245
4 crop_coveryes                       0.0229   -0.330    0.375
5 managementyes                       0.257    -0.164    0.678
6 yearwet:crop_coveryes              -0.0717   -0.482    0.338
7 yearwet:managementyes              -0.157    -0.631    0.317
8 crop_coveryes:managementyes        -0.276    -0.843    0.290
9 yearwet:crop_coveryes:managementyes  0.331   -0.312    0.974
```

We can see that only the 95% confidence interval for the difference in logit(μ_Y) between maize and soybean did not include the null difference, since the plausible values ranged from –0.358 to –0.059. Using Gelman et al.'s (2014) rule of four applicable in models with a logit link function (Chapter 9), we can see that –0.209 / 4 = –0.052 means that the effect of switching from maize to soybean is a maximum reduction in `prop_above_biom` of 5.2%, with plausible ranges of reduction between –0.358 / 4 = –0.0895 and –0.059 / 4 = –0.0147. For all other terms in the model (i.e., single effects, pairwise interactions, and third-order interaction), the magnitudes of their estimated effects were smaller than for the difference between crops.

We gather the random effect estimates and their variances for model `m4` with `rand.eff.m4 = as.data.frame(ranef(m4, condVar = T))`. After observing `rand.eff.m4`, we gather that the variable `grpvar` differentiates the three levels of the random effects that we use to select the data for each of the three plots in Fig. 15.25.

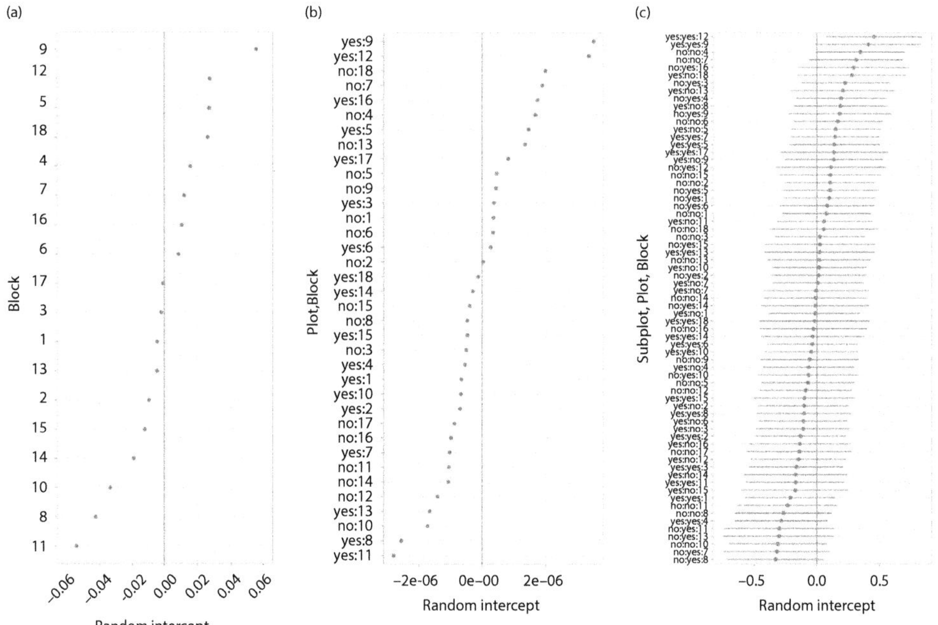

Fig. 15.25 Estimated coefficients for the hierarchy of random effects of the beta frequentist GLM `m4` for the blocks (a), plots within blocks (b), and sub-plots within plots and within blocks (c). The coefficients are shown in terms of their increasing magnitude. The standard errors in plots (a) and (b) are not shown because their magnitude was much higher than the point estimates and thus hindered the visualization.

```
> unique(rand.eff.m4$grpvar)
[1] "crop_cover:management:block" "management:block"          "block"
```

The point estimates of the random effects shown in Fig. 15.25 denote the changes in logit(μ_Y) related to the heterogeneity of the sub-plots within each plot and block (Fig. 15.25c), of plots within each block (Fig. 15.25b), and among blocks (Fig. 15.25a). These random shifts are above and beyond the "global effect" of the explanatory variables shown in the fixed effects. These point estimates range from –0.333 to +0.455 (Fig. 15.25c) or, equivalently, to changes in `prop_above_biom` ranging from a decrease of –0.333 / 4 = –0.083 to an increase of +0.455 / 4 =0.114. This range of changes in `prop_above_biom` related to different levels of the random effects is comparable in magnitude of the effects of the explanatory variables modeled as fixed effects.

To assess the goodness of fit of model `m4` to the data, we start by gathering the data for the plots with `resid.m4 = augment(m4)` from package broom.mixed. We use the randomized quantile residuals (Chapter 8) generated by the package `DHARMa`: `res.m4 = simulateResiduals(fittedModel = m4, n = 1e3, integerResponse = F, refit = F, plot = F)`. These simulated residuals are converted to having a normal distribution with `resid.m4$.std.resid = residuals(res.m4, quantileFunction = qnorm)`. We finally use the function `Cookdist.glmmTMB` to obtain the Cook distances for each data point: `resid.m4$Cook = Cookdist.glmmTMB(m4)`.

Figure 15.26 suggests that model `m4` has a barely acceptable fit to the data, as there are some problems. Figure 15.26a suggests an upward trend in the cloud of points. Figure 15.26b and c indicate that while the medians of the residuals are all close to zero (good!), the ranges of the three groups (no.dry, no.yes, and maize.yes) are clearly smaller

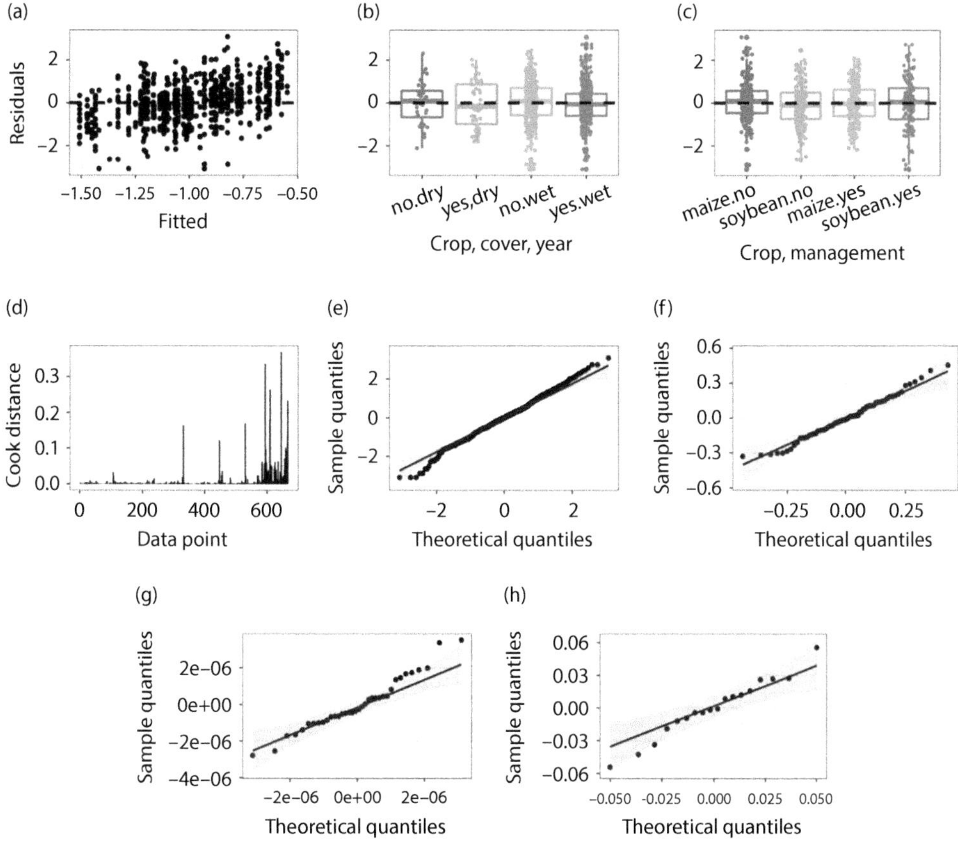

Fig. 15.26 Residual analysis of the frequentist beta GLMM `m4` based on randomized quantile residuals. Plots (a)–(c) show the relation between the residuals and the fitted values and to the combinations of explanatory variables. Plot D shows the Cook distances for each data point. Plots (e)–(h) are the quantile-quantile plots (and the 95% confidence band) for the residuals, and the estimates of the random intercepts for the three levels of the hierarchy of random effects, namely `crop_cover:management:block`, `management:block`, and `block`, respectively.

than those of the other groups. Figure 15.26d shows that that there are four points (that can be identified with `resid.m4[resid.m4$Cook > 0.2,])`) that stand out with large Cook distances and hence with possible large influence on parameter estimates. The Q–Q plot in Fig. 15.26e shows a slight deviation in the lower tail of the residuals, which is not too worrisome. Figure 15.26f, g, and h suggest that the random effect estimates at all levels of the hierarchy have normal distributions. What can we do to improve the model fit to the data? Besides adding additional explanatory variables (none available), we might consider modeling the dispersion parameter ϕ of the beta distribution as we did in Chapter 12. Because this parameter governs the relation between the mean and variance of a beta distribution, it might help dealing with the trend and unevenness of the residuals just discussed. The price to pay is that this would make the beta GLMM much harder to fit. In the meantime, we only accord some tentative credence to the estimated parameters of model `m4`.

We use the package `ggeffects` to obtain the conditional plots predicted by `m4`. We start by generating data frames with the predicted values taking into account only the fixed effects: `pred.m4.f2 = ggpredict(m4, type = "fixed", terms = c("crop", "crop_cover", "year", "management"))`. The command `plot(pred.m4.f2)` produces a basic plot of two panels whose main aesthetic attributes we could not change at will. This is why we saved the plots in `plot.m4.f2 = plot(pred.m4.f2)`, to manipulate each of them separately with `plot.m4.f2[[1]]` and `plot.m4.f2[[2]]` and reformat them at will to generate Fig. 15.27.

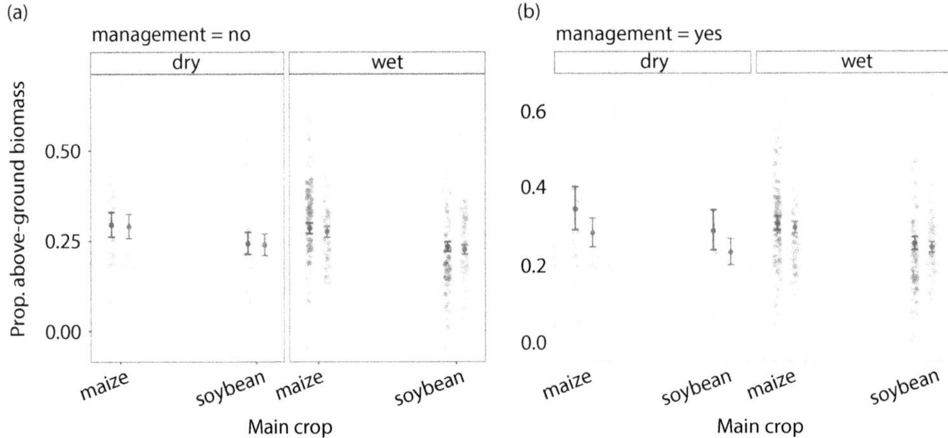

Fig. 15.27 Conditional plots of the frequentist beta GLMM `m4` including only the fixed effects. Plots (a) and (b) separate the predictions of the mean of the response variable for different combinations of the explanatory variables, with `crop_over` = yes and no shown in blue and red. All points are vertically jittered by adding a random value to allow better visualization. The bands correspond to the 95% confidence intervals of each predicted mean.

Figure 15.27 displays two main features. First, there are minor differences in the predicted average values of `prop_above_biom` and their 95% confidence intervals among the groups being compared. Second, including only the fixed effects in the predictions does a poor job in accounting for the wide scatter of the data. Yes, we could have used the option `type = "random"` in the `ggpredict` command to also include the random effects in the predictions. The resulting 95% confidence intervals of the eight groups formed by the $2 \times 2 \times 2$ combinations of levels of the explanatory variables all ranged between zero and one, which is a total defeat in the prediction of a proportion whose potential range of values we knew beforehand. But this defeat may serve a good purpose: highlighting the importance that the often dismissed random effects may have on the predictive ability of a GLMM.

15.4.2 *Beta GLMM with a split-plot design: Bayesian model*

We yet again use the package `brms` to fit this model. We start by defining the model formula: `formula.m4.brms = bf(prop_above_biom ~ crop + year*crop_cover*management + (1|block/management/crop_cover))`. We visualize the default weakly informative model priors:

```
> get_prior(formula.m4.brms, data=DF3,family='Beta')
               prior     class                              coef                           group resp dpar nlpar lb ub       source
              (flat)         b                                                                                              default
              (flat)         b                     crop_coveryes                                                       (vectorized)
              (flat)         b      crop_coveryes:managementyes                                                       (vectorized)
              (flat)         b                       cropsoybean                                                       (vectorized)
              (flat)         b                     managementyes                                                       (vectorized)
              (flat)         b                           yearwet                                                       (vectorized)
              (flat)         b             yearwet:crop_coveryes                                                       (vectorized)
              (flat)         b yearwet:crop_coveryes:managementyes                                                       (vectorized)
              (flat)         b             yearwet:managementyes                                                       (vectorized)
 student_t(3, 0, 2.5) Intercept                                                                                             default
   gamma(0.01, 0.01)       phi                                                                                 0            default
 student_t(3, 0, 2.5)        sd                                                                                0            default
 student_t(3, 0, 2.5)        sd                                                            block               0       (vectorized)
 student_t(3, 0, 2.5)        sd                         Intercept                          block               0       (vectorized)
 student_t(3, 0, 2.5)        sd                                                 block:management               0       (vectorized)
 student_t(3, 0, 2.5)        sd                         Intercept               block:management               0       (vectorized)
 student_t(3, 0, 2.5)        sd                                       block:management:crop_cover               0       (vectorized)
 student_t(3, 0, 2.5)        sd                         Intercept block:management:crop_cover               0       (vectorized)
```

From top to bottom, there are three sets of priors that need to be defined: the population-level effects involving the effects of the explanatory variables, the dispersion parameter of the beta distribution, and the standard deviations of the intercepts of the nested hierarchy of group-level effects. Let's assume that we expect that the single variable and interactive effects of the explanatory variables (the class b parameters) lead to at most a ±30% change in the average `prop_above_biom`. Because the priors in GLM(M) are defined in the scale of the logit link function, we need to "translate" this range of ±30% change into the parameters of the normal distribution that we will use for these priors. We set the mean of these prior normal distributions to zero because the expected effects are as likely to be positive as negative. We use Wan et al.'s (2014) approximation of the standard deviation from a range, (max – min) / 4, and then transform the value to obtain the standard deviation of the priors in the logit scale. Recall (Chapter 9) that, contrary to intuition, priors in the logit scale are more vague as their standard deviation decreases (Hobbs and Hooten 2015, p. 96).

```
> logit(0.3-(-0.3))/4
[1] 0.101
```

We then set `prior.m4.brms = prior(normal(0, logit(0.3-(-0.3))/4), class = b)`, and leave the other priors to their weakly informative defaults.

We run the model: `m4.brms = brm(prop_above_biom ~ crop + year*crop_cover*management + (1|block/management/ crop_cover), data = DF3, prior = prior.m4.brms, family = 'Beta', control = list(adapt_delta = 0.99), warmup = 1000, iter = 5000, chains = 3, future = T)`; its summary is:

```
> summary(m4.brms)
 Family: beta
  Links: mu = logit; phi = identity
Formula: prop_above_biom ~ crop + year * crop_cover * management + (1 | block/management/crop_cover)
   Data: DF3 (Number of observations: 666)
  Draws: 3 chains, each with iter = 5000; warmup = 1000; thin = 1;
         total post-warmup draws = 12000

Group-Level Effects:
~block (Number of levels: 18)
              Estimate Est.Error l-95% CI u-95% CI Rhat Bulk_ESS Tail_ESS
sd(Intercept)     0.08      0.05     0.00     0.20 1.00     2468     4708

~block:management (Number of levels: 36)
              Estimate Est.Error l-95% CI u-95% CI Rhat Bulk_ESS Tail_ESS
sd(Intercept)     0.09      0.06     0.00     0.21 1.00     1940     4817

~block:management:crop_cover (Number of levels: 69)
              Estimate Est.Error l-95% CI u-95% CI Rhat Bulk_ESS Tail_ESS
sd(Intercept)     0.22      0.04     0.16     0.30 1.00     3965     7078

Population-Level Effects:
                                   Estimate Est.Error l-95% CI u-95% CI Rhat Bulk_ESS Tail_ESS
Intercept                             -0.94      0.09    -1.11    -0.77 1.00    11803     9292
cropsoybean                           -0.12      0.07    -0.25     0.02 1.00     9313     8471
yearwet                               -0.04      0.08    -0.19     0.11 1.00    14703     9656
crop_coveryes                         -0.02      0.07    -0.16     0.11 1.00    13183     8898
managementyes                          0.06      0.07    -0.08     0.20 1.00    15112    10315
yearwet:crop_coveryes                 -0.02      0.07    -0.16     0.13 1.00    13166     8480
yearwet:managementyes                  0.02      0.08    -0.12     0.18 1.00    12838     9180
crop_coveryes:managementyes            0.00      0.08    -0.15     0.16 1.00    14286    10103
yearwet:crop_coveryes:managementyes    0.03      0.08    -0.14     0.19 1.00    16195     9780

Family Specific Parameters:
    Estimate Est.Error l-95% CI u-95% CI Rhat Bulk_ESS Tail_ESS
phi    27.30      1.60    24.26    30.54 1.00    16936     8707
```

We first notice that the `Rhat` values are all equal to 1.00, and that the ESS values are very large (> 1,000), thus providing an initial suggestion that the Hamiltonian Monte Carlo algorithm converged to a stationary distribution. The ESS values for the standard deviations of the group-level intercepts are smaller than those of the other model parameters. This is because these standard deviations are always much harder to estimate than parameters related to the mean of the response variable. The mean of the standard deviation of the group-level intercepts at the level of `block:management` was much bigger than the tiny point estimate in the frequentist beta GLMM (model `m4`). While we cannot say which estimate is correct or even better, that of `m4.brms` seems more credible. It is clear that estimating standard deviations of small magnitude close to their lowest feasible boundary of zero is always problematic and plagued with uncertainties. The marginal means of the population-level parameters were all globally similar to `summary(m4)`, which the exception of `cropsoybean` that was more than 50% smaller. Again, we cannot possibly know if the point estimate of `m4` is "more correct" than the marginal mean of `m4.brms`. May the disparity between the frequentist and Bayesian models be due to the undue influence of the priors used? This can be evaluated by changing the standard deviations of the hyperpriors in `prior.m4.brms` as we did in Chapter 4. Doing so (results not shown) does not reveal any relevant difference between models' outputs due to the priors. Because there is an infinite set of possible priors to compare in separate Monte Carlo runs, we can never be positively sure that the posterior distributions are not unduly influenced by priors for which we may lack strong empirical justification. A comparable sensitivity analysis in the frequentist model would involve using different starting parameter values and different optimization algorithms to solve the maximum likelihood equations. But this is very rarely done by most data analysts. It is by no means wrong that previous empirical knowledge informs and constrains our findings. After all, the world's knowledge does not start anew with each additional statistical analysis. The magnitudes of the means of the population-level parameters were all comparatively small. We had to create a three-dimensional figure (Fig. 15.24) to get close to the actual interpretation of the parameters of a model involving several pairwise interactions and a third-order interaction. We will not repeat Fig. 15.24 using the means of the marginal posterior distributions of model `m4.brms`. Using Gelman et al.'s (2014) "rule of four," we can state that the explanatory variables and their interactions had a modest effect on the average `prop_above_biom`.

We start by making a better evaluation of the convergece of the Hamiltonian Monte Carlo algorithm. Figure 15.28 shows the convergence of the algorithm to a stationary posterior distribution for the population-level parameters of model `m4.brms`. The sampled estimates of the model parameters also had very low autocorrelation (figure not shown, but the code can be found on the companion website), meaning that we have a very large number of statistically independent estimates in the marginal posterior distributions of each model parameter. A better depiction of the posterior distribution of the parameters of model `m4.brms` can be found in Fig. 15.29.

We use Gelman et al.'s (2018) metric to obtain the proportion of variation of the response variable accounted for by model `m4.brms`:

```
> bayes_R2(m4.brms)
   Estimate Est.Error  Q2.5 Q97.5
R2    0.284    0.0284 0.227 0.338
```

It is a modest proportion, meaning that the data contains a larger heterogeneity than the one that can be associated to the explanatory variables included as population- and group-level effects.

As always, we evaluate the goodness of fit of model `m4.brms` with residual analysis. We use the randomized quantile residuals appropriate for a response variable with beta distribution (Chapter 8) based on the model's posterior predictive distribution, `post.pred.m4.brms = predict(m4.brms, ndraws = 1e3, summary = F)`, which is then used with the package DHARMa to obtain the residuals:

```
qres.m4.brms=createDHARMa(simulatedResponse=t(post.pred.m4.brms),
   observedResponse = DF3$
   fittedPredictedResponse=apply(post.pred.m4.brms,2,median),integerResponse=F)
```

These simulated residuals, which have a uniform distribution when a model has a perfect fit, are converted to a normal distribution with `res.m4.brms = data.frame(res = qnorm(residuals(qres.m4.brms)))`, and stored in a data frame where we will incorporate other components of the plots of the residual analysis such as the Pareto k of LOO-CV:

```
res.m4.brms=cbind(res.m4.brms,
                  DF3[,c("year","crop","block","management","crop_cover")],
                  fitted=fitted(m4.brms, ndraws=1000)[,1],
                  pareto=loo(m4.brms, pointwise=T)$diagnostics$pareto_k)
```

To evaluate whether the means of the group-level effects have a normal distribution as hypothesized in the GLMM, we must first extract them: `stats.ranef.m4.brms = (ranef(m4.brms))`. The resulting object, `stats.ranef.m4.brms`, is a list composed of three matrices of different sizes containing the mean, standard deviation, and upper and lower limits of the 95% credible intervals for the three levels of the group-level intercepts:

```
> names(stats.ranef.m4.brms)
[1] "block"                    "block:management"            "block:management:crop_cover"
```

The three matrices have different sizes because each level in the hierarchy of the group-level effects has a different number of combinations of levels of the explanatory variables

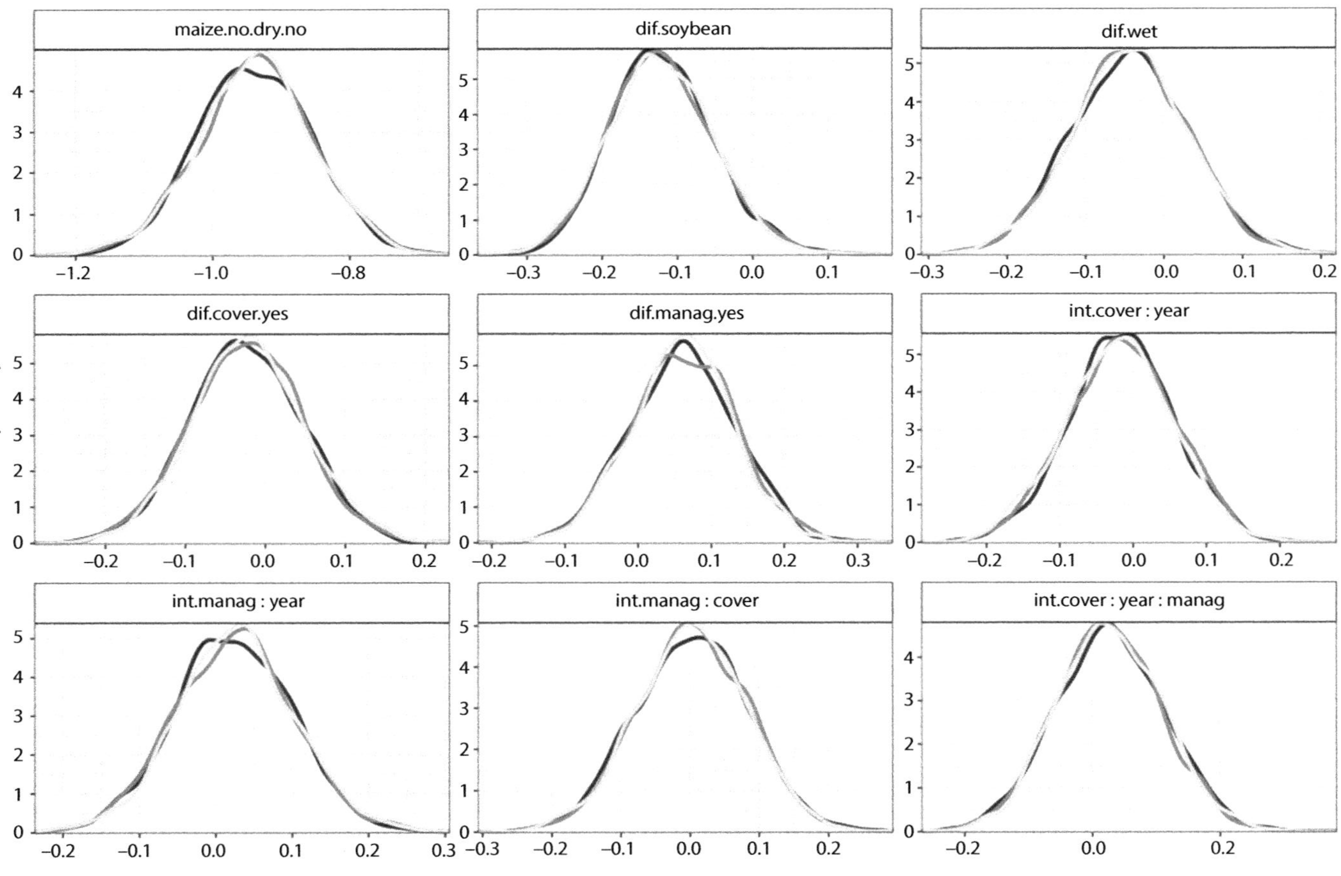

Fig. 15.28 Convergence of the three chains to a common stationary posterior distribution for the population-level parameters of the Bayesian beta GLMM `m4.brms`.

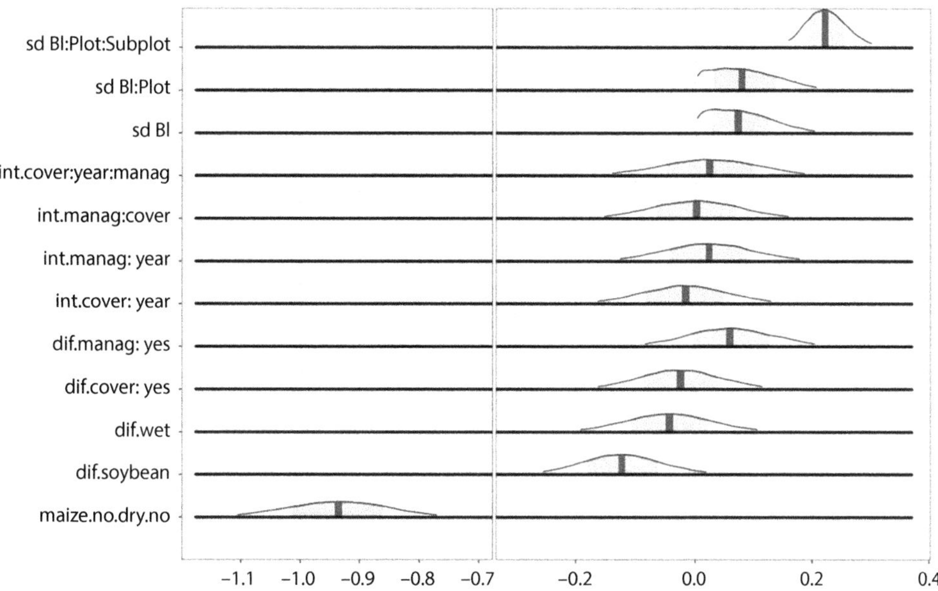

Fig. 15.29 Posterior distributions of the population- and group-level parameters of the Bayesian beta GLMM `m4.brms` showing their 62% (dark blue) and 95% credible intervals. All parameters denote effects in the scale of the logit link function. maize.no.dry.no is the mean for the reference levels of main crop (maize), crop_cover (no), management (no), and year (dry); dif.soybean, dif.wet, dif.cover.yes, and dif.manag.yes are the differentials between soybean, wet, no, and no and the reference levels of all other variables; int.cover:year, int.cover:year, int.manag:year, int.manag:cover denote the pairwise interactions; int.cover:year:manag, the third-order interaction; sd Bl, sd Bl:Plot, and sd Bl:plot:Subplot are the standard deviations of the intercepts of the hierarchy of group-level effects.

involved. We can convert each of these matrices into separate data frames to make the appropriate plots with:

```
int.block.m4.brms=as.data.frame(stats.ranef.m4.brms$block)
int.block.plot.m4.brms=as.data.frame(stats.ranef.m4.brms$"block:management")
int.block.plot.subplot.m4.brms=
as.data.frame(stats.ranef.m4.brms$"block:management:crop_cover")
```

The quotes in the second and third line are needed to force R to take the variable name with the colons.

We can now make Fig. 15.30, which shows that model `m4.brms` has a reasonably good fit to the data overall, with a few glitches. The residuals have a seemingly random scatter without any visible trends (Fig. 15.30a), and their variation among combinations of levels of the explanatory variables (Fig. 15.30b and c) is reasonably homogeneous, although slightly smaller for the dry year (Fig. 15.30b). The Pareto LOO-CV *k* statistics are all in the "good" range, expect for three of them that lie in the "OK" range (Chapter 7). The Q–Q plots (Fig. 15.30e to h) assessing whether the residuals and the mean of the estimated intercepts in the group-level hierarchy are very good, except for three blocks (Fig. 15.30f) that are well outside the 95% credible band. We can identify the extreme blocks with:

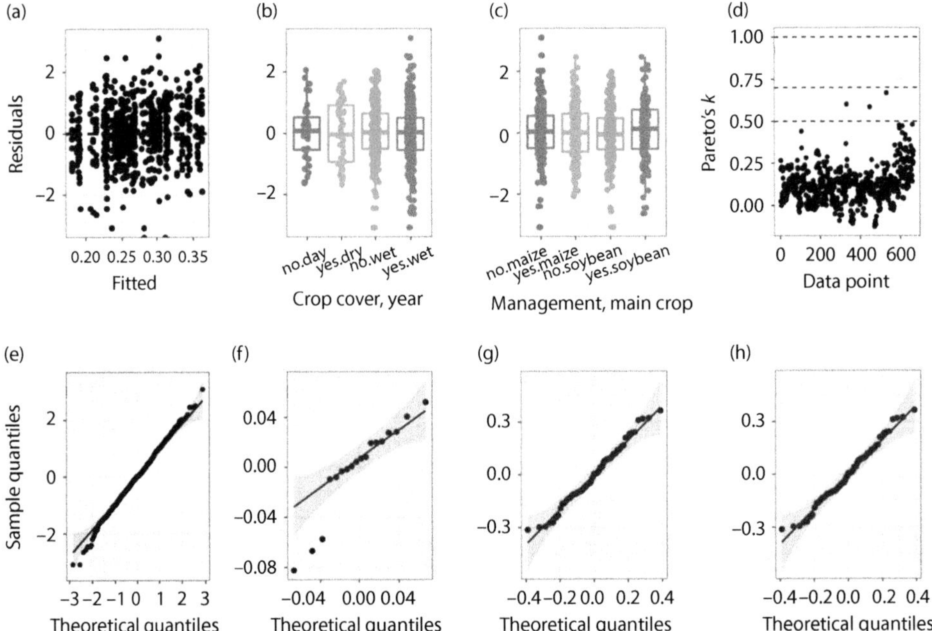

Fig. 15.30 Residual analysis of the beta GLMM `m4.brms` based on randomized quantile residuals. Plots (a)–(c) show the relation between the residuals and the fitted values and to the combination of the explanatory variable. Plot (d) is the Pareto *k* cross-validation statistic for each data point. Plots (e)–(h) are the quantile–quantile plots (and their 95% credible band) for the residuals, and the estimates of the random intercepts for the blocks, plots within blocks, and sub-plots within plots, respectively.

```
> int.block.m4.brms
   Estimate.Intercept
1            0.004142
2           -0.003359
3            0.006923
4            0.028196
5            0.040516
6            0.019161
7            0.000653
8           -0.066789
9            0.052099
10          -0.057360
11          -0.082520
12           0.020529
13           0.008064
14          -0.007997
15          -0.001821
16           0.019602
17          -0.009661
18           0.027574
```

They are the blocks 8, 10, and 11. We can compare the summary statistics for the data of these three blocks with those of the entire data set:

```
> summary(DF3[DF3$block==c(8,10,11),])
  year        crop         block        management crop_cover prop_above_biom
 dry: 0    maize  : 0    Min.   : 8.00   no :27     no :21     Min.   :0.071
 wet:51    soybean:51    1st Qu.: 8.00   yes:24     yes:30     1st Qu.:0.153
                         Median :10.00                         Median :0.246
                         Mean   : 9.47                         Mean   :0.230
                         3rd Qu.:11.00                         3rd Qu.:0.282
                         Max.   :11.00                         Max.   :0.444
> summary(DF3)
  year          crop          block         management crop_cover prop_above_biom
 dry: 83    maize  :358    Min.   : 1.00   no :377     no :298    Min.   :0.040
 wet:583    soybean:308    1st Qu.: 4.00   yes:289     yes:368    1st Qu.:0.200
                           Median : 7.00                          Median :0.264
                           Mean   : 7.38                          Mean   :0.269
                           3rd Qu.:10.00                          3rd Qu.:0.323
                           Max.   :18.00                          Max.   :0.581
```

We can see that the three blocks are 51 data points from the wet year and soybean. Comparing the summary statistics of the response variable for these three blocks with the entire data set, we cannot spot any obvious main difference between them. While the robustness of GLMMs to departures from the normal distribution is still an unresolved issue (Chapter 13), there are no grounds to imagine fitting the model without these data. We have no grounds for suspecting 51 / 666 = 0.076 of the data, and are nowhere near suggesting discarding them for the sake of attaining a better fit in one of the eight plots of Fig. 15.30.

We can use the package `ggeffects` to obtain the conditional curves predicted by model `m4.brms`. We first obtain a data frame with the predicted values for all combinations of the explanatory variables: `pred.m4.f = ggpredict(m4, type = "fixed", terms = c("crop", "crop_cover", "year", "management"))`. These predictions include only the population-level effects. Then, we generate a plot, `plot.m4.f2 = plot(pred.m4.f, add.data = T, jitter = 0.1, dot.alpha = 0.10)`, whose two parts we are going to edit separately to obtain Fig. 15.31. This figure highlights that no combination of effects of the explanatory variables generated much difference in the predicted means of the response variable `prop_above_biom`, in agreement with our comments about `summary(m4.brms)`.

To display the posterior distributions of the group-level effects, we must first obtain them with `post.gr_lev.m4.brms = posterior_samples(m4.brms, pars = "∧r")`. The resulting data frame has variable names (`names(post.gr_lev.m4.brms)`) starting with `r_block[...]`, `r_:block:management[...]`, and `r_:block:management:crop_cover[...]` that are in columns 1–18, 19–54, and 55–123, respectively. We can then separate the three sets of group-level intercepts into separate data frames with:

```
post.block.m4.brms $=$ post.gr_lev.m4.brms[,1:18] # for blocks
post.block.plot.m4.brms $=$ post.gr_lev.m4.brms[,19:54]
 # for blocks:plots
post.block.plot.subplot.m4.brms $=$ post.gr_lev.m4.brms[,55:123]
 # for blocks:plots:subplots
```

We now need to reformat these data frames into the "long format" by stacking the columns on top of each other using the package `reshape2`, `post.block.m4.brms = melt (post.block.m4.brms, value.name = "int_block")`, and likewise with the other two data frames.

Figure 15.32 displays the 123 posterior distributions of the group-level effects in the three levels of the hierarchy of effects induced by the split-plot design. Note that the range

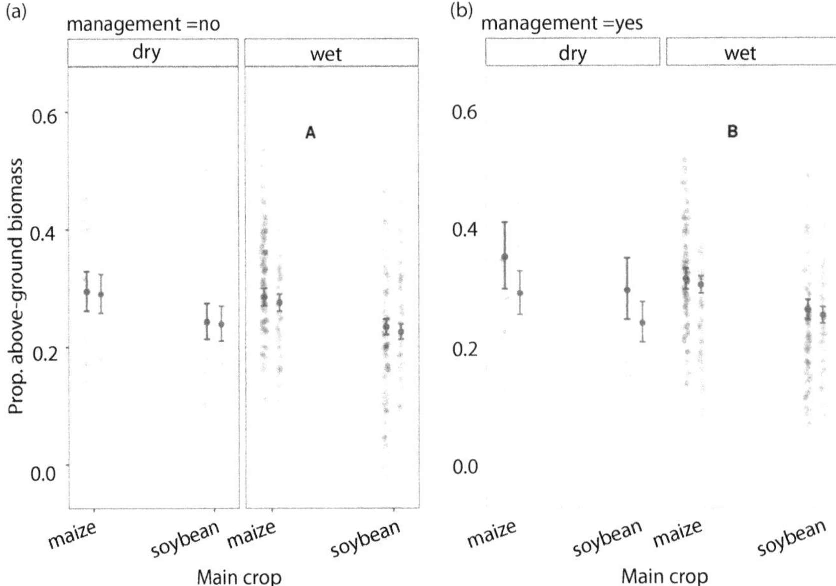

Fig. 15.31 Conditional plots of the Bayesian beta GLMM `m4.brms` including only the population-level effects. Plots (a) and (b) separate the predictions of the mean of the response variable for different combinations of the explanatory variables, with `crop_over` = yes and no shown in blue and red. All points are vertically jittered by adding a random value to allow better visualization. The bands correspond to the 95% credible intervals of each predicted mean.

of Fig. 15.32c is larger than for the other two plots, in agreement with the differences in the means of their posterior standard deviations shown in `summary(m4.brms)`.

Finally, we can use the posterior predictive distribution (Chapter 7) of model `m4.brms` to assess its capacity to reproduce observed features in the data as a "typical" prediction (and not a lucky or coincidential match). To that end, we use the versatile commands of package `bayesplot` in Fig. 15.33. We can use the posterior predictive distribution in the posterior prediction checks to generate the overall density distribution of the response variable `prop_above_biom` that the model `m4.brms` can faithfully generate as a "typical" prediction (Fig. 15.33a). We can also aim for the joint distribution of any two summary statistics calculated from the data (Fig. 15.33b). Each dot in Fig. 15.33b is the joint prediction of the median and the standard deviation from the posterior predictive distribution. Again, because the observed pair of values is found in the middle of the predicted cloud of points, we can see that `m4.brms` can reproduce the observed pair of summary statistics. We can also test whether the model can predict a summary statistic for groups across the levels of a single, or a combination of, explanatory variable(s) as in Fig. 15.34.

We see (Fig. 15.34) that while the predictive capacity of the model tends to decrease as we use more stringent summary statistics for certain groups, it retains its ability to reproduce the observed features of the data. We can then give some credence to a statistical model capable of faithfully reproducing the evidence. There is no equivalence of the posterior predictive checks in the frequentist framework. We can create any function (see the examples in Chapter 9) to be calculated both in the data and in the posterior predictive distribution for the entire data set or for any subset of it. We stress again that

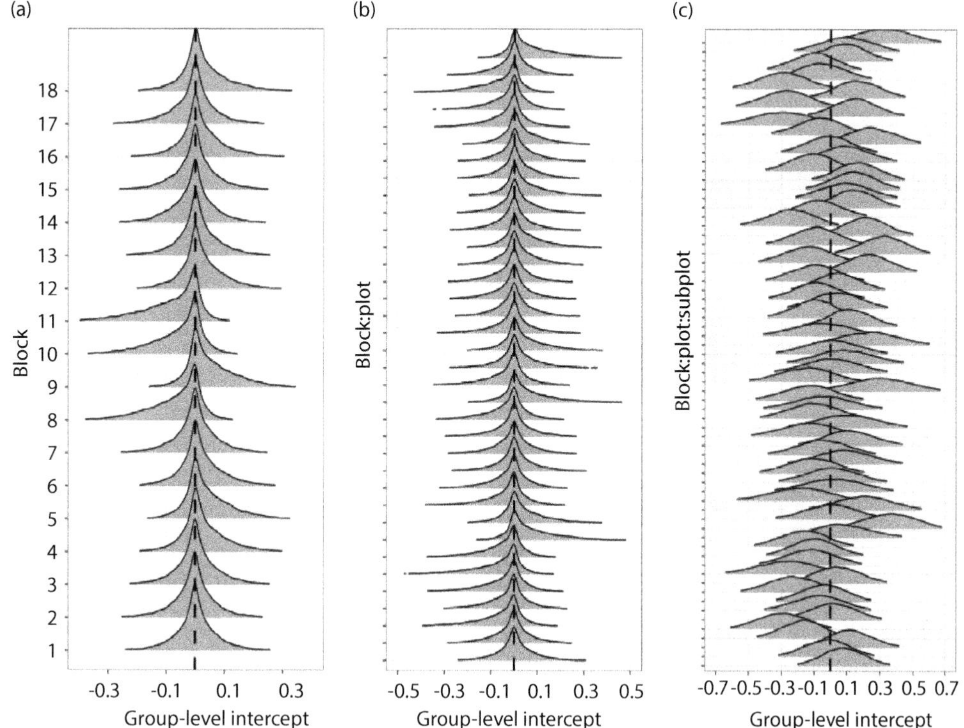

Fig. 15.32 Posterior distributions of the intercepts at each level of the hierarchy of group-level effects of the Bayesian beta GLMM `m4.brms`, showing the blocks (a), plots within blocks (b), and sub-plots within plots within blocks (c). Note that the scales are different for each plot.

any realistic statistical model should at some point break under the pressure of requiring it to make stringent predictions of complex summary statistics for small subsets of data. However, the limit of what can be done with posterior predictive checks to assess the predictive capacity of a model just depends on your inventiveness and your good sense, as always.

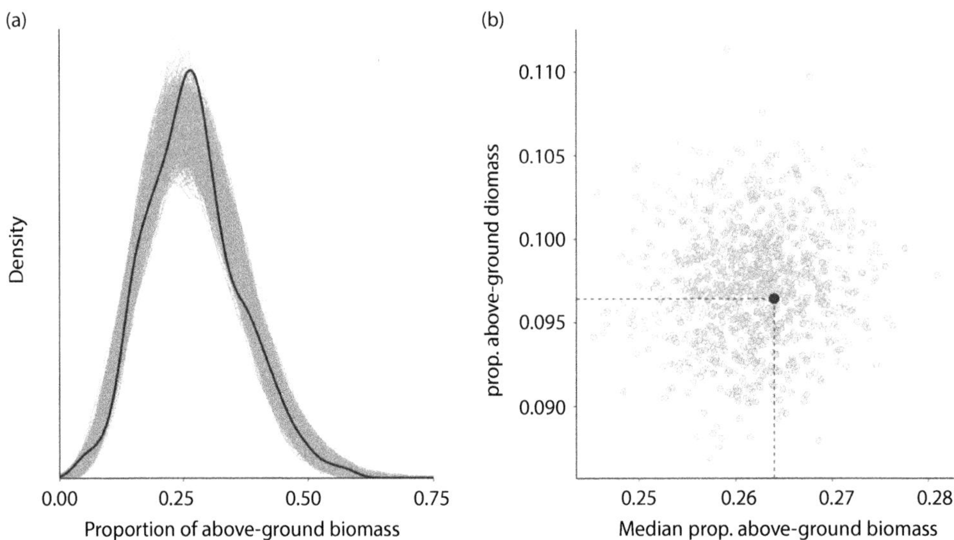

Fig. 15.33 Use of the posterior predictive distribution of the Bayesian beta GLMM `m4.brms` to predict the observed density curve of the response variable `prop_above_biom` (a), and a joint pair of summary statistics (here, the median and standard deviation) for the entire data (b). The gray lines (a) are the density curves calculated over the posterior predictive distribution. Each dot (b) corresponds to the pair of values of summary statistics calculated over the posterior predictive distribution.

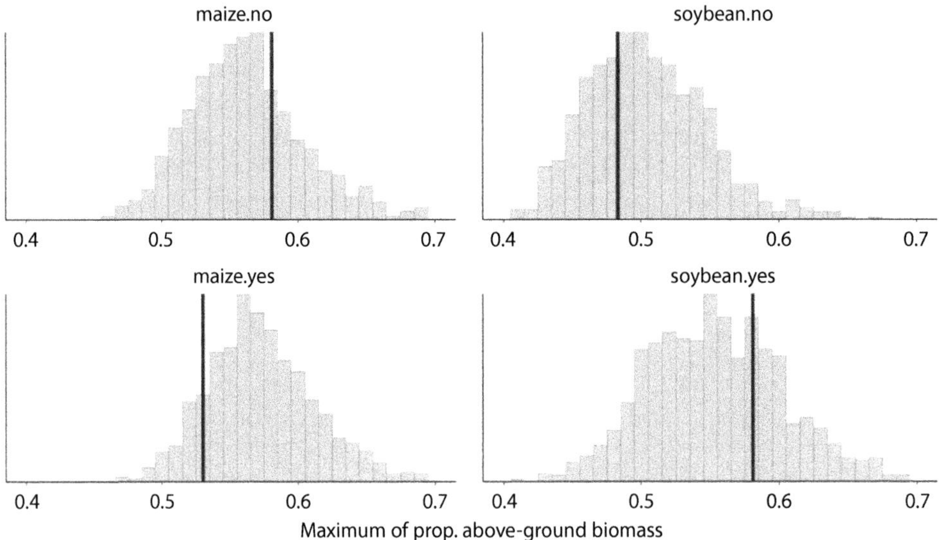

Fig. 15.34 Posterior predictive check of the Bayesian beta GLMM `m4.brms` to predict the maximum of the response variable `prop_above_biom` for combinations of two explanatory variables. The top row corresponds to `management` = no, and the two crops.

15.5 Problems

15.1 McGoldrick and Mac Nally (1998) wanted to compare the effects the dominant Eucalyptus species (ironbark-dominated vs. stringybark-dominated forests) and a regional contrast (north vs. south) on the temporal dynamics of avian nectarivores in south-central Victoria, Australia. They defined eight large (80–100 ha) study sites, two of each habitat type, in both north and south parts of the mountain range. Each site was visited once a month for one year. The total number of birds was estimated by walking a strip transect of 7 ha (1 km × 70 m) at constant speed under favorable conditions for observation to maximize the probability of observing forest birds. File: Pr15-1.csv

15.2 Campana and Yahdjian (2021) studied whether grazing (top-down control) and nutrient availability (bottom-up control) interactively modified above-ground plant biomass in Argentinian grasslands devoted to livestock production. They applied the two experimental factors at different spatial scales: grazing/enclosure in six 20 m × 50 m plots, and within each plot they randomly placed four 25 m^2 sub-plots where they fertilized/control. The response variable was the dry weight of above-ground biomass (live_biom; g/m^2). File: Pr15-2.csv

15.3 Sáez et al. (2014) assessed pollination and drupelet set (i.e., the number of drupelets per fruit) in 16 raspberry fields in southern Argentina. Using pollen supplementation, they also tested whether drupelet set was pollen limited in a subset of six large, relatively homogeneous, fields. They supplemented flowers in the same stems where they, at the same time, tagged open-pollinated flowers to control for individual stem variation. The pollen used for supplementation was collected from flowers on at least five other stems in the same field. Each experimental flower was marked with a paper tag for later harvesting. They measured the number of drupelets per fruit (drupelets) as the response variable, the amount of volcanic ash in the soil (cm_ash) from a recent eruption, and the experimental treatment. File: Pr15-3.csv

References

Aranda, M., Tognetti, P., and Mazía, N. (2021) Are field crops refuge for woody invaders? Rainfall, crop type and management shaped tree invasion in croplands. *Agriculture, Ecosystems and Environment*, 319, 107564.

Bolker, B. (2015). Generalized linear mixed models. In G. Fox, A. Negrete-Yankelevich, and V. Sosa (eds), *Ecological Statistics: Contemporary Theory and Applications*, pp. 309–333. Oxford University Press, Oxford.

Campana, S. and Yahdjian, L. (2021). Plant quality and primary productivity modulate plant biomass responses to the joint effects of grazing and fertilization in a mesic grassland. *Applied Vegetation Science*, 24, e12588.

Efron, B. and Tibshirani, R. (1993). *An Introduction to the Bootstrap*. CRC Press, New York.

Franco-Trecu, V., Aurioles-Gamboa, D., and Inchausti, P. (2014) Individual trophic specialisation and niche segregation explain the contrasting population trends of two sympatric otariids. *Marine Biology*, 161, 609–618.

Gelman, A., Carlin, J., Stern, H. et al. (2014). *Bayesian Data Analysis*, 3rd edn. Chapman and Hall, New York.

Gelman, A., Goodrich, B., Gabry, J. et al. (2018). R-squared for Bayesian regression models. *The American Statistician*, 73, 307–309.

Heinze, G. and Schemper, M. (2002). A solution to the problem of separation in logistic regression. *Statistics in Medicine*, 21, 2049–2419.

Hobbs, N. and Hooten, M. (2015). *Bayesian Models: A Statistical Primer for Ecologists*. Princeton University Press, Princeton.

McGoldrick, J. and Mac Nally, R. (1998). Impact of flowering on bird community dynamics in some central Victorian eucalypt forests. *Ecological Research*, 13, 125–139.

Nakagawa, S., Johnson, P., and Schielzeth, H. (2017). The coefficient of determination R^2 and intra-class correlation coefficient from generalized linear mixed-effect models revisited and expanded. *Journal of the Royal Society Interface*, 14, 20170213.

Sáez, A., Morales, C., Ramos, L. et al. (2014). Extremely frequent bee visits increase pollen deposition but reduce drupelet set in raspberry. *Journal of Applied Ecology*, 51, 1603–1612.

Singmann, H. and Keller, D. (2019). An introduction to mixed models for experimental psychology. In D. Speier and E. Schumacher (eds), *New Methods in Cognitive Psychology*, pp. 4–31. Psychology Press, New York.

Vehtari, A., Gelman, A., and Gabry, J. (2017). Practical Bayesian model evaluation using leave-one-out cross validation and WAIC. *Statistical Computing*, 27, 1413–1432.

Vicente, J., Fernández de Mera, I., and Cortázar, C. (2006). Epidemiology and risk factors analysis of elaphostrongylosis in red deer (*Cervus elaphus*) from Spain. *Parasitology*, 98, 77–85.

Wan, X., Wang, W., Liu, J. et al. (2014). Estimating the sample mean and standard deviation from the sample size, median, range and/or interquartile range. *BMC Medical Research Methodology*, 14, 135–147.

Afterword

This is the end of the book. This book aimed to show that we can view the analysis of a large array of univariate data commonly arising in the biological sciences as the fitting of statistical models with the goal of estimating their parameters. This is to be contrasted with viewing data analysis as the matching of data with statistical tests aiming to detect statistical significance under the constraints imposed by the assumptions of the tests. We have striven to show that statistical modeling is a far more interesting and revealing endeavor than striving to find a test that happens to give a statistically significant result to "save the blushes" of our research project.

We can employ either the Bayesian or frequentist framework to estimate the parameters of statistical models as we please. Regardless of your framework of choice, we must always start by writing a statistical model postulating specific relations between its parameters and the explanatory variables. Only by knowing the statistical model we are fitting and how it is implemented in R can we avoid being mesmerized by its summary output. Prior to anything else, we must always perform the fundamental quality-control step of assessing the model's goodness of fit to the available data. We can only interpret the parameters of a satisfactorily fitting model in terms of the questions and scientific hypotheses that motivated the gathering of the evidence. We have striven to provide predicted model outputs as graphics of conditional curves in the same scale as the data. These are often far more enlightening for our understanding of the implications of the statistical findings than tables of coefficients and their confidence/credible intervals that we tend to report in writing anyway.

We trust that by including the presentation of simple study cases that you probably knew in advance (e.g., the *t*-test!), you progressively gained an understanding of the fundamental concepts and methods used by the Bayesian or frequentist frameworks to tackle the same statistical problems. For the comparatively simple and well-trodden problems considered here their answers may coincide, but this coincidence need not always be the case. There are more complex data analysis problems that can be more easily or only tackled with one of these frameworks. But the choice of the framework is likely to depend on the nature of the specific problem and on our familiarity with the methods, in order to cast it as either the optimization of maximum likelihood equations or the uncovering of posterior distributions with MCMC algorithms. For these more complex problems, you might need to tweak optimizers to numerically solve complex likelihood equations, or use and adapt the latest clever MCMC algorithm with some programming. At the end, what matters most to us is to have provided you with a fundamental understanding of the main ideas underlying the Bayesian and frequentist frameworks that make up what statistics really is in the twenty-first century.

APPENDIX A

List of R Packages Used in This Book

We have attempted as far as possible to use much the same packages in every chapter. Unless explicitly noted, all the plots in this book were made with the package `ggplot2`. The R packages used are referenced here to shorten each chapter's bibliography.

Almeida, L. and Hofmann, H. (2018). ggplot2 compatible quantile–quantile plots in R. *The R Journal*, 10, 248–261. (`qqplotr`)

Auguie, B. (2017). `gridExtra`: Miscellaneous functions for "grid" graphics. R package, version 2.3.

Bolker, B. and Robinson, D. (2020). `broom.mixed`: Tidying methods for mixed models. R package, version 0.2.6.

Bolker, B. and R Development Core Team (2020). `bbmle`: Tools for general maximum likelihood estimation. R package, version 1.0.23.1.

Brooks, M., Kristensen, K., van Benthem, K. et al. (2017). `glmmTMB` balances speed and flexibility among packages for zero-inflated generalized linear mixed modeling. *The R Journal*, 9, 378–400.

Bürkner, P. (2017). `brms`: An R package for Bayesian multilevel models using Stan. *Journal of Statistical Software*, 80, i80.

Bürkner, P., Gabry. J., Kay, M., et al. (2021). `posterior`: Tools for working with posterior distributions. R package, version 1.0.1.

Delignette-Muller, M. and Dutang, C. (2015). `fitdistrplus`: An R package for fitting distributions. *Journal of Statistical Software*, 64, 1–34.

Fox, J. and Weisberg, S. (2019). *An R Companion to Applied Regression*, 3rd edn. Sage, Thousand Oaks. (`car`)

Gabry, J. and Mahr, T. (2021). `bayesplot`: Plotting for Bayesian models. R package, version 1.8.1.

Gelman, A. and Su, Y. (2020). *Data Analysis Using Regression and Multilevel/Hierarchical Models*. Cambridge University Press, Cambridge. (`arm`)

Genz, A., Bretz, F., Miwa, T. et al. (2021). `mvtnorm`: Multivariate normal and *t* distributions. R package, version 1.1-2.

Guangchuang, Y. and Shuangbin, X. (2021). `ggbreak`: Set axis break for ggplot2. R package, version 0.0.5.

Halekoh, U. and Højsgaard, S. (2014). A Kenward–Roger approximation and parametric bootstrap methods for tests in linear mixed models – the R package `pbkrtest`. *Journal of Statistical Software*, 59, 1–30.

Hartig, F. (2021). `DHARMa`: Residual diagnostics for hierarchical (multi-level/mixed) regression models. R package, version 0.4.1.

Højsgaard, S. and Halekoh, U. (2021). `doBy`: Groupwise statistics, LSmeans, linear contrasts, utilities. R package, version 4.6.10.

Hothhorn, H., Bretz, F., and Westfall, P. (2008). Simultaneous inference in general parametric models. *Biometrical Journal*, 50, 346–363. (`multcomp`)

Jeppson, H., Hofmann, H., and Cook, D. (2021). `ggmosaic`: Mosaic plots in the "ggplot2" framework. R package, version 0.3.3.

Kay, M. (2021). `ggdist`: Visualizations of distributions and uncertainty. R package, version 3.0.0.

Kuznetsova, A, Brockhoff, P., and Christensen, R. (2017). `lmerTest` package: Tests in linear mixed effects models. *Journal of Statistical Software*, 82, 1–26.

Loy, A., and Hofmann, H. (2014). `HLMdiag`: A suite of diagnostics for hierarchical linear models in R. *Journal of Statistical Software*, 56, 1–28.

Lüdecke, D. (2018). `ggeffects`: Tidy data frames of marginal effects from regression models. *Journal of Open Source Software*, 3, 772.

Lüdecke, D., Ben-Shachar, M., Patil, I. et al. (2021). `performance`: An R package for assessment, comparison and testing of statistical models. *Journal of Open Source Software*, 6, 3139.

Nieuwenhuis, R., Grotenhuis, M., and Pelzer, B. (2012). `influence.ME`: Tools for detecting influential data in mixed effects models. *R Journal*, 4, 38–47.

Pinheiro, J., Bates, D., DebRoy, S., et al. (2021). `nlme`: Linear and nonlinear mixed effects models. R package, version 3.1-153.

Robin, X., Turck, N., Hainard, A., et al. (2011). `pROC`: An open-source package for R and S+ to analyze and compare ROC curves. *BMC Bioinformatics*, 12, 77.

Robinson D., Hayes, A., and Couch, S. (2021). `broom`: Convert statistical objects into tidy tibbles. R package, version 0.7.8.

Schloerke, B., Cook, D., Larmarange, J., et al. (2021). `GGally`: Extension to ggplot2. R package, version 2.1.2.

Singmann, H., Bolker, B., Westfall, J., et al. (2021). `afex`: Analysis of factorial experiments. R package, version 1.0-1.

Stasinopoulos, M. and Rigby, R. (2021). `gamlss.dist`: Distributions for generalized additive models for location scale and shape. R package, version 5.3-2.

Vehtari, A., Gelman, A., and Gabry, J. (2020). `Loo`: Efficient leave-one-out cross-validation and WAIC for Bayesian models. R package, version 2.4.1.

Venables, W. and Ripley, B. (2002). *Modern Applied Statistics with S*, 4th edn. Springer, New York. (`MASS`)

Wickham, H. (2007). Reshaping data with the reshape package. *Journal of Statistical Software*, 21, 1–20. (`reshape2`)

Wickham, H. (2016). `ggplot2`: *Elegant Graphics for Data Analysis*. Springer, New York.

Wickham, H. and Seidel, D. (2020). `scales`: Scale functions for visualization. R package, version 1.1.1.

Wilke, C. (2021). `ggridges`: Ridgeline plots in ggplot2. R package, version 0.5.3.

Index

Note: Tables, figures, and notes are indicated by *t*, *f*, and *n*.